Physics of Neural Networks

Springer
Berlin
Heidelberg
New York
Barcelona
Budapest
Hong Kong
London
Milan
Paris
Santa Clara
Singapore
Tokyo

E. Domany J. L. van Hemmen
K. Schulten (Eds.)

Models of Neural Networks I

Second Updated Edition
With 79 Figures, 3 in colour

Springer

Series and Volume Editors:

Professor Dr. J. Leo van Hemmen
Institut für Theoretische Physik
Technische Universität München
D-85747 Garching bei München
Germany

Professor Eytan Domany
Department of Electronics
Weizmann Institute of Science
76100 Rehovot, Israel

Professor Klaus Schulten
Department of Physics
and Beckman Institute
University of Illinois
Urbana, IL 61801, USA

Cataloging-in-Publication Data applied for

Die Deutsche Bibliothek - CIP-Einheitsaufnahme

Models of neural networks. - New York ; Berlin ; Heidelberg ; London ; Paris ; Tokyo ; Hong Kong ; Barcelona ; Budapest : Springer.

1. E. Domany ... - 2., updated ed. - 1995
(Physics of neural networks)
ISBN 3-540-59403-5
NE: Domany, Eytan

ISBN 3-540-59403-5 2nd Ed. Springer-Verlag Berlin Heidelberg New York

ISBN 3-540-51109-1 1st Ed. Springer-Verlag Berlin Heidelberg New York

Typesetting: Data conversion by Springer-Verlag
SPIN 10484141 54/3144 – 5 4 3 2 1 0 – Printed on acid-free paper

This volume is dedicated to the memory of Elizabeth Gardner. Her outstanding scientific achievements, quiet unassuming manners, and personality have set the highest standards to all of us who work in this field. *

* An appreciation of Elizabeth Gardner can be found at the end of this book.

Foreword

> One of the great intellectual challenges for the next few decades is the question of brain organization. What is the basic mechanism for storage of memory? What are the processes that serve as the interphase between the basically chemical processes of the body and the very specific and nonstatistical operations in the brain? Above all, how is concept formation achieved in the human brain? I wonder whether the spirit of the physics that will be involved in these studies will not be akin to that which moved the founders of the "rational foundation of thermodynamics".
>
> *C.N. Yang*[1]

The human brain is said to have roughly 10^{10} neurons connected through about 10^{14} synapses. Each neuron is itself a complex device which compares and integrates incoming electrical signals and relays a nonlinear response to other neurons. The brain certainly exceeds in complexity any system which physicists have studied in the past. Nevertheless, there do exist many analogies of the brain to simpler physical systems. We have witnessed during the last decade some surprising contributions of physics to the study of the brain. The most significant parallel between biological brains and many physical systems is that both are made of many tightly interacting components. Physics, more so than any other discipline, has devoted itself to the systematic study of multicomponent systems, and some of the most important concepts developed in physics have emerged from such studies, in particular, the ideas connected with the phenomena of phase transitions. The activities which sprang from applying these and other concepts and methods of physics to the study of information processing in neural networks have developed at an enormous pace recently, and it is felt that in this emerging multidisciplinary field, timely reviews and monographs integrating physics, neurobiology, and computer science are urgently needed. In our opinion, integrating these three disciplines is the most promising path to new insights in many aspects of the brain's functioning and to efficient architectural designs and algorithms for massively parallel computers.

We intend the new series *Physics of Neural Networks* to be aimed at the level of beginning graduate students, that is, to be accessible to a wide readership. This holds both for monographs to come and for the present and future collections of in-depth reviews, written by internationally renowned experts. After all, who can better explain relevant new ideas than the person who has introduced and

[1] Introductory Note to *Phase Transitions and Critical Phenomena*, Vol. I, edited by C. Domb and M.S. Green (Academic, New York 1972)

exploited them extensively in his/her own research? Though we shall not hesitate to invite papers, we encourage prospective authors to contact us.

We hope that readers of the series *Physics of Neural Networks* will derive ideas, motivation, and enjoyment from its pages.

Rehovot, *E. Domany*
Munich, *J.L. van Hemmen*
Urbana, December 1990 *K. Schulten*

Preface to the Second Edition

A steady demand for Volume 1 of our series *Models of Neural Networks* has invited us to provide a revision after the corrected 2nd printing of 1992 went out of stock last year. Most notably, Chap. 1 has been expanded so as to include a full treatment of graded-response (analog) neurons. We have also inserted some new data concerning the storage capacity of the Hopfield model and replica symmetry breaking.

The Editors

Preface to the First Edition

This book offers a multifaceted presentation of several main issues in the theory of neural networks that have recently witnessed considerable progress: statistics and dynamics of Hopfield-type nets with symmetric and asymmetric couplings, learning algorithms, temporal association, structured data (software), and structured nets (hardware). We consider these issues in turn.

Each review collection that is to appear in the series *Physics of Neural Networks* is intended to begin with a longer paper that puts together the theoretical foundations of the articles that follow. Here the introductory chapter, authored by van Hemmen and Kühn, concentrates on the long neglected *Collective Phenomena in Neural Networks*. It shows that the physical insights and techniques that have been obtained from both equilibrium and nonequilibrium statistical mechanics and the theory of spin glasses can be extremely fruitful. It also shows that Yang's prophetic remarks (see the Foreword), which were written two decades ago, have been vindicated in a rather surprising way.

The present series of review papers, then, starts with Braitenberg's thought-provoking *Information from Structure: A Sketch of Neuroanatomy*. In our opinion, theory needs feedback from experimental neurobiology.

A learning algorithm should solve the problem of coding the information presented to a network. Forrest and Wallace review in *Storage Capacity and Learning in Ising-Spin Neural Networks* the beautiful work of the Edinburgh group, in particular that of the late Elizabeth Gardner, and show how the different learning algorithms are implemented. In their paper *Dynamics of Learning*, Kinzel and Opper emphasize supervised learning (adaline and perceptron). They also estimate the learning time and provide additional information on forgetting.

The data that are offered to a network and have to be learned are usually assumed to be random. This kind of data is convenient to generate, often allows an analytical treatment, and avoids specific assumptions. It is known from real-life situations, though, that information can also be hierarchically structured, being tree-like with branches that become finer as we proceed. Here is an academic example: physics (an overall notion), theoretical physics (a category), the theory of neural networks (a class), the theory of learning (a family of notions) to which, say, the perceptron algorithm belongs. In their paper *Hierarchical Organization of Memory*, Feigel'man and Ioffe describe how this type of information can be modeled and how the memory's performance can be optimized.

Both software and hardware determine the performance of a network. In the present context, a hardware setup means either specifying the synapses, which contain the information, or devising a layered structure. At the moment, two types of synaptic organization of a single network (without hidden units) allow for an exact analytic treatment of the dynamical response and the storage capacity. One treats either a fully connected network with finitely many patterns or a net which is so diluted that dynamical correlations between different neurons need not be taken into account. In their paper *Asymmetrically Diluted Neural Networks*, Kree and Zippelius analyze this second type of network, which has proven useful in several different contexts.

Compared with stationary data, the storage and retrieval of patterns that change in space *and time* require a new type of synaptic organization. Various approaches to *Temporal Association* are analyzed by Kühn and van Hemmen. Not only do they examine the performance but they also compare theoretical predictions of the timing with simulation results, i.e. experiment. It seems that Hebbian learning, for long a venerable abstract issue but revived recently both in experimental work and in theory through appropriate techniques and mathematical implementation, gives rise to unexpected possibilities.

Only a part of the brain's architecture is genetically specified. Most of the synaptic connections are achieved through self-organization in response to input percepts. In their paper *Self-organizing Maps and Adaptive Filters*, Ritter, Obermayer, Schulten, and Rubner study competitive learning networks and show how self-organizing feature maps *à la* Kohonen are able to generate connections between, for example, the retina and the visual cortex. They provide an analysis of the formation of "striped" projections, well known from the work of Hubel and Wiesel, and present simulations of a system containing $2^{16} = 65\,536$ neurons in the (primary) cortex. Though the system size surpasses most of what has been done so far, the performance of the brain is even more impressive, if one realizes that several hours on the Connection Machine, one of the world's most powerful parallel computers, were required to simulate a system with about as many neurons as are contained in $1\,\text{mm}^3$ of the brain.

It has been known for a long time that the cerebral cortex has a specific "hardware" structure in that it consists of six layers. Regrettably, present-day theory falls short of appropriately explaining this extremely complicated system. In spite of that, it is a challenging, and rewarding, effort to try to catch the essence of information processing in layered feed-forward structures. In their essay *Layered Neural Networks*, Domany and Meir present an in-depth analysis.

Several promising developments in the theory of neural networks, in particular the idea of abstractly analyzing a task in the space of interactions, were initiated by the late Elizabeth Gardner. Her ideas permeate several chapters of this book. She died in 1988 at the age of thirty. It is with great respect and admiration that we dedicate the first volume of *Models of Neural Networks* to her.

The Editors

Contents

Contributors

Braitenberg, Valentino
MPI für Biologische Kybernetik, Spemannstrasse 38,
D-72076 Tübingen, Germany

Domany, Eytan
Department of Electronics, Weizmann Institute of Science,
76100 Rehovot, Israel

Feigel'man, Michail V.
Landau Institute for Theoretical Physics,
117940 Moscow, Russia

Forrest, Bruce M.
Institut für Polymere, ETH Zentrum,
CH-8092 Zürich, Switzerland

van Hemmen, J. Leo
Institut für Theoretische Physik, Physik-Department,
Technische Universität München,
D-85747 Garching bei München, Germany

Ioffe, Lev B.
Department of Physics, Rutgers University,
New Brunswick, NJ 08855-0849, USA

Kinzel, Wolfgang
Lehrstuhl für Theoretische Physik, Universität Würzburg,
Am Hubland, D-97074 Würzburg, Germany

Kree, Reiner
Institut für Theoretische Physik, Universität Göttingen,
Bunsenstrasse 9, D-37073 Göttingen, Germany

Kühn, Reimer
Institut für Theoretische Physik, Universität Heidelberg,
Philosophenweg 19, D-69120 Heidelberg, Germany

Meir, Ronny
Department of Electrical Engineering, Technion,
32000 Haifa, Israel

Obermayer, Klaus
Technische Universität Berlin, Fachbereich Informatik,
Franklinstrasse 28/29, D-10587 Berlin, Germany

Opper, Manfred
Institut für Theoretische Physik, Universität Würzburg,
Am Hubland, D-97074 Würzburg, Germany

Ritter, Helge
Technische Fakultät, Universität Bielefeld,
Universitätsstrasse 25, D-33615 Bielefeld, Germany

Rubner, Jeanne
Süddeutsche Zeitung, Abt. Wissenschaft,
Sendlinger Strasse 8, D-80331 München, Germany

Schulten, Klaus
Beckman Institute and Department of Physics,
University of Illinois at Urbana Champaign,
1110 W. Green Street, Urbana, IL 61801, USA

Wallace, David J.
Loughborough University of Technology, Ashby Road,
Loughborough, LE11 3TU, United Kingdom

Zippelius, Annette
Institut für Theoretische Physik, Universität Göttingen,
Bunsenstrasse 9, D-37073 Göttingen, Germany

1. Collective Phenomena in Neural Networks

J. Leo van Hemmen and Reimer Kühn

With 15 Figures

Synopsis and Note. In this paper we review some central notions of the theory of neural networks. In so doing we concentrate on collective aspects of the dynamics of large networks. The neurons are usually taken to be formal but this is not a necessary requirement for the central notions to be applicable. Formal neurons just make the theory simpler.

There are at least two ways of reading this review. It may be read as a self-contained introduction to the theory of neural networks. Alternatively, one may regard it as a *vade mecum* that goes with the other articles in the present book and may be consulted if one needs further explanation or meets an unknown idea.. In order to allow the second approach as well we have tried to keep the level of redundancy much higher than is strictly necessary. So the attentive reader should not be annoyed if (s)he notices that some arguments are repeated.

Equations are labeled by (x.y.z). Referring to an equation within a subsection we only mention (z), within a section (y.z), and elsewhere in the paper (x.y.z). The chapter number is ignored.

The article also contains some new and previously unpublished results.

1.1 Introduction and Overview

In this section some basic notions are introduced, a few hypotheses and underlying ideas are discussed, and a general outline of the paper is given.

1.1.1 Collective Phenomena in a Historical Perspective

Learning and recollection of data, such as intricate patterns, rhythms, and tunes, are capabilities of the brain that are used (surprisingly) fast and efficiently. The underlying neural processes are of a distributive and collective nature. As early as the 1940s, there was a vague feeling that understanding these collective phenomena *per se* is a key challenge to any theory that aims at modeling the brain. That this feeling was vague and could not yet be made precise is nicely illustrated by Chap. II of Wiener's *Cybernetics* [1.1]. The book appeared in 1948. Its second chapter is devoted to "Groups and statistical mechanics". Intuitively, Wiener must have realized that statistical mechanics is ideally suited to analyzing collective phenomena in a network consisting of very many, relatively simple, constituents. Explicitly, he starts out by introducing Gibbs as one of the founding fathers of statistical mechanics, credits him[1] for the ergodic hypothesis (time average = phase average), and proceeds to stress the relevance of Birkhoff's ergodic theorem (the mere existence of the time average) in conjunction with the existence of many (!) *ergodic components* [1.1: p. 69]. But he then finishes the chapter by treating entropy and ... Maxwell's demons.

Entropy is related to thermal disorder, noise, and the notion of temperature (T). In modern terms, ergodic components are energy ($T = 0$) or "free-energy" ($T > 0$) valleys in phase space. For a large system and at low enough temperatures, the barriers are so high that the system will stay in a specific valley and explore it "completely". In the context of a neural network, valleys in phase space are to be associated with patterns: specific configurations that have been stored.

It is somewhat surprising and in any case deplorable that, though Wiener did realize the correct interpretation of the ergodic decomposition [1.1: p. 69, bottom], he did not grasp the physical significance of the ideas he had just expounded. It was more than thirty years before Hopfield [1.2] took up the lead left by Wiener. It has to be admitted, though, that the idea of putting each pattern at the bottom of a valley (ergodic component) and allowing a dynamics that minimizes the (free) energy so that the valley becomes a domain of attraction of a pattern is novel and probably too radical to have been accessible to Wiener.[2]

The Hopfield picture [1.2] can be explained most easily, but not exclusively [1.5], in terms of formal neurons, a concept dating back to McCulloch and Pitts [1.6]. In a paper of unusual rigor these authors have shown that a task performed by a network of "ordinary" neurons can be executed equally well by a collection of formal neurons: two-state, "all-or-none" variables. This kind of variable can be modeled by Ising spins $S_i = \pm 1$, where i labels the neurons and ranges between 1 and N, the size of the network. A pattern is now a *specific* Ising-spin configuration, which we denote by $\{\xi_i^\mu; 1 \leq i \leq N\}$. So the label of the pattern is μ, and ranges, say, between 1 and q.

Let J_{ij} be the synaptic efficacy of j operating on i. Then the postsynaptic potential or, in physical terms, the local field h_i equals $\sum_j J_{ij} S_j$. Hopfield took random patterns where $\xi_i^\mu = \pm 1$ with probability 1/2, assumed

$$J_{ij} = N^{-1} \sum_\mu \xi_i^\mu \xi_j^\mu \equiv N^{-1} \boldsymbol{\xi}_i \cdot \boldsymbol{\xi}_j \; , \tag{1.1.1}$$

and allowed a (sequential) dynamics of the form

$$S_i(t + \Delta t) = \mathrm{sgn}\,[h_i(t)] \; , \tag{1.1.2}$$

[1] instead of Boltzmann

[2] We will not review here the work of Rosenblatt and Widrow et al. (see the articles by Forrest and Wallace, and Kinzel and Opper in this book) but only refer the reader to an index composed by Posch [1.3] and the excellent analysis and critique of the early perceptron theory by Minsky and Papert [1.4].

where $\mathrm{sgn}(x)$ is the sign of x. It then turns out (see below) that the dynamics (2) is equivalent to the rule that the state of a neuron is changed, or a spin is flipped, if and only if the *energy*

$$H_N = -\tfrac{1}{2} \sum_{i \neq j} J_{ij} S_i S_j \tag{1.1.3}$$

is lowered. That is, the Hamiltonian H_N is a so-called Lyapunov functional for the dynamics (2), which therefore converges to a (local) minimum or ground state of H_N. If q is not too large, the patterns are identical with, or very near to, ground states of H_N, i.e., each of them is at the bottom of a valley – as predicted.

1.1.2 The Role of Dynamics

The above argument in fact holds for any *symmetric* coupling $J_{ij} = J_{ji}$, as was stressed by Hopfield himself [1.2]. Furthermore, the search for ground states typically belongs to the realm of equilibrium statistical mechanics [1.7]. One might object, though, that the notion of temperature is still absent. It is, however, easily included [1.8]: for positive temperatures, we introduce a *stochastic* dynamics. In so doing we lift for a moment the above symmetry condition.

Let $\beta = 1/k_{\mathrm{B}}T$ denote the universe temperature. Here k_{B} is Boltzmann's constant and $k_{\mathrm{B}}T$ has the dimension of energy. Our units are chosen in such a way that $k_{\mathrm{B}} = 1$. We now specify the probability that the spin at i is flipped. There are two canonical ways of doing so [1.9]. We first compute the local field $h_i(t) = \sum_{j(\neq i)} J_{ij} S_j(t)$. Then the Glauber dynamics [1.10] is obtained through the transition probability

$$\mathrm{Prob}\,\{S_i \rightarrow -S_i\} = \tfrac{1}{2}[1 - \tanh(\beta h_i S_i)]\ , \tag{1.2.1}$$

while the Monte Carlo dynamics [1.11] is specified by

$$\mathrm{Prob}\,\{S_i \rightarrow -S_i\} = \begin{cases} 1, & \text{if } h_i S_i \leq 0 \\ \exp(-2\beta h_i S_i), & \text{if } h_i S_i > 0 \end{cases}\ . \tag{1.2.2}$$

In theoretical work the use of Glauber dynamics is recommended since it gives rise to simpler formulae. Furthermore, Peretto [1.8: p. 57] has given convincing arguments why a Glauber dynamics with Δt scaling as N^{-1} provides a fair description of the neural dynamics itself. This specific scaling with the system size N guarantees that each spin (formal neuron) is updated a finite number of times per second.

Given the local fields $h_i(t)$, one can use either dynamics *sequentially*, updating one spin after the other, or *in parallel*, updating all spins at the same time. As $\beta \rightarrow \infty$, either dynamics requires that h_i and S_i have the same sign. If they do, then $h_i(t)S_i(t) > 0$. If they do not, then $h_i(t)S_i(t) < 0$. In both cases the assertion follows from (1) and (2). It can be summarized by saying $S_i(t + \Delta t) = \mathrm{sgn}\,[h_i(t)]$.

In the case of *symmetric* couplings, the energy may be written $-1/2 \sum_j h_j S_j$ and the energy change after flipping a single spin, say at i, is $\Delta E = 2h_i S_i$. If we use sequential dynamics, the discussion of the previous paragraph directly implies that, as $\beta \to \infty$, a spin is flipped only if $\Delta E < 0$. (The reader may decide for himself what to do if $\Delta E = 0$.)

Both the Glauber and the Monte Carlo dynamics are Markov processes and, provided one performs them sequentially, both converge [1.9], as time proceeds, to an equilibrium distribution: the Gibbs distribution $Z_N^{-1} \exp(-\beta H_N)$, where

$$Z_N = \mathrm{Tr}\left\{e^{-\beta H_N}\right\} \tag{1.2.3}$$

is the partition function, widely studied in equilibrium statistical mechanics. The trace in (3) is a sum over all 2^N Ising spin configurations. As $\beta \to \infty$ the Gibbs distribution singles out the minima of H_N and, as we have just seen, these are precisely the states to which both dynamics converge.

As an approximation to a very large but finite network one takes [1.7] the limit $N \to \infty$ and studies the free energy $f(\beta)$,

$$-\beta f(\beta) = \lim_{N\to\infty} N^{-1} \ln Z_N \ , \tag{1.2.4}$$

as a function of β and other parameters, such as the number of stored patterns. The advantages of the thermodynamic or bulk limit $N \to \infty$ are well known [1.7]. If transitions occur as β or other parameters are varied, they usually show up as nonanalyticities in $f(\beta)$. And they do so clearly only in the limit $N \to \infty$. Furthermore, in all cases where $f(\beta)$ can be evaluated analytically, the very same limit is instrumental.

In general [1.7], $f(\beta) = u(\beta) - Ts(\beta)$, where $u(\beta)$ is the energy and $s(\beta)$ is the *entropy* (per spin or formal neuron). Entropy is *not* a function on phase space but measures the "degeneracy" of the energy levels which are sampled at inverse temperature β. Only at zero temperature are we left with the energy and is the dynamics such that H_N as given by (1.3) is minimized. We will see later on that the evaluation of $f(\beta)$ through (4) provides us with all the information which we need to determine the stability of the ergodic components, the retrieval quality, and the storage capacity of the network. Moreover, it will turn out that for the models under consideration we can find a free-energy functional which (i) is minimized under dynamics and (ii) reduces to "$N^{-1} H_N$" as $\beta \to \infty$. In the context of neural networks the efficacy of this procedure was demonstrated in a series of beautiful papers [1.13] by Amit, Gutfreund, and Sompolinsky, whose ideas we will discuss at length in Sect. 1.3.

In summary, a well-chosen dynamics such as Glauber or Monte Carlo converges to an equilibrium characterized by the Gibbs distribution (3). In this way one reconciles statics and dynamics and makes the behavior of the latter accessible to techniques devised to analyze the former.

1.1.3 Universality, Locality, and Learning

One might complain, though, that formal neurons, random patterns, and symmetric interactions are quite far from reality. Are these serious objections? The symmetry condition can be left out without destroying the associative memory. As to formal neurons, random patterns, and other simplifications which will arise in due course, the reply is the following. It is well known from physics [1.7], in particular critical phenomena [1.12], that in studying *collective* aspects of a system's behavior one may introduce several, even severe, simplifications without essentially altering the conclusions one arrives at. This is called *universality*. For example, random patterns certainly do not resemble the ones we are used to, but they allow analytic progress and insight which are hard to obtain in any other way, and, as we will see shortly, they capture many aspects of reality much better than one might naively expect.

There is another topic which has not been discussed yet but whose explanation becomes much simpler if one assumes symmetric couplings. We want a memory to be *content*-addressable or autoassociative, i.e., the memory should not operate by label, as most computers do, but by content. Suppose that in one way or another we have stored several (stationary) patterns in the J_{ij} and that the system is offered a noisy version of one of them. If the noise was not too strong, the system remains in the valley associated with that pattern and under its natural dynamics it will relax *by itself* to the (free-) energy minimum where the stored patterns live. See Fig. 1.1.

That is, the system has recalled the pattern. If the procedure were by label, the system would have to go through *all* the stored patterns, compare them with the noisy one, and choose the patterns which give the best fit. With human memory, such a procedure does not seem plausible.

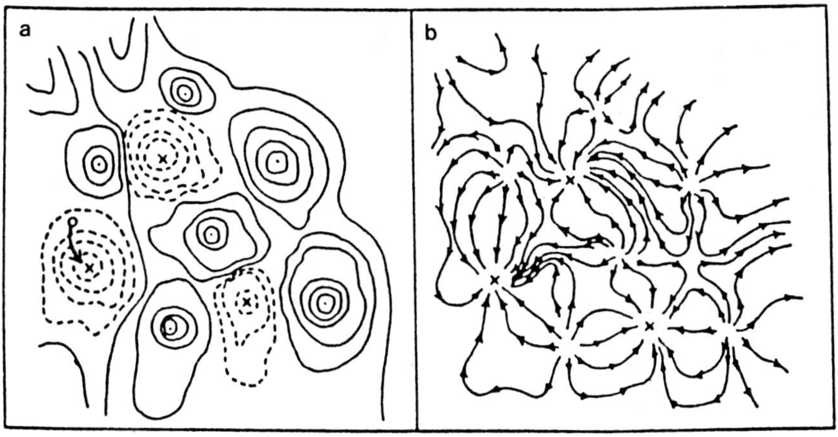

Fig. 1.1. (**a**) Energy landscape. The phase space is two-dimensional and the contour lines indicate the height of the energy surface. *Solid lines* indicate a hill and *dashed lines* a valley. From a noisy pattern (o, to the left) the system relaxes back to the original (x) at the bottom of the valley. (**b**) Flow lines of the dynamics, a down-hill motion, in the energy landscape of a. After [1.14]

In statistical mechanics, one is usually given the J_{ij} and one of the first tasks consists in finding the minima of the Hamiltonian H_N. In the theory of neural networks, however, one is given the patterns and one is asked to solve the *inverse* problem: finding J_{ij} such that the patterns are minima of H_N. Learning, then, is related to solving this inverse problem.

But how does the system learn? Or in computer-science terms, how does it program the J_{ij}? There is an important physiological requirement, *locality*, which can be considered as a generalization of an old rule dating back to Hebb [1.15]. Locality means that J_{ij} is determined by the information available to the pre- and postsynaptic neurons j and i; e.g., one requires that J_{ij} depend only on $\boldsymbol{\xi}_i = (\xi_i^\mu; 1 \leq \mu \leq q)$ and $\boldsymbol{\xi}_j = (\xi_j^\mu; 1 \leq \mu \leq q)$. In more mathematical terms [1.16, 17],

$$J_{ij} = N^{-1} Q(\boldsymbol{\xi}_i; \boldsymbol{\xi}_j) \tag{1.3.1}$$

for some synaptic kernel Q defined on $\mathrm{R}^q \times \mathrm{R}^q$. The learning procedure is now hidden in Q. For stationary patterns and $Q(\boldsymbol{\xi}_i; \boldsymbol{\xi}_j) = \boldsymbol{\xi}_i \cdot \boldsymbol{\xi}_j$, one recovers the Hebb rule (1.1) as it was used by Hopfield [1.2]. In what follows, however, we will frequently have occasion to use *non*symmetric Qs.

1.1.4 Outline of this Paper

The aim of this paper is rather modest. We want to isolate some unifying concepts and indicate some important ideas and techniques. Of course, the authors – as well as the readers – have a bias, but we think that it is not too hard to acquire a multifaceted approach by browsing through or studying the other articles in this book. There one can also find many applications (the flesh) of the theoretical structures (the bones) which are to be explored here.

In Sect. 1.2 we treat some basic notions, viz., large deviations, sublattice magnetizations, and the replica method. In Sect. 1.3, we apply these techniques to the Hopfield model, first with finitely many, then with extensively many patterns ($q = \alpha N$). We then turn to nonlinear neural networks in equation Sect. 1.4, where $Q(\boldsymbol{\xi}_i; \boldsymbol{\xi}_j)$ is not of the simple linear form $\boldsymbol{\xi}_i \cdot \boldsymbol{\xi}_j$ as in the Hopfield model. A typical example is provided by *clipped* synapses with $Q(\boldsymbol{\xi}_i; \boldsymbol{\xi}_j) = \mathrm{sgn}\,(\boldsymbol{\xi}_i \cdot \boldsymbol{\xi}_j)$. Plainly, the sign function clips quite a bit of information, but in spite of that the storage capacity and retrieval quality are hardly reduced. One may consider this another indication of universality.

In Sect. 1.5, we treat Hebbian learning in some detail and sketch how nonstationary patterns such as cycles (rhythms and tunes like BACH) can be memorized by taking advantage of delays. Forgetting can also be modeled satisfactorily as an *intrinsic* property of the memory. Through forgetting the memory creates space for new patterns.

The types of model which have been referred to so far and will be discussed in this paper can be described by the following set of equations.

(i) The Hopfield model,

$$NJ_{ij} = \sum_{\mu} \xi_i^{\mu} \xi_j^{\mu} \equiv \boldsymbol{\xi}_i \cdot \boldsymbol{\xi}_j \, , \tag{1.4.1}$$

where the ξ_i^{μ} are independent, identically distributed random variables which assume the values ± 1 with equal probability. If $q = \alpha N$ and $\alpha > \alpha_c \simeq 0.14$, the network has lost its memory completely.

(ii) Nonlinear neural networks of the *inner-product* type,

$$NJ_{ij} = \sqrt{q}\phi(\boldsymbol{\xi}_i \cdot \boldsymbol{\xi}_j / \sqrt{q}) \tag{1.4.2}$$

for some odd function ϕ. The linear $\phi(x) = x$ reduces (2) to (1), and clipping means that $\phi(x) = \text{sgn}\,(x)$. Here we also allow $q = \alpha N$ for some $\alpha > 0$.

(iii) General nonlinearity *à la* Hebb,

$$NJ_{ij} = Q(\boldsymbol{\xi}_i ; \boldsymbol{\xi}_j) \, , \tag{1.4.3}$$

can be solved exactly only if the number of patterns q is finite. However, a weak invariance condition already allows q to increase with N – without destroying the stability of the model.

(iv) Forgetful memories which are characterized by the iterative prescription

$$J_{ij}^{(\mu)} = \phi(\varepsilon_N \xi_i^{\mu} \xi_j^{\mu} + J_{ij}^{(\mu-1)}) \, . \tag{1.4.4}$$

If $\phi(x)$ saturates as $|x| \to \infty$, the memory creates storage capacity for new patterns by forgetting the old ones. We parenthetically note that (iv) is a special case of (iii).

(v) Temporal association. Now the problem is to store and retrieve a *sequence* of patterns. This is done, for instance, through

$$NJ_{ij} = \sum_{\mu} \xi_i^{\mu} \xi_j^{\mu} + \varepsilon \sum_{\mu} \xi_i^{\mu+1} \xi_j^{\mu} \, , \tag{1.4.5}$$

where the second term on the right is associated with a temporal *delay* τ. One imagines that the second term "pushes" the system through an energy landscape created by the first. We will see, though, that this picture is not quite right, that Hebbian learning with a broad distribution of delays encompasses (5), and that a sequence of patterns should be interpreted as a single, spatiotemporal one.

Hierarchically structured information is easier to remember – at least, to us. We will study this type of information in Sect. 1.6 and apply the results to low-activity patterns. The final section (1.7) is an outlook to future developments.

1.2 Prerequisites

In this section we outline a few unifying theoretical concepts and techniques such as large deviations, sublattices, and the replica method.

Large-deviation techniques have become a powerful tool for efficiently analyzing the equilibrium statistical mechanics of a neural network. Before treating the general principles (Sect. 1.2.2) we therefore start by critically examining a simple case (Sect. 1.2.1). The mathematical proofs, though relatively simple and elementary, have been relegated to a separate section (Sect. 1.2.3), which can be skipped on a first reading. In Sect. 1.2.4 we turn to the notion of sublattice and its associated order parameter, the sublattice magnetization. Finally, in Sect. 1.2.5, we present a short discussion of the replica method, which is – as we will see in Sects. 1.3 and 1.4 – an elegant and effective method for treating randomness analytically.

1.2.1 Large Deviations: A Case Study

Large-deviation theory [1.17–20] is concerned with determining the probability distribution of sums of random variables of the form

$$N^{-1}\sum_{i=1}^{N}\eta_i \equiv N^{-1}W_N \tag{2.1.1}$$

as $N \to \infty$. Determining a probability distribution means that one has to evaluate probabilities such as $\mathrm{Prob}\{N^{-1}W_N \geq \varepsilon\}$. When the theory was introduced in the 1930's, the η_i were independent random variables with mean zero. The event $\{N^{-1}W_N \geq \varepsilon\}$ is nothing but $\{W_N \geq \varepsilon N\}$, and it then means that the sum *deviates* from its mean (= 0) by εN, which is proportional to N – whence "large" compared to the $\sqrt{N}$ of the central-limit theorem [1.22]. So what is important is the scaling of W_N by N.

In the context of neural-network models, large deviations occur very naturally. Let us take, for instance, the Hopfield model. Combining (1.1.1) and (1.1.3) we can write the Hamiltonian

$$H_N = -\tfrac{1}{2}N\sum_{\mu}\left(N^{-1}\sum_{i=1}^{N}\xi_i^{\mu}S_i\right)^2 \equiv -\tfrac{1}{2}N\sum_{\mu}m_{\mu}^2\,, \tag{2.1.2}$$

where the quantities

$$m_{\mu} = N^{-1}\sum_{i=1}^{N}\xi_i^{\mu}S_i \tag{2.1.3}$$

are the so-called *overlaps*. We first list some simple properties of the overlaps and then show that they are of the form (1).

A pattern is a specific Ising-spin configuration. So $(\xi_i^{\mu})^2 = 1$. If $S_i = \xi_i^{\mu}$ for all i, then $m_{\mu} = 1$. Conversely, if $m_{\mu} = 1$, then $S_i = \xi_i^{\mu}$. In all other cases,

$m_\mu < 1$. In all other cases, $m_\mu < 1$ by the Cauchy–Schwarz inequality. In fact, if ξ_i^μ and S_i are uncorrelated, we may expect m_μ to be of the order of $N^{-1/2}$, since the sum consists of N terms, each containing a ξ_i^μ. In the Hopfield case, the ξ_i^μ are independent random variables with mean zero, and we may therefore apply the central-limit theorem [1.22]. On the other hand, if the S_i are positively correlated with the ξ_i^μ, then m_μ is of the order of unity. So the overlaps give precisely the *global* information we are interested in and hence are good order parameters [1.23]. They are also nice quantities to plot.

The Hamiltonian (2) depends only on the overlaps m_μ. Let us assume for the moment that their number (q) is fixed and finite. In evaluating the free energy (1.2.4) we have to perform the trace (1.2.3), i.e., the S_i have to be traced over. The ξ_i^μ are *fixed* random numbers since they belong to specific Ising-spin configurations which represent the memories of the system. We therefore interpret the S_i as independent random variables which assume the values ± 1 with equal probability. The trace $\mathrm{Tr}\{\ldots\}$, here a sum over all the 2^N Ising-spin configurations, then becomes a mathematical expectation by dividing it by 2^N so as to get the *normalized* trace, which we denote by $\mathrm{tr}\{\ldots\}$. And we have the logical chain

$$S_i \text{ random} \to \xi_i^\mu S_i \text{ random} \to m_\mu \text{ is of the form (1)} .$$

Since several m_μ occur in (2), the theory of large deviations should be – and is – able to specify the *common* probability distribution of at least finitely many m_μ. If so, performing the (normalized) trace is trivial because it is reduced to an integration over the m_μ with respect to a given density. We start, however, by studying the simpler case of a single overlap.

Since we will take the trace, we first perform the gauge transformation $\xi_i^\mu S_i \to S_i$ and are left with the *Curie–Weiss* Hamiltonian

$$H_N = -\tfrac{1}{2}JN\left(N^{-1}\sum_{i=1}^{N} S_i\right)^2 \equiv -\tfrac{1}{2}JNm_N^2 , \tag{2.1.4}$$

where the N in m_N explicitly denotes the dependence upon the system size N. To get a perfect agreement with (2), one has to put $J = 1$.

The free energy associated with (2.1.4) is

$$-\beta f(\beta) = \lim_{N\to\infty} N^{-1}\ln[\mathrm{tr}\{\exp(-\beta H_N)\}] + \ln 2 , \tag{2.1.5}$$

where tr is, as agreed, the normalized trace. We henceforth drop $\ln 2$. There are at least two ways of computing $f(\beta)$ which we consider in turn.

The first, which does not use the large-deviations philosophy, takes advantage of the *linearization trick*

$$\exp\left(\frac{1}{2}\lambda a^2\right) = \int_{-\infty}^{+\infty}\frac{\mathrm{d}z}{\sqrt{2\pi}}\exp\left(-\frac{1}{2}z^2 + \sqrt{\lambda}az\right), \tag{2.1.6}$$

an idea which dates back at least to Kac [1.24]. Combining (4) and (6) we have that the partition function $Z_N = \mathrm{tr}\{\exp(-\beta H_N)\}$ is given by

$$Z_N = \int \frac{\mathrm{d}z}{\sqrt{2\pi}} \mathrm{e}^{-z^2/2} \mathrm{tr}\left\{\exp\left(z\sqrt{\beta J N} m_N\right)\right\} \tag{2.1.7}$$

Since we have "linearized" m_N^2 in the exponent, the trace factorizes into a product over N sites and we find

$$Z_N = \int \frac{\mathrm{d}z}{\sqrt{2\pi}} \mathrm{e}^{-z^2/2} \left[\cosh\left(\sqrt{\frac{\beta J}{N}} z\right)\right]^N \tag{2.1.8}$$

Inserting (8) into (5) and substituting $z/\sqrt{N} := \sqrt{\beta J} y$ we obtain

$$\begin{aligned} -\beta f(\beta) = \lim_{N\to\infty} N^{-1} \ln \int_{-\infty}^{+\infty} \mathrm{d}y \\ \times \exp\left\{N\left[-y^2/2 + \ln\cosh(\sqrt{\beta J} y)\right]\right\} , \end{aligned} \tag{2.1.9}$$

which gives, through a Laplace argument,

$$-\beta f(\beta) = \sup_y \left[-y^2/2 + \ln\cosh(\sqrt{\beta J} y)\right] . \tag{2.1.10}$$

Two elements of the above derivation deserve to be noted: (i) the linearization in (7), which allows us to perform the trace, and (ii) the Laplace argument in (9), which is nothing but the statement that under rather general conditions [1.25]

$$\lim_{N\to\infty} N^{-1} \ln \int \mathrm{d}y\, \mathrm{e}^{N g(y)} = \sup_y g(y) . \tag{2.1.11}$$

So far so good. However, the argument is rather special because it *only* works if H_N is a *perfect square* or a sum of perfect squares. What can be said if, say, $-\beta H_N = NF(m_N)$ for some continuous function F? This question is quite sensible for, as we will see in Sects. 1.3 and 1.4, there are quite a few cases which cannot be "linearized". We therefore imagine that we are to derive the free energy of the Curie–Weiss Hamiltonian (4) *without* using the linearization trick (6). This directly leads us to a typical large-deviations argument [1.16, 21, 26].

To evaluate the (normalized) trace in (5), we note that the whole expression depends only on the magnetization m_N. It therefore seems reasonable to perform a coordinate transformation from the $S_i, 1 \le i \le N$, to m_N as a new "integration" variable with values between -1 and 1. Suppose we had found the corresponding density, to be called $\mathcal{D}_N(m)$. Then, as $N \to \infty$,

$$\mathrm{tr}\{\exp(N\beta J m_N^2/2)\} = \int_{-\infty}^{+\infty} \mathrm{d}m\, \mathcal{D}_N(m) \exp[N(\beta J m^2/2)] . \tag{2.1.12}$$

$\mathcal{D}_N(m)$ is easily found. We have

$$\mathrm{tr}\left\{\exp\left[N(\beta J m_N^2/2)\right]\right\} = \sum_{k=0}^{N} 2^{-N} \binom{N}{k} \exp\left[N\left(\beta J m_N^2(k)/2\right)\right] , \tag{2.1.13}$$

where $m_N(k) = N^{-1}[-(N-k)+k] = N^{-1}[2k-N]$ is the magnetization for $(N-k)$ spins down and k spins up. Hence $k = N(1+m)/2$ and, by Stirlings's formula,

$$\mathcal{D}_N(m) \sim 2^{-N}\binom{N}{N(1+m)/2} = \exp[-Nc^*(m)] \ , \tag{2.1.14}$$

where

$$c^*(m) = \tfrac{1}{2}[(1+m)\ln(1+m)+(1-m)\ln(1-m)] \tag{2.1.15}$$

if $|m| \leq 1$, and $+\infty$ elsewhere. Combining (5) and (12–15) we get, using a Laplace argument,

$$\begin{aligned} -\beta f(\beta) &= \lim_{N\to\infty} N^{-1}\ln\int_{-\infty}^{+\infty} \mathrm{d}m \exp\left[N\left(\beta J m^2/2 - c^*(m)\right)\right] \\ &= \sup_m \left[\beta J m^2/2 - c^*(m)\right] \ . \end{aligned} \tag{2.1.16}$$

The supremum is realized for those m which satisfy the fixed-point equation

$$\beta J m = \frac{\mathrm{d}}{\mathrm{d}m}c^*(m) = \tanh^{-1}(m) \Longrightarrow m = \tanh(\beta J m) \ . \tag{2.1.17}$$

We leave it to the reader to show that (10) and (16) are equivalent (see also below). Two elements of the above derivation deserve to be noted: (i) the fact that we have derived the probability distribution of m_N itself as $N\to\infty$ and in so doing have replaced the trace by an integration over m with density $\mathcal{D}_N(m)$ and (ii) the Laplace argument in (16). As we pointed out at the beginning of this subsection, deriving a probability distribution for m_N is at the heart of the large-deviations philosophy.

So far so good. We now have two ways of asymptotically evaluating the trace in (5). But once again one can complain. This time, that it was the simple discrete nature of the Ising spins which allowed a straightforward combinatorial evaluation of the density $\mathcal{D}_N(m)$ in (14). That is correct. An analogous derivation for, say, XY or Heisenberg spins would turn out to be prohibitively complicated. It is here, however, that the *theory* of large deviations, which we have not used until now, comes to our aid. In order not to obliterate the key ideas involved, we shall first isolate the general principles and relegate the proofs to a separate section.

1.2.2 Large Deviations: General Principles

Suppose we have a sequence of stochastic vectors $\boldsymbol{W}_N = \left(W_N^{(1)},\dots,W_N^{(q)}\right)$, where q is fixed and finite. For example, one might take $W_N^{(\mu)} = Nm_\mu, 1\leq\mu\leq q$, with the overlaps m_μ defined by (1.3). Large-deviations theory now provides us with the theorem that, as $N\to\infty$,

$$\begin{aligned} &\mathrm{Prob}\{m_\mu \leq N^{-1}W_N^{(\mu)} < m_\mu + \mathrm{d}m_\mu; 1\leq\mu\leq q\} \\ &\quad\sim \exp[-Nc^*(\boldsymbol{m})]\mathrm{d}\boldsymbol{m} \ , \end{aligned} \tag{2.2.1}$$

where $\mathrm{d}\boldsymbol{m} = \prod_\mu \mathrm{d}m_\mu$ and

$$c^*(\boldsymbol{m}) = \sup_{\boldsymbol{t}} [\boldsymbol{m} \cdot \boldsymbol{t} - c(\boldsymbol{t})] \tag{2.2.2}$$

is the *Legendre transform* [1.27] of the cumulant generating function or, for short, *c-function*

$$c(\boldsymbol{t}) = \lim_{N\to\infty} N^{-1} \ln \mathbb{E}_N \exp(\boldsymbol{t} \cdot \boldsymbol{W}_N) \,. \tag{2.2.3}$$

The symbol $\mathbb{E}_N$ denotes a mathematical expectation. For the theorem to hold, one needs some conditions. Here we assume that $c(\boldsymbol{t})$ exists and is differentiable. Since $c(\boldsymbol{t})$ is a limit of convex functions, it is convex itself and, by general theory [1.27], so is $c^*(\boldsymbol{m})$. Note that no explicit reference is made to the nature of the $\boldsymbol{W}_N$. They can be anything as long as (3) exists and is differentiable. We first show how to apply and simplify (1). Its proof can be found in the next section.

The expression (1) is pretty general. Let us therefore specialize to the Ising spins of the Curie–Weiss model. Then $\mathbb{E}_N$ is the normalized trace over the S_i and

$$c(t) = \lim_{N\to\infty} N^{-1} \ln \left[\mathrm{tr} \left\{ \exp\left(t \sum_{i=1}^{N} S_i \right) \right\} \right] = \ln[\cosh(t)] \,, \tag{2.2.4}$$

which is convex, as claimed, and differentiable. One easily verifies that $c^*(m)$ is given by (1.15). So (1), the assertion of the theorem, and (1.15), the result of a combinatorial argument, agree – as they should.

In the context of neural networks, a typical application of (1) is the evaluation of the free energy $f(\beta)$, say for finitely many patterns as in (1.2) or, more generally [1.28], for a Hamiltonian of the form $-\beta H_N = NF(m_\mu; 1 \le \mu \le q)$, where F is some function on $\mathbb{R}^q$. Note that the ξ_i^μ characterize the patterns and, hence, are fixed random numbers. We first compute the c-function:

$$\begin{aligned} c(\boldsymbol{t}) &= \lim_{N\to\infty} N^{-1} \ln \left[\mathrm{tr} \left\{ \exp\left(N \sum_\mu t_\mu m_\mu \right) \right\} \right] \\ &= \lim_{N\to\infty} N^{-1} \sum_{i=1}^{N} \ln \left[\cosh\left(\sum_\mu t_\mu \xi_i^\mu \right) \right] \\ &= \left\langle \ln \left[\cosh\left(\sum_\mu t_\mu \xi^\mu \right) \right] \right\rangle , \end{aligned} \tag{2.2.5}$$

where the angular brackets denote an average over the q random variables $\xi^\mu, 1 \le \mu \le q$, and the last equality follows from the strong law of large numbers [1.22]. Because $c(\boldsymbol{t})$ is differentiable, we can apply the theorem, i.e., (1), and write

$$-\beta f(\beta) = \lim_{N\to\infty} N^{-1} \ln \mathbb{E}_N \{\exp[NF(m_\mu; 1 \le \mu \le q)]\}$$

$$= \lim_{N\to\infty} N^{-1} \ln \int \mathrm{d}\boldsymbol{m} \exp\{N[F(\boldsymbol{m}) - c^*(\boldsymbol{m})]\}$$
$$= \sup_{\boldsymbol{m}} [F(\boldsymbol{m}) - c^*(\boldsymbol{m})] . \quad (2.2.6)$$

An explicit comment is in order. Equation (5) tells us that $c(\boldsymbol{t})$ does *not* depend on the specific random configurations we started with. Thus the density $\mathcal{D}_N(\boldsymbol{m})$ and the free energy itself do not depend on them either. That is, the free energy is a *self-averaging* quantity. This is a valuable property. In the present context, it is not self-evident since all spins (neurons) interact with each other so that the system cannot be split up into many independent subsystems (plus a neglible boundary term) and hence the strong law of large numbers cannot be applied directly [1.29].

Equation (6) suggests that one has to compute the Legendre transform $c^*(\boldsymbol{m})$ explicitly. This, however, is not the case. Life is much simpler. We will show [1.28] that if $c(\boldsymbol{t})$ is a (strictly) convex function with Legendre transform $c^*(\boldsymbol{m})$ and $F(\boldsymbol{m})$ is smooth, then

$$\sup_{\boldsymbol{m}} [F(\boldsymbol{m}) - c^*(\boldsymbol{m})] \quad (2.2.7)$$

may be written

$$\max_{\boldsymbol{\mu}} [F(\boldsymbol{\mu}) - \boldsymbol{\mu} \cdot \nabla F(\boldsymbol{\mu}) + c(\nabla F(\boldsymbol{\mu}))] , \quad (2.2.8)$$

where $\boldsymbol{\mu}$ satisfies the fixed-point equation

$$\boldsymbol{\mu} = \nabla c(\nabla F(\boldsymbol{\mu})) . \quad (2.2.9)$$

The proof is simple. We note that $\boldsymbol{t} \to \nabla c(\boldsymbol{t})$ is a mapping of $\mathbb{R}^n$, say, into $\mathbb{R}^n$. Its inverse exists and equals ∇c^* [1.27]. See, for instance, (1.17). (As a hint for the proof, note that the inverse of ∇c exists because c is strictly convex and finish the argument by using (11) and the implicit-function theorem.) Since the supremum in (7) is realized among those $\boldsymbol{m}$ which satisfy the relation $\nabla F(\boldsymbol{m}) = \nabla c^*(\boldsymbol{m})$, we immediately get the fixed-point equation

$$\boldsymbol{m} = \nabla c(\nabla F(\boldsymbol{m})) . \quad (2.2.10)$$

Its solutions are denoted by $\boldsymbol{\mu}$.

We now evaluate $c^*(\boldsymbol{\mu})$. By definition,

$$c^*(\boldsymbol{\mu}) = \sup_{\boldsymbol{t}} [\boldsymbol{\mu} \cdot \boldsymbol{t} - c(\boldsymbol{t})] . \quad (2.2.11)$$

To get the supremum we have to find a $\boldsymbol{t}$ so well behaved that

$$\nabla c(\boldsymbol{t}) = \boldsymbol{\mu} . \quad (2.2.12)$$

But $\boldsymbol{\mu}$ satisfies (10). By comparison we see that $\boldsymbol{t} = \nabla F(\boldsymbol{\mu})$. If we substitute this into (11) and return to (7), then (8) follows directly. The equations (8) and (9) will be used repeatedly [1.53]. It is a simple corollary of (8) that (1.10) and (1.16) are equivalent.

1.2.3 A Mathematical Detour

In view of its apparent simplicity one may wonder how (2.1) comes about. There exist some intricate proofs [1.18, 20] and we therefore would like to present a more intuitive and completely elementary argument [1.21]. The expert will easily notice that our probability estimates to be derived below are surprisingly sharp and, compared to [1.18, 20], are certainly good enough to provide an alternative proof of (2.6), including finite-N corrections. It may satisfy the reader, though, just to know the main results, (2.6–9). In that case, (s)he can proceed directly to the next section.

Instead of considering a sequence of random vectors $\boldsymbol{W}_N$ we take a single W_N. This simplifies the argument, which is, however, easily extended to the vector case. Let $\{W_N, \mu_N\}$ be a sequence of pairs of random variables W_N and probability measures μ_N. As before we require the sequence $\{W_N, \mu_N\}$ to be such that the corresponding c-function,

$$c(t) = \lim_{N\to\infty} N^{-1} \ln \mathrm{E}_N(\mathrm{e}^{tW_N}) \equiv \lim_{N\to\infty} c_N(t) \tag{2.3.1}$$

exists and is differentiable. We remind the reader that $\mathrm{E}_N(\ldots)$ is the mathematical expectation with respect to the probability distribution or measure μ_N. Instead of $\mathrm{E}_N(g) = \int \mathrm{d}\mu_N(\omega) g(\omega)$ we will also write $\mu_N(g)$. As always, $\mu_N(A)$ is the probability or measure of the set (event) A with respect to μ_N. Furthermore, $\mu_N(g; B)$ is the integral of g with respect to μ_N but *restricted* to these points ω where B holds; cf. (4) below.

It will now be shown that, under the aforementioned differentiability condition,

$$\mathrm{Prob}\{m \le N^{-1} W_N < m + \mathrm{d}m\} \sim \exp[-N c^*(m)] \mathrm{d}m \ , \tag{2.3.2}$$

where $c^*(m) = \sup_t \{mt - c(t)\}$ is the Legendre transform of $c(t)$. To verify (2) we use a key lemma, whose proof will be given at the very end of this section.

Lemma. If $c(t)$ in (1) is differentiable at $t = 0$, then whatever $\delta > 0$ there exists a constant $a(\delta) > 0$ such that, as $N \to \infty$,

$$\mathrm{Prob}\{|N^{-1} W_N - c'(0)| \ge \delta\} \le \mathrm{e}^{-N a(\delta)} \ . \tag{2.3.3}$$

Let us now consider a small open interval (a, b) surrounding $[m, m + \mathrm{d}m)$. We have

$$\begin{aligned} &\mu_N(a < N^{-1} W_N < b) \\ &\quad = \mu_N\left(\frac{\mathrm{e}^{tW_N}}{\mu_N(\mathrm{e}^{tW_N})} \mathrm{e}^{-tW_N} \mu_N(\mathrm{e}^{tW_N}); a < N^{-1} W_N < b\right) \\ &\quad \equiv \mu_N^*(\mathrm{e}^{-tW_N} \mu_N(\mathrm{e}^{tW_N}); a < N^{-1} W_N < b) \ . \end{aligned} \tag{2.3.4}$$

Since $Na < W_N < Nb$, we can easily estimate $\exp(-tW_N)$ in (4) and obtain

$$\begin{aligned} &\mathrm{e}^{-tNa}\mu_N(\mathrm{e}^{tW_N})\mu_N^*(a < W_N/N < b) \geq \ldots \\ &\quad \geq \mathrm{e}^{-tNb}\mu_N(\mathrm{e}^{tW_N})\mu_N^*(a < W_N/N < b)\,, \end{aligned} \tag{2.3.5}$$

where ... comprises the probability (2) we are interested in. We will show shortly that $\mu_N^*(a < W_N/N < b)$ converges to 1 as $N \to \infty$. Taking this for granted, applying logarithms, and dividing by N, we then get from (5)

$$ta - \frac{1}{N}\ln \mu_N(\mathrm{e}^{tW_N}) \leq -\frac{1}{N}\ln \mathrm{Prob}\left\{a < \frac{1}{N}W_N < b\right\} \leq tb - \frac{1}{N}\ln \mu_N(\mathrm{e}^{tW_N})$$

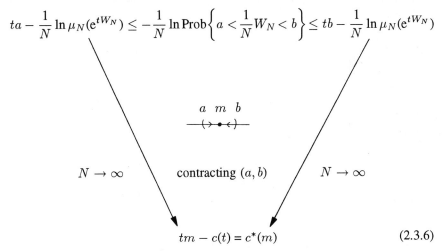

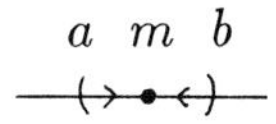

$N \to \infty$ contracting (a, b) $N \to \infty$

$$tm - c(t) = c^*(m) \tag{2.3.6}$$

where (6) holds provided t is chosen in such a way that $c'(t) = m$. By the lemma, $\mu_N^*(a < N^{-1}W_N < b) \to 1$. To see this, we calculate the c-function corresponding to $\{W_N, \mu_N^*\}$,

$$c_1(s) = \lim_{N\to\infty} N^{-1}\ln \mu_N^*(\mathrm{e}^{sW_N}) = c(s+t) - c(t)\,, \tag{2.3.7}$$

and note that $c_1'(0) = c'(t) = m$. So we are done.

The upshot of the above considerations is that to estimate the probability of a large deviation which is a rare event with respect to the original measure you shift or "translate" this measure so that your large deviation gets probability one with respect to the new measure. This is the *translation principle*, which goes back to Lanford [1.19]. The estimate (5) is extremely precise. It straightforwardly allows the computation of finite-size corrections to (6) and (2.6).

As promised, we now prove the lemma [1.29]. We are given that $c(t)$ is differentiable at $t = 0$ and we have to estimate $\mathrm{Prob}\{|N^{-1}W_N - c'(0)| \geq \delta\}$. For any $t > 0$,

$$\begin{aligned} &\mathrm{Prob}\left\{\frac{1}{N}W_N - c'(0) \geq \delta\right\} \\ &\quad = \mathrm{Prob}\{W_N - Nc'(0) \geq N\delta\} \\ &\quad \leq \int_{W_N - Nc'(0)\geq N\delta} \mathrm{d}\mu_N \exp[t(W_N - Nc'(0) - N\delta)] \end{aligned}$$

$$\leq \exp[-tN(c'(0)+\delta)] \int d\mu_N \exp(tW_N)$$
$$= \exp[tN(-\delta + t^{-1}c_N(t) - c'(0))] , \qquad (2.3.8)$$

where $c_N(t)$ has been defined in (1). Since $c(0) = 0$ and, by hypothesis, $c(t)$ is differentiable at $t = 0$, we choose a t_0 so small that $|t_0^{-1}c(t_0) - c'(0)| < \delta/4$. Because of (1), we next choose N so large that $t_0^{-1}|c_N(t_0) - c(t_0)| < \delta/4$. Then the last term in (8) can be majorized by $\exp[-(t_0\delta/2))N]$. The probability $\text{Prob}\{N^{-1}W_N - c'(0) \leq -\delta\}$ can be handled similarly, so the result is established.

1.2.4 Sublattice Magnetizations

According to Hebb [1.15], the synaptic efficacy J_{ij} is determined by the pre- and postsynaptic data at j and i, i.e., by the vectors $\boldsymbol{\xi}_j = (\xi_j^\mu; 1 \leq \mu \leq q)$ and $\boldsymbol{\xi}_i = (\xi_i^\mu; 1 \leq \mu \leq q)$. Accordingly we have [1.16, 17]

$$J_{ij} = N^{-1}Q(\boldsymbol{\xi}_i; \boldsymbol{\xi}_j) \qquad (2.4.1)$$

for some synaptic kernel Q on $\mathbb{R}^q \times \mathbb{R}^q$. Let q be fixed. The vector $\boldsymbol{\xi}_i$ varies as i travels from 1 to N but it is always on a corner of the hypercube $[-1, 1]^q$. The 2^q corners are denoted by C^q. We now want to introduce the sublattice idea [1.16, 17, 29, 31], which makes explicit the *homogeneity* in the system even though the $\boldsymbol{\xi}_i$ may be random vectors themselves.

Let $\boldsymbol{x} \in C^q$ be a specific corner. Then the *sublattice* belonging to $\boldsymbol{x}$ is

$$I(\boldsymbol{x}) = \{i; \boldsymbol{\xi}_i = \boldsymbol{x}\} . \qquad (2.4.2)$$

In words, $I(\boldsymbol{x})$ consists of all i with $\boldsymbol{\xi}_i = \boldsymbol{x}$. As $\boldsymbol{x}$ varies through the corners C^q of the hypercube we get all i between 1 and N and in this way obtain a (disjoint) partition of the index set $\{1, 2, \ldots, N\}$. The homogeneity we alluded to shows up in a twofold sense. First, by definition, $\boldsymbol{\xi}_i = \boldsymbol{x}$ for *all* $i \in I(\boldsymbol{x})$. Second, from (1) and (2), we obtain

$$J_{ij} = N^{-1}Q(\boldsymbol{x}; \boldsymbol{y}) \text{ for } \textit{all } i \in I(\boldsymbol{x}) \text{ and } \textit{all } j \in I(\boldsymbol{y}) . \qquad (2.4.3)$$

As shown in Sects. 1.4 and 1.5, these observations are the (indispensable) key to several analytic results.

In most models which are analytically soluble, the $\boldsymbol{\xi}_i$ are independent random vectors that assume $\boldsymbol{x}$ with probability $p(\boldsymbol{x})$. Then we have for the size $|I(\boldsymbol{x})|$ of the sublattice $I(\boldsymbol{x})$

$$N^{-1}|I(\boldsymbol{x})| \to p(\boldsymbol{x}) \quad \text{as } N \to \infty \qquad (2.4.4)$$

by the strong law of large numbers [1.22]. So $|I(\boldsymbol{x})| \propto N$ for all $\boldsymbol{x}$ which occur with nonzero probability $p(\boldsymbol{x}) > 0$. For nonrandom vectors one simply *requires* that (4) be true. Otherwise there is no well-defined limit behavior of the dynamics as $N \to \infty$. See Sect. 1.5 and Chap. 7.

The notion of sublattice $I(\boldsymbol{x})$ naturally leads to an order parameter associated with it, the *sublattice magnetization*

$$m(\boldsymbol{x}) = |I(\boldsymbol{x})|^{-1} \sum_{i \in I(\boldsymbol{x})} S_i \,. \tag{2.4.5}$$

It is clear that each sublattice magnetization $m(\boldsymbol{x})$ is of the same form as the total magnetization m_N in (1.4), except that N is to be replaced by the sublattice size $|I(\boldsymbol{x})|$. So we can apply all the considerations of Sect. 1.2.1 concerning the probability distribution of the $m(\boldsymbol{x})$. In particular, for Ising spins (1.14) holds if $|I(\boldsymbol{x})|$ increases with N as $N \to \infty$. This then provides us with a powerful tool for evaluating the free energy of a neural network with *arbitrary* symmetric kernel Q as long as q is finite; see Sect. 1.4.1.

In passing, we also note another interesting consequence of the homogeneity. The overlap with pattern μ, as defined by (1.3), may be written

$$\begin{aligned} m_\mu &= N^{-1} \sum_{i=1}^{N} \xi_i^\mu S_i = N^{-1} \sum_{\boldsymbol{x}} \sum_{i \in I(\boldsymbol{x})} \xi_i^\mu S_i \\ &= \sum_{\boldsymbol{x}} (N^{-1} |I(\boldsymbol{x})|) |I(\boldsymbol{x})|^{-1} \sum_{i \in I(\boldsymbol{x})} x_\mu S_i \\ &= \sum_{\boldsymbol{x}} p(\boldsymbol{x}) x_\mu m(\boldsymbol{x}) \equiv \langle x_\mu m(\boldsymbol{x}) \rangle \,. \end{aligned} \tag{2.4.6}$$

In obtaining this result, we used the definition of m_μ, the fact that the sublattices induce a partition of $\{1, 2, \ldots, N\}$, the homogeneity on each sublattice, and (4). So the m_μ follow once we know the $m(\boldsymbol{x})$.

1.2.5 The Replica Method

Patterns are *specific* phase-space configurations which are chosen once and for all. For analytical purposes, one usually takes random patterns where the ξ_i^μ are independent, identically distributed random variables. This is a theoretical formulation, an abstraction, of the fact that many real-life data hardly resemble each other. It is also hard to imagine how a true theory could be developed without the ansatz that patterns are random.

The ξ_i^μ being *fixed* random numbers, there does not seem any reason to average over them. Some quantities, however, such as the free energy in Sect. 1.2.3, converge to a nonrandom number as $N \to \infty$, even though they explicitly depend on the ξ_i^μ. That is, they are *self*-averaging. If so, we could have averaged them at the very beginning without altering the final result since it was nonrandom (deterministic). On the other hand, if we are not sure that for large N we are left with a deterministic quantity, e.g., in studying extensively many patterns, then averaging at least gives a unique answer.[3]

[3] Even in this case, the bulk limit need not exist, but in applying the replica method we will find that its average does.

Accordingly, it may be advantageous to average the free energy (1.2.4). This average must be computed *after* taking the logarithm but *before* taking the thermodynamic limit, giving

$$-\beta f(\beta) = \lim_{N\to\infty} N^{-1}\langle \ln Z_N \rangle \, . \tag{2.5.1}$$

Here the symbol $\langle \dots \rangle$ denotes averaging over the randomness. However, performing an average over many random variables seems hopeless because the logarithm prevents any useful factorization into a product of one-variable averages. A way out is offered by the replica method. The presentation below closely follows [1.32].

The conventional replica trick employs the relation

$$\langle \ln Z_N \rangle = \lim_{n\to 0} \frac{\langle Z_N^n \rangle - 1}{n} \, , \tag{2.5.2}$$

but we find it more convenient to use the equivalent identity

$$\langle \ln Z_N \rangle = \frac{\mathrm{d}}{\mathrm{d}n} \ln \langle Z_N^n \rangle |_{n=0} \, . \tag{2.5.3}$$

There are several advantages of this formulation, as will soon become apparent. Defining

$$\phi_N(n) = N^{-1} \ln \langle Z_N^n \rangle \tag{2.5.4}$$

we must calculate

$$-\beta f(\beta) = \lim_{N\to\infty} N^{-1}\langle \ln Z_N \rangle = \lim_{N\to\infty} \left(\frac{\mathrm{d}}{\mathrm{d}n} \phi_N(n)|_{n=0} \right) . \tag{2.5.5}$$

For positive integer n, apart from a factor $-\beta$, the quantity $\phi_N(n)$ is just the free energy per site of a replica system of N sites with n spins at each site. The replicas $1, 2, \dots, n$ all have the same random parameter set, but are uncoupled before averaging. The average $\langle \dots \rangle$ in (4) is to be computed *before* taking the logarithm. We therefore expect $\phi_N(n)$ to be meaningful in the thermodynamic limit, at least for n a positive integer, and define

$$\phi(n) = \lim_{N\to\infty} \phi_N(n) \, . \tag{2.5.6}$$

We also *assume* for the present that the limit in (6) exists for all *real* n. The function $\phi_N(n)$ is defined, by (4), for real n and all finite N, but it is not obvious that it tends to a limit at $N \to \infty$ unless n is a positive integer. The fact that it does indeed have a limit is central to the replica trick, and is essentially proved in [1.32].

For positive integer n, we may expect to be able to evaluate $\phi(n)$ explicitly for a specific system, as we shall do in Sect. 1.3 and 1.4. The use of the thermodynamic limit will be essential in that evaluation, and in general our hopes for exact evaluation may be high for $\phi(n)$ but are vanishingly low for $\phi_N(n)$.

Unfortunately, the prescription (5) requires the differentiation at $n = 0$ *before* use of the thermodynamic limit. This naturally raises the question: Under what circumstances can we interchange the order of the limit $N \to \infty$ and the differentiation with respect to n at $n = 0$? Or, more formally,

$$\lim_{N\to\infty} \left(\frac{\mathrm{d}}{\mathrm{d}n} \phi_N(n)|_{n=0} \right) = \frac{\mathrm{d}}{\mathrm{d}n} \left(\lim_{N\to\infty} \phi_N(n) \right) \Big|_{n=0} \equiv \phi'(0) \; ? \tag{2.5.7}$$

We note that both sides of (7) are well defined in the thermodynamic limit, in contrast to previous prescriptions, which involve $\langle Z_N^n \rangle$ rather than $\langle Z_N^n \rangle^{N^{-1}}$.

The definition (4), and an examination of the computations required, gives us little confidence in the possibility of explicitly evaluating $\phi_N(n)$ for anything but positive integer n, and then only in the limit $N \to \infty$. We must therefore ask: Under what conditions can we extend $\phi(n)$ from positive integer n to real or complex n in the neighborhood of $n = 0$? Symbolically,

$$\phi(n)\,, \quad n \in \mathbb{N} \to \phi(n)\,, \qquad\qquad n \in \mathbb{R} \text{ or } \mathbb{C} \; ? \tag{2.5.8}$$

There are many techniques that might appear useful in answering questions (7) and (8), but it seems in practice that these reduce to just two fundamental ideas. One is *analyticity* and the other is *convexity*; both are treated in [1.32]. Perhaps surprisingly, convexity turns out to be the more powerful tool.

Postponing a detailed discussion to subsequent sections, we see that we have effectively broken down the computation of $f(\beta)$ into a three-step process:

a) Calculate $\phi(n)$ for positive integer n.
b) Find an extension of $\phi(n)$ to $n \approx 0$, and then compute $\phi'(0)$, showing that the value so obtained is unique.
c) Prove that equation (7) is true, and hence that the result $\phi'(0)$ is indeed the required $-\beta f(\beta)$.

This procedure might justly be called the replica *method*, as opposed to the basic replica *trick* of (2) or (3).

We implement step (a) in the Sects. 1.3, 4, and obtain $\phi(n)$ for positive integer n as an *explicit* function of n. It then will turn out that *an* extension (step b) can be obtained easily by considering n as a real variable. In the present context, this is the so-called replica-symmetric extension of ϕ, since all replicas are treated on an equal footing. The extension need not be unique, however [1.33]; by Carlson's theorem – see the discussion in [1.32] – it only is, if $|\phi(n)| \leq a \exp[b\,Re(n)]$, where a and b are positive constants. Since numerical simulations have confirmed the analytical predictions based on steps (a) and (b) to high accuracy, the final step (c) will not be implemented here.

1.3 The Hopfield Model

The Hopfield model has been studied extensively. In this section we first turn to the case of finitely many patterns, which is exactly soluble (Sect. 1.3.1). We discuss various states which may occur at low temperatures and study their stability (Sect. 1.3.2). Then we carefully analyze the Hopfield model with extensively many patterns (Sect. 1.3.3), paying due attention to the underlying mathematical structures. It is also indicated how the noise generated by the *extensively* many patterns modifies the stability of the retrieval states (as compared to the finite-q case) and why something like a spin-glass state ought to appear in the phase diagram (Sect. 1.3.4). Section 1.3.5 is devoted to the internal consistency of the theory as illustrated by the zero-temperature entropy and a comparison of theoretical predictions with numerical simulations. Using the idea of duplicate spins we sketch in Sect. 1.3.6 how the very same model with parallel instead of sequential dynamics (the Little model [1.43]) can be treated analytically. In Sect. 1.3.7 we show how theory has to be modified if one replaces the Ising spins, which have a discrete distribution at ± 1, by continuous spins or, what is the same, graded-response neurons. Owing to the flexibility of our large-deviation method, the modification turns out to be straightforward.

1.3.1 The Hopfield Model with Finitely Many Patterns

As we have seen in Sect. 2.1, the Hopfield model is given by the Hamiltonian (2.1.2)

$$-\beta H_N = \frac{1}{2}\beta N \sum_{\mu=1}^{q} m_\mu^2 \equiv N F(\boldsymbol{m}) \,, \tag{3.1.1}$$

where the $m_\mu = N^{-1}\sum_i \xi_i^\mu S_i$ are the overlaps. In the case of finitely many patterns, q is finite and fixed as $N \to \infty$. The ξ_i^μ are random numbers, which have been chosen once and for all. Since the Ising spins S_i have to be traced over to obtain the free energy (1.2.4), we interpret them as random variables. The exact solution to the free energy $f(\beta)$ can then be derived in a few lines. According to Sect. 1.2.2,

$$\begin{aligned} -\beta f(\beta) &= \lim_{N\to\infty} N^{-1} \ln \operatorname{tr}\{\exp(-\beta H_N)\} \\ &= \lim_{N\to\infty} N^{-1} \ln \int \mathrm{d}\boldsymbol{m} \exp\{N[F(\boldsymbol{m}) - c^*(\boldsymbol{m})]\} \\ &= \sup_{\boldsymbol{m}} \,[F(\boldsymbol{m}) - c^*(\boldsymbol{m})] \,, \end{aligned}$$

where the second equality directly follows from (2.2.1) and $c^*(\boldsymbol{m})$ is the Legendre transform (2.2.2) of the c-function (2.2.5),

$$c(\boldsymbol{t}) = \left\langle \ln \left[\cosh \left(\sum_\mu t_\mu \xi^\mu \right) \right] \right\rangle . \tag{3.1.2}$$

The expression (2) does not depend on the specific random patterns we started with. Here – and elsewhere – angular brackets denote an average over random variables ξ^μ; in the present case, $1 \leq \mu \leq q$. Because $\nabla F(\boldsymbol{m}) = \beta \boldsymbol{m}$, the corresponding fixed-point equation is, by virtue of (2.2.9),

$$\boldsymbol{m} = \left\langle \boldsymbol{\xi} \tanh \left[\beta \left(\sum_\mu m_\mu \xi^\mu \right) \right] \right\rangle . \tag{3.1.3}$$

Finally, from (2.2.8), the free energy $f(\beta)$ may be written

$$-\beta f(\beta) = \max_{\boldsymbol{m}} \left[-\tfrac{1}{2} \beta \boldsymbol{m}^2 + c(\beta \boldsymbol{m}) \right] , \tag{3.1.4}$$

where $\boldsymbol{m}$ satisfies (3). We are done. Well, are we? The simplicity of (3) and (4) is a bit specious. In particular, (3) gives rise to more intricate solutions than one might guess at first sight.

The m_μs are the overlaps defined by (2.1.3) and their values are, as $N \to \infty$, determined by the fixed-point equation (3). This means that, for N large, the equilibrium or Gibbs state $Z_N^{-1} \exp(-\beta H_N)$ *lives on spin configurations* $\{S_i; 1 \leq i \leq N\}$ which are such that each of the m_μ has the value assigned to it by the fixed-point equation (3). An $m_\mu \neq 0$ indicates a nonzero overlap with pattern μ. We now list the solutions to (3).

i) *Retrieval states* (or patterns) constitute what we are after. They are solutions to (3) which have a nonzero overlap *with a single pattern only*. If the system is in pattern ν, than $m_\mu = \delta_{\mu\nu}$. So we make the ansatz $m_\mu = m\delta_{\mu\nu}$ in (3), check that this ansatz is consistent with (3), and find

$$m = \tanh(\beta m) . \tag{3.1.5}$$

For $\beta > 1$, we have a solution $m > 0$, which corresponds to an absolute maximum of the free-energy functional (4). As $\beta \to \infty$, m approaches 1 at an exponential rate.

ii) *n-symmetric states* [1.13a] are mixture states which correspond to the ansatz $\boldsymbol{m} = m_n(1, 1, \ldots, 1, 0, \ldots, 0)$ or *permutations thereof* with the $1n$ times. One easily verifies that this ansatz is consistent with (3) and

$$m_n = n^{-1} \left\langle \left(\sum_{\mu=1}^{n} \xi^\mu \right) \tanh \left[\beta m_n (\sum_{\mu=1}^{n} \xi^\mu) \right] \right\rangle . \tag{3.1.6}$$

For $n = 1$, (6) reduces to (5). As to stability (see below), it turns out [13a] that the states with n odd are stable at sufficiently low temperatures, whereas those with n even are unstable. At very low temperatures, $m_n \approx n^{-1}\langle | \sum_\mu \xi^\mu | \rangle$; it is of the order of $n^{-1/2}$ as n becomes large. The corresponding microscopic state is

$$S_i = \text{sgn}\left(\sum_\mu \xi_i^\mu\right), \tag{3.1.7}$$

where μ ranges through the labels of the n nonzero components (n odd).

iii) *Asymmetric (mixture) states* [1.13a] correspond for example to the ansatz $\boldsymbol{m} = (m_1, \ldots, m_1, m_2, \ldots, m_2, 0, \ldots, 0)$ or permutations thereof where m_i occurs n_i times, $i = 1, 2$. This specific state may also be called (n_1, n_2)-symmetric. A generalization to the case with $i > 2$ is plain. Once again it is easy to verify that the ansatz is consistent with (3). At zero temperature, one can characterize the *stable* mixture states explicitly [1.34]. Microscopically we have

$$S_i = \text{sgn}\left(\sum_\mu m_\mu \xi_i^\mu\right), \tag{3.1.8}$$

where $\boldsymbol{m} = (m_\mu)$ is a fixed real vector with the following properties:

a) $\boldsymbol{m}$ has a finite number of nonzero components. [This condition is empty as long as $q < \infty$ but it also holds if $q = \alpha N$ and $\alpha \leq \alpha_c$.]
b) $\pm m_1 \pm m_2 \pm \ldots \neq 0$ for any choice of $\pm$ signs.
c) For each nonzero m_μ, the sum $\pm m_1 \pm m_2 \pm \ldots$ has the same sign as $\pm m_\mu$ (this latter $\pm$ is the same choice as the one appearing before m_μ in the sum) for exactly a fraction $(1 + m_\mu)/2$ out of all possible sign choices.

For random patterns, a condition equivalent to (c) is
c′) $\langle S_i \xi_i^\mu \rangle = m_\mu$ for each μ. Here S_i is given by (8).

Examples are $(\frac{1}{2}, \frac{1}{2}, \frac{1}{2}, 0, \ldots)$, $(\frac{3}{8}, \frac{3}{8}, \frac{3}{8}, \frac{3}{8}, \frac{3}{8}, 0, \ldots)$ and $(\frac{1}{2}, \frac{1}{2}, \frac{1}{4}, \frac{1}{4}, \frac{1}{4}, 0, \ldots)$, representing a 3-symmetric, a 5-symmetric, and a (2,3)-symmetric state, respectively. Newman's elegant characterization [1.34] of the mixture states encompasses all previously known examples [1.13].

The retrieval states bifurcate from the paramagnetic state $\boldsymbol{m} = 0$ at $\beta = 1$. So a natural question is: When do the mixture states appear as we lower the temperature? It can be shown [1.13a, 16] that the n-symmetric states are the *only* ones that bifurcate from $\boldsymbol{m} = 0$ at $\beta = 1$. If n is even, they never become stable, whereas for n odd they do at temperatures T_n. In fact, one has [1.13a] in decreasing order $T_1 = T_c = 1$, $T_3 = 0.46$, $T_5 = 0.39$, $T_7 = 0.35$, and so on. Amit et al. [1.13a] argue – but do not prove – that the highest temperature where asymmetric states appear is below 0.58. A general proof that produces this kind of number seems to be extremely hard.

How many mixture states are there? Quite a few. Let us take, for instance, the n-symmetric states. Since the distribution of the ξ^μ is invariant under permutation of the μs and inversion of the ξs, we get $2^n \binom{q}{n}$ n-symmetric states. Summing over all n, we find 3^q symmetric states. Roughly half of them are stable at low temperatures. We leave it to the reader to try an analogous calculation for the asymmetric states.

1.3.2 Stability

If q is finite, the overlaps m_μ induce a *parametrization* of phase space. Each spin configuration $\{S_i; 1 \le i \le N\}$ determines a specific $\boldsymbol{m} = (m_\mu; 1 \le \mu \le q)$. Let us now consider the right-hand side of (1.4) more closely. We show that $R(\boldsymbol{m}) \equiv -\beta \boldsymbol{m}^2/2 + c(\beta \boldsymbol{m})$ is a Lyapunov function for the Glauber dynamics, i.e., $\mathrm{d}R(\boldsymbol{m}(t))/\mathrm{d}t \ge 0$, so that the dynamics relaxes to a *maximum* of $R(\boldsymbol{m})$, which is what we are after in (1.4). Under a Glauber dynamics and as $N \to \infty$ [see (5.4.13) with $\Gamma = 1$ and $\varepsilon = 0$], the m_μ satisfy the differential equation

$$\dot{m}_\mu = -m_\mu + \langle \xi^\mu \tanh[\beta(\boldsymbol{m} \cdot \boldsymbol{\xi})] \rangle \, , \tag{3.2.1}$$

where $\dot{m}_\mu$ denotes a differentiation of m_μ with respect to time. Hence we find

$$\frac{\mathrm{d}}{\mathrm{d}t} R(\boldsymbol{m}) = \beta[-\boldsymbol{m} + \nabla c(\beta \boldsymbol{m})] \cdot \dot{\boldsymbol{m}} = \beta \dot{\boldsymbol{m}}^2 \ge 0 \, , \tag{3.2.2}$$

as claimed. Therefore, dynamic stability and thermodynamic stability are equivalent.

A global maximum of $R(\boldsymbol{m})$ in (1.4) corresponds thermodynamically to a stable phase and a local maximum to a *meta*stable phase whereas a saddle point or a minimum is to be related to an unstable phase. Owing to (2), this distinction is of particular relevance to Glauber (and Monte Carlo) dynamics: once the system is in a stable or metastable phase it will never get out in a finite amount of time[4] ($N \to \infty$). On the other hand, unstable phases are left at a finite speed. A maximum of the free-energy functional corresponds to a minimum of "$f(\beta)$ itself". So we can say that the asymptotic behavior of the dynamics is determined by the structure of the *ergodic components* or, more loosely formulated, the free-energy valleys. These are labeled by the solutions m_μ of the fixed-point equation (1.3) and their thermodynamic stability tells us that a valley is "really a valley", i.e., concave *upwards*. So once the system is in a valley, it will never get out since the barriers have a height proportional to N and $N \to \infty$.

In summary, stability of a state corresponding to a solution $\boldsymbol{m}$ of (3) requires that the second derivative of R evaluated at $\boldsymbol{m}$ be negative-definite, i.e., the matrix with elements [1.28]

$$-\delta_{\mu\nu} + \beta \langle \xi^\mu \xi^\nu \cosh^{-2}[\beta(\boldsymbol{m} \cdot \boldsymbol{\xi})] \rangle \tag{3.2.3}$$

should have negative eigenvalues only. For instance, the retrieval states are stable for $\beta > 1$.

1.3.3 The Hopfield Model with Extensively Many (Weighted) Patterns

The Hopfield model with extensively many patterns ($q = \alpha N$) was first solved by Amit et al. [1.13b, c]. Their key idea was to single out *finitely* many patterns,

---

[4] If N is really infinite, the dynamics is governed by (1), which is completely deterministic, and the system will never get out. For finite N, the probability is exponentially small in N; the system has to mount a (free-) energy barrier of a height proportional to the system size N.

which we denote by μ, to assume that the system is in a state, i.e., a part of phase space where these patterns "live", and to treat the remaining, extensively many patterns, to be denoted by ν, as a *noise term* in the context of the replica method. To treat the noise generated by the ν-terms analytically, the second important idea of Amit et al. was to take it as Gaussian. As we will see, one has to be quite courageous to do so. The analysis below mainly follows [1.35], supplemented by some more recent ideas [1.36].

We consider the Hamiltonian (1.1.3) with exchange couplings

$$J_{ij} = N^{-1} \sum_{\lambda} \varepsilon_\lambda \xi_i^\lambda \xi_j^\lambda \,. \tag{3.3.1}$$

Each pattern λ has been given a weight ε_λ. For the time being the weights are arbitrary. If desired, one may assume $0 \le \varepsilon_\lambda \le 1$. We recover the original Hopfield model by putting $\varepsilon_\lambda = 1$ for $0 \le \lambda \le q = \alpha N$ and $\varepsilon_\lambda = 0$ for $\lambda > \alpha N$.

For later purposes it may be convenient to add external fields h_μ that single out the finitely many μ-patterns. Then the Hamiltonian may be written

$$\begin{aligned} -\beta H_N = & \left[\frac{\beta}{2N} \sum_\mu \varepsilon_\mu \left(\sum_{i=1}^N \xi_i^\mu S(i) \right)^2 + \sum_\mu \beta h_\mu \left(\sum_{i=1}^N \xi_i^\mu S(i) \right) \right] \\ & + \frac{\beta}{2N} \sum_\nu \varepsilon_\nu \left(\sum_{i=1}^N \xi_i^\nu S(i) \right)^2 \equiv -\beta H_N^{(1)} - \beta H_N^{(2)} \,, \end{aligned} \tag{3.3.2}$$

where $H_N^{(1)}$ and $H_N^{(2)}$ refer respectively to the μ- and the ν-patterns. Since the spins will soon carry replica indices, we have denoted them by $S(i)$ instead of the usual S_i. Moreover, compared to (1.1.3), we have included self-interactions J_{ii}. These are completely harmless. They just give the constant term $-\beta/2 \sum_\lambda \varepsilon_\lambda$, which has been dropped from (2) but will be included later on [in (30)].

The Gaussian Approximation. In the context of the replica method (Sect. 2.5), one first determines

$$\phi_N(n) = \frac{1}{N} \ln \langle Z_N^n \rangle \tag{3.3.3}$$

for positive integer n, takes the thermodynamic limit $N \to \infty$ so as to arrive at $\phi(n)$, and obtains an extension (here the replica-symmetric one) to a neighborhood of $n = 0$. Then $\phi'(0)$ is supposed to give $-\beta f(\beta)$, where $f(\beta)$ is the free energy per spin at inverse temperature β. As usual, $Z_N = 2^{-N} \mathrm{Tr}\{\exp(-\beta H_N)\} \equiv \mathrm{tr}\{\exp(-\beta H_N)\}$ is the partition function, a normalized sum over all Ising-spin configurations.

The angular brackets in (3) denote an average over the disorder, here the patterns ξ_i^λ. Since we first integrate out the ν-patterns, we leave aside $H_N^{(1)}$ for a while and instead of $\langle Z_N^n \rangle$ concentrate on the n-fold replicated $H_N^{(2)}$,

$$\left\langle \exp\left[\frac{\beta}{2N}\sum_{\nu,\sigma}\varepsilon_\nu\left(\sum_{i=1}^{N}\xi_i^\nu S_\sigma(i)\right)^2\right]\right\rangle$$

$$= \prod_\nu \left\langle \exp\left[\frac{\beta}{2}\varepsilon_\nu \sum_{\sigma=1}^{n} X_\sigma^2(\nu)\right]\right\rangle , \qquad (3.3.4)$$

where

$$X_\sigma(\nu) = \frac{1}{\sqrt{N}}\sum_{i=1}^{N}\xi_i^\nu S_\sigma(i) . \qquad (3.3.5)$$

The $1 \le \sigma \le n$ label the n replicas. The equality in (4) follows from the independence of the different patterns. For the moment we fix ν and drop it from $X_\sigma(\nu)$.

The $S_\sigma(i)$ in (4) and (5) refer, replicated or not, to the state of the system. Now, it was assumed that the system lives in that part of phase space which is correlated with the μ-patterns or, in other words [1.13c], where the μ-patterns are "condensed". If so, the $S_\sigma(i)$ are *not* correlated with the ν-patterns[5]. It therefore seems reasonable to use the multidimensional central-limit theorem [1.37] and take the X_σ as *Gaussian* random variables with mean zero, covariance matrix

$$C_{\sigma\sigma'} = \langle X_\sigma X_{\sigma'}\rangle = N^{-1}\sum_{i=1}^{N} S_\sigma(i)S_{\sigma'}(i) , \qquad (3.3.6)$$

and common distribution

$$\left(\frac{1}{\sqrt{2\pi}}\right)^n (\det C)^{-1/2}\exp\left(-\frac{1}{2}\boldsymbol{X}\cdot C^{-1}\boldsymbol{X}\right)\mathrm{d}\boldsymbol{X} . \qquad (3.3.7)$$

The meaning of the multidimensional central-limit theorem is easily understood. By the above assumption, each of the X_σ has a Gaussian distribution with mean zero and variance one. However, the n random variables X_σ are not independent but are correlated with covariance matrix (6). The *multi*dimensional Gaussian distribution which reproduces these mutual correlations among the X_σ is (7).

Let us use, then, the Gaussian distribution (7) in performing the averages in (4). For each ν we obtain a contribution

$$\left\langle \exp\left[\frac{\beta}{2}\varepsilon_\nu\sum_{\sigma=1}^{n} X_\sigma^2\right]\right\rangle = \det C^{-1/2}\det(C^{-1} - \beta\varepsilon_\nu \mathbb{1})^{-1/2} . \qquad (3.3.8)$$

This leads to

$$\det(\mathbb{1} - \beta\varepsilon_\nu C)^{-1/2} \equiv \det(Q_\nu)^{-1/2} , \qquad (3.3.9)$$

where Q_ν is a *symmetric* $n \times n$ matrix with elements

[5] We will see later on that this assumption is self-consistent.

$$(Q_\nu)_{\sigma\sigma'} = \delta_{\sigma\sigma'} - \beta\varepsilon_\nu \left[N^{-1} \sum_{i=1}^{N} S_\sigma(i) S_{\sigma'}(i) \right] . \tag{3.3.10}$$

Here and elsewhere $\delta_{\sigma\sigma'}$ is the Kronecker delta. Collecting terms we get

$$\prod_\nu \det(Q_\nu)^{-1/2} = \exp\left[-\tfrac{1}{2} \sum_\nu \mathrm{Tr}\{\ln Q_\nu\} \right] . \tag{3.3.11}$$

Stepping back for a first overview we see [1.38] that something must be wrong. For (9) to make sense the matrix Q_ν has to be positive-definite, and therefore its diagonal elements *must* be positive. However, $(Q_\nu)_{\sigma\sigma} = 1 - \beta\varepsilon_\nu$ is negative for β large enough. The reason for this can be traced back to the transition from (4) to (8), i.e., to the Gaussian ansatz. One might think that one could do better, but this does not appear to be the case – as will be explained at the end of this subsection. We first proceed, only noting that Q_ν becomes positive in the replica limit $n \to 0$.

The term (11), which still depends on the spins but no longer contains any randomness, represents the *noise* produced by the other patterns and has to be added to the remaining part of the Hamiltonian, which contains the μ-patterns. This complex can be treated exactly, modulo the difficulty we just noted. We have, by (3) and (11),

$$\begin{aligned} &\phi_N(n) \\ &= N^{-1} \ln \left\langle \mathrm{tr}_{S_\sigma} \exp \left\{ N \left[\frac{1}{2}\beta \sum_{\mu,\sigma} \varepsilon_\mu \left(N^{-1} \sum_{i=1}^{N} \xi_i^\mu S_\sigma(i) \right)^2 + \beta \sum_{\mu,\sigma} h_\mu \right.\right.\right. \\ &\quad \left.\left.\left. \times \left(N^{-1} \sum_{i=1}^{N} \xi_i^\mu S_\sigma(i) \right) - \frac{1}{2} N^{-1} \sum_\nu \mathrm{Tr}\{\ln Q_\nu\} \right] \right\} \right\rangle . \end{aligned} \tag{3.3.12}$$

The first trace is a normalized sum over all 2^{nN} Ising-spin configurations of the n replicas and the second one is an ordinary trace. Let us define the order parameters

$$\begin{aligned} m_{\mu\sigma} &= N^{-1} \sum_{i=1}^{N} \xi_i^\mu S_\sigma(i) , \qquad && 1 \le \sigma \le n , \\ q_{\sigma\sigma'} &= N^{-1} \sum_{i=1}^{N} S_\sigma(i) S_{\sigma'}(i) , \qquad && 1 \le \sigma < \sigma' \le n . \end{aligned} \tag{3.3.13}$$

Then the expression between the square brackets in (12) may be written

$$\begin{aligned} F(\boldsymbol{m}, \boldsymbol{q}) = &\beta \sum_{\mu,\sigma} \left(\varepsilon_\mu m_{\mu\sigma}^2/2 + h_\mu m_{\mu\sigma} \right) \\ &- \tfrac{1}{2} N^{-1} \sum_\nu \mathrm{Tr}\{\ln[Q_\nu(\boldsymbol{q})]\} . \end{aligned} \tag{3.3.14}$$

As explained in Sects. 1.2.1, 2, we now perform a coordinate transformation from the $S_\sigma(i), 1 \le i \le N$ and $1 \le \sigma \le n$, to $m_{\mu\sigma}$ and $q_{\sigma\sigma'}$ as new integration variables. According to (2.2.1–2.2.3), the corresponding density ("Jacobian") is

$$\mathcal{D}_N(\boldsymbol{m}, \boldsymbol{q}) = \exp\{-Nc^*(\boldsymbol{m}, \boldsymbol{q})\} , \tag{3.3.15}$$

where

$$c^*(\boldsymbol{m}, \boldsymbol{q}) = \sup_{(\boldsymbol{x},\boldsymbol{y})} \{\boldsymbol{m} \cdot \boldsymbol{x} + \boldsymbol{q} \cdot \boldsymbol{y} - c(\boldsymbol{x}, \boldsymbol{y})\} \tag{3.3.16}$$

is the Legendre transform of the (strictly) convex c-function

$$c(\boldsymbol{x}, \boldsymbol{y}) = \left\langle \ln \operatorname{tr} \left\{ \exp \left(\sum_{\mu,\sigma} x_{\mu\sigma} \xi^\mu S_\sigma + \sum_{(\sigma,\sigma')} y_{\sigma\sigma'} S_\sigma S_{\sigma'} \right) \right\} \right\rangle . \tag{3.3.17}$$

The second term in (17) is over pairs (σ, σ') only. The trace refers to n Ising spins $S_\sigma, 1 \le \sigma \le n$, and in the outer average each ξ^μ appears only once; there are finitely many of them. The derivation proceeds exactly as in (2.2.5). Furthermore, as in (2.2.5), the function $c(\boldsymbol{x}, \boldsymbol{y})$ does not depend on the specific random configuration as $N \to \infty$. Hence $\mathcal{D}_N(\boldsymbol{m}, \boldsymbol{q})$ does not depend on it either. For that reason we may drop the angular brackets from (12). In passing we note that an extra set of order parameters $r_{\varrho\sigma}$ as introduced by Amit et al. [1.13c] can simply be dispensed with.

Given (12), (14), and (17), $\phi(n)$ directly follows from (2.2.8) and (2.2.9). The matrix $Q_\nu(\boldsymbol{q})$ has elements:

$$(Q_\nu)_{\sigma\sigma'} = \delta_{\sigma\sigma'}(1 - \beta\varepsilon_\nu) - \beta\varepsilon_\nu q_{\sigma\sigma'} = (Q_\nu)_{\sigma'\sigma} . \tag{3.3.18}$$

Using the relation

$$\frac{\partial}{\partial Q_{\sigma\sigma'}} \operatorname{Tr}\{\ln Q\} = 2(Q^{-1})_{\sigma\sigma'} \tag{3.3.19}$$

one easily verifies that, with $\boldsymbol{\mu} = (\boldsymbol{m}, \boldsymbol{q})$,

$$\nabla F(\boldsymbol{\mu}) = \begin{pmatrix} \beta(\varepsilon_\mu m_{\mu\sigma} + h_\mu) \\ N^{-1} \sum_\nu \beta\varepsilon_\nu (Q_\nu^{-1})_{\sigma\sigma'} \end{pmatrix} . \tag{3.3.20}$$

This has to be used in (2.2.8) and inserted into the fixed-point equation (2.2.9), viz., $\boldsymbol{\mu} = \nabla c(\nabla F(\boldsymbol{\mu}))$. For small enough β (2.2.8) is exact, but it becomes formal as soon as $\beta \max_\mu(\varepsilon_\mu) > 1$. In spite of that we proceed and perform the extension of $\phi(n)$ to $n = 0$ by assuming replica symmetry.

Before doing so, however, we first return[6] to the transition from (4) to (8). Amit et al. [1.13c] used a different argument. They linearized the $X_\sigma^2(\nu)$ through (2.1.6), so that (4) reappears as

[6] The arguments below are not needed in what follows. At a first reading the reader should proceed directly to the next subsection (Replica Symmetry) in order to avoid any interruption of the argument.

$$\left\langle \int \prod_{\nu,\sigma} \frac{\mathrm{d}z_{\nu\sigma}}{\sqrt{2\pi}} \exp\left\{ -\frac{1}{2} \sum_{\nu,\sigma} z_{\nu,\sigma}^2 + \sum_{\nu,\sigma} z_{\nu,\sigma} \left[\sqrt{\frac{\beta\varepsilon_\nu}{N}} \sum_{i=1}^{N} \xi_i^\nu S_\sigma(i) \right] \right\} \right\rangle , \quad (3.3.21)$$

and then took the average with respect to the ξ_i^ν exactly so as to find

$$\int \prod_{\nu,\sigma} \frac{\mathrm{d}z_{\nu\sigma}}{\sqrt{2\pi}} \exp\left(-\frac{1}{2} \sum_{\nu,\sigma} z_{\nu\sigma}^2 + \sum_{i,\nu} \ln\left\{ \cosh\left[\sqrt{\frac{\beta\varepsilon_\nu}{N}} \sum_{\sigma=1}^{n} z_{\nu\sigma} S_\sigma(i) \right] \right\} \right) . \quad (3.3.22)$$

It was then argued that the terms between the square brackets in (22) were "small" so that one could replace $\ln[\cosh(x)]$ for "small" x by $x^2/2$. This gives

$$\int \prod_{\nu,\sigma} \frac{\mathrm{d}z_{\nu\sigma}}{\sqrt{2\pi}} \exp\left\{ -\frac{1}{2} \sum_{\nu,\sigma} z_{\nu\sigma}^2 + \sum_{\sigma,\sigma',\nu} \tfrac{1}{2}\beta\varepsilon_\nu z_{\nu\sigma} z_{\nu\sigma'} \left[N^{-1} \sum_{i=1}^{N} S_\sigma(i) S_{\sigma'}(i) \right] \right\} , \quad (3.3.23)$$

which can be integrated exactly (fix ν). The answer is (11), which we know to be pretty formal for large β. So what went wrong? Since the $z_{\nu\sigma}$ are integration variables which vary between $-\infty$ and $+\infty$, it is hard to see why the argument of the hyperbolic cosine in (22) should be "small". Whereas (22) itself is well defined for all β, since $\ln[\cosh(x)] \sim |x|$ as $|x| \to \infty$, (23) is not for large β. Simply fix ν and take the diagonal elements of the quadratic form in the exponent. These must be negative if the integral converges, but they are not for large enough β.

Is there an alternative? One could, for example [1.39], argue that the argument of the hyperbolic cosine in (22) – or X_ν in (24) below – contains a sum over σ with upper limit n which is sent to zero at the end of the calculation. Since we cannot do this calculation, we just take the limit $n \to 0$ a bit earlier. This makes the transition from (22) to (23) plausible – but not more than that. It is clearer, and also more physical, to use the argument which has led us from (4) to (8). Let us fix ν and put

$$X_\nu = \sum_{\sigma=1}^{n} z_{\nu\sigma} \left[\frac{1}{\sqrt{N}} \sum_{i=1}^{N} \xi_i^\nu S_\sigma(i) \right] , \quad (3.3.24)$$

which is $\sum_\sigma z_{\nu\sigma} X_\sigma(\nu)$ in the notation of (5). By assumption, the ξ_i^ν are not correlated with the $S_\sigma(i)$ so that the expressions between the square brackets are taken to be Gaussian and X_ν, as a linear combination of Gaussians, is Gaussian itself, whence

$$\langle \exp(tX_\nu) \rangle = \exp(\tfrac{1}{2}t^2 \langle X_\nu^2 \rangle) . \quad (3.3.25)$$

Equation (23) then follows directly.

Can we do better? – Say, by replacing $\frac{1}{2}t^2\langle X_\nu^2\rangle$ by a polynomial $P(t)$ of order exceeding 2 and with coefficients determined by the moments of X_ν? No, we cannot. According to a theorem of Marcinkiewicz [1.40], the representation $\langle \exp(tX) \rangle = \exp[P(t)]$ implies that $P(t) = \alpha t + \gamma t^2/2$ for some real numbers α and γ. So X is Gaussian (with mean α and variance γ). In a sense, the only consistent approximation is the Gaussian one, viz. (25). Of course, an exact explicit evaluation of (4) or (22) would be highly desirable, but no one has done so yet.

Replica Symmetry. We assume that all the replicas are equal so that the symmetry between them is not broken[7]: $m_{\mu\sigma} = m_\mu$ and $q_{\sigma\sigma'} = \hat{q}\ (\sigma \neq \sigma')$. This assumption is consistent with the fixed-point equation $\boldsymbol{\mu} = \nabla c(\nabla F(\boldsymbol{\mu}))$. Moreover,

$$Q_\nu(\hat{q}) = (1 - \beta\varepsilon_\nu + \beta\varepsilon_\nu\hat{q})\mathbb{1} - \beta\varepsilon_\nu\hat{q}n \left| \frac{1}{\sqrt{n}}\mathbf{1} \right\rangle \left\langle \frac{1}{\sqrt{n}}\mathbf{1} \right|$$
$$\equiv a(n) - nb(n)P \ , \tag{3.3.26}$$

where $\mathbb{1}$ is the unit matrix, $\mathbf{1}$ is the vector $(1, 1, \dots, 1) \in \mathbb{R}^n$, and $P = P^2$ is a projection operator. Through the ansatz $Q^{-1} = c - dP$ the inverse of Q is easily obtained and $(\sigma \neq \sigma')$

$$(Q^{-1})_{\sigma\sigma'} = -b(n)[a(n)(nb(n) - a(n))]^{-1} \ . \tag{3.3.27}$$

Hence we can write

$$N^{-1}\sum_\nu \beta\varepsilon_\nu(Q_\nu^{-1})_{\sigma\sigma'} \equiv \beta^2\hat{q}r(n) \ , \tag{3.3.28}$$

and by (20)

$$\nabla F(\boldsymbol{\mu}) = \begin{pmatrix} \beta(\varepsilon_\mu m_\mu + h_\mu) \\ \beta^2\hat{q}r(n) \end{pmatrix} \ . \tag{3.3.29}$$

Using (26) one directly verifies that the eigenvalues of $Q_\nu(\hat{q})$ are $1 - \beta\varepsilon_\nu(1 - \hat{q}) - \beta\varepsilon_\nu\hat{q}n$, which is simple, and $1 - \beta\varepsilon_\nu(1 - \hat{q})$, which is $(n-1)$-fold degenerate. In the limit $n \to 0$ one is left with $1 - \beta\varepsilon_\nu(1 - \hat{q})$, which has to be positive; cf. the discussion in the previous subsection and (32) below. In the Hopfield model one can verify that this limiting value is indeed positive (Sect. 1.3.5).

Combining the expression for $\phi(n)$ with (26–29) and adding the constant we dropped from (2), we get

$$\phi(n) = -\tfrac{1}{2}\beta n \left(N^{-1}\sum_\lambda \varepsilon_\lambda \right) - \tfrac{1}{2}\beta n \left(\sum_\mu \varepsilon_\mu m_\mu^2 \right)$$
$$- \frac{1}{2}\frac{1}{N}\sum_\nu \{\ln[1 - \beta\varepsilon_\nu(1 - \hat{q}) - \beta\varepsilon_\nu\hat{q}n]$$

[7] For instance, in the case of the Sherrington–Kirkpatrick model of a spin glass it has been shown [1.32] that for positive integer n this is indeed true. In the present situation, the proof is expected to proceed in a similar vein.

$$+(n-1)\ln[1-\beta\varepsilon_\nu(1-\hat{q})]\} - \tfrac{1}{2}n(n-1)(\beta\hat{q})^2 r(n)$$

$$+\left\langle \ln \mathrm{tr}_{S_\sigma}\left\{\exp\left[\beta\sum_{\mu,\sigma}(\varepsilon_\mu m_\mu + h_\mu)\xi^\mu S_\sigma + \tfrac{1}{2}\beta^2\sum_{\sigma\neq\sigma'}\hat{q}r(n)S_\sigma S_{\sigma'}\right]\right\}\right\rangle . \tag{3.3.30}$$

If we use the linearization trick (2.1.6) and carry out the trace, we can rewrite the last term of (30) in the form

$$-\frac{1}{2}n\beta^2\hat{q}r(n) + \left\langle \ln \int_{-\infty}^{+\infty}\frac{\mathrm{d}z}{\sqrt{2\pi}}\mathrm{e}^{-z^2/2} \right.$$
$$\left. \times \cosh^n\left\{\beta\left[(\varepsilon\boldsymbol{m}+\boldsymbol{h})\cdot\boldsymbol{\xi} + \sqrt{\hat{q}r(n)}z\right]\right\}\right\rangle , \tag{3.3.31}$$

where $\varepsilon = \mathrm{diag}\,(\varepsilon_\mu)$ is a diagonal matrix. The right-hand sides of (30) and (31) exhibit $\phi(n)$ explicitly as a function of n. Taking the "evident" real-variable extension of $\phi(n)$ to a neighborhood of $n = 0$ we then get, as argued in Sect. 1.2.5,

$$\begin{aligned}
-\beta f(\beta) &= \lim_{n\to 0} n^{-1}\phi(n) \\
&= -\frac{1}{2}\beta\left(N^{-1}\sum_\lambda \varepsilon_\lambda\right) - \frac{1}{2}\beta\left(\sum_\mu \varepsilon_\mu m_\mu^2\right) \\
&\quad -\frac{1}{2}N^{-1}\sum_\nu\{\ln[1-\beta\varepsilon_\nu(1-\hat{q})] - \beta\varepsilon_\nu\hat{q}[1-\beta\varepsilon_\nu(1-\hat{q})]^{-1}\} \\
&\quad -\frac{1}{2}\beta^2\hat{q}r(1-\hat{q}) \\
&\quad +\left\langle\int\frac{\mathrm{d}z}{\sqrt{2\pi}}\mathrm{e}^{-z^2/2}\ln\{\cosh[\beta((\varepsilon\boldsymbol{m}+\boldsymbol{h})\cdot\boldsymbol{\xi}+\sqrt{\hat{q}r}z)]\}\right\rangle
\end{aligned} \tag{3.3.32}$$

with N very large and

$$r = \lim_{n\to 0} r(n) = N^{-1}\sum_\nu \varepsilon_\nu^2[1-\beta\varepsilon_\nu(1-\hat{q})]^{-2} \geq 0 . \tag{3.3.33}$$

Furthermore, one should choose that solution of the fixed-point equations,

$$\boldsymbol{m} = \langle\langle \boldsymbol{\xi}\tanh\{\beta[(\varepsilon\boldsymbol{m}+\boldsymbol{h})\cdot\boldsymbol{\xi}+\sqrt{\hat{q}r}z]\}\rangle\rangle , \tag{3.3.34a}$$

$$\hat{q} = \langle\langle \tanh^2\{\beta[(\varepsilon\boldsymbol{m}+\boldsymbol{h})\cdot\boldsymbol{\xi}+\sqrt{\hat{q}r}z]\}\rangle\rangle , \tag{3.3.34b}$$

which maximizes the right-hand side of (32). The double angular brackets in (34) denote an average with respect to both the finitely many ξ^μ and the Gaussian distribution of z. If $\varepsilon_\nu = 1$ for $1 \leq \nu \leq \alpha N$ and $\varepsilon_\nu = 0$ for $\nu > \alpha N$, then (32–34) reproduce the result of Amit et al. [1.13b, c]. Throughout what follows we put $\boldsymbol{h} = 0$. In view of (33) and (34) we interpret $r = r(\hat{q})$ as a renormalization constant that rescales the order parameter $\hat{q}$. It is a consequence of the noise generated by the "infinitely many" other patterns ($N \to \infty$).

1.3.4 The Phase Diagram of the Hopfield Model

In this section we concentrate on the original Hofield model [1.2], study the solutions to the fixed-point equations (3.34), i.e., in the present case

$$\boldsymbol{m} = \left\langle\left\langle \boldsymbol{\xi} \tanh\left[\beta\left(\sum_\mu m_\mu \xi^\mu + \sqrt{\hat{q}r}z\right)\right]\right\rangle\right\rangle , \tag{3.4.1a}$$

$$\hat{q} = \left\langle\left\langle \tanh^2\left[\beta\left(\sum_\mu m_\mu \xi^\mu + \sqrt{\hat{q}r}z\right)\right]\right\rangle\right\rangle , \tag{3.4.1b}$$

and provide them with a physical interpretation. Here $r = r(\hat{q})$ has the simple form

$$r(\hat{q}) = \alpha[1 - \beta(1 - \hat{q})]^{-2} , \tag{3.4.2}$$

which follows directly from (3.33). Note the dependence upon the storage ratio, the fraction of stored patterns $\alpha = q/N$. For $\alpha = 0$ we recover (1.3). Moreover, (3.32) reduces to (1.4). So a natural question is whether the states we have found in Sect. 1.3.1 survive for $\alpha > 0$.

We have seen that for vanishing α the n-symmetric states with n odd are stable, though only at temperatures below T_n, and that T_n decreases monotonically in n. In other words, the higher n, the more loosely these states are bound and thus the more readily they are destabilized by thermal noise. A similar effect occurs if we put, say, $T = 0$ and increase α, i.e., the statistical noise generated by the other patterns. Then the n-symmetric states with n odd are stable only if α is less than α_n where α_n decreases with n. Furthermore, it turns out that the transition at α_n is *discontinuous* (first order), whatever n. For example, the 3-symmetric states cease to exist at $\alpha_3 \approx 0.03$ whereas the retrieval states are – luckily – much more stable and disappear at $\alpha_1 \equiv \alpha_c = 0.138$. We will return to this first-order transition at α_c in the more general context of nonlinear inner-product models (Sect. 1.4.3). As expected, the n-symmetric states with n even are always unstable.

We now derive the equations from which the above results are obtained. The ansatz for an n-symmetric solution of (1) is that it has n equal components m while the other ones vanish. We then arrive at the expression

$$m = n^{-1}\left\langle\left\langle \left(\sum_\mu \xi^\mu\right) \tanh\left\{\beta\left[m\left(\sum_\mu \xi^\mu\right) + \sqrt{\hat{q}r}z\right]\right\}\right\rangle\right\rangle . \tag{3.4.3}$$

Equation (3) is a generalization of (1.6), to which it reduces in the limit $\alpha \to 0$. As $\beta \to 0$, $\tanh(\beta x)$ converges to $\operatorname{sgn}(x)$; furthermore, by (1.b), $\hat{q} \to 1$. Assuming that $r(\beta, \hat{q})$ has a well-defined limit r as $\beta \to \infty$ (it has: see below), the integration over the Gaussian z can be performed in terms of the error function (erf), and (3) reappears in the form

$$m = n^{-1}\left\langle\left(\sum_\mu \xi^\mu\right) \operatorname{erf}\left[m\left(\sum_\mu \xi^\mu\right)/\sqrt{2r}\right]\right\rangle . \tag{3.4.4}$$

We are left with determining r and thus, in view of (2),

$$\begin{aligned} C(\beta) &= \beta(1-\hat{q}) \\ &= \left\langle\int_{-\infty}^{+\infty} \frac{\mathrm{d}z}{\sqrt{2\pi}} \mathrm{e}^{-(z/\beta)^2/2} \cosh^{-2}\left[\beta\left(\sum_\mu m_\mu \xi^\mu\right) + \sqrt{\hat{q}r}z\right]\right\rangle \end{aligned} \tag{3.4.5}$$

as β becomes large $(T \to 0)$. In this limit we obtain after the substitution $y := \beta(\sum_\mu m_\mu \xi^\mu) + \sqrt{\hat{q}r}z$,

$$C = \lim_{\beta\to\infty} C(\beta) = \sqrt{\frac{2}{\pi r}}\left\langle \exp\left[-\left(\sum_\mu m_\mu \xi^\mu\right)^2 /2r\right]\right\rangle \tag{3.4.6}$$

so that $\hat{q} = 1 - CT$ as $T \to 0$. In the very same limit, (2) is $r = \alpha(1-C)^{-2}$. For n-symmetric states, the nonzero m_μ in (5) and (6) can be replaced by m. We then have three equations, viz., (4), (6), and the one for r, which determine m, C, and r. The solutions can be obtained numerically. Their nature has already been indicated. In passing we note that by (5) we have $C(\beta) < 1$ for all β.

In addition to the retrieval and mixture states, there is also another one which only occurs for $\alpha > 0$: the spin-glass state. It is correlated with *none* of the patterns but, in spite of that, is generated by all of them. That is to say, in the context of the present theory it is characterized by $m_\mu = 0$ for all μ while $\hat{q}$ is a *non*zero solution of

$$\hat{q} = \int_{-\infty}^{+\infty} \frac{\mathrm{d}z}{\sqrt{2\pi}} \mathrm{e}^{-z^2/2} \tanh^2(\beta\sqrt{\hat{q}r}z) , \tag{3.4.7}$$

which has to be combined with $r = \alpha[1-C(\beta)]^{-2}$ and (5). The bifurcation from $\hat{q} = 0$ is continuous, so a simple Taylor expansion of (7) around $\hat{q} = 0$ suffices to give the critical value of β, to be called β_g,

$$\hat{q} \approx [\beta\sqrt{\hat{q}r(0)}]^2\langle z^2\rangle \quad \Rightarrow \quad \beta_\mathrm{g}^2\alpha(1-\beta_\mathrm{g})^{-2} = 1 . \tag{3.4.8}$$

Hence $T_\mathrm{g} = 1 + \sqrt{\alpha}$ and below T_g the spin-glass phase exists. The name "spin glass" is due to the "frustration" in the system [1.38,41,42].

We now turn to the phase diagram, Fig. 1.2, which has been obtained by Amit et al. [1.13c]. As a function of α, there are three main critical lines, namely, $T_\mathrm{g}(\alpha)$, $T_\mathrm{M}(\alpha)$, and $T_\mathrm{c}(\alpha)$. As we lower the temperature, we first pass T_g and then T_M, provided that $\alpha < \alpha_\mathrm{c}$. Directly below T_M the retrieval states are metastable. If $\alpha < 0.051$, we also pass a third line, T_c, below which the retrieval states are globally stable. In the present context, however, metastability already suffices to guarantee associative memory since it implies the existence of an ergodic

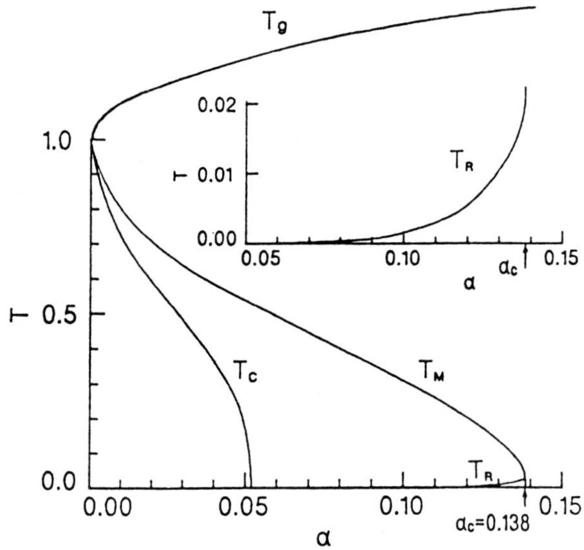

Fig. 1.2. Phase diagram of the Hopfield model. At T_g, the spin-glass phase appears. Directly below T_M, the retrieval states (patterns) are metastable so that the network can function as an associative memory. The retrieval states attain global stability below T_c. Below T_R, replica symmetry is broken. As the inset shows, this is restricted to a tiny region

component, i.e., a real (free-) energy valley surrounded by barriers of a height proportional to N, the size of the system, and with the pattern "at the bottom" of the valley. See also the discussion of this issue in Sect. 1.3.2.

In the neighborhood of $\alpha = 0$ and $T = 1$, one finds that $T_M \approx 1 - 1.95\sqrt{\alpha}$ and $T_c \approx 1 - 2.6\sqrt{\alpha}$. So all three critical temperatures exhibit a square-root singularity as $\alpha \to 0$.

Near α_c there is another critical line, $T_R(\alpha)$, below which the retrieval states do not give rise to *maxima* of the free-energy functional (3.32), so that replica symmetry is broken. As the inset of the phase diagram shows, the instability is restricted to a fairly small region: at α_c one has $T_R(\alpha_c) = 0.07$ and, as $\alpha \to 0$, T_R decreases at an exponential rate according to $T_R(\alpha) \approx (8\alpha/9\pi)^{1/2} \exp(-1/2\alpha)$. The reader can find a sketch of the algebra, which is somewhat involved, in [1.13c: Appendix B].

1.3.5 Discussion

The derivation of (3.32–3.34) is based on the Gaussian approximation (7) and the assumption that we have "replica symmetry" in the limit $n \to 0$. So one may wonder how good all this is. Well, the proof of the pudding is in the eating. There are two excellent checks. First, we examine the internal consistency of the theory by studying the zero-temperature entropy, which is bound to be nonnegative for Ising spins. Second, we compare the predictions of the theory, such as the critical storage ratio α_c, with numerical simulations.

The entropy $s(\beta)$ is obtained most easily through the relation[8] $s(\beta) = \beta^2 \partial f(\beta)/\partial\beta$. Taking advantage of (3.22) and the fact that $r = \alpha[1 - C(\beta)]^{-2}$ we find

$$s(\beta) = -\frac{\alpha}{2}\{\ln[1 - C(\beta)] + C(\beta)[1 - C(\beta)]^{-1}\} + \langle\langle\{\ln[2\cosh(x)] - x\tanh(x)\}\rangle\rangle , \tag{3.5.1}$$

where $x = \beta(\boldsymbol{m}\cdot\boldsymbol{\xi} + \sqrt{\hat{q}r}z)$. Moreover, we have included $\ln 2$ so as to obtain the usual entropy. (We had lost the $\ln 2$ because of our using the normalized trace.) To arrive at (1) we need not differentiate $f(\beta)$ with respect to $m, \hat{q}$, and r, since these contributions vanish because of (3.33) and (3.34). What remains is the explicit dependence upon β.

The contributions of the terms between the curly brackets in (1) are positive. The first is easily treated. Let $\psi(C) = \ln(1 - C) + C/(1 - C)$. The function ψ is meaningful *only if* $0 \le C < 1$. (For $C > 1$ we get an entropy with an imaginary part, a possibility which has to be rejected both on physical grounds and because, by (4.5), C is bound to be less than 1. If the fixed-point equations allow, say, two positive solutions for C, we are consequently bound to choose the one which is less than one.) Since $\psi(0) = 0$ and $\psi'(C) = C(1 - C)^{-2} \ge 0$, ψ is positive on $[0, 1)$. The reason the second contribution is positive is somewhat subtler. We have

$$\ln[2\cosh(x)] - x\tanh(x) = -\left[\left(\frac{1+t}{2}\right)\ln\left(\frac{1+t}{2}\right) + \left(\frac{1-t}{2}\right)\ln\left(\frac{1-t}{2}\right)\right] , \tag{3.5.2}$$

where $t = \tanh(x)$. Hence $0 \le (1 \pm t)/2 < 1$ and the assertion follows.

The upshot of the above considerations is that the right-hand side of (1) consists of a negative and a positive term. In the limit $\beta \to \infty$, the latter vanishes, since $t \to \pm 1$, $C(\beta) \to C$ as given by (4.6), and

$$s(\infty) \equiv s = -\frac{\alpha}{2}[\ln(1 - C) + C/(1 - C)] < 0 . \tag{3.5.3}$$

So we have to determine C.

In the spin-glass state, all the m_μ vanish and (4.6) gives

$$C = \sqrt{\frac{2}{\alpha\pi}}|1 - C| . \tag{3.5.4}$$

We choose the solution $C = [1 + (\alpha\pi/2)^{1/2}]^{-1} < 1$ and disregard the other one. As $\alpha \to 0$, we use (4) so as to get $C/(1 - C) = (2/\alpha\pi)^{1/2}$ and substituting this into (3) we find $s \to 0^-$, as should be the case. On the other hand, as $\alpha \to \alpha_c = 0.138$ we obtain $C = 0.682$ and $s = -0.069 \approx -0.07$, which is comparable to the $-1/2\pi = -0.159 \approx -0.16$ of the Sherrington–Kirkpatrick

[8] Alternatively, one can use $f(\beta) = u(\beta) - Ts(\beta)$ where the energy $u(\beta)$ follows from $u(\beta) = \partial\beta f(\beta)/\partial\beta$.

model of a spin glass [1.41]. Apparently, the low-temperature state of the spin-glass phase is not described very well by the present theory. There is no reason for despair, however. After all, the theory was devised for something else: the retrieval states.

In a retrieval state, $m_\mu = m$ for some μ while the other overlaps vanish. Then the dependence upon ξ in (4.6) has gone and

$$C = \sqrt{\frac{2}{\alpha\pi}}|1 - C| \exp\{-[m(1 - C)]^2/2\alpha\} . \tag{3.5.5}$$

As $\alpha \to 0$, C converges to zero at an exponential rate. With m very close to one we immediately get from (5) that $C = (2/\alpha\pi)^{1/2} \exp(-1/2\alpha)$ and thus, via (3),

$$s \approx -\frac{1}{4}\alpha C^2 = -\frac{1}{2\pi} \exp(-1/\alpha) . \tag{3.5.6}$$

For example, for $\alpha = 0.02$ we would get $s \approx -3 \times 10^{-23}$, which is extremely near to zero indeed. On the other hand, as $\alpha \to \alpha_c = 0.138$, we have $m_c = 0.967, C = 0.182$, and $s = -1.49 \times 10^{-3}$, which is also quite close to zero, the exact value.

We now consider the critical storage ratio α_c at $T = 0$. The fixed-point equations (4.1) give $\alpha_c \simeq 0.1379$, with $m_c \simeq 0.967$ [1.13c]. This allows for for a small percentage of errors (about 1.5%), which indeed do occur. Numerical simulations yielded $\alpha_c = 0.145 \pm 0.009$ [1.13c] and $\alpha_c = 0.143 \pm 0.001$ [1.144], the latter being based on *much* larger system sizes. A first one-step replica-symmetry breaking calculation (1RSB) based on a different extension of $\phi(n)$ from the integers to $n \approx 0$ gave $\alpha_c^{\text{1RSB}} \simeq 0.144$ [1.44] in very good agreement with the simulations. However, this result is now known to be in error, the correct value being $\alpha_c^{\text{1RSB}} \simeq 0.138\,186$ [1.145]. A 2RSB calculation and analogies with the SK model suggest that the *exact* value does not exceed $\alpha_{\max} \simeq 0.138\,189$ (see [1.145] for details), a result which raises critical questions regarding the analysis of the simulation data.

Summarizing, the replica-symmetry theory describes both the zero-temperature entropy and the critical α_c and the nonzero error percentage of the retrieval states at α_c rather well. And precisely that is what we were after.

Finally, we turn to the so-called *signal-to-noise-ratio analysis*, which provides a qualitative but more physical understanding of the finite value of α_c. Let us assume that the system is in pattern μ, i.e., $S_j = \xi_j^\mu$, and consider the local field at site i,

$$h_i^\mu = \sum_{j(\neq i)} J_{ij}\xi_j^\mu = \xi_i^\mu + N^{-1} \sum_{\substack{j(\neq i)\\ \nu(\neq \mu)}} \xi_i^\nu \xi_j^\nu \xi_j^\mu . \tag{3.5.7}$$

The first term on the right is the signal which stabilizes the state μ whereas the second is the (unwanted) noise generated by the other patterns, which destabilizes μ. It is a sum of independent, identically distributed random variables with mean zero and variance one. By DeMoivre–Laplace or, more generally, the central-limit theorem [1.22, Sects. 14 and 15], the noise has a Gaussian distribution with mean zero and standard deviation (the "typical size" or square root of the variance) $\sqrt{\alpha}$. As long as α is small, the signal term will dominate. Errors are allowed – but rare. If, however, α becomes *of the order of one*, then the noise term will frequently overcome ("wash out") the signal ξ_i^μ so that the assumption of a stable state μ breaks down. The present argument cannot say more than that because (i) at site i the noise has a probability distribution and (ii) noise terms at different sites are correlated. This rather subtle correlation, a collective effect, is represented by $r(\hat{q})$ and shows up in a pronounced way in the first-order transition as a function of α at α_c.

Until now we have studied in detail the fully symmetric case $J_{ij} = J_{ji}$ and one might wonder whether all this, in particular, the associative memory, breaks down if the symmetry condition is dropped. It turns out, however, that this is not the case [1.45–48]. One may perform, for instance, asymmetric dilution by assigning a direction to both $J_{ij}(j \to i)$ and $J_{ji}(i \to j)$ and randomly deleting either of the bonds. The ability of associative recall is *not* seriously modified. Only the spin-glass phase may be strongly suppressed or disappear, which seems to be an advantage. For details, the reader is referred to the literature.

1.3.6 Parallel Dynamics (the Little Model)

The statistical mechanical analysis of the Hopfield model has provided us with information about the thermodynamic and, hence (Sect. 1.3.2), about the dynamical stability of the retrieval states. The only proviso was the use of a *sequential* Glauber (or Monte Carlo) dynamics. A natural question therefore is what can be said about parallel dynamics. Though the very notion of (free-) energy landscape with the patterns as minima and the system performing a downhill motion under its own (sequential) dynamics has to be abandoned, a thermodynamic substitute can be provided in the case of parallel *Glauber* dynamics, i.e., the Little model [1.43].

In 1974 Little published a seminal paper with the title "The existence of persistent states in the brain". There he already noticed that "persistent states are distinguished by the property that a coherence or correlation exists between the neurons *throughout* [Little's italics] the entire brain or large portions of it." In our terms, recalling a memory is a collective phenomenon. Little also observed that "the transformation from the uncorrelated to the correlated state ... occurs in a manner closely similar to the phase transition in the analogous spin system" and he accordingly introduced Ising spins to describe the neurons. However, no critical temperature is mentioned yet and the connection to statistical mechanics is specific in that time translation is to operate as a transfer matrix [1.7], each time slice $\{S_i(t); 1 \le i \le N\}$ being a row.

For a Glauber dynamics, the probability of getting the value σ_i at time $t+\Delta t$, given $\boldsymbol{S} = \{S_i(t); 1 \le i \le N\}$ and hence $h_i(\boldsymbol{S}) = \sum_j J_{ij} S_j$, may be written

$$\frac{1}{2}\{1 + \tanh[\beta h_i(\boldsymbol{S})\sigma_i]\} = \frac{\exp[\beta h_i(\boldsymbol{S})\sigma_i]}{2\cosh[\beta h_i(\boldsymbol{S})]} \, . \tag{3.6.1}$$

This follows directly from (1.2.1). Parallel updating means that, given $\boldsymbol{S}$, all the spins are updated independently of each other according to the rule (1). The transition probability $W(\boldsymbol{\sigma}|\boldsymbol{S})$ from $\boldsymbol{S}$ to $\boldsymbol{\sigma}$ is the product of the individual transition probabilities (1),

$$W(\boldsymbol{\sigma}|\boldsymbol{S}) = \frac{\exp[\beta \sum_{i,j} J_{ij}\sigma_i S_j]}{\prod_i 2\cosh[\beta h_i(\boldsymbol{S})]} . \tag{3.6.2}$$

By construction, $\sum_{\boldsymbol{\sigma}} W(\boldsymbol{\sigma}|\boldsymbol{S}) = 1$.

To describe the time evolution of a stochastic process, one has to specify how a *probability distribution* ϱ evolves as time proceeds. Here we have a homogeneous Markov chain where the state at time t and the transition probabilities W determine ϱ at time $t + \Delta t$,

$$\varrho(\boldsymbol{\sigma}, t + \Delta t) = \sum_{\boldsymbol{S}} W(\boldsymbol{\sigma}|\boldsymbol{S})\varrho(\boldsymbol{S}, t) . \tag{3.6.3}$$

Since the Ws are normalized we have $W(\boldsymbol{\sigma}|\boldsymbol{\sigma}) = 1 - \sum' W(\boldsymbol{S}|\boldsymbol{\sigma})$ where the primed sum is over all $\boldsymbol{S} \neq \boldsymbol{\sigma}$. Singling out the term $\boldsymbol{S} = \boldsymbol{\sigma}$ in (3) and using the above expression for $W(\boldsymbol{\sigma}|\boldsymbol{\sigma})$ we then find

$$\varrho(\boldsymbol{\sigma}, t + \Delta t) = \varrho(\boldsymbol{\sigma}, t) + \sum_{\boldsymbol{S}} [W(\boldsymbol{\sigma}|\boldsymbol{S})\varrho(\boldsymbol{S}, t) - W(\boldsymbol{S}|\boldsymbol{\sigma})\varrho(\boldsymbol{\sigma}, t)] . \tag{3.6.4}$$

If ϱ satisfies the so-called *detailed-balance* condition,

$$W(\boldsymbol{\sigma}|\boldsymbol{S})\varrho(\boldsymbol{S}) = W(\boldsymbol{S}|\boldsymbol{\sigma})\varrho(\boldsymbol{\sigma}) , \tag{3.6.5}$$

then ϱ is time invariant by (4). Such a ϱ need not exist. It is a theorem [1.49], however, that the Markov process converges to the (unique) limit ϱ if (i) the process is *regular* or *irreducible*, i.e., any two configurations communicate in a finite number of steps, and (ii) ϱ satisfies (5).

For finite β, the regularity (or irreducibility) of our Markov chain is guaranteed by (2). Furthermore, as was first noticed by Peretto [1.8], $\varrho(\boldsymbol{S}) = C \exp[-\beta \bar{H}(\boldsymbol{S})]$ with[9]

$$-\beta \bar{H}(\boldsymbol{S}) = \sum_{i=1}^{N} \ln \left[\cosh \left(\beta \sum_j J_{ij} S_j \right) \right] \tag{3.6.6}$$

and C a normalizing constant satisfies (5) provided $J_{ij} = J_{ji}$. Note that the "Hamiltonian" (6) is used to define an equilibrium *distribution*. As in the Hopfield case, it is *also* a Lyapunov functional for the zero-temperature (parallel) dynamics [1.143]. Taking the limit $\beta \to \infty$ in (6), we obtain

[9] Referring to [1.8], we have dropped a trivial $N \ln 2$.

$$\bar{H}(\boldsymbol{S}) = -\sum_{i=1}^{N}\left|\sum_{j=1}^{N} J_{ij}S_j\right| . \tag{3.6.7}$$

If $\boldsymbol{S}_0$ minimizes $\bar{H}$, so does $-\boldsymbol{S}_0$, and there is nothing against the system oscillating between $\boldsymbol{S}_0$ and $-\boldsymbol{S}_0$ (a 2-cycle). In fact, this kind of oscillation does occur – also at finite β, though not as a perfect 2-cycle. To see why (7) is a Lyapunov functional and why we either have fixed points or period-two limit cycles as $t \to \infty$, we make a small detour.

Keeping in mind the dependence of $\boldsymbol{S}$ upon t we rewrite (7):

$$\bar{H}(t) = -\sum_{i=1}^{N} |h_i(t)| = -\sum_{i=1}^{N} S_i(t+1)h_i(t)$$

using the fact that the local field at i is $h_i = \sum_j J_{ij}S_j$ and, for parallel dynamics at zero temperature, we have $S_i(t+\Delta t) = \operatorname{sgn}[h_i(t)]$ *for all* i. There is no harm in taking $\Delta t = 1$. Owing to the symmetry $J_{ij} = J_{ji}$ we can also write

$$\begin{aligned}\bar{H}(t) &= -\sum_{i,j} S_i(t+1)J_{ij}S_j(t) = -\sum_j S_j(t)\Big[\sum_i J_{ji}S_i(t+1)\Big] \\ &= -\sum_j S_j(t)h_j(t+1) .\end{aligned}$$

Subtracting this from $\bar{H}(t+1) = -\sum_i |h_i(t+1)|$ we obtain, using $h_j(t+1) = |h_j(t+1)|\operatorname{sgn}[h_j(t+1)]$ in tandem with $\operatorname{sgn}[h_j(t+1)] = S_j(t+2)$,

$$\Delta\bar{H}(t) = \bar{H}(t+1) - \bar{H}(t) = -\sum_i |h_i(t+1)|\big[1 - S_i(t)S_i(t+2)\big] \leq 0$$

as $|S_i(t)S_i(t+2)| \leq 1$. So $\bar{H}$ is a Lyapunov functional. For finite N, $\bar{H}$ is bounded from below. After finitely many steps $\Delta\bar{H}$ therefore has to vanish identically. This can be realized only if for all i we have $S_i(t)S_i(t+2) = 1$, i.e., $S_i(t+2) = S_i(t)$. Accordingly, the dynamics converges either to a fixed point or to a period-two limit cycle, as claimed. Note that, in contrast to sequential dynamics, nonvanishing diagonal terms J_{ii} do not modify the above argument. We now leave aside the dynamics and determine the thermodynamics associated with (6), using the idea of duplicate spins [1.28].

As usual, the S_i are Ising spins whose normalized trace is denoted by $\operatorname{tr}_S\{\ldots\}$. Since the J_{ij} are as in the Hopfield case we define overlaps

$$m_\mu^{(1)} = N^{-1}\sum_{i=1}^{N} \xi_i^\mu S_i . \tag{3.6.8}$$

To each site i we also assign a *duplicate spin* $\sigma_i = \pm 1$ with normalized trace $\operatorname{tr}_\sigma(\ldots)$ and overlaps

$$m_\mu^{(2)} = N^{-1} \sum_{i=1}^{N} \xi_i^\mu \sigma_i \,. \tag{3.6.9}$$

For the moment, the σ_i have no relation to those in (1–5).

Using the duplicate spins we rewrite the Hamiltonian (6),

$$\begin{aligned} -\beta \bar{H}_N &= \ln\left\{2^{-N} \prod_{i=1}^{N} \left[\exp\left(\beta \sum_\mu \xi_i^\mu m_\mu^{(1)}\right) + \exp\left(-\beta \sum_\mu \xi_i^\mu m_\mu^{(1)}\right)\right]\right\} \\ &= \ln\left[\mathrm{tr}_\sigma \prod_{i=1}^{N} \exp\left(\beta \sigma_i \sum_\mu \xi_i^\mu m_\mu^{(1)}\right)\right] \\ &= \ln\left[\mathrm{tr}_\sigma \exp\left(\beta N \sum_\mu m_\mu^{(1)} m_\mu^{(2)}\right)\right] \end{aligned} \tag{3.6.10}$$

so that

$$\begin{aligned} -\beta f(\beta) &= \lim_{N\to\infty} N^{-1} \ln\left[\mathrm{tr}_S \left\{\exp(-\beta \bar{H}_N)\right\}\right] \\ &= \lim_{N\to\infty} N^{-1} \ln\left[\mathrm{tr}_S \left\{\mathrm{tr}_\sigma \left\{\exp(\beta N \boldsymbol{m}^{(1)} \cdot \boldsymbol{m}^{(2)})\right\}\right\}\right] \,. \end{aligned} \tag{3.6.11}$$

In (11) we in fact have a new Hamiltonian

$$H(\boldsymbol{\sigma}|\boldsymbol{S}) = -\sum_{\mu=1}^{q} \boldsymbol{m}_\mu^{(1)} \boldsymbol{m}_\mu^{(2)} = -\boldsymbol{m}^{(1)} \cdot \boldsymbol{m}^{(2)} \,, \tag{3.6.12}$$

where $m_\mu^{(1)}$ is defined in terms of the S_i and $m_\mu^{(2)}$ in terms of the σ_i. Before proceeding we quickly pause to return to (2) and obtain the appealing representation

$$W(\boldsymbol{\sigma}|\boldsymbol{S}) = \frac{\exp[-\beta H(\boldsymbol{\sigma}|\boldsymbol{S})]}{2^N \mathrm{tr}_\sigma\{\exp[-\beta H(\boldsymbol{\sigma}|\boldsymbol{S})]\}} \,. \tag{3.6.13}$$

However, we will not pursue this issue here any further.

For finite q, the large-deviation technique of Sect. 1.2.2 (see also Sect. 1.3.1) allows a straightforward and exact evaluation of the free energy $f(\beta)$,

$$-\beta f(\beta) = \sup_{\boldsymbol{m}} [\beta \boldsymbol{m}^{(1)} \cdot \boldsymbol{m}^{(2)} - c^*(\boldsymbol{m})] \,. \tag{3.6.14}$$

We can now apply (2.2.7–2.2.9) with $F(\boldsymbol{m}) = \boldsymbol{m}^{(1)} \cdot \boldsymbol{m}^{(2)}$. Because we have duplicated the spins, the dimension has increased from q to $2q$ as compared to the Hopfield model and the c-function is $c(\boldsymbol{t}) = c(\boldsymbol{t}_1) + c(\boldsymbol{t}_2)$ where $\boldsymbol{t} = (\boldsymbol{t}_1, \boldsymbol{t}_2) \in \mathbb{R}^{2q}$ and the $c(\boldsymbol{t}_i)$ are given by (1.2). The fact that $c(\boldsymbol{t})$ is simply the sum of two Hopfield c-functions and $F(\boldsymbol{m}) = \boldsymbol{m}^{(1)} \cdot \boldsymbol{m}^{(2)}$ suggests that the symmetry between the S- and the σ-variables is not broken and that $f_{\mathrm{Little}}(\beta) = 2 f_{\mathrm{Hopfield}}(\beta)$. This is indeed the case [1.13a, 28].

That is to say, owing to (2.2.8) with $F(\boldsymbol{m}) = \boldsymbol{m}^{(1)} \cdot \boldsymbol{m}^{(2)}$, the fixed-point equation for $\boldsymbol{m} = (\boldsymbol{m}^{(1)}, \boldsymbol{m}^{(2)})$ is

$$m^{(1)} = \langle \xi \tanh(\beta m^{(2)} \cdot \xi) \rangle \, ,$$
$$m^{(2)} = \langle \xi \tanh(\beta m^{(1)} \cdot \xi) \rangle \, . \tag{3.6.15}$$

Subtracting $m^{(2)}$ from $m^{(1)}$ and taking the inner product with $(m^{(1)} - m^{(2)})$ one obtains [1.13a]

$$\begin{aligned} 0 &\leq (m^{(1)} - m^{(2)}) \cdot (m^{(1)} - m^{(2)}) \\ &= \langle (m^{(1)} \cdot \xi - m^{(2)} \cdot \xi)[\tanh(\beta m^{(2)} \cdot \xi) - \tanh(\beta m^{(1)} \cdot \xi)] \rangle \leq 0 \, , \end{aligned} \tag{3.6.16}$$

since $(x - y)[\tanh(\beta y) - \tanh(\beta x)] \leq 0$. So $m^{(1)} = m^{(2)} \equiv m$ and the $S - \sigma$ symmetry is not broken. Furthermore, by (2.2.9), the free energy directly follows,

$$-\beta f(\beta) = 2 \max_{m} \left[-\beta m^2/2 + c(\beta m) \right] \, , \tag{3.6.17}$$

where the q-dimensional vector m satisfies (1.3). Comparing (17) with (1.4) we see that $f_{\text{Little}}(\beta) = 2 f_{\text{Hopfield}}(\beta)$, as claimed. With a bit more effort [1.28] it can also be shown that the stability analysis for the Little model is identical with that of the Hopfield model.

If the number of stored patterns q increases with N, we can exploit the idea of duplicate spins once again, start with (10), and proceed as in Sect. 1.3.3. Instead of a single spin-glass order parameter $\hat{q}$ related to (3.13) we now get three, corresponding to $S - S, S - \sigma$, and $\sigma - \sigma$ combinations. For explicit formulae and phase diagrams, the reader may consult Fontanari and Köberle [1.50]. These authors have also added a diagonal term $J_{ii} = J_0$ so as to control (or even suppress) the 2-cycles (for large enough $J_0 > 0$). Note that J_0 gives rise to a self-interaction $J_0 S_i(t)$ in the local field $h_i(t)$ and occurs as such in the Hamiltonian $\bar{H}_N$ for the Little model. For $J_0 > -1$, the critical α_c turns out to be nearly identical with that of the Hopfield model.

1.3.7 Continuous-Time Dynamics and Graded-Response Neurons

With respect to dynamics, the Little and the Hopfield model represent two extreme cases. In the Little model, all neurons are updated at every time step in complete synchrony. In the Hopfield model, the updating is sequential, that is, at each time step only a single neuron redefines its state in response to its post-synaptic potential.

From a biological point of view, both types of dynamics have obvious deficiencies. In biological nerve nets, there appears to be no agent that would enforce a global synchrony of updatings as required for the parallel Little dynamics. In asynchronous dynamics, on the other hand, neurons do not determine their state strictly in response to their post-synaptic potential, but *only* when the updating procedure decides it is their turn.

Here we discuss a third variant: continuous-time dynamics. It is introduced along with a different modelling at the single-neuron level. In the context of continuous-time dynamics, a single neuron is conceived of as a capacitance which is charged by afferent currents fed into it via synaptic junctions or from external current sources, and which is discharged by leakage across the neuronal membrane. The neuron responds with an output *firing rate* which depends on the current value of its trans-membrane potential in a nonlinear but smooth way, i.e., in a graded fashion. The dynamics of a network of graded-response neurons is then described by a set of RC charging equations according to [1.5]

$$C_i \frac{\mathrm{d}U_i}{\mathrm{d}t} = \sum_{j=1}^{N} J_{ij} V_j - \frac{U_i}{R_i} + I_i \ , \tag{3.7.1}$$

with

$$V_j = g_j(\gamma_j U_j) \ . \tag{3.7.2}$$

In (1), C_i denotes the input capacitance of the ith neuron, R_i is its transmembrane resistance, U_i its postsynaptic potential, and V_i its instantaneous output. The input-output characteristics of a neuron is encoded in its transfer (gain) function g_j as in (2), γ_j denoting a gain parameter. The I_i represent the input from external (current) scources and the synaptic weights are as usual denoted by J_{ij}

The continuous-time dynamics (1) has been proposed by Hopfield [1.5], who put the main emphasis on probing the degree of robustness of collective network-based computation. There are beyond this, of course, also a number of *specific* points to be put forward in favour of (1). First, the continuous-time dynamics (1) allows the inclusion of neurophysiological detail into formal neural network models, which is not available in the standard models using two-state neurons with (stochastic) synchronous or asynchronous dynamics. For example, capacitive input delays and trans-membrane leakages are explicitly taken into account in (1) — input delays, though, perhaps not as detailed as the variability of synaptic-dendritic information transport would require. Moreover, it can be shown that the shape of the gain functions $g_j(x)$ in (2) encodes neural behaviour during relative refractory periods [1.146,147]. To be honest, a drawback of (1) is that it is a *rate* description and, therefore, does not apply to situations where a high temporal resolution is needed. For that, one would have to take recourse to a description in terms of spiking neurons; for a review, see [1.148].

Alternatively, the Eqs. (1) and (2) provide a quantitative description of the dynamics of networks of resistively coupled nonlinear amplifiers [1.5], in which case the V_j denote output voltages of amplifiers with gain functions g_j. Such networks have been suggested [1.149] as real-time solvers of hard optimization tasks, so a quantitative theoretical understanding of their performance will be of use as a guide to improving the efficiency of devices of this type.

The purpose of the present subsection is to describe the long-time behaviour of networks of graded-response neurons with continuous-time dynamics, using a statistical mechanics approach proposed in [1.147,150,151].

For this approach to be applicable, the dynamics (1) must be governed by a Lyapunov function, a condition that is satisfied, if the J_{ij} are symmetric and neural gain functions are monotonically increasing functions of their argument. These conditions were identified by Cohen and Grossberg [1.51] and by Hopfield [1.5], and the Lyapunov function was shown to be of the form

$$\mathcal{H}_N = -\frac{1}{2}\sum_{i,j=1}^{N} J_{ij} V_i V_j + \sum_{i=1}^{N} \frac{1}{\gamma_i R_i} G_i(V_i) - \sum_{i=1}^{N} I_i V_i \; , \tag{3.7.3}$$

where G_i denotes the integrated inverse input-output relation

$$G_i(V) = \int^{V} \mathrm{d}V' \; g_i^{-1}(V') \; . \tag{3.7.4}$$

The value of the lower integration limit in (4) is arbitrary. It can be used to define the zero of the energy scale in (3). In terms of (3), the dynamics (1) reads

$$C_i \frac{\mathrm{d}U_i}{\mathrm{d}t} = -\frac{\partial \mathcal{H}_N}{\partial V_i} \; , \tag{3.7.5}$$

so that

$$\frac{\mathrm{d}\mathcal{H}_N}{\mathrm{d}t} = \sum_{i=1}^{N} \frac{\partial \mathcal{H}_N}{\partial V_i}\frac{\mathrm{d}V_i}{\mathrm{d}t} = -\sum_{i=1}^{N} C_i \gamma_i g_i'(\gamma_i U_i) \left(\frac{\mathrm{d}U_i}{\mathrm{d}t}\right)^2 \leq 0 \; , \tag{3.7.6}$$

with equality in (6) *only* at stationary points of (1). If $\mathcal{H}_N$ is bounded from below, the dynamical flow generated by (1) will *always* converge to a fixed point which is a global or local minimum of $\mathcal{H}_N$. The required boundedness is guaranteed if, for large $|U_i|$, the g_i do not increase faster than linearly with U_i [1.152].

The statistical-mechanical way of locating (and characterizing) the minima of $\mathcal{H}_N$, hence the attractors of the dynamics (1), is to compute the zero-temperature ($\beta \to \infty$) limit of the free energy

$$\begin{aligned} f_N(\beta) &= -(\beta N)^{-1} \ln \mathrm{Tr}_{\boldsymbol{V}} \exp[-\beta \mathcal{H}_N(\boldsymbol{V})] \\ &= -(\beta N)^{-1} \ln \int \prod_i \mathrm{d}\rho(V_i) \exp[-\beta \mathcal{H}_N(\boldsymbol{V})] \end{aligned} \tag{3.7.7}$$

and to investigate the nature of its stable and metastable phases [1.147,150,151]. Here, $\mathrm{d}\rho(V_i)$ denotes an *a priori* measure on the space of neural output states which is taken to be uniform (though not normalized) on the range of g_i so as to avoid encoding hidden assumptions about the system's behaviour already at the level of this output measure.

To be specific, and to allow for a comparison with the standard model, we demonstrate how this approach works for networks of analog neurons with Hebbian couplings

$$J_{ij} = \frac{1}{N}\sum_{\nu=1}^{q} \xi_i^{\nu} \xi_j^{\nu} \quad , \quad i \neq j \; , \tag{3.7.8}$$

which have been designed to store a set of q unbiased binary random patterns $\xi_i^\nu \in \{\pm 1\}$, $1 \le \nu \le q$. For the sake of simplicity, we assume that all neurons have the same input-output relation $g_i = g$, with gain parameters $\gamma_i = \gamma$. Moreover we take $R_i = C_i = 1$ in suitable units. As shown in [1.150,151], these simplifying homogeneity assumptions can be dropped, if desired.

For networks with synaptic couplings given by (8), the free energy (7) may be expressed as

$$f_N(\beta) = -(\beta N)^{-1} \ln \int \prod_i \mathrm{d}\tilde{\rho}(V_i) \exp\left(\frac{N\beta}{2} \sum_\nu m_\nu^2\right) , \tag{3.7.9}$$

where we have introduced the overlaps

$$m_\nu = \frac{1}{N} \sum_{i=1}^N \xi_i^\nu V_i , \tag{3.7.10}$$

and where the integrated inverse input-output relation G as well as a term correcting for the absence of self-interactions have been absorbed in the single-site measure

$$\mathrm{d}\tilde{\rho}(V) = \mathrm{d}\rho(V) \exp[-\alpha\beta V^2/2 - \beta\gamma^{-1} G(V)] , \tag{3.7.11}$$

with $\alpha = q/N$.

In the limit of extensively stored patterns ($\alpha > 0$) the free energy is evaluated by the replica method [1.13c]. As for the details, the computations follow those for the Hopfield model, which have been discussed at length in Sect. 3.3. Here too, a *finite* subset of patterns (labeled by μ) is singled out, and one assumes that the system "lives" in a region of phase space which is uncorrelated with the remaining patterns (labeled by ν). This justifies (see Sect. 3.3) treating the

$$X_\sigma(\nu) = \frac{1}{\sqrt{N}} \sum_{i=1}^N \xi_i^\nu V_i^\sigma \tag{3.7.12}$$

as Gaussian random variables with mean zero and covariance matrix

$$\langle X_\sigma(\nu) X_{\sigma'}(\nu') \rangle = \delta_{\nu,\nu'} \frac{1}{N} \sum_{i=1}^N V_i^\sigma V_i^{\sigma'} \equiv \delta_{\nu,\nu'} q_{\sigma\sigma'} , \tag{3.7.13}$$

and to perform the averaging of the replicated partition sum over the ν-patterns as an average over the Gaussians $X_\sigma(\nu)$. This yields, with weights $\varepsilon_\nu = 1$ and thus $Q_\nu = Q = (\mathbb{1} - \beta \boldsymbol{q})$,

$$\phi_N(n) = \frac{1}{N} \ln \left\langle \int \prod_{i,\sigma} \mathrm{d}\tilde{\rho}(V_i^\sigma) \exp\left[\frac{N}{2}\left(\beta \sum_{\mu,\sigma} m_{\mu\sigma}^2 - \alpha \mathrm{Tr} \ln Q(\boldsymbol{q})\right)\right]\right\rangle \tag{3.7.14}$$

where the $m_{\mu\sigma}$ denote replicated overlaps ($1 \le \sigma \le n$), $\boldsymbol{q} = (q_{\sigma\sigma'})$ is the matrix of Edwards-Anderson order parameters introduced in (13), and the angular brackets indicate an average over the μ-patterns according to their distribution. Formally, this is the same as (3.12-14), except that the normalized spin-trace tr_{S_σ} is replaced by an integration over the unnormalized measure $\prod_{i,\sigma} \mathrm{d}\tilde{\rho}(V_i^\sigma)$. This measure can, however, be normalized, so that the large-deviations techniques described in Secs 2.1-3 and used in Sec 3.3 to analyse the Hopfield model are directly applicable in the present case. All formal manipulations of Sect. 3.3 go through virtually unaltered, with the c-function (3.17) being replaced by

$$c(\boldsymbol{x},\boldsymbol{y}) = \left\langle \ln \int \prod_\sigma \mathrm{d}\hat{\rho}(V^\sigma) \exp\left[\sum_{\mu,\sigma} x_{\mu\sigma}\xi^\mu V^\sigma + \sum_{\sigma\le\sigma'} y_{\sigma\sigma'} q_{\sigma\sigma'}\right]\right\rangle . \quad (3.7.15)$$

Here $\mathrm{d}\hat{\rho}$ denotes the normalized measure corresponding to the unnormalized $\mathrm{d}\tilde{\rho}$. A second, last, and notable difference is related to the fact that the matrix of Edwards-Anderson order parameters has *non*trivial diagonal entries, $q_{\sigma\sigma} \neq 1$, since $V_i^2 = 1$ uniquely characterizes Ising spins.

With these observations at hand, it should not be overly difficult to derive the following results in the replica-symmetric approximation. The free energy is given by [1.150,151]

$$\begin{aligned} f(\beta) = & \frac{1}{2}\sum_\mu m_\mu^2 + \frac{\alpha}{2}\left\{\beta^{-1}\ln[1-\beta(q_0-q_1)] + (q_0-q_1)\tilde{r} + \beta(q_0-q_1)r\right\} \\ & - \beta^{-1}\left\langle\!\left\langle \ln \int \mathrm{d}\tilde{\rho}(V)\exp\left\{\beta\left[\left(\boldsymbol{\xi}\cdot\boldsymbol{m} + \sqrt{\alpha r}z\right)V + \tfrac{1}{2}\alpha\tilde{r}V^2\right]\right\}\right\rangle\!\right\rangle , \end{aligned} \quad (3.7.16)$$

where $\boldsymbol{\xi}\cdot\boldsymbol{m} = \sum_\mu \xi^\mu m_\mu$, and where double angular brackets designate a combined average over the finitely many ξ^μ with which the system is macroscopically correlated and a Gaussian random variable z with zero mean and unit variance. In (16),

$$r = \frac{q_1}{[1-\beta(q_0-q_1)]^2} \quad \text{and} \quad \tilde{r} = \frac{1}{1-\beta(q_0-q_1)} . \quad (3.7.17)$$

The m_μ and the diagonal and off-diagonal elements q_0 and q_1 of the matrix $\boldsymbol{q}$ of Edwards-Anderson order parameters must be chosen in such a way that they satisfy the fixed-point equations

$$m_\mu = \langle\langle \xi^\mu [V]_{\xi,z}\rangle\rangle , \quad (3.7.18a)$$

$$q_0 = \langle\langle [V^2]_{\xi,z}\rangle\rangle , \quad (3.7.18b)$$

$$q_1 = \langle\langle [V]_{\xi,z}^2\rangle\rangle . \quad (3.7.18c)$$

In (18), $[\ldots]_{\xi,z}$ denotes the 'thermal' average

$$
\begin{aligned}
[\ldots]_{\xi,z} &= \frac{\int \mathrm{d}\hat{\rho}(V)(\ldots)\exp\left[\beta(\boldsymbol{\xi}\cdot\boldsymbol{m}+\sqrt{\alpha r}z)V+\frac{\alpha\beta}{2}\tilde{r}V^2\right]}{\int \mathrm{d}\hat{\rho}(V)\exp\left[\beta(\boldsymbol{\xi}\cdot\boldsymbol{m}+\sqrt{\alpha r}z)V+\frac{\alpha\beta}{2}\tilde{r}V^2\right]} \\
&= \frac{\int \mathrm{d}\rho(V)(\ldots)\exp[-\beta H_{\xi,z}(V)]}{\int \mathrm{d}\rho(V)\exp[-\beta H_{\xi,z}(V)]} ,
\end{aligned}
\tag{3.7.19}
$$

where, using (11), we have introduced an effective single-site Hamiltonian

$$
H_{\xi,z}(V) = \gamma^{-1}G(V) - \left(\boldsymbol{\xi}\cdot\boldsymbol{m}+\sqrt{\alpha r}z\right)V - \frac{\alpha}{2}(\tilde{r}-1)V^2 . \tag{3.7.20}
$$

In order to obtain information about the nature of the local and global minima of $\mathcal{H}_N$, the averages (19) should be evaluated in the deterministic limit $\beta \to \infty$. In this limit one obtains [1.147,150,151]

$$
[V^k]_{\xi,z} = \hat{V}^k(\boldsymbol{\xi}, z) \tag{3.7.21}
$$

for $k = 1, 2$, where $\hat{V}(\boldsymbol{\xi}, z)$ is a point which minimizes $H_{\xi,z}(V)$ and thus dominates the integrals in (19). It must be determined among the solutions of the transcendental fixed-point equation

$$
\hat{V} = \hat{V}(\boldsymbol{\xi}, z) = g\left(\gamma\left[\boldsymbol{\xi}\cdot\boldsymbol{m}+\sqrt{\alpha r}z+\alpha(\tilde{r}-1)\hat{V}\right]\right) \tag{3.7.22}
$$

on the support of $\mathrm{d}\rho$. For numerical purposes, it is advantageous to rewrite the fixed-point equations (18) in terms of the variables m_μ, q_1 and $C \equiv \beta(q_0 - q_1)$, so as to get (as $\beta \to \infty$)

$$
m_\mu = \left\langle\!\left\langle \xi^\nu \hat{V}(\boldsymbol{\xi}, z)\right\rangle\!\right\rangle , \quad C = \frac{1}{\sqrt{\alpha r}}\left\langle\!\left\langle z\hat{V}(\boldsymbol{\xi}, z)\right\rangle\!\right\rangle , \quad q_1 = \left\langle\!\left\langle \hat{V}^2(\boldsymbol{\xi}, z)\right\rangle\!\right\rangle , \tag{3.7.23}
$$

with $r = q_1/(1-C)^2$, $\tilde{r} = 1/(1-C)$, and $\hat{V}(\boldsymbol{\xi}, z)$ being determined from (22). Thus, we have obtained a macroscopic characterization of the attractors of (1) through the solutions of a set of fixed-point equations (23) which in turn require the solution of a single transcendental fixed-point equation, viz. (22), for *all* real z. This nesting of fixed-point equations constitutes the main difference between the present theory and that of the standard model.

It is perhaps worth noting that up to this point, the statistical-mechanical theory could be developed in *complete generality* with respect to input-output relations. Formally, this is due to the fact that input-output relations affect *only* single-site measures in (9-18), and not terms related to synaptic interactions.

Let us now briefly state the main results for the model described above. First, the case of finitely many stored patterns is recovered by taking the limit $\alpha \to 0$ in Eqs. (16-23). In this limit, the m_μ alone are sufficient to describe the state of the system. They must be chosen so as to satisfy (18a). The solution $\hat{V}$ of (22) can be determined explicitly in this case. It is given by $\hat{V} = g(\gamma\sum_\mu \xi^\mu m_\mu)$, so that in the limit $\beta \to \infty$ Eqs. (18a) take the form $m_\mu = \langle \xi^\mu g(\gamma\sum_\mu \xi^\mu m_\mu)\rangle$. These equations bear a strong formal similarity to those describing the stochastic

Hopfield model, with the hyperbolic tangent replaced by a general input-output relation, and inverse temperature β by the gain parameter γ. Needless to say that we get a true formal equivalence for the choice $g(x) = \tanh(x)$, which does not, however, persist at extensive levels of loading ($\alpha > 0$), as we shall presently see.

We restrict our attention to conventional, odd, sigmoid input-output relations g, normalized such that they have asymptotes $\lim_{x\to\pm\infty} g(x) = \pm 1$, and maximum slope equal to 1 at the origin, $g'(x) \le g'(0) = 1$. Examples are the hyperbolic tangent, $g(x) = \tanh(x)$, or the piecewise linear function $g(x) = \mathrm{sgn}(x)\min(|x|, 1)$. For such gain functions, the topology of the phase diagram is similar to that of the stochastic model, inverse gain playing a role similar to that of temperature in the Hopfield model; see Fig. 1.3. In particular, at $\alpha = 0$ solutions with non-zero m_μ exist for $\gamma > \gamma_c = 1$. The transition at γ_c may be continuous or discontinuous, depending on the shape of $g(x)$ near $x = 0$. The instability of the "paramagnetic" null solution with respect to spin-glass ordering occurring at $\gamma_g^{-1} = 1 + 2\sqrt{\alpha}$ and the storage capacity $\alpha_c \simeq 0.138$ at $\gamma^{-1} = 0$, too, are *universal* properties of networks having a normalized sigmoid input-output relation as defined above. As for the case of finitely many patterns, the *order* of the spin-glass transition at $\gamma_g(\alpha)$ will depend on properties of $g(x)$ in the vicinity of $x = 0$, and there will be hysteresis effects, if the transition is discontinuous [1.147,150,151].

In an enlarged portion of the phase diagram (Fig. 1.3b), we also show the curve, satisfying $\alpha\gamma(\tilde{r} - 1) = 1$, that indicates the opening of a gap in the local field distribution for the retrieval phase (see below). The other additional curve signals an instability of the replica symmetric retrieval solution against replica symmetry breaking [1.153]. It can be shown [1.151] that replica symmetry is definitely broken in regions of parameter space where the local field distribution computed in the replica symmetric approximation has a gap.

At this point, we call attention to the fact that the mean-field theory presented above provides more than just a tool for identifying stable stationary states of the network dynamics and for characterizing them macroscopically in terms of order parameters. The order parameters constitute, in fact, a *parametrization* of the local field distributions pertaining to the various types of attractor. This is true for the standard model too but there the distribution is trivial (a Gaussian) and, more importantly, no useful information can be derived from it because of the all-or-none response of the McCulloch-Pitts type neurons employed there. Networks of graded-response neurons, on the other hand, have the distinct advantage that local fields or membrane potentials $\hat{U}$ and firing rates $\hat{V}$ are in one-to-one correspondence through $\hat{V} = g(\gamma\hat{U})$. Thus, local field distributions can be translated into firing-rate distributions (and vice versa), the latter being directly accessible to experimental techniques of the neurophysiologist — *in contrast* to the order parameters m_ν, C, and q_1 or the storage capacity α_c.

The computation of local field distributions [1.151] starts from (22). Using $\hat{V} = g(\gamma\hat{U})$, one rewrites (22) in terms of $\hat{U}$ so as to read $\hat{U} = \boldsymbol{\xi}\cdot\boldsymbol{m} + \sqrt{\alpha r}z + \alpha(\tilde{r} - 1)g(\gamma\hat{U})$. As before, in case of several solutions one has to choose the one

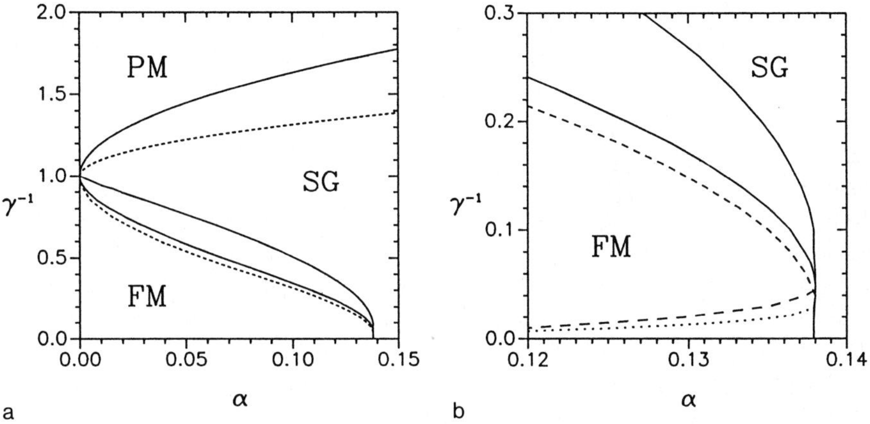

Fig. 1.3a, b. Phase diagram for soft-neuron versions of the Hopfield model. PM, SG, and FM denote the paramagnetic (null-), spin-glass, and retrieval phases, respectively. Results for the stochastic Ising network are shown as *dashed* lines (upon identification of inverse gain and temperature scales) for comparison. Retrieval phase boundaries are shown for neurons with hyperbolic tangent ($g(x) = \tanh(x)$; *lower full curve*) and piecewise linear ($g(x) = \mathrm{sgn}(x)\min(|x|, 1)$; *upper full curve*) response. **(b)** Enlarged portion of the phase diagram (a). It also shows the "gap-line" (*dotted*) below which the local field distribution for the retrieval phase has a gap at small $\hat{U}$, and the AT line (*long-dashed*), both for neurons with hyperbolic tangent response. Taken from [1.151].

minimizing $H_{\xi,z}(V) = H_{\xi,z}(g(\gamma U))$. From this, one obtains the $\hat{U}$ distribution for a given $\boldsymbol{\xi} = (\xi^\mu)_\mu$ as follows. Instead of determining $\hat{U}(z)$ for given z, one proceeds the other way round and solves for $z = z(\hat{U})$ which gives

$$z = z(\hat{U}) = \frac{1}{\sqrt{\alpha r}}\left(\hat{U} - \boldsymbol{\xi}\cdot\boldsymbol{m} - \alpha(\tilde{r}-1)g(\gamma\hat{U})\right) . \tag{3.7.24}$$

In cases where (22) has multiple solutions, several different $\hat{U}$ will give rise to the same z in which case one has to choose that representation which minimizes $H_{\xi,z}(g(\gamma\hat{U}))$. The $\hat{U}$ chosen to represent $z(\hat{U})$ in (22) may occasionally jump, namely when two minima of $H_{\xi,z}(g(\gamma\hat{U}))$, i.e., two solutions of (22), exchange their relative depth — as order parameters do at a first-order phase transition. Locally, however $\hat{U}(z)$ is smooth and invertible, implying that $z = z(\hat{U})$ is continuous and differentiable. Hence, knowledge of $\mathcal{P}(z)\mathrm{d}z = \frac{\mathrm{d}z}{\sqrt{2\pi}}\exp(-z^2/2)$ allows one to obtain the local field distribution

$$\begin{aligned}\mathcal{P}_\xi(\hat{U}) &\equiv \mathcal{P}(z(\hat{U}))\frac{\mathrm{d}z}{\mathrm{d}\hat{U}}\\ &= \frac{1}{\sqrt{2\pi\alpha r}}\exp\left\{-\frac{1}{2\alpha r}\left(\hat{U} - \boldsymbol{\xi}\cdot\boldsymbol{m} - \alpha(\tilde{r}-1)g(\gamma\hat{U})\right)^2\right\}\\ &\quad\times\left(1 - \alpha\gamma(\tilde{r}-1)g'(\gamma\hat{U})\right)\chi(\hat{U}) ,\end{aligned} \tag{3.7.25}$$

where $\chi(\hat{U}) = 1$ for those $\hat{U}$ where $\hat{U}(z)$ is locally smooth and invertible, whereas $\chi(\hat{U}) = 0$ on those intervals across which the solution $\hat{U}(z)$ to (22) *jumps* due to the minimality criterion for $H_{\xi,z}(g(\gamma\hat{U}))$. Note that $\alpha\gamma(\tilde{r}-1)g'(\gamma\hat{U}) < 1$ for a solution to (22) that represents a (local) minimum of $H_{\xi,z}(g(\gamma U))$, so that $\mathcal{P}_\xi(\hat{U})$ is indeed nonnegative. If jumps of the solution $\hat{U}(z)$ do occur, they create gaps in the local field distribution $\mathcal{P}_\xi(\hat{U})$. For the "conventional" sigmoid input-output relations introduced above, jumps were shown to occur [1.151] when $\alpha\gamma(\tilde{r}-1) > 1$, a condition that is realized for sufficiently high gains and loading levels. Note that the determination of local field distributions can be used to speed up the numerical solution of Eqs. (23) considerably, since it circumvents the problem of solving nested fixed-point eqations by reformulating the integrations over the Gaussian z in Eqs. (23) as integrations over the $\hat{U}$ distributions (25); they are known explicitly once the gaps, if any, have been located [1.151].

Gaps or no gaps, Eq. (25) clearly shows that local field distributions deviate from the Gaussian form obtained for networks of stochastic two-state neurons in the context of the replica approach, and that their precise form will also depend on details of the neural gain function. Using the gain function to transform to the $\hat{V}$, firing rate distributions are readily obtained from (25).

We refrain here from reviewing the by now extensive literature on graded reponse neurons. The reader will find an account of it in [1.151].

1.4 Nonlinear Neural Networks

According to Hebb's principle [1.15], the local data at i and j, i.e., $\boldsymbol{\xi}_i = (\xi_i^\mu; 1 \leq \mu \leq q)$ and $\boldsymbol{\xi}_j = (\xi_j^\mu; 1 \leq \mu \leq q)$, determine the synaptic efficacies J_{ij} $(j \to i)$ and J_{ji} $(i \to j)$. So we may write $J_{ij} = N^{-1}Q(\boldsymbol{\xi}_i; \boldsymbol{\xi}_j)$ for some synaptic kernel Q defined on $\mathbb{R}^q \times \mathbb{R}^q$. The Hopfield case is rather special in that Q is symmetric, $Q(\boldsymbol{x};\boldsymbol{y}) = Q(\boldsymbol{y};\boldsymbol{x})$, and in addition has the form $Q(\boldsymbol{x};\boldsymbol{y}) = \boldsymbol{x}\cdot\boldsymbol{y}$, a scalar product which is *linear* both in $\boldsymbol{x}$ and in $\boldsymbol{y}$. This linearity greatly simplifies the analytical work. In this section we concentrate on *non*linear neural networks. We start by presenting an exact solution of the free energy for arbitrary symmetric Qs and finitely many patterns (Sect. 1.4.1). In so doing we also obtain explicit criteria for the dynamic stability of the patterns. Section 1.4.2 is a short interlude devoted to the spectral theory of the matrix $Q(\boldsymbol{x};\boldsymbol{y})$, where the labels $\boldsymbol{x}$ and $\boldsymbol{y}$ range through the corners C^q of the (hyper)cube $[-1,1]^q$ and Q itself satisfies a weak invariance condition. The spectral theory can be applied both to the analysis of the fixed-point equations, whose solutions determine the (thermo)dynamic equilibrium states, and the statistical mechanics of nonlinear neural networks with *extensively* many patterns (Sect. 1.4.3).

A special, but for hardware realizations relevant, case is provided by *clipped synapses* where we "clip" the $\boldsymbol{\xi}_i\cdot\boldsymbol{\xi}_j$ so as to retain their sign only. Since $\operatorname{sgn}(\boldsymbol{\xi}_i\cdot\boldsymbol{\xi}_j) = \pm 1$ with probability 1/2, one can store one bit per bond. If the bonds J_{ij} were independent (they are not), one could store there $N(N-1)/2$ bit since there are $N(N-1)/2$ bonds. Because they are not independent the actual storage capacity is always less than that. On the other hand, the amount of information contained in the data, i.e., the patterns, is $qN = \alpha N^2$ bits. Thus we arrive at the theoretical upper bound $\alpha_c^{\text{clipped}} \leq 0.5$. It will be shown analytically that $\alpha_c^{\text{clipped}} = 0.10 < 0.14 = \alpha_c^{\text{Hopfield}}$. Moreover we will find that moderate dilution improves the performance of a network with clipped synapses: by removing about half of the bonds one can increase α_c so as to get 0.12. The retrieval quality is then increased as well. This improvement of the performance is a rather surprising result.

1.4.1 Arbitrary Synaptic Kernel and Finitely Many Patterns

Let us suppose for the moment that the vectors $\boldsymbol{\xi}_i$ assume the corners $\boldsymbol{x}$ of the (hyper)cube $[-1,1]^q$ with probability $p(\boldsymbol{x})$. We write $\boldsymbol{x} \in \mathcal{C}^q$. The case of more general probability distributions will be considered later on. According to Sect. 2.4, the index set $\{1,2,\ldots,N\}$ can be decomposed into $M = 2^q$ disjoint sublattices $I(\boldsymbol{x}) = \{i; \boldsymbol{\xi}_i = \boldsymbol{x}\}$ whose sizes $|I(\boldsymbol{x})|$ become deterministic as $N \to \infty$,

$$N^{-1}|I(\boldsymbol{x})| = N^{-1}\sum_{i=1}^{N} \delta_{\boldsymbol{\xi}_i,\boldsymbol{x}} \longrightarrow \text{Prob}\{\boldsymbol{\xi} = \boldsymbol{x}\} = p(\boldsymbol{x}) \,. \tag{4.1.1}$$

See in particular (2.4.2) and (2.4.4). With each $I(\boldsymbol{x})$ we associate a sublattice magnetization

$$m(\boldsymbol{x}) = |I(\boldsymbol{x})|^{-1} \sum_{i \in I(\boldsymbol{x})} S_i \,, \tag{4.1.2}$$

as in (2.4.5). If we performed a trace over the S_i with $i \in I(\boldsymbol{x})$ *only*, then $m(\boldsymbol{x})$ would be distributed according to the density $\mathcal{D}_N(m) \sim \exp[-|I(\boldsymbol{x})|c^*(m)]$, where, for Ising-spins, $c^*(m)$ is given by (2.1.15). Compared to (2.1.14), one only has to replace N by $|I(\boldsymbol{x})|$, which increases with N as $p(\boldsymbol{x})N$.

To obtain the free energy and the (thermo)dynamic stability of a nonlinear neural network with arbitrary symmetric Q and finitely many patterns, the above notions of sublattice and sublattice magnetization are instrumental [1.16, 17]. To see this, we begin by rewriting the Hamiltonian

$$-\beta H_N = \tfrac{1}{2}\beta N^{-1} \sum_{i,j} S_i Q(\boldsymbol{\xi}_i;\boldsymbol{\xi}_j) S_j \tag{4.1.3}$$

in terms of the sublattice magnetizations (2), exploiting (1) and the fact that the sublattices induce a partition of $\{1,2,\ldots,N\}$,

$$\begin{aligned}
-\beta H_N &= \frac{\beta}{2N} \sum_{\boldsymbol{x},\boldsymbol{y}} \sum_{\substack{i \in I(\boldsymbol{x}) \\ j \in I(\boldsymbol{y})}} S_i Q(\boldsymbol{\xi}_i;\boldsymbol{\xi}_j) S_j \\
&= \frac{\beta}{2N} \sum_{\boldsymbol{x},\boldsymbol{y}} \left(\sum_{i \in I(\boldsymbol{x})} S_i \right) Q(\boldsymbol{x};\boldsymbol{y}) \left(\sum_{j \in I(\boldsymbol{y})} S_j \right) \\
&= \tfrac{1}{2} N\beta \sum_{\boldsymbol{x},\boldsymbol{y}} m(\boldsymbol{x}) \left[p(\boldsymbol{x}) Q(\boldsymbol{x};\boldsymbol{y}) p(\boldsymbol{y}) \right] m(\boldsymbol{y}) \equiv N F(\boldsymbol{m}) \,.
\end{aligned} \tag{4.1.4}$$

Here it is understood that $\boldsymbol{m} = (m(\boldsymbol{x}); \boldsymbol{x} \in \mathcal{C}^q)$ and $N \to \infty$.

Using once more the observation that the $I(\boldsymbol{x})$, $\boldsymbol{x} \in C^q$, induce a partition of $\{1, 2, \ldots, N\}$ we now perform the trace,

$$\operatorname{tr}\{\exp(-\beta H_N)\} = 2^{-N} \sum_{\{S_i = \pm 1; 1 \le i \le N\}} \exp(-\beta H_N)$$
$$= \left[\prod_{\boldsymbol{x}} 2^{-|I(\boldsymbol{x})|} \sum_{\{S_i; i \in I(\boldsymbol{x})\}}\right] \exp(-\beta H_N) .$$

In view of the large-deviation philosophy of Sects. 1.2.1, 2, it seems (and is!) natural to take the $m(\boldsymbol{x})$ as new integration variables. For $\boldsymbol{x} \neq \boldsymbol{x}'$ the summations over $\{S_i; i \in I(\boldsymbol{x})\}$ and $\{S_i; i \in I(\boldsymbol{x}')\}$ can be performed *independently*. Accordingly, the integrations over the $m(\boldsymbol{x}), \boldsymbol{x} \in C^q$, can also be performed independently, and we obtain, taking advantage of (2.1.14) and (2.1.15) in conjunction with (1) and (4)

$$\operatorname{tr}\{\exp(-\beta H_N)\} \sim \left[\prod_{\boldsymbol{x}} \int \mathrm{d}m(\boldsymbol{x}) \mathrm{e}^{-|I(\boldsymbol{x})| c^*(m(\boldsymbol{x}))}\right] \mathrm{e}^{N F(\boldsymbol{m})}$$
$$= \int \mathrm{d}\boldsymbol{m} \exp\left\{N\left[F(\boldsymbol{m}) - \sum_{\boldsymbol{x}} p(\boldsymbol{x}) c^*(m(\boldsymbol{x}))\right]\right\} \tag{4.1.5}$$

and thus, by another Laplace argument,

$$-\beta f(\beta) = \lim_{N \to \infty} N^{-1} \ln[\operatorname{tr}\{\exp(-\beta H_N)\}]$$
$$= \sup_{\boldsymbol{m}} \left[F(\boldsymbol{m}) - \sum_{\boldsymbol{x}} p(\boldsymbol{x}) c^*(m(\boldsymbol{x}))\right] . \tag{4.1.6}$$

We are done.

The maximum in (6) is realized among the $\boldsymbol{m}$ that satisfy the fixed-point equation

$$m(\boldsymbol{x}) = \tanh\left[\beta \sum_{\boldsymbol{y}} Q(\boldsymbol{x}; \boldsymbol{y}) p(\boldsymbol{y}) m(\boldsymbol{y})\right] \equiv \tanh[\beta h(\boldsymbol{x})] , \tag{4.1.7}$$

which generalizes (2.1.17). A fixed-point $\boldsymbol{m}$ gives rise to a stable phase (cf. Sect. 3.2) if and only if the second derivative of (6) at $\boldsymbol{m}$ is negative-definite. As before we will say that we "maximize" the free-energy functional (6) though, of course, the free energy $f(\beta)$ itself is minimized. By (6), (7), and the inverse-function theorem applied to $\mathrm{d}c^*(m)/\mathrm{d}m = \tanh^{-1}(m)$ we then find that the matrix with elements

$$\beta p(\boldsymbol{x}) Q(\boldsymbol{x}; \boldsymbol{y}) p(\boldsymbol{y}) - p(\boldsymbol{x}) \delta_{\boldsymbol{x},\boldsymbol{y}} [1 - m^2(\boldsymbol{x})]^{-1} \tag{4.1.8}$$

should have negative eigenvalues only.

The expression (6) can be simplified,

$$-\beta f(\beta) = -\tfrac{1}{2}\beta \sum_{\boldsymbol{x},\boldsymbol{y}} m(\boldsymbol{x})p(\boldsymbol{x})Q(\boldsymbol{x};\boldsymbol{y})p(\boldsymbol{y})m(\boldsymbol{y}) + \sum_{\boldsymbol{x}} p(\boldsymbol{x})c(\beta h(\boldsymbol{x})) , \quad (4.1.9)$$

where we take the solution(s) $\boldsymbol{m} = (m(\boldsymbol{x}); \boldsymbol{x} \in C^q)$ which maximize(s) (9). The function $h(\boldsymbol{x})$ has been defined by (7). In fact, $h(\boldsymbol{x})$ is nothing but the local field experienced by all spins (neurons) which belong to the sublattice $I(\boldsymbol{x})$; see also Sect. 5.4. To obtain the result (9), which also holds for more general c-functions corresponding to n-component, Potts, or soft spins (graded-response neurons), we can either use (2.2.8) directly or exploit the fact that the $m(\boldsymbol{x})$ satisfy (7) and, by explicit calculation, $c^*(m(\boldsymbol{x})) = \beta m(\boldsymbol{x})h(\boldsymbol{x}) - c(\beta h(\boldsymbol{x}))$.

What should be modified in the above reasoning if we allow other probability distributions? As long as the distribution is discrete we need not change a single word. For a general distribution μ, e.g., a continuous one, we have to make the substitution

$$\sum_{\boldsymbol{x}} p(\boldsymbol{x})g(\boldsymbol{x}) \longrightarrow \int d\mu(\boldsymbol{x})g(\boldsymbol{x}) \quad (4.1.10)$$

in (6), (7), and (9). The proof [1.30] requires only a mild regularity condition on Q.

1.4.2 Spectral Theory

Given a synaptic kernel Q, some knowledge of the spectrum of the matrix $Q(\boldsymbol{x};\boldsymbol{y})$ with $\boldsymbol{x}$ and $\boldsymbol{y}$ ranging as labels through $C^q = \{-1,1\}^q$ is quite useful. To wit, we consider the fixed-point equation (1.7) and, after having developed the relevant spectral theory for Q, we derive a new representation of the Hamiltonian (1.3), which will allow the analytic treatment of extensively many patterns in the next section.

The case we are mainly interested in is $p(\boldsymbol{x}) = 2^{-q}$ for all $\boldsymbol{x} \in C^q$. This we assume for the moment. For β small enough, the only solution to (1.7) is the trivial one, viz., $\boldsymbol{m} = 0$. Postponing uniqueness, we first check its stability and to this end turn to (1.8). In the present case, (1.8) leads to the requirement that $\beta 2^{-q}Q - \mathbb{1}$ should have negative eigenvalues only. (Since Q is symmetric, all its eigenvalues are real.) Let us denote by $\Lambda_{\max}$ the *largest* (positive) eigenvalue of $2^{-q}Q$. Then we see that $\boldsymbol{m} = 0$ is stable as long as $\beta\Lambda_{\max} < 1$.

We now show that under the condition $\beta\Lambda_{\max} < 1$ the trivial solution $\boldsymbol{m} = 0$ is also the *only* one. The argument [1.17] is simple. Suppose there were a (real) solution $\boldsymbol{m} \neq 0$ of (1.7). Then at least one of the $m(\boldsymbol{x})$ is nonzero and

$$\begin{aligned} \boldsymbol{m}\cdot\boldsymbol{m} &= \sum_{\boldsymbol{x}} |m(\boldsymbol{x})|^2 \\ &= \sum_{\boldsymbol{x}} |m(\boldsymbol{x})|\,|\tanh[\beta 2^{-q}(Q\boldsymbol{m})(\boldsymbol{x})]| \\ &< \beta 2^{-q} \sum_{\boldsymbol{x}} |m(\boldsymbol{x})|\,|(Q\boldsymbol{m})(\boldsymbol{x})| . \end{aligned}$$

Here we have used the fact that $|\tanh(u)| < |u|$ for $u \neq 0$. Because $m(\boldsymbol{x})$, $\tanh[\beta 2^{-q}(Q\boldsymbol{m})(\boldsymbol{x})]$, and $(Q\boldsymbol{m})(\boldsymbol{x})$ all have the same sign, we conclude that

$$\begin{aligned} \boldsymbol{m} \cdot \boldsymbol{m} &< \beta 2^{-q} \sum_{\boldsymbol{x}} m(\boldsymbol{x})(Q\boldsymbol{m})(\boldsymbol{x}) = \beta \boldsymbol{m} \cdot 2^{-q} Q\boldsymbol{m} \\ &\leq \beta \Lambda_{\max} \boldsymbol{m} \cdot \boldsymbol{m} \end{aligned} \tag{4.2.1}$$

and thus $\beta\Lambda_{\max} > 1$. For $\beta\Lambda_{\max} < 1$, $\boldsymbol{m} = 0$ is the only solution to (1.7) and it is a stable one.

At β_c determined by $\beta_c\Lambda_{\max} = 1$, the trivial solution $\boldsymbol{m} = 0$ loses its stability, and nontrivial solutions to (1.7) branch off [1.17] into the direction of the eigenvector(s) belonging to $\Lambda_{\max}$. The reason is that the bifurcation is continuous, so that for β just above β_c the $m(\boldsymbol{x})$ are small and (1.7) may be written

$$\boldsymbol{m} = \beta_c 2^{-q} Q\boldsymbol{m} \quad \Rightarrow \quad 2^{-q} Q\boldsymbol{m} = T_c \boldsymbol{m} \tag{4.2.2}$$

where $T_c = \Lambda_{\max}$ is the critical temperature. So it is desirable to know both $\Lambda_{\max}$ itself and the corresponding eigenvector(s). In general, this is a very hard problem. Notwithstanding that, it can be solved for all Qs which satisfy a certain, rather weak, invariance condition [1.16, 17]. To understand the underlying rationale fully, we must make a small detour.

We equip $C^q = \{-1, 1\}^q$ with a group structure. Let $(\boldsymbol{x})_i = x_i$ denote the ith component of the vector $\boldsymbol{x}$. Through the operation $\boldsymbol{x} \circ \boldsymbol{y}$ defined by

$$(\boldsymbol{x} \circ \boldsymbol{y})_i = x_i y_i \,, \quad 1 \leq i \leq q \,, \tag{4.2.3}$$

the elements of C^q form an Abelian group with $\boldsymbol{e} = (1, 1, \ldots, 1)$ as unit element. Equation (3) tells us that we multiply $\boldsymbol{x}$ and $\boldsymbol{y}$ componentwise. Every $\boldsymbol{x}$ is its own inverse, i.e., $\boldsymbol{x} \circ \boldsymbol{x} = \boldsymbol{e}$. We now require that Q be invariant under the group action,

$$Q(\boldsymbol{x} \circ \boldsymbol{y}; \boldsymbol{x} \circ \boldsymbol{z}) = Q(\boldsymbol{y}; \boldsymbol{z}) \tag{4.2.4}$$

for all $\boldsymbol{x}, \boldsymbol{y}$, and $\boldsymbol{z}$ in C^q.

There are quite a few models which satisfy (4). First, we have the Hopfield model with $Q(\boldsymbol{x}; \boldsymbol{y}) = \boldsymbol{x} \cdot \boldsymbol{y}$. Then

$$Q(\boldsymbol{x} \circ \boldsymbol{y}; \boldsymbol{x} \circ \boldsymbol{z}) = \sum_i (x_i y_i)(x_i z_i) = \sum_i y_i z_i = Q(\boldsymbol{y}; \boldsymbol{z}) \,,$$

since for all components $x_i^2 = 1$. In fact, the very same argument also works for *scalar-product models* which are defined by

$$Q(\boldsymbol{x}; \boldsymbol{y}) = \sqrt{q}\,\phi(\boldsymbol{x} \cdot \boldsymbol{y} / \sqrt{q}) \,. \tag{4.2.5}$$

The scaling by $\sqrt{q}$ in (5) will become clear later. As long as q is fixed and finite, it is not essential. If ϕ is linear, i.e., $\phi(x) = x$, then (5) reduces to the Hopfield model.

The Hopfield model and, as we will see shortly, all scalar-product models have a critical α_c such that for $q = \alpha N$ patterns with $\alpha > \alpha_c$ no retrieval is possible any more. The system should – but cannot – create room for new patterns by *forgetting* the old ones. There are at least two ways to induce forgetting and, from a neurophysiological viewpoint, it is at the moment unclear which of them is the correct one. First, one imagines that forgetting is due to a chemical process which degrades certain molecules. Degrading is usually described by an exponential decay so it seems reasonable to give [1.35, 54, 55] each pattern μ a weight $\varepsilon_\mu = A\exp(-a\mu)$ where A and a are positive constants and μ ranges through the natural numbers; cf. (3.3.1). The larger μ, the older the pattern is. Toulouse and coworkers [1.54, 55] called this scheme "marginalist learning".

Alternatively, forgetting can be modeled as an *intrinsic* property of the network [1.56–58]. If $J_{ij}(\mu-1)$ contains the information up to and including pattern $\mu - 1$, then we put

$$J_{ij}(\mu) = \phi(\varepsilon_N \xi_i^\mu \xi_j^\mu + J_{ij}(\mu - 1)) \tag{4.2.6}$$

for some odd function $\phi(x)$ which saturates as $|x| \to \infty$. The prefactor ε_N depends on the model [1.58]. The prescription (6) is an iterative procedure which a priori does not single out any pattern. If, then, forgetting occurs, it is an intrinsic property of the network. One easily verifies that the J_{ij} defined by (6) satisfy (4). Furthermore, $\phi(x) = \mathrm{e}^{-a}x$ with $a > 0$ and $\varepsilon_N = A$ in (6) is identical with the marginalist learning scheme but does not saturate as $|x| \to \infty$. We will return to forgetting and all that in the next chapter.

The Qs obeying (4) all have the *same* set of eigenvectors though the eigenvalues may, and in general will, be different. This is most easily seen as follows [1.17]. Let ϱ be one of the 2^q subsets of $\{1, 2, \ldots, q\}$ and define

$$v_\varrho(\boldsymbol{x}) = \prod_{i \in \varrho} x_i \,. \tag{4.2.7}$$

Take $v_\emptyset(\boldsymbol{x}) = 1$ for the empty subset $\varrho = \emptyset$. Plainly,

$$v_\varrho(\boldsymbol{x} \circ \boldsymbol{y}) = v_\varrho(\boldsymbol{x}) v_\varrho(\boldsymbol{y}) \,, \tag{4.2.8}$$

so v_ϱ is a group character. Moreover,

$$\sum_{\boldsymbol{x}} v_\varrho(\boldsymbol{x}) v_{\varrho'}(\boldsymbol{x}) = 2^q \delta_{\varrho, \varrho'} \,, \tag{4.2.9}$$

so the v_ϱs are orthogonal. Finally, from (8), the invariance property (4) which implies that $Q(\boldsymbol{x} \circ \boldsymbol{y}; \boldsymbol{z}) = Q(\boldsymbol{x} \circ \boldsymbol{x} \circ y; \boldsymbol{x} \circ \boldsymbol{z}) = Q(\boldsymbol{y}; \boldsymbol{x} \circ \boldsymbol{z})$, and the group property of C^q, the v_ϱs are eigenvectors of Q, we obtain

$$\begin{aligned}\sum_{y} Q(\boldsymbol{x};\boldsymbol{y})v_\varrho(\boldsymbol{y}) &= \sum_{y} Q(\boldsymbol{x}\circ \mathrm{e};\boldsymbol{y})v_\varrho(\boldsymbol{x}\circ\boldsymbol{y})v_\varrho(\boldsymbol{x}) \\ &= \left[\sum_{y} Q(\mathrm{e};\boldsymbol{x}\circ\boldsymbol{y})v_\varrho(\boldsymbol{x}\circ\boldsymbol{y})\right]v_\varrho(\boldsymbol{x}) \\ &= \left[\sum_{z} Q(\mathrm{e};\boldsymbol{z})v_\varrho(\boldsymbol{z})\right]v_\varrho(\boldsymbol{x}) \\ &= \lambda_\varrho v_\varrho(\boldsymbol{x})\,, \end{aligned} \tag{4.2.10}$$

where

$$\lambda_\varrho = \sum_{x} Q(\mathrm{e};\boldsymbol{x})v_\varrho(\boldsymbol{x}) \tag{4.2.11}$$

is an explicit representation of the eigenvalue λ_ϱ corresponding to the eigenvector v_ϱ. The λ_ϱ depend on Q, as is evident from (11).

If Q is odd (even) in the sense that

$$Q(\mathrm{e};-\boldsymbol{x}) = \pm Q(\mathrm{e};\boldsymbol{x})\,, \tag{4.2.12}$$

where the minus sign on the right stands for odd (and plus for even), then $\lambda_\varrho = 0$ for $|\varrho|$, the number of elements in ϱ, being even (odd). To see this, change $\boldsymbol{x}$ in (11) into $-\boldsymbol{x}$ and use (12),

$$\lambda_\varrho = \sum_{x} Q(\mathrm{e};-\boldsymbol{x})v_\varrho(-\boldsymbol{x}) = \pm(-1)^{|\varrho|}\sum_{x} Q(\mathrm{e};\boldsymbol{x})v_\varrho(\boldsymbol{x})\,,$$

so that $\lambda_\varrho = \pm(-1)^{|\varrho|}\lambda_\varrho$. The assertion follows.

Another interesting consequence of (9) and (10) is the following. According to the spectral theorem, we can write any Hermitean operator (in brak-ket notation) as

$$\mathrm{O} = \sum_{\lambda} \lambda|\lambda\rangle\langle\lambda|\,,$$

where we sum over all the eigenvalues λ. The corresponding eigenvectors $|\lambda\rangle$ are normalized to one. Taking matrix elements we get

$$\langle\boldsymbol{x}|\mathrm{O}|\boldsymbol{y}\rangle = \sum_{\lambda} \lambda\langle\boldsymbol{x}|\lambda\rangle\langle\lambda|\boldsymbol{y}\rangle\,.$$

Writing out what this means for our Q, we immediately arrive at

$$Q(\boldsymbol{x};\boldsymbol{y}) = \sum_{\varrho} \lambda_\varrho 2^{-q} v_\varrho(\boldsymbol{x})v_\varrho(\boldsymbol{y})\,. \tag{4.2.13}$$

The extra factor 2^{-q} in (13) arises from the normalization of the v_ϱ; see (9). It will turn out to be convenient to put $\Lambda_\varrho = 2^{-q}\lambda_\varrho$.

Recalling that $J_{ij} = N^{-1}Q(\boldsymbol{\xi}_i; \boldsymbol{\xi}_j)$ we obtain a useful representation of the Hamiltonian (1.3) as a sum of perfect squares,

$$H_N = -\frac{1}{2N} \sum_{\varrho} \Lambda_\varrho \left(\sum_{i=1}^{N} v_\varrho(\boldsymbol{\xi}_i) S_i \right)^2 . \tag{4.2.14}$$

This holds for *any* Q satisfying the invariance condition (4). To appreciate what we have gained, we split up the sum in (14) into one with $|\varrho| = 1$ and the rest, note that $v_\varrho(\boldsymbol{\xi}_i)$ with $|\varrho| = 1$ equals ξ_i^μ for $1 \le \mu \le q$, and find

$$H_N = -\frac{1}{2N} \sum_{\mu} \Lambda_\mu \left(\sum_{i=1}^{N} \xi_i^\mu S_i \right)^2 - \frac{1}{2N} \sum_{|\varrho| \neq 1} \Lambda_\varrho \left(\sum_{i=1}^{N} v_\varrho(\boldsymbol{\xi}_i) S_i \right)^2 . \tag{4.2.15}$$

The first term on the right represents a generalized Hopfield model with weights Λ_μ as in (3.3.1). It contains overlaps only and we henceforth assume that Q has been scaled in such a way (e.g., as in (5): see below) that the Λ_μ corresponding to the patterns converge to a finite, nonzero limit as $q \to \infty$. The remaining terms in (15), which represent the *non*linearity of Q, just produce extra noise compared to the extensively many patterns in the first sum. Naively one might therefore expect that among the models satisfying (4) the generalized Hopfield model is *optimal*. This is indeed the case [1.39]. In passing we note that even Qs are no good for pattern retrieval since the λ_μ vanish. It has been shown, however, that they are quite useful for storing the outcomes of logical operations on patterns [1.60, 61].

Before turning to the case of extensively many patterns it may be worthwhile considering the finite-q case in some detail. Here we restrict ourselves to inner-product models, which are defined by (5). By (11) we directly obtain

$$\Lambda_\varrho = \sqrt{q} 2^{-q} \sum_{\boldsymbol{x}} \phi \left(\sum_{i=1}^{q} x_i / \sqrt{q} \right) \prod_{i \in \varrho} x_i . \tag{4.2.16}$$

Since the sum is invariant under a permutation of the coordinates x_i, it only depends on the *size* $|\varrho|$ of the set ϱ so that $\Lambda_\varrho = \Lambda_{|\varrho|}$. Furthermore, since ϕ is taken to be odd, the Λ_ϱ with $|\varrho|$ even vanish. Specializing, e.g., to clipped synapses with $\phi(x) = \mathrm{sgn}(x)$, one finds [1.17] that the $\Lambda_{|\varrho|}$ have alternating signs and that their absolute values decrease monotonically in $|\varrho|$ until $|\varrho| \approx q/2$. Accordingly, $\Lambda_1 = \Lambda_{\max}, \Lambda_3 < 0, \Lambda_5 > 0$ and Λ_5 is the second largest eigenvalue; and so on. With the scaling as in (5) and (16), Λ_1 is of the order of one as q becomes large whereas Λ_5 *decreases* as q^{-x} with $x = 1$ or $x = 2$. We will see shortly in (3.13) and (3.14) that this is no accident.

The above results have interesting implications for the bifurcation diagram of (1.7), Fig. 1.4, and hence for the existence of phases related to the *non*linearity of Q.

Here $T_c = \Lambda_1, T_2 = \Lambda_5$, and we get a sequence of decreasing temperatures T_n with $T_n = \Lambda_{4n-3}$. Each time when T passes downwards through a T_n, states

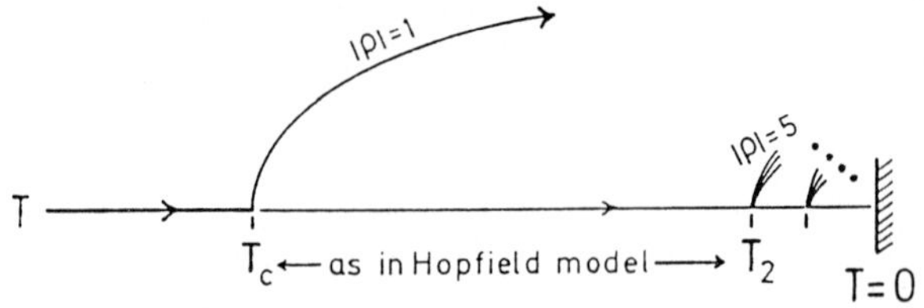

Fig. 1.4. Bifurcation of the fixed-point equation (1.7) for clipped synapses with q fixed and finite. At $T_c = \Lambda_1$, we have a bifurcation at a q-fold degenerate eigenvalue which is associated with the stored patterns *only* ($|\varrho| = 1$). The patterns themselves and all the other solutions associated with the (linear) Hopfield model, symmetric states, bifurcate from $\boldsymbol{m} \equiv 0$ [1.17: pp. 271ff.]. At $T_2 = \Lambda_5$, states corresponding to v_ϱ with $|\varrho| = 5$ bifurcate from $\boldsymbol{m} \equiv 0$. More and more states associated with the nonlinearity of the model branch off as T is lowered and most of them become stable as $T \to 0$ (indicated by the *black dots* near $T = 0$) However, $T_2/T_c \to 0$ as q becomes large

corresponding to v_ϱ with $|\varrho| = 4n-3$ and mixtures thereof bifurcate from $\boldsymbol{m} \equiv 0$ and most of them become stable as $T \to 0$. (This is easily seen [1.17: Part II, Sects. 2.3 and 2.4] for $m(\boldsymbol{x}) = \alpha_\varrho v_\varrho(\boldsymbol{x})$.) Then $\alpha_\varrho = \tanh(\beta\Lambda_\varrho\alpha_\varrho)$ approaches 1 at an exponential rate as $\beta \to \infty$ and stability directly follows from (1.8).] For $n \geq 2$, these states are associated with the nonlinearity of Q and they are not wanted as long as they are attractors in phase space.

However, in the region $[T_2, T_c]$ no such states occur and the system behaves *as if it were linear* [1.17]. The fact that $T_2/T_c \to 0$ as $q \to \infty$ might suggest that the nonlinearity becomes irrelevant[10] for large q. In a sense, this is correct. It will be shown in the next section that scalar-product models reduce to the (linear) Hopfield model *plus* an extra noise term generated by the nonlinearity.

1.4.3 Extensively Many Patterns

The motivation to study nonlinear neural networks with $q = \alpha N$ patterns is at least threefold. First, we would like to know whether the results obtained for the generalized Hopfield model in Sect. 1.3.3 are generic or whether they constitute a singular case. For Qs satisfying the invariance condition (2.4) it is shown that the Hopfield model is both generic and optimal. In addition, we will elaborate on the *universality* in the class of scalar-product models, to which the Hopfield model also belongs. Second, clipped synapses and also dilution are highly relevant to hardware realizations. A surprising result of our analysis is that clipping hardly deteriorates the network and that its performance is even *improved* by moderate dilution. Third, as argued in the previous section, most forgetful memories are nonlinear. They do satisfy the invariance condition (2.4) and are therefore amenable to the theory developed below. The condition (2.4) is assumed throughout what follows.

[10] There is a problem of limits. Here we first take $N \to \infty$ for *fixed* q, then allow q to increase without bound.

The starting point of our analysis [1.39] is (2.14),

$$-\beta H_N = \frac{\beta}{2N} \sum_{\varrho} \Lambda_{\varrho} \left(\sum_{i=1}^{N} v_{\varrho}(\boldsymbol{\xi}_i) S(i) \right)^2 \tag{4.3.1}$$

where, in view of replica indices to come, S_i has been replaced by $S(i)$. To find the free-energy valleys of the model (1) we closely follow the procedure of Sect. 1.3.3, single out *finitely* many patterns, labeled by μ, and use the replica method to average over the remaining, extensively many patterns ν. We split up the index set $\{1, 2, \ldots, q\} = I_\mu \cup I_\nu$ and divide the sum in (1) into two parts. One part, $-\beta H_N^{(1)}$, is a sum over subsets of I_μ only and need not be averaged. The other part, $-\beta H_N^{(2)}$, is a sum over subsets ϱ of the form $\varrho = A \cup B$ with $A \subseteq I_\mu$ and $B \subseteq I_\nu$ with B *non*empty (otherwise ϱ would belong to the first group). Let Z_N be the partition function $\mathrm{tr}\{\exp(-\beta H_N)\}$. Instead of studying the average $\langle Z_N^n \rangle$ we note that $\exp(-\beta H_N) = \exp(-\beta H_N^{(1)}) \exp(-\beta H_N^{(2)})$ and that in the present case we need only average the replicated $\exp(-\beta H_N^{(2)})$ over the ξ_i^ν,

$$\begin{aligned}\left\langle \exp\left\{ -\beta \sum_{\sigma} H_N^{(2)}(\sigma) \right\} \right\rangle &= \left\langle \exp\left\{ \frac{\beta}{2N} \sum_{\varrho,\sigma} \Lambda_{\varrho} \left[\sum_{i=1}^{N} v_{\varrho}(\boldsymbol{\xi}_i) S_{\sigma}(i) \right]^2 \right\} \right\rangle \\ &= \left\langle \exp\left[\frac{\beta}{2} \sum_{\varrho,\sigma} \Lambda_{\varrho} X_{\sigma}^2(\varrho) \right] \right\rangle ,\end{aligned} \tag{4.3.2}$$

where

$$X_{\sigma}(\varrho) = \frac{1}{\sqrt{N}} \sum_{i=1}^{N} v_{\varrho}(\boldsymbol{\xi}_i) S_{\sigma}(i) \tag{4.3.3}$$

and the $1 \leq \sigma \leq n$ label the n replicas.

As in Sect. 1.3.3, it is assumed that the $S_\sigma(i)$ live in μ-space and are *not* correlated with the ν-patterns. Since we average over the latter only, we take the $X_\sigma(\varrho)$ as *Gaussian* random variables with mean zero ($\varrho = A \cup B, v_{\varrho}(\boldsymbol{\xi}_i) = v_A(\boldsymbol{\xi}_i) v_B(\boldsymbol{\xi}_i)$, and $B \neq \emptyset$), covariance matrix

$$\begin{aligned}C_{\varrho\sigma;\varrho'\sigma'} &= \langle X_{\sigma}(\varrho) X'_{\sigma}(\varrho') \rangle \\ &= N^{-1} \sum_{i=1}^{N} S_{\sigma}(i) S_{\sigma'}(i) \langle v_{\varrho}(\boldsymbol{\xi}_i) v_{\varrho'}(\boldsymbol{\xi}_i) \rangle ,\end{aligned} \tag{4.3.4}$$

and common distribution (3.3.7). For $\varrho = A \cup B$ and $\varrho' = A' \cup B'$, the average

$$\langle v_{\varrho}(\boldsymbol{\xi}_i) v'_{\varrho}(\boldsymbol{\xi}_i) \rangle = \delta_{B,B'} v_A(\boldsymbol{\xi}_i) v_{A'}(\boldsymbol{\xi}_i) \tag{4.3.5}$$

easily follows from (2.7–2.9). We order the ϱ antialphabetically: for fixed B, we let A range through the subsets of I_μ. From (5), C is then a block-diagonal matrix. Each block, to be called C_B, is labeled by B and has dimension $2^{|I_\mu|} n$.

If $D = \mathrm{diag}(\Lambda_\varrho)$ is a diagonal matrix where each Λ_ϱ occurs n times, then (2) gives

$$\det C^{-1/2} \det(C^{-1} - \beta D)^{-1/2} = \det(\mathbb{1} - \beta D^{1/2} C D^{1/2})^{-1/2} . \tag{4.3.6}$$

This expression is completely analogous to (3.3.8), the only difference being D and the increased dimension. The determinant factorizes into a product over the B-blocks. The restriction of D to a B-block is called D_B.

Combining (6) with the replicated $\exp(-\beta H_N^{(1)})$ we get

$$\langle Z_N^n \rangle = \mathrm{tr} \left\{ \exp\left[-\beta \sum_\sigma H_N^{(1)}(\sigma) - \tfrac{1}{2} \sum_{B \subseteq I_\nu} \mathrm{Tr}\{\ln \mathcal{Q}_B\} \right] \right\} , \tag{4.3.7}$$

where $\mathcal{Q}_B = \mathbb{1} - \beta D_B^{1/2} C_B D_B^{1/2}$. Being interested in the stability of a *single* pattern we take $I_\mu = \{\mu\}$. This assumption simplifies the algebra substantially. To see why, we note that (i) given B, A is either $\emptyset$ or $\{\mu\}$ and $\mathcal{Q}_B$ consists of four $n \times n$ subblocks (ϱ, ϱ'), each of them multiplied by $\sqrt{\Lambda_\varrho \Lambda_{\varrho'}}$, (ii) ϕ is odd so that Λ_ϱ vanishes for even $|\varrho|$, and (iii) ϱ and ϱ' differ by at most *one* element. If then $\varrho \neq \varrho'$, either Λ_ϱ or $\Lambda_{\varrho'}$ vanishes and only one of the diagonal subblocks survives. Hence $\mathcal{Q}_B$ *de facto* reduces to an $n \times n$ matrix with elements

$$(\mathcal{Q}_B)_{\sigma,\sigma'} = \delta_{\sigma,\sigma'} - \beta \Lambda(B) \left(N^{-1} \sum_{i=1}^{N} S_\sigma(i) S_{\sigma'}(i) \right) \tag{4.3.8}$$

and

$$\Lambda(B) = \sum_{A \subseteq I_\mu} \Lambda_{A \cup B} = \begin{cases} \Lambda_B , & \text{if } |B| \text{ is odd,} \\ \Lambda_{\{\mu\} \cup B} , & \text{if } |B| \text{ is even.} \end{cases} \tag{4.3.9}$$

By assumption, $B \subseteq I_\nu$ is nonempty. $A \subseteq I_\mu$ may be empty though.

For a single pattern μ, we have reduced the problem to a generalized Hopfield model so that we can proceed exactly as from (3.3.10) to (3.3.34). We then find for N very large ($N \to \infty$)

$$\begin{aligned} -\beta f(\beta) = &-\frac{1}{2}\beta \left(\sum_{A \subseteq I_\mu} \Lambda_A m_A^2 \right) - \frac{1}{2N} \sum_{B \subseteq I_\nu} \Big\{ \ln\big[1 - \beta\Lambda(B)(1-\hat{q})\big] \\ &- \beta \Lambda(B) \hat{q} \big[1 - \beta \Lambda(B)(1-\hat{q})\big]^{-1} \Big\} - \frac{1}{2}\beta^2 \hat{q} r (1-\hat{q}) \\ &+ \left\langle \int \frac{\mathrm{d}z}{\sqrt{2\pi}} \mathrm{e}^{-z^2/2} \ln\Big\{ \cosh\big[\beta(\Lambda_\mu m_\mu \xi + \sqrt{\hat{q} r}\, z)\big] \Big\} \right\rangle \end{aligned} \tag{4.3.10}$$

with $r = r(\hat{q})$ given by

$$r(\hat{q}) = N^{-1} \sum_{B \subseteq I_\nu} \Lambda(B)^2 \big[1 - \beta \Lambda(B)(1-\hat{q})\big]^{-2} . \tag{4.3.11}$$

In addition, one should choose that solution of the fixed-point equations

$$m_\mu = \ll \xi \tanh\left[\beta(\Lambda_\mu m_\mu \xi + \sqrt{\hat{q}r}\,z)\right] \gg , \tag{4.3.12a}$$

$$\hat{q} = \ll \tanh^2\left[\beta(\Lambda_\mu m_\mu \xi + \sqrt{\hat{q}r}\,z)\right] \gg , \tag{4.3.12b}$$

which maximizes the right-hand side of (10). The m_μ determines the retrieval quality of the μ-pattern while the spin-glass order parameter $\hat{q}$ comes from (8). The angular brackets in (12) denote an average over ξ^μ (which may be dropped) and the Gaussian z. If I_μ consists of more than one element, the first term of the right-hand side of (10) remains the same but the other terms change and become more complicated. In passing we note that by adding $\ln 2$ to (10) one obtains the conventional entropy; cf. (3.5.1).

The inner-product models (2.5) provide an interesting application of the general formulae (10–12). These models have two additional, distinctive features. First, the eigenvalues λ_ϱ and thus Λ_ϱ depend only on the *size* $|\varrho|$ of the set ϱ; cf. (2.16). Moreover, $\Lambda_1 = 2^{-q}\lambda_1(|\varrho| = 1)$ converges to a finite limit as $q \to \infty$. This follows from (2.16) and the central-limit theorem [1.22],

$$\begin{aligned} \Lambda_1 &= 2^{-q} \sum_{\boldsymbol{x}} \left(q^{-1/2} \sum_\gamma x_\gamma \right) \phi \left(q^{-1/2} \sum_\gamma x_\gamma \right) \\ &\longrightarrow \int_{-\infty}^{+\infty} \frac{dx}{\sqrt{2\pi}} e^{-x^2/2} x \phi(x) . \end{aligned} \tag{4.3.13}$$

Second, Λ_ϱ vanishes as $q \to \infty$ for *all* ϱ with $|\varrho| \neq 1$. To see this, let us assume that $|\varrho| = 3$. By (2.16) we get ($\alpha \neq \beta \neq \gamma$)

$$\Lambda_3 = \sqrt{q}\, 2^{-q} \sum_{\boldsymbol{x}} x_\alpha x_\beta x_\gamma \phi \left(q^{-1/2} \sum_\delta x_\delta \right) \tag{4.3.14}$$

and besides four terms ($\alpha = \beta \neq \gamma, \ldots, \alpha = \beta = \gamma$) of order q^{-1} or less we end up with

$$q^{-1} 2^{-q} \sum_{\boldsymbol{x}} \left(q^{-1/2} \sum_\gamma x_\gamma \right)^3 \phi \left(q^{-1/2} \sum_\gamma x_\gamma \right) \longrightarrow q^{-1} \int \frac{dx}{\sqrt{2\pi}} e^{-x^2/2} x^3 \phi(x)$$

which is $O(q^{-1})$ too. And so on.

Let us now return to (7) and consider $-\beta H_N^{(1)}$, which refers to the μ-pattern(s). As $q \to \infty$, only the Λ_ϱ with $|\varrho| = 1$ survive and, up to $\Lambda_1, H_N^{(1)}$ therefore reduces to the original Hopfield Hamiltonian (3.1.1). Absorbing Λ_1 in β by putting $\beta' = \beta\Lambda_1$, we get a *perfect correspondence*. The last term in (7) is a noise term, which we now study in more detail.

In the case of a single pattern, with $I_\mu = \{\mu\}$, we note that for odd Q the sum in (9) has only one term (the other one vanishes) and that $\Lambda(B)$ in (11) may be replaced by Λ_ϱ with ϱ ranging through all subsets of $\{1, \ldots, q\}$. Using the above observation that $\Lambda_\varrho \to 0$ for $|\varrho| \neq 1$ we can simplify (11) even further so as to get

$$r = \alpha \Lambda_1^2 \left[1 - \beta \Lambda_1 (1 - \hat{q})\right]^{-2} + N^{-1} \sum_{|\varrho| \neq 1} \Lambda_\varrho^2 . \tag{4.3.15}$$

The last term in (15) is nothing but

$$N^{-1}\left[2^{-2q}\mathrm{Tr}\,\{Q^2\} - q\Lambda_1^2\right] = \alpha\left[\langle\phi^2(z)\rangle - \langle z\phi(z)\rangle^2\right] , \tag{4.3.16}$$

which we rewrite as $\alpha[\Lambda_Q^2 - \Lambda_1^2]$; as before, z is a Gaussian with mean zero and variance one. Taking the limit $\beta' = \beta\Lambda_1 \to \infty$ one can reduce (11) and (12) to a single equation of the form $\sqrt{2\alpha}x = F(x)$, where

$$F(x) = \left[\left(\mathrm{erf}\,(x) - \frac{2}{\sqrt{\pi}} x \mathrm{e}^{-x^2}\right)^{-2} + C/\mathrm{erf}^2(x)\right]^{-1/2} \tag{4.3.17}$$

with $C = [(\Lambda_Q/\Lambda_1)^2 - 1]$. This determines the storage ratio α_c, as explained in Fig. 1.5. The retrieval quality is given by $m = \mathrm{erf}\,(x)$. The function F is *universal* in that choosing another model, and thus another ϕ, only modifies the constant C. For instance, the original Hopfield model has $C = 0$ since $\Lambda_\varrho = 0$ for $|\varrho| \neq 1$. In fact, it is uniquely characterized by $C = 0$ and a glance at Fig. 1.5 now suffices to tell us that among the inner-product models it is optimal: For any other inner-product model we have $C > 0$ and thus $\alpha_c < \alpha_c^{\mathrm{Hopfield}}$. Clipped synapses with $\phi(x) = \mathrm{sgn}\,(x)$ have $\Lambda_1 = \sqrt{2/\pi}, \Lambda_Q = 1$ and thus $C = (\pi/2 - 1) = 0.571$; this gives $m_c = 0.948$ at $\alpha_c = 0.102$. See Fig. 1.5. The present data agree with Sompolinsky's estimates [1.61, 62], which were obtained through a kind of signal-to-noise analysis where the "signal" is produced by the Hopfield model and the noise by the nonlinearity.

Deterioration of a network usually means that synaptic efficacies $\boldsymbol{\xi}_i \cdot \boldsymbol{\xi}_j$ with values near zero no longer function [1.63]. This gives rise to dilution and can be modeled by deleting all bonds with $|\boldsymbol{\xi}_i \cdot \boldsymbol{\xi}_j| \leq a\sqrt{q}$. For instance, in the case of clipped synapses we get[11] $\phi(x) = \mathrm{sgn}\,(x)\Theta(|x| - a)$ and

$$C(a) = \frac{\pi}{2} \exp(a^2)\mathrm{erfc}\,(a/\sqrt{2}) - 1 , \tag{4.3.18}$$

where $\mathrm{erfc} = 1 - \mathrm{erf}$ is the complementary error function; $\mathrm{erf}\,(a/\sqrt{2})$ is the fraction of bonds that have been deleted. The insert of Fig. 1.5 shows a plot of $C(a)$. Surprisingly, the performance of the network is *improved* by moderate dilution. The best value of C is obtained for $a = 0.612$. Then $m_c = 0.959$ at $\alpha_c = 0.120$ and about half of the bonds have been deleted.

With the benefit of hindsight the physical explanation of the above improvement is simple. We have seen that the nonlinearity – and $\mathrm{sgn}\,(x)$ is extremely

[11] $\Theta(x) = [\mathrm{sgn}\,(x) + 1]/2$ is the Heaviside function.

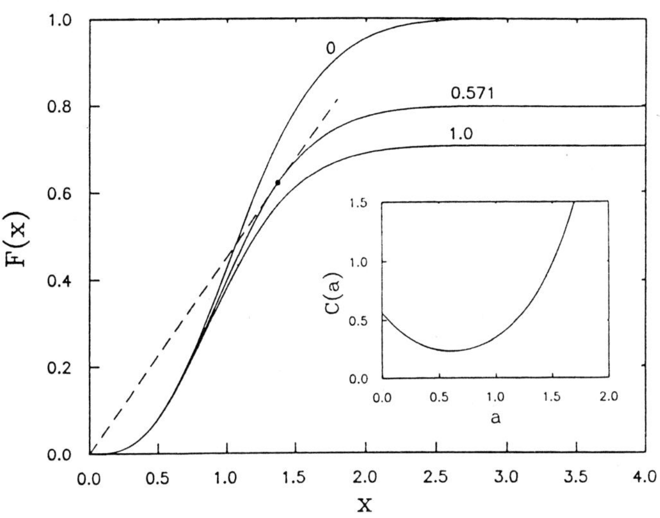

Fig. 1.5. $F(x)$ as given by (17) for $C = 0$ (Hopfield case), $C = 0.571$ (clipped synapses), and $C = 1$. The equation $\sqrt{2\alpha}x = F(x)$ possesses a nontrivial solution ($x \neq 0$) only for $\alpha \leq \alpha_c$, thus fixing α_c. The *dashed line* represents $\sqrt{2\alpha_c}x$ for $C = 0.571$. For $\alpha < \alpha_c$, there are two nontrivial solutions, of which the larger is the physical one. The insert shows $C(a)$ as a function of the dilution parameter a; cf. (18). It has a minimium $0.235 < C(0) = 0.571$. The smaller C, the better the performance of the network

nonlinear! – induces extra noise. Deleting some bonds means deleting some information but also reducing the noise generated by the nonlinearity. Apparently, the latter effect dominates at the beginning, before the former takes over as a increases and the network deteriorates.

In summary, we have obtained the free energy of a neural network with a nonlinearity of the form (2.4) and extensively many ($q = \alpha N$) patterns. The inner-product models are thus fully understood. In the limit $\alpha \to 0$, the solution joins onto the one for a finite but *large* number of patterns; cf. Sect. 1.4.1 and [1.17]. Replica-symmetry breaking is not expected to become important since the zero-temperature entropy, though negative, is quite small. External noise is also easily included [1.62, 64]. The first-order transition at α_c is physiologically not satisfying. However, the general equations (10–12) open up the way to studying more complicated but highly interesting nonlinear memories, such as the ones which gradually forget.

1.5 Learning, Unlearning, and Forgetting

Learning, unlearning, and forgetting are three main themes in the theory of neural networks. All three are concerned with handling the data stream, and as complementary tasks they belong closely together. Learning aims at programming the synaptic efficacies J_{ij} in such a way that a maximal amount of information can be retrieved. Unlearning allows an efficient *Hebbian* coding of correlated data with *varying* activity. Forgetting creates space for new data and is modeled as an intrinsic property of the network. Before turning to these subjects proper we first scan the underlying difficulties. A preview of this chapter can be found at the end of Sect. 1.5.1.

1.5.1 Introduction

As we have seen in the previous chapters, the Hopfield model and, more generally, inner-product models such as clipped synapses have the property that *no* retrieval is possible any more as soon as the number of stored patterns q exceeds a critical boundary $\alpha_c N$ where α_c is of the order 0.10. For $q > \alpha_c N$, the network has lost its ability to function as an associative memory. In terms of short-term memory one may interpret this to advantage [1.65] but in general it seems worthwhile to look for alternatives that improve the network's performance.

There is also another circumstance, viz., correlation between the patterns, that can severely hamper the retrieval ability of the network. More precisely, if we have two patterns, $\boldsymbol{\xi}^\mu = (\xi_i^\mu;\, 1 \le i \le N)$ and $\boldsymbol{\xi}^\nu = (\xi_i^\nu;\, 1 \le i \le N)$, then we can define their mutual overlap or correlation coefficient

$$C_{\mu\nu} = N^{-1} \sum_{i=1}^{N} \xi_i^\mu \xi_i^\nu \,. \tag{5.1.1}$$

In the case of the Hopfield model we have *unbiased* patterns, i.e., independent identically distributed random variables with mean zero and variance one so that $C_{\mu\nu}$ is of the order $N^{-1/2}$, vanishing as $N \to \infty$. In other words, these patterns are uncorrelated. Most patterns, however, are not, their overlap is of order one as $N \to \infty$, and the original Hebb rule (1.1.1) won't work. This is explained most easily in terms of *biased* random patterns which have $\xi_i^\mu = +1$ with probability p and $\xi_i^\mu = -1$ with probability $(1-p)$, whatever i and μ are. Then $\langle \xi^\mu \rangle = 2p - 1 \equiv a$, which is nonzero for $p \neq 0.5$; the quantity a is called the *activity*. Until now we have worked with a vanishing activity ($p = 0.5$). As before, the ξ_i^μ are independent random variables. For the more general kind of pattern with arbitrary a we obtain the correlation ($\mu \neq \nu$)

$$C_{\mu\nu} = N^{-1} \sum_{i=1}^{N} \xi_i^\mu \xi_i^\nu \quad \longrightarrow \quad \langle \xi^\mu \rangle \langle \xi^\nu \rangle = a^2 \tag{5.1.2}$$

as $N \to \infty$ by the strong law of large numbers [1.22]. If $p \neq 0.5$, then $a \neq 0$, the patterns are correlated, and no extensive storage capacity is possible. To wit, we calculate the local field at site i for a Hebbian synaptic efficacy of the form (1.1.1) and a specific pattern μ,

$$\begin{aligned} h_i^\mu &= N^{-1} \sum_{j(\neq i)} \left(\sum_\nu \xi_i^\nu \xi_j^\nu \right) \xi_j^\mu = \xi_i^\mu + N^{-1} \sum_{\substack{j(\neq i) \\ \nu(\neq \mu)}} \xi_i^\nu \xi_j^\nu \xi_j^\mu \\ &= \xi_i^\mu + (q-1)a^3 + \mathrm{O}(\sqrt{q/N}) . \end{aligned} \tag{5.1.3}$$

The signal term ξ_i^μ is dominated by $(q-1)a^3$ for q large enough *but finite*. In fact, the leading contribution to J_{ij} itself is $(q/N)a^2$, which is ferromagnetic, and for q large enough the system can only store a single bit of information (*all* spins up or *all* spins down). From the point of view of information storage, the ferromagnetic state is somewhat restricted.

To improve the network's performance, one can either promote a more efficient learning algorithm (Sects. 1.5.2–4) or try to weed out the spurious states through unlearning (Sect. 1.5.6). Moreover, to maintain the associative capabilities of the network while ever acquiring new data one should let the system forget the old data so as to create room for new ones (Sect. 1.5.7). Though this listing suggests an antibiosis, the relation between the three is symbiotic. In the *Intermezzo* (Sect. 1.5.5) we derive a highly efficient coding for correlated patterns.

As to learning proper, we treat the pseudoinverse method, discuss the perceptron learning rule and the associated convergence theorem (the PCT of the theory of learning algorithms), and then turn to Hebbian learning, a mainstay of neurophysiologically motivated models of associative memory. We also present an analytic treatment of the dynamics of the overlaps – including an amusing application to the original Hopfield model.

1.5.2 The Pseudoinverse Learning Rule

We are given q patterns $(\xi_i^\mu;\ 1 \le i \le N)$ with $1 \le \mu \le q$. For the moment, these patterns are arbitrary and need not be random whatsoever. Finding J_{ij} so that the network functions as an associative memory is a truly difficult task and the first simplification therefore consists in requiring that the patterns be "only" fixed points of the dynamics, i.e.,

$$\xi_i^\mu = \operatorname{sgn}\left(\sum_j J_{ij} \xi_j^\mu + \theta_i \right) \tag{5.2.1}$$

for all μ and some given set of thresholds θ_i, $1 \le i \le N$. The condition (1) is highly nonlinear and the second simplification now consists in reducing it to linear algebra: instead of (1) we impose the stronger condition

$$\xi_i^\mu = \sum_{j=1}^{N} J_{ij}\xi_j^\mu + \theta_i \tag{5.2.2}$$

for some matrix J. Plainly, (2) implies (1). Let $\boldsymbol{\xi}^\mu = (\xi_i^\mu;\ 1 \le i \le N) \in \mathbb{R}^N$ denote an N-dimensional column vector. Then (2) can be written

$$\boldsymbol{\xi}^\mu = J\boldsymbol{\xi}^\mu + \boldsymbol{\theta}\,, \quad 1 \le \mu \le q\,, \tag{5.2.3}$$

where $\boldsymbol{\theta} = (\theta_i;\ 1 \le i \le N)$. Our task is to find a linear transformation J that maps the known vectors $\boldsymbol{\xi}^\mu$ onto the known vectors $\boldsymbol{\delta}^\mu = \boldsymbol{\xi}^\mu - \boldsymbol{\theta}$,

$$J\boldsymbol{\xi}^\mu = \boldsymbol{\delta}^\mu\,, \quad 1 \le \mu \le q\,. \tag{5.2.4}$$

In other terms, given the $N \times q$ matrices

$$\Xi = (\boldsymbol{\xi}^\mu;\ 1 \le \mu \le q) \quad \text{and} \quad \Delta = (\boldsymbol{\delta}^\mu;\ 1 \le \mu \le q)\,, \tag{5.2.5}$$

find a matrix J so that

$$J\Xi = \Delta\,. \tag{5.2.6}$$

If Ξ were a square matrix with inverse Ξ^{-1}, then (6) could be solved immediately so as to give $J = \Delta\Xi^{-1}$. In general, however, Ξ is an $N \times q$ matrix with $q < N$ so that no inverse exists. In that case we put $J = \Delta\Xi^{\mathrm{I}}$ where Ξ^{I} is the Moore–Penrose *pseudoinverse* [1.66–68] of Ξ, for short also called just the pseudoinverse. It represents the "best possible" solution to (6) and, as such, is unique. In general, however, there are several solutions to (6).

To see what the pseudoinverse does for us, we make a small detour and turn to a simpler problem: solving $A\boldsymbol{x} = \boldsymbol{y}$ for $\boldsymbol{x}$, if A is an $m \times n$ matrix with $m > n$. Let us denote the columns of A by $\boldsymbol{a}_i$, $1 \le i \le n$. Then solving $A\boldsymbol{x} = \boldsymbol{y}$ is nothing but finding coefficients x_i such that $\sum_i x_i\boldsymbol{a}_i = \boldsymbol{y}$. This can be done, if $\boldsymbol{y}$ is in the linear span of the $\boldsymbol{a}_i$, but the solution need not be unique. What do we do however, if $\boldsymbol{y}$ is *not* in the linear span of the $\boldsymbol{a}_i$? Well, the "best" we can do, following Penrose [1.66], is to project $\boldsymbol{y}$ onto $P_A\boldsymbol{y}$ in the linear span of the $\boldsymbol{a}_i$ and solve $A\boldsymbol{x} = P_A\boldsymbol{y}$. This, then, is what the pseudoinverse does. If the $\boldsymbol{a}_i$ are linearly independent, one easily verifies (see below) that $P_A = A(A^{\mathrm{t}}A)^{-1}A^{\mathrm{t}}$ is an explicit representation of the orthogonal projection onto the $\boldsymbol{a}_i$, $1 \le i \le n$, and that $\boldsymbol{x} = A^{\mathrm{I}}\boldsymbol{y} = (A^{\mathrm{t}}A)^{-1}A^{\mathrm{t}}\boldsymbol{y}$ gives the coefficients x_i we are looking for. Here A^{t} denotes the transpose of A. Plainly, P_A is also well-defined if the $\boldsymbol{a}_i$ are linearly dependent. To get the above explicit representation one has to restrict the $\boldsymbol{a}_i$ to a maximally independent subset. In practice, the numerics of calculating P_A boils down to a Gram–Schmidt orthonormalization [1.66].

With the benefit of hindsight, there is a second way of seeing what the pseudoinverse does. We want to choose the x_i in such a way that the remaining error $\boldsymbol{Y} = \boldsymbol{y} - \sum_j x_j\boldsymbol{a}_j$ is minimal. To this end, it is necessary and sufficient that $\boldsymbol{Y}$ is orthogonal to the linear span of the $\boldsymbol{a}_i$. That is, we have to require that

$\sum_j \boldsymbol{a}_i \cdot \boldsymbol{a}_j x_j = \boldsymbol{a}_i \cdot \boldsymbol{y}$ for $1 \leq i \leq n$. For linearly independent $\boldsymbol{a}_i$ (see also below) we then recover $\boldsymbol{x} = (A^t A)^{-1} A^t \boldsymbol{y}$. We now return to our original problem.

If the columns $\boldsymbol{\xi}^\mu$, $1 \leq \mu \leq q$, of the $N \times q$ matrix Ξ are linearly independent, then Ξ^{I} can also be calculated explicitly,

$$\Xi^{\mathrm{I}} = (\Xi^t \Xi)^{-1} \Xi^t \Rightarrow J = \Delta \Xi^{\mathrm{I}} . \tag{5.2.7}$$

As a check we compute

$$J\Xi = \Delta(\Xi^t \Xi)^{-1} \Xi^t \Xi = \Delta,$$

as desired. We only have to verify that $\Xi^t \Xi$ is invertible. And that's easy. Its matrix elements are

$$(\Xi^t \Xi)_{\mu\nu} = \sum_i \xi_i^\mu \xi_i^\nu = \boldsymbol{\xi}^\mu \cdot \boldsymbol{\xi}^\nu \tag{5.2.8}$$

so $\Xi^t \Xi$ is the Gramian [1.69] of the vectors $\boldsymbol{\xi}^\mu$, $1 \leq \mu \leq q$. A Gramian is invertible if and only if the $\boldsymbol{\xi}^\mu$ are linearly independent. [Let $G = \Xi^t \Xi$ and assume that the $\boldsymbol{\xi}^\mu$ are independent. Then G is invertible since $\det(G) \neq 0$. Otherwise we could find $a_\nu \neq 0$ such that

$$\sum_\nu a_\nu (\boldsymbol{\xi}^\mu \cdot \boldsymbol{\xi}^\nu) = \boldsymbol{\xi}^\mu \cdot \left(\sum_\nu a_\nu \boldsymbol{\xi}^\nu \right) = 0$$

for all μ. Hence

$$\sum_\mu a_\mu \boldsymbol{\xi}^\mu \cdot \left(\sum_\nu a_\nu \boldsymbol{\xi}^\nu \right) = \left\| \sum_\mu a_\mu \boldsymbol{\xi}^\mu \right\|^2 = 0$$

so that $\sum_\mu a_\mu \boldsymbol{\xi}_\mu = 0$. Contradiction. The other way around is even simpler.] We parenthetically note that $J\boldsymbol{\eta} = 0$ for all $\boldsymbol{\eta}$ in the orthogonal complement of $\boldsymbol{\xi}^\mu$. This directly follows from (7) and $\Xi^t \boldsymbol{\eta} = 0$.

The above mathematics becomes more transparent if the thresholds θ_i all vanish. Then (4) tells us that we have to find a J such that

$$J\boldsymbol{\xi}^\mu = \boldsymbol{\xi}^\mu , \quad 1 \leq \mu \leq q . \tag{5.2.9}$$

The projection operator P onto the linear span of the $\boldsymbol{\xi}^\mu$, $1 \leq \mu \leq q$, certainly does the job. But it is by no means unique: any projection P' with $P \leq P' \leq \mathbb{1}$ satisfies (9). In a sense, however, P is the "best possible" solution – both mathematically [1.66] and physically: mathematically, since P projects onto the *minimal* subspace containing the $\boldsymbol{\xi}^\mu$, $1 \leq \mu \leq q$; physically, P being symmetric, we have the Hamiltonian $H(\boldsymbol{S}) = -\|P\boldsymbol{S}\|^2/2$ whose *sole minima* are the patterns $\boldsymbol{S} = \boldsymbol{\xi}^\mu$, $1 \leq \mu \leq q$, and linear combinations thereof (if any). There is a subtle difference, though, between the Hamiltonian dynamics and (1), which we will discuss shortly.

If the $\boldsymbol{\xi}^\mu$ are linearly independent, then we have, from (7), the explicit representation

$$J = \Xi(\Xi^{\mathrm{t}}\Xi)^{-1}\Xi^{\mathrm{t}} . \tag{5.2.10}$$

Since $J = J^{\mathrm{t}} = J^2$, the real matrix J represents a projection operator. It maps the orthogonal complement of the $\boldsymbol{\xi}^\mu$ onto zero. And by construction $J\Xi = \Xi$ so that J *is* the projection onto the $\boldsymbol{\xi}^\mu$, $1 \le \mu \le q$.

Before turning to the applications (see also Sect. 1.5), we summarize the procedure as follows. We define the *correlation matrix* C,

$$C_{\mu\nu} = N^{-1} \sum_{i=1}^{N} \xi_i^\mu \xi_i^\nu , \tag{5.2.11}$$

and suppose that the $\boldsymbol{\xi}^\mu$ are linearly independent. Then, because of (8) and (10), the matrix J has the elements

$$J_{ij} = N^{-1} \sum_{\mu,\nu} \xi_i^\mu (C^{-1})_{\mu\nu} \xi_j^\nu = J_{ji} . \tag{5.2.12}$$

By (11), this is a nonlocal prescription and therefore physiologically not very plausible. One might compute the inverse through an infinite series of local operations though [1.70].

As a first application of the pseudoinverse formalism we take unbiased random patterns having $\xi_i^\mu = \pm 1$ with probability 0.5. For the correlation matrix (11) we get $C_{\mu\mu} = 1$ while for $\mu \neq \nu$ we find fluctuating terms of the order of $N^{-1/2}$. If we drop these, then $C_{\mu\nu} = \delta_{\mu\nu}$ and we recover the Hopfield model with $\alpha_{\mathrm{c}} = 0.14$. This is far below the theoretical upper bound $\alpha_{\mathrm{c}}^{\mathrm{theory}} = 1$ which is suggested by (11) and (12). Apparently, the "fluctuating terms" ($\mu \neq \nu$) are important. But even if one includes them, *associative* capabilities are not guaranteed up to $\alpha_{\mathrm{c}}^{\mathrm{theory}} = 1$. We finish this section by analyzing why.

Given a Hamiltonian $H = -1/2 \sum_{i,j} J_{ij} S_i S_j$, its zero-temperature dynamics is the rule that a spin be flipped only if the energy is lowered or equivalently (Sect. 1.1.2) that $S_i(t + \Delta t) = \mathrm{sgn}\,[h_i(t)]$, where

$$h_i(t) = \sum_{j(\neq i)} J_{ij} S_j(t) . \tag{5.2.13}$$

Since the diagonal terms in the sum defining H are just constants, they *de facto* drop out and (13) contains no self-interaction.

On the other hand, the zero-temperature dynamics (1) can also be written $S_i(t + \Delta t) = \mathrm{sgn}\,[\tilde{h}_i(t)]$, but here the local field

$$\tilde{h}_i(t) = \sum_{j=1}^{N} J_{ij} S_j(t) \tag{5.2.14}$$

does include a self-interaction $J_{ii}S_i(t)$. The J_{ii} is needed in (12) for the matrix inversion and it is nonzero. We now want to determine the effect of the self-interaction in some detail [1.71].

The argument leading to (11) and (12) guarantees that the patterns $\boldsymbol{\xi}^\mu$ are fixed points – but no more than that. In fact, it turns out that the patterns are not stable, not even under a single spin flip, if $\alpha = q/N$ exceeds 0.5 ($N \to \infty$). The reason is the self-interaction. To see why we have to estimate J_{ii}.

First,

$$\begin{aligned}\sum_j J_{ij}^2 &= N^{-2} \sum_j \sum_{\substack{\mu,\nu\\ \gamma,\delta}} \xi_i^\mu \xi_i^\gamma \xi_j^\nu \xi_j^\delta (C^{-1})_{\mu\nu} (C^{-1})_{\gamma\delta} \\ &= N^{-1} \sum_{\substack{\mu,\nu\\ \gamma,\delta}} \xi_i^\mu \xi_i^\gamma (C^{-1})_{\mu\nu} C_{\nu\delta} (C^{-1})_{\delta\gamma} \\ &= N^{-1} \sum_{\mu,\delta} \xi_i^\mu (C^{-1})_{\mu\delta} \xi_i^\delta = J_{ii} \end{aligned} \tag{5.2.15}$$

so that for all i

$$J_{ii} - J_{ii}^2 = \sum_{j(\neq i)} J_{ij}^2 \geq 0 \tag{5.2.16}$$

and thus $0 \leq J_{ii} \leq 1$. Furthermore,

$$\sum_i J_{ii} = \sum_{\mu,\nu} (C^{-1})_{\mu\nu} (N^{-1} \sum_i \xi_i^\mu \xi_i^\nu) = q = \alpha N \tag{5.2.17}$$

so that $\langle J_{ii} \rangle = \alpha$. It is believed [1.71] that the fluctuations of the J_{ii} around their mean α are small, typically $O(N^{-1/2})$, so that $J_{ii} \approx \alpha$, whatever i is.

The self-interaction restricts the size of the basins of attraction of the individual patterns quite severely. Let us fix, for instance, S_1 at time t and assume that $S_i = \xi_i^\mu$ for all $i > 1$. Then (14) gives

$$\begin{aligned} S_i(t + \Delta t) &= \mathrm{sgn} \left(\sum_{j(\neq 1)} J_{1j} \xi_j^\mu + J_{11} S_1 \right) \\ &= \mathrm{sgn} \left[(1 - \alpha) \xi_1^\mu + \alpha S_1 \right] . \end{aligned} \tag{5.2.18}$$

If $\alpha > 0.5$, then both $S_1 = \xi_1^\mu$ and $S_1 = -\xi_1^\mu$ are stable and for $S_1 = -\xi_1^\mu$ error correction no longer occurs. Equation (18) suggests that the critical storage capacity for *associative* memory is $\alpha_c = 0.5$ – as is confirmed by a more careful argument [1.71], which also shows that a Hamiltonian formulation with a vanishing self-interaction leads to $\alpha_c = 1.0$. In passing we note the rigorous result [1.72] that unbiased random patterns $\boldsymbol{\xi}^\mu$, $1 \leq \mu \leq q$, are linearly independent with probability one as long as $q = \alpha N$ ($N \to \infty$) and $\alpha < 1$. Thus C^{-1} exists for $\alpha < \alpha_c = 1$.

For additional information concerning the pseudoinverse method the reader may also consult Personnaz et al. [1.73].

1.5.3 The Perceptron Convergence Theorem

If one is willing to give up the symmetry condition $J_{ij} = J_{ji}$ (why not?) and allow J_{ij} and J_{ji} to be determined independently of each other, then a truly local learning rule becomes feasible. It dates back to the early 1960s and was conceived for Rosenblatt's perceptron [1.74]. A perceptron [1.4, 74] is a single layer of formal neurons whose unidirectional couplings J_{0j} have to be determined in such a way that, given a certain input, their common output at the element 0 is a Boolean yes or no. In the present context, the perceptron learning rule may be formulated as follows.

Let us assume that we start with a set of couplings J_{ij}, $1 \le i, j \le N$. We fix i, which now plays the role of the output of a perceptron, and check whether the input $\boldsymbol{\xi}^\mu = (\xi_j^\mu; 1 \le j \le N)$ reproduces ξ_i^μ at site i, i.e., whether $\xi_i^\mu = \operatorname{sgn}(\sum_{j(\neq i)} J_{ij}\xi_j^\mu) \equiv \operatorname{sgn}(h_i^\mu)$. If so, then $\{J_{ij}\}$ is a *solution.* As yet, this does not differ from the argument leading to the pseudoinverse procedure. The next step, however, exploits the nonlinearity or, more precisely, the Boolean character of the sign function. If, that is, $h_i^\mu \xi_i^\mu > 0$, then we go to the next pattern $\mu + 1 (\operatorname{mod} q)$. Otherwise we first perform the change

$$J_{ij} \to J_{ij} + N^{-1}\xi_i^\mu \xi_j^\mu \tag{5.3.1}$$

and then turn to pattern $\mu + 1$. The modification (1) is quite sensible since $(\xi_i^\mu \xi_j^\mu)\xi_j^\mu = \xi_i^\mu$ improves the performance with respect to the pattern μ. A priori it is not clear, though, that (1) is also good for other patterns.

More formally, we define a *mask*

$$\varepsilon_i^\mu = \Theta\left(-\xi_i^\mu \sum_{j(\neq i)} J_{ij}\xi_j^\mu\right), \tag{5.3.2}$$

where $\Theta(x) = 1$ for $x \ge 0$ and $\Theta(x) = 0$ if $x < 0$. Then the complete learning rule is

$$J_{ij} \to J_{ij} + N^{-1}\varepsilon_i^\mu \xi_i^\mu \xi_j^\mu, \tag{5.3.3}$$

where it is implicitly understood that all patterns are taken in some order, say $\mu \to \mu + 1 (\operatorname{mod} q)$. The procedure (3) is iterated until it stops. Well, does it? It is a theorem, the *perceptron convergence theorem* [1.74, 75], that it does stop, *provided* a solution $\{J_{ij}^*\}$ exists.

In an abstract setting, the convergence theorem is a statement about finding a hyperplane such that a *given* set of unit vectors is on the same side of it, i.e., in the same half-space supported by the hyperplane. More precisely [1.75], suppose we have a set of unit vectors $\boldsymbol{\eta}^\mu$, $1 \le \mu \le q$, in $\mathbb{R}^N$. Then the perceptron convergence theorem says:

If there exists a unit (!) vector $\boldsymbol{A}^*$ and a $\delta > 0$ such that for all μ we have[12] $\boldsymbol{A}^* \cdot \boldsymbol{\eta}^\mu > \delta$, then the algorithm below terminates after finitely many steps ($\le \delta^{-2}$

[12] This is a kind of stabilility condition. Compare the discussion at the end of the previous section and (9) below.

as $\delta \to 0$). Let us assume that we start with an arbitrary $\boldsymbol{A}$ in the unit sphere (this is immaterial).

Algorithm

sweep: `for` $\mu := 1$ `to` q `do`
update: `if` $\boldsymbol{A} \cdot \boldsymbol{\eta}^\mu \leq 0$ `then` $\boldsymbol{A} := \boldsymbol{A} + \boldsymbol{\eta}^\mu$;
`if` "$\boldsymbol{A}$ has been changed during sweep" `goto` *sweep*
`stop`

According to the theorem, the algorithm will converge to a solution $\boldsymbol{A}$ with $\boldsymbol{A} \cdot \boldsymbol{\eta}^\mu > 0$, $1 \leq \mu \leq q$, in finitely many steps. We now turn to the proof.

Let $\boldsymbol{A}^{(n)}$ denote the vector $\boldsymbol{A}$ after n updates have been performed. So n times we had $\boldsymbol{A} \cdot \boldsymbol{\eta}' \leq 0$ for some $\boldsymbol{\eta}'$ and added $\boldsymbol{\eta}'$ so that $\boldsymbol{A} \to \boldsymbol{A} + \boldsymbol{\eta}'$. By the Cauchy–Schwarz inequality

$$\boldsymbol{A}^* \cdot \boldsymbol{A}^{(n)} / \|\boldsymbol{A}^{(n)}\| \leq 1 \,. \tag{5.3.4}$$

Equation (4) leads to an upper bound for n. To show this, we estimate the numerator and the denominator in (4) separately. For the numerator we have, since $\boldsymbol{A}^* \cdot \boldsymbol{\eta}^\mu > \delta$ for all μ,

$$\begin{aligned} \boldsymbol{A}^* \cdot \boldsymbol{A}^{(n)} = \boldsymbol{A}^* \cdot (\boldsymbol{A}^{(n-1)} + \boldsymbol{\eta}') > \boldsymbol{A}^* \cdot \boldsymbol{A}^{(n-1)} + \delta > \dots \\ > \boldsymbol{A}^* \cdot \boldsymbol{A}^{(0)} + n\delta = a + n\delta \,, \end{aligned} \tag{5.3.5}$$

where $-1 \leq a \leq 1$. The denominator is estimated by

$$\begin{aligned} \|\boldsymbol{A}^n\|^2 &= \boldsymbol{A}^n \cdot \boldsymbol{A}^n = (\boldsymbol{A}^{n-1} + \boldsymbol{\eta}') \cdot (\boldsymbol{A}^{n-1} + \boldsymbol{\eta}') \\ &= \|\boldsymbol{A}^{n-1}\|^2 + 2\boldsymbol{A}^{n-1} \cdot \boldsymbol{\eta}' + \|\boldsymbol{\eta}'\|^2 \\ &\leq \|\boldsymbol{A}^{n-1}\|^2 + 1 \leq \dots \leq n + 1 \end{aligned} \tag{5.3.6}$$

because $\boldsymbol{A}^{n-1} \cdot \boldsymbol{\eta}' \leq 0$; otherwise $\boldsymbol{\eta}'$ would not have been added – and so on. Inserting (5) and (6) into (4) we get

$$\frac{a + n\delta}{\sqrt{n+1}} \leq \frac{\boldsymbol{A}^* \cdot \boldsymbol{A}^n}{\|\boldsymbol{A}^n\|} \leq 1 \,. \tag{5.3.7}$$

So the process stops after at most $n_{\max}$ iterations and

$$n_{\max} \leq \frac{1}{2\delta^2}\{1 - 2a\delta + [(1 - 2a\delta)^2 + 4\delta^2(1 - a^2)]^{1/2}\} \sim \delta^{-2} \tag{5.3.8}$$

as $\delta \to 0$.

Note the important role played by the existence of a solution $\boldsymbol{A}^*$ and by the stability requirement $\boldsymbol{A}^* \cdot \boldsymbol{\eta}^\mu > \delta > 0$ for all μ. In our original problem we take $\boldsymbol{\eta}^\mu = \xi_i^\mu \boldsymbol{\xi}^\mu / \sqrt{N}$ and $\boldsymbol{A} = (J_{ij};\ 1 \leq j \leq N)$ with $J_{ii} = 0$. For unbiased random patterns, it has been shown by Gardner [1.76, 77] that the set of solutions $\{J_{ij}^*\}$ has a *positive* volume in the space of interactions $\{J_{ij}\}$ as long as $\alpha < 2.0$, whence $\alpha_c = 2.0$ – in agreement with known results [1.78].

Stability is improved if one uses the masks [1.76, 77]

$$\varepsilon_i^\mu = \Theta\left(\kappa\sqrt{\sum_{j(\neq i)} J_{ij}^2} - \xi_i^\mu \sum_{j(\neq i)} J_{ij}\xi_j^\mu\right) \tag{5.3.9}$$

with $\kappa > 0$. Given $\alpha = q/N$, one can optimize κ [1.79]. For unbiased random patterns, the equation for the storage capacity $\alpha_c(\kappa)$ is [1.76]

$$1 = \alpha_c(\kappa)\int_{-\kappa}^{+\infty} \frac{dz}{\sqrt{2\pi}} e^{-z^2/2}(z+\kappa)^2 \tag{5.3.10}$$

and, once again, $\alpha_c(0) = 2$. Gardner [1.76] has also computed the storage capacity α_c for correlated, or biased, random patterns and shown that in the low-p limit α_c diverges as

$$\alpha_c \sim -\frac{1}{(1+a)\ln(1+a)} \tag{5.3.11}$$

for $\kappa = 0$. Here a is the activity: $a = \langle\xi\rangle = 2p - 1$.

Through a simple modification of the learning rule (3) one can also construct symmetric $J_{ij} = J_{ji}$,

$$J_{ij} \to J_{ij} + N^{-1}(\varepsilon_i^\mu + \varepsilon_j^\mu)\xi_i^\mu\xi_j^\mu \ . \tag{5.3.12}$$

Because of the symmetry, the algorithm (12) has to be executed *in parallel* over the sites. The convergence proof has to be adapted accordingly [1.76, 77].

Returning to (9), we have to face another, and important, question: What is the learning time? That is to say, how long should we iterate (3) before we have found a solution $\{J_{ij}^*\}$ satisfying $\varepsilon_i^\mu \equiv 0$ in (9)? The perceptron convergence theorem gives us (8) as an upper bound for $n_{\max}$, so we have to rewrite (9) in terms of the theorem. We have $\boldsymbol{A} = (J_{ij};\, 1 \le j \le N)$, $\boldsymbol{\eta} = \xi_i^\mu \boldsymbol{\xi}^\mu/\sqrt{N}$ and, if there were a solution with an identically vanishing mask (9), it would satisfy

$$\boldsymbol{\eta}\cdot\boldsymbol{A}/\|\boldsymbol{A}\| > \frac{\kappa}{\sqrt{N}} \equiv \delta \ . \tag{5.3.13}$$

According to Gardner [1.76, 77], there *is* a solution $\boldsymbol{A}^* = \boldsymbol{A}/\|\boldsymbol{A}\|$ to (13) provided certain conditions on α and κ are met. Note that the perceptron convergence theorem requires $\boldsymbol{A}^*$ to be a unit vector. Then $\delta = \kappa/\sqrt{N}$ is the δ of the theorem in (8) and

$$n_{\max} \le \kappa^{-2}N \ . \tag{5.3.14}$$

Since there are N operations per vector $(J_{ij};\, 1 \le j \le N)$ and N sites i, we obtain N^3 as the size dependence of the number of operations needed to compute all the bonds. In Sect. 1.5.6 we will see that a similar estimate holds for unlearning.

Another hot question is how we can *shorten* the learning time. Equation (14) gives an upper bound for the perceptron algorithm, but there might be – and are smarter learning rules. So what could be a smart way of learning? Naively, we would like to add "lots of" $\xi_i^\mu \xi_j^\mu$ if $\xi_i^\mu h_i^\mu$ is strongly negative (since the answer h_i^μ is so bad) whereas "hardly any" $\xi_i^\mu \xi_j^\mu$ seems to be needed if $\xi_i^\mu h_i^\mu > 0$ (because the answer is all right anyway). A first suggestion then is to take [1.99]

$$J_{ij} \to J_{ij} + \frac{\varepsilon}{N} \xi_i^\mu \xi_j^\mu f(\xi_i^\mu h_i^\mu) \,, \tag{5.3.15}$$

where $\varepsilon > 0$ and $f(x)$ is a nonnegative nonincreasing function of x. The Hopfield model has $f(x) \equiv 1$, there is no convergence theorem, and for correlated patterns its performance is poor. The perceptron algorithm is characterized by $f(x) = \Theta(-x)$ where $\Theta(x) = 1$ for $x \geq 0$ and $\Theta(x) = 0$ if $x < 0$. Convergence is guaranteed provided there exists a solution. A candidate for $f(x)$ which appeals to the above philosophy of strongly correcting the J_{ij} if the system performs badly is $f(x) = \exp(-\beta x)$ for some $\beta > 0$. This function, however, leads to *overshooting*, i.e., it increases too fast as $x \to -\infty$ and therefore corrects too much. A modification, $f(x) = \exp(-\beta x)$ for $x > 0$ and $f(x) = -\beta x + 1$ for $x < 0$, gives a smooth fit at $x = 0$ and, according to the literature [1.99], does improve the learning time. Disadvantages of this choice are that a convergence proof is not available yet and that a "solution" is not a true fixed point under (15), in contrast to the perceptron algorithm where a solution is a fixed point under (3).

An interesting alternative, for which a convergence theorem can be proved, has been put forward by Abbott and Kepler [1.100]. They start by rewriting (9) in the form $\varepsilon_i^\mu = \Theta(\kappa - \gamma_i^\mu)$, where $\gamma_i^\mu = \xi_i^\mu h_i^\mu / \|J_i\|$ with $\|J_i\| = (\sum_j J_{ij}^2)^{1/2}$ and, as usual, $J_{ii} = 0$. They argue that (i) the step size should scale with $\|J_i\|$ and (ii) one should include a function $f(x)$ as in (15). They then end up with

$$J_{ij} \to J_{ij} + N^{-1} \xi_i^\mu \xi_j^\mu f(\gamma_i^\mu) \|J_i\| \Theta(\kappa - \gamma_i^\mu) \,, \tag{5.3.16}$$

where $f(x) = \kappa + \delta - x + [(\kappa + \delta - x)^2 - \delta^2]^{1/2}$ appears to be optimal. Because of $\Theta(\kappa - \gamma_i^\mu)$ in (16), it suffices to consider $f(x)$ for $\kappa - x > 0$. It is assumed that there exists a solution $\{J_{ij}^*\}$ with identically vanishing ε_i^μ in (9) and κ replaced by $(\kappa + \delta)$ for some $\delta > 0$; typically, $\delta \approx 0.01$. Such a solution is a fixed point to (16). The advantages of (16) are twofold. First, there is a convergence theorem, which gives an $n_{\max}$ proportional to N. Second, the performance of (16) is much better [1.100] than that of the perceptron algorithm (3–9). To see why this is so, we note that $f(\kappa) = \delta \ll 1$, while $f(x) \sim 2(\kappa - x)$ if $\kappa - x \gg \delta$. Thus we get a small correction or none if we need not correct, whereas the correction is proportional to $\kappa - \gamma_i^\mu$ if the performance is really bad in that $\kappa - \gamma_i^\mu \gg \delta$; that is, the worse the performance, the larger the correction.

1.5.4 Hebbian Learning

Sloppily formulated, Hebbian learning is learning by repetition. The organism repeats the very same pattern or sequence of patterns several times and in so doing trains the synapses. (A nice example is learning to play an allegro or a presto on a musical instrument.) It has turned out that Hebbian learning is robust and faithful.

In this section, we first concentrate on the encoding and decoding procedure, then study a versatile analytic description of temporal association in terms of sublattice magnetizations (Sect. 1.2.4), and finally analyze how an external signal can trigger the eigendynamics of the network.

Encoding and Decoding. In 1949 Hebb published his classic *The Organization of Behavior – A Neurophysiological Theory* [1.15]. On p. 62 of this book one can find the now famous "neurophysiological postulate": *When an axon of cell A is near enough to excite a cell B and* repeatedly *or* persistently *takes part in firing it, some growth process or metabolic change takes place in one or both cells such that A's efficiency, as one of the cells firing B, is increased.*

One then may wonder, of course, where the above "metabolic change" might take place. Hebb continued by suggesting that "synaptic knobs develop" and on p. 65 he states very explicitly: "I have chosen to assume that the growth of synaptic knobs, with or without neurobiotaxis, is the basis of the change of facilitation[13] from one cell on another, and this is not altogether implausible". No, as we now know, it is not. It is just more complicated [1.80].

Hebb's postulate has been formulated in plain English – but no more than that – and here our main question is how to implement it mathematically. Most of the information which is presented to a network varies in space *and* time. So what is needed is a *common* representation of both the spatial and the temporal aspects. As a pattern changes, the system should be able to measure and store this change. How can it do that?

Until now we have assumed that a neuron, say i, instantaneously notices what happens at j so that the action of j via the synaptic efficacy J_{ij} as experienced by i at time t is $J_{ij}S_j(t)$. Strictly speaking, however, this is incorrect. Once a spike has left the soma of neuron j it travels through the axon to neuron i, whose dendritic tree is reached only after a *delay* τ which may, and in general will, depend on i and j [1.81]. The range of the axonal delay is wide [1.82–84], up to 100–200 ms, so that we may assume a *broad* distribution of delays τ. This is one of the key ingredients of Hebbian learning [1.85, 86].

At the synapse, information is stored in the synaptic efficacy J_{ij} only if neuron i's activity is concurrent with (or slightly after) the arrival of the pulse coming from j [1.87, 88]. If the axonal delay is τ, we have to pair $S_i(t)$ with $S_j(t-\tau)$ so as to satisfy this simultaneity requirement.

---

[13] From *Webster's Ninth New Collegiate Dictionary*: facilitation = the increasing of the ease or intensity of response by repeated stimulation.

We now implement the Hebb rule for the case of unbiased random patterns [1.89]. Let J_{ij} be the value of the synaptic efficacy before the learning session, whose duration is denoted by T. To avoid any confusion with the temperature, we will denote the latter by its inverse value β. After the learning session, J_{ij} is to be changed into $J_{ij} + \Delta J_{ij}$ with [1.85, 86]

$$\Delta J_{ij}(\tau) = N^{-1} \varepsilon_{ij}(\tau) \frac{1}{T} \int_0^T \mathrm{d}t\, S_i(t)\, S_j(t - \tau) \,. \tag{5.4.1}$$

This equation provides us with a *local* encoding of the data at the synapse $j \to i$. The N^{-1} is the well-known scaling factor, $\varepsilon_{ij}(\tau)$ is a neurophysiological factor which is at our disposal, the prefactor T^{-1} in front of the integral takes saturation into account, and $S_i(t)$ is combined with the signal that arrives at i at time t, i.e., $S_j(t - \tau)$ where τ is the axonal delay. Here $S_i(t)$, $1 \leq i \leq N$, denotes the pattern as it is taught to the network during the learning session, $0 \leq t \leq T$. If the pattern is stationary, which means that $S_i(t) = \xi_i^\mu$ for some μ and $T \to \infty$, then $\Delta J_{ij}(\tau) \propto N^{-1} \xi_i^\mu \xi_j^\mu$ and, neglecting $\varepsilon_{ij}(\tau)$, we recover the customary Hebb rule *à la* Cooper [1.90] and Hopfield [1.2]. In general, however, the J_{ij} are asymmetric, as is clearly brought out by (1).

The expression (1) is a time correlation function, so from an abstract point of view there is nothing against (i) its storing time-dependent effects and (ii) its doing so faithfully if the distribution of the τs is broad. Be that as it may, the form of (1) alone does not guarantee yet that the *retrieval* is faithful as well. (It is [1.85, 86].) Decoding during retrieval proceeds as follows. Owing to (1.2.1), we only have to specify the local field

$$h_i(t) = \sum_{j(\neq i)} J_{ij}(\tau_{ij}) S_j(t - \tau_{ij}) \,. \tag{5.4.2}$$

The very same delays which occurred during the learning session also occur during the retrieval session and they do so in exactly the same way as they did during learning. This is the second key ingredient of Hebbian learning (in addition to the broad distribution of delays). Of course, more general models than (1) and (2) are possible [1.81, 85, 86].

Given an encoding and a decoding procedure, there are several questions which have to be answered. Is the performance of (1) and (2) model dependent? (No, hardly.) Does the performance depend critically on the distribution of the delays τ? (No, it either works well or it does not work at all.) And if Hebbian learning works, why does it? As to the why, the succinct answer [1.86] is that synaptic representations are selected according to their resonance with the input percept; the stronger the resonance, the larger $\Delta J_{ij}(\tau)$. Full details can be found elsewhere in this book [1.93]. In the next subsection we turn to a simpler system with a single delay, show what delays are good for, and how a dynamics which incorporates them can be described analytically.

An Analytic Description of Temporal Association. Suppose that we have q unbiased random patterns with a certain order $1 \leq \mu \leq q$ and that we want the system to stay in each of them during a time Δ before going to the next $[\mu \to \mu + 1 (\mathrm{mod}\, q)]$. In short, we want the system to perform a cycle. If Δ were immaterial, one might think [1.2] that

$$J_{ij} = N^{-1} \sum_{\mu} \xi_i^\mu \xi_j^\mu + \varepsilon N^{-1} \sum_{\mu} \xi_i^{\mu+1} \xi_j^\mu \equiv J_{ij}^{(1)} + \varepsilon J_{ij}^{(2)} \tag{5.4.3}$$

would do. The idea is simply that the first term on the right ($J_{ij}^{(1)}$) generates a (free-) energy landscape with hills and valleys and that, for large enough ε, the second term ($J_{ij}^{(2)}$) "pushes" the system from one valley to the next, i.e., induces transitions. However, (3) does not perform as desired. The reason is that $J_{ij}^{(2)}$ wants to induce yet another transition, $\mu \to \mu + 1$, *as soon as* μ *appears*, so that the system has no time to stabilize in state μ. Stabilization, then, can be provided if the signals transmitted through the $J_{ij}^{(2)}$ are delayed in time – even a *single* delay will do [1.91, 92].

Let the characteristic delay time be τ and define

$$\bar{S}_i(t) = \int_0^\infty \mathrm{d}s\, w(s) S_i(t-s) \tag{5.4.4}$$

for some memory kernel $w(s) \geq 0$, normalized to one, $\int_0^\infty \mathrm{d}s\, w(s) = 1$. A δ-function delay has $w(s) = \delta(s - \tau)$ and exponential (typically RC) delay has $w(s) = \tau^{-1} \exp(-s/\tau)$. Using (3) and (4) we write the local field

$$h_i(t) = \sum_j J_{ij}^{(1)} S_j(t) + \varepsilon \sum_j J_{ij}^{(2)} \bar{S}_j(t) \,, \tag{5.4.5}$$

which gives in terms of the overlaps (2.1.3)

$$h_i(t) = \sum_\mu \xi_i^\mu m_\mu(t) + \varepsilon \sum_\mu \xi_i^{\mu+1} \bar{m}_\mu(t) \,. \tag{5.4.6}$$

To see that (6) may give rise to a cycle, we take a zero-temperature parallel dynamics (Sect. 1.1.2) with δ-function delay and assume that the system makes the transition $\nu \to \nu + 1$ at time $t = 0$. From (6), we then obtain that for t small the second term stabilizes the first and $h_i(t) = (1 + \varepsilon)\xi_i^{\nu+1}$. As time proceeds, the delay term will notice the transition at $t = \tau$ and we then get $h_i(\tau^+) = \xi_i^{\nu+1} + \varepsilon \xi_i^{\nu+2}$. If $\varepsilon > 1$, the system makes the transition to $\nu + 2$. And so on. The duration Δ of each pattern is (about) τ and the latter is *fixed* by the hardware. For other delay mechanisms such as exponential delay, Δ is a more elaborate function of ε and τ.

The above argument has made plausible that a limit cycle can exist but it gives no indication of how to describe the dynamics analytically for a general delay and finite β and how to determine the stability of a fixed point or limit cycle. This is what we now want to do [1.94].

We start by taking

$$J_{ij}^{(1)} = N^{-1}Q^{(1)}(\boldsymbol{\xi}_i;\boldsymbol{\xi}_j) = J_{ji}^{(1)} \text{ and } J_{ij}^{(2)} = N^{-1}Q^{(2)}(\boldsymbol{\xi}_i;\boldsymbol{\xi}_j) \neq J_{ji}^{(2)} ,$$

where $\boldsymbol{\xi}_i = (\xi_i^\mu;\ 1 \le \mu \le q)$ is the local information available to neuron i. For example,

$$Q^{(1)}(\boldsymbol{\xi}_i;\boldsymbol{\xi}_j) = \phi(\sum_\mu \xi_i^\mu \xi_j^\mu) , \quad Q^{(2)}(\boldsymbol{\xi}_i;\boldsymbol{\xi}_j) = \phi(\sum_\mu \xi_i^{\mu+1} \xi_j^\mu)$$

for some odd function ϕ. The case studied until now has $\phi(x) = x$ and is therefore called linear. However, *double clipping* [1.94] with $\phi(x) = \mathrm{sgn}(x)$ can be shown to work equally well. It is relevant to electronic hardware realizations.

Given q binary patterns, there are 2^q different positions available to the vectors $\boldsymbol{\xi}_i$, viz., the 2^q corners C^q of the q-dimensional hypercube $[-1,1]^q$. As in Sect. 1.2.4, we assign to each corner $\boldsymbol{x}$ a sublattice $I(\boldsymbol{x}) = \{i;\ \boldsymbol{\xi}_i = \boldsymbol{x}\}$ and a sublattice magnetization

$$m(\boldsymbol{x};t) = |I(\boldsymbol{x})|^{-1} \sum_{i \in I(\boldsymbol{x})} S_i(t) . \tag{5.4.7}$$

Taking a single Q and forgetting about the time dependence for a moment, we find

$$\begin{aligned} h_i &= N^{-1} \sum_j Q(\boldsymbol{\xi}_i;\boldsymbol{\xi}_j) S_j \\ &= \sum_{\boldsymbol{y}} (N^{-1}|I(\boldsymbol{y})|)|I(\boldsymbol{y})|^{-1} \sum_{j \in I(\boldsymbol{y})} Q(\boldsymbol{\xi}_i;\boldsymbol{\xi}_j) S_j \\ &= \sum_{\boldsymbol{y}} p_N(\boldsymbol{y}) Q(\boldsymbol{\xi}_i;\boldsymbol{y}) |I(\boldsymbol{y})|^{-1} \sum_{j \in I(\boldsymbol{y})} S_j \\ &= \sum_{\boldsymbol{y}} p_N(\boldsymbol{y}) Q(\boldsymbol{\xi}_i;\boldsymbol{y}) m(\boldsymbol{y}) . \end{aligned} \tag{5.4.8}$$

For unbiased random patterns, $p_N(\boldsymbol{y}) = N^{-1}|I(\boldsymbol{y})|$ converges to $p(\boldsymbol{y}) = 2^{-q}$ with probability one as $N \to \infty$ [1.22]. If i belongs to $I(\boldsymbol{x})$, then (8) depends only on $\boldsymbol{x}$ since $\boldsymbol{\xi}_i = \boldsymbol{x}$ for all i in $I(\boldsymbol{x})$. In the bulk limit the local field $h_i(t)$ produced by the instantaneous and delayed signals therefore converges to

$$h(\boldsymbol{x};t) = \sum_{\boldsymbol{y}} [Q^{(1)}(\boldsymbol{x};\boldsymbol{y}) m(\boldsymbol{y};t) + \varepsilon Q^{(2)}(\boldsymbol{x};\boldsymbol{y}) \bar{m}(\boldsymbol{y};t)] p(\boldsymbol{y}) , \tag{5.4.9}$$

whatever i in $I(\boldsymbol{x})$.

Suppose now that after each elementary time step Δt a parallel updating is performed according to (1.2.1). For *all* i in $I(\boldsymbol{x})$ the thermal average of $S_i(t + \Delta t)$ is $\tanh[\beta h(\boldsymbol{x};t)]$. The argument is simple. The probability of getting S_i is $[1 + \tanh(\beta h_i S_i)]/2$ so that $\mathrm{Prob}\,\{S_i = +1\} - \mathrm{Prob}\,\{S_i = -1\} = \tanh(\beta h_i)$ with $h_i(t) = h(\boldsymbol{x};t)$ for all i in $I(\boldsymbol{x})$, as claimed. The size $|I(\boldsymbol{x})| \sim p(\boldsymbol{x})N$ grows with

N so that by (7), (9), and the strong law of large numbers [1.22], $m(\boldsymbol{x};t+\Delta t)$ converges to

$$m(\boldsymbol{x};t+\Delta t) = \tanh[\beta h(\boldsymbol{x};t)] \tag{5.4.10}$$

as $N \to \infty$. This is a recursion relation, which has to be iterated. Equation (10) also tells us that, in the bulk limit $N \to \infty$, the stochastic dynamics of the sublattice magnetizations becomes fully deterministic. (This holds as long as $p_N(\boldsymbol{y})$ converges to a limit, $p(\boldsymbol{y})$, and the initial conditions converge as well.)

If we use sequential updating, only one spin per elementary time step Δt is updated. To get each spin updated a finite number of times per second as N becomes large one has to rescale time so that $\Delta t \propto N^{-1}$. Then one obtains a set of 2^q coupled nonlinear ordinary differential equations [1.93, 94]

$$\dot{m}(\boldsymbol{x};t) = -\Gamma\{m(\boldsymbol{x};t) - \tanh[\beta h(\boldsymbol{x};t)]\}\ , \tag{5.4.11}$$

where Γ is the mean attempt rate per (rescaled) unit time. Once again $m(\boldsymbol{x};t)$ has become fully deterministic as $N \to \infty$. Because of the delay contained in $h(\boldsymbol{x};t)$, (11) represents a so-called *functional differential equation*. Both (10) and (11) are exact, whatever the neural activity level and the synaptic kernels Q.

Only in the case of the *linear* model (3–6) can we reduce (10) or (11) to a set of q equations for the overlaps themselves. To this end we exploit (2.4.6) and find

$$m_\mu(t+\Delta t) = \left\langle x_\mu \tanh\left\{\beta \sum_\nu x_\nu [m_\nu(t) + \varepsilon \bar{m}_{\nu-1}(t)]\right\}\right\rangle \tag{5.4.12}$$

for parallel updating and

$$\dot{m}_\mu = -\Gamma\left[m_\mu(t) - \left\langle x_\mu \tanh\left\{\beta \sum_\nu x_\nu [m_\nu(t) + \varepsilon \bar{m}_{\nu-1}(t)]\right\}\right\rangle\right] \tag{5.4.13}$$

for sequential updating. The angular brackets have the same meaning as in (2.4.6). As is shown in Fig. 1.6, sequential and parallel updating give rise to a different performance. In addition, double clipping works well. In passing, we note [1.94] that for $\varepsilon > 1$ (but not "too" large) the cycle runs both at low temperatures and at high temperatures which may even *exceed* the critical temperature $T_c = 1$ of the Hopfield model. For $\varepsilon < 1$, a moderate amount of thermal noise is needed to help the system to "surmount the hills which separate the valleys (=patterns)".

Because of the delay, the stability analysis of a solution to (10–13) is somewhat intricate [1.95, 96]. For a cycle of length two ($q = 2$), an exact phase diagram in the (β, ε) plane could be obtained [1.94, 97, 98] but for larger cycles one has to resort to approximations [1.92, 93].

Though the model (5) with a single delay τ is capable of producing stable temporal sequences, its performance in some association tasks is poor. To see why, let Δ and Δ_R be the time the system is spending in each pattern during the learning and the retrieval session, respectively. It is important to realize that

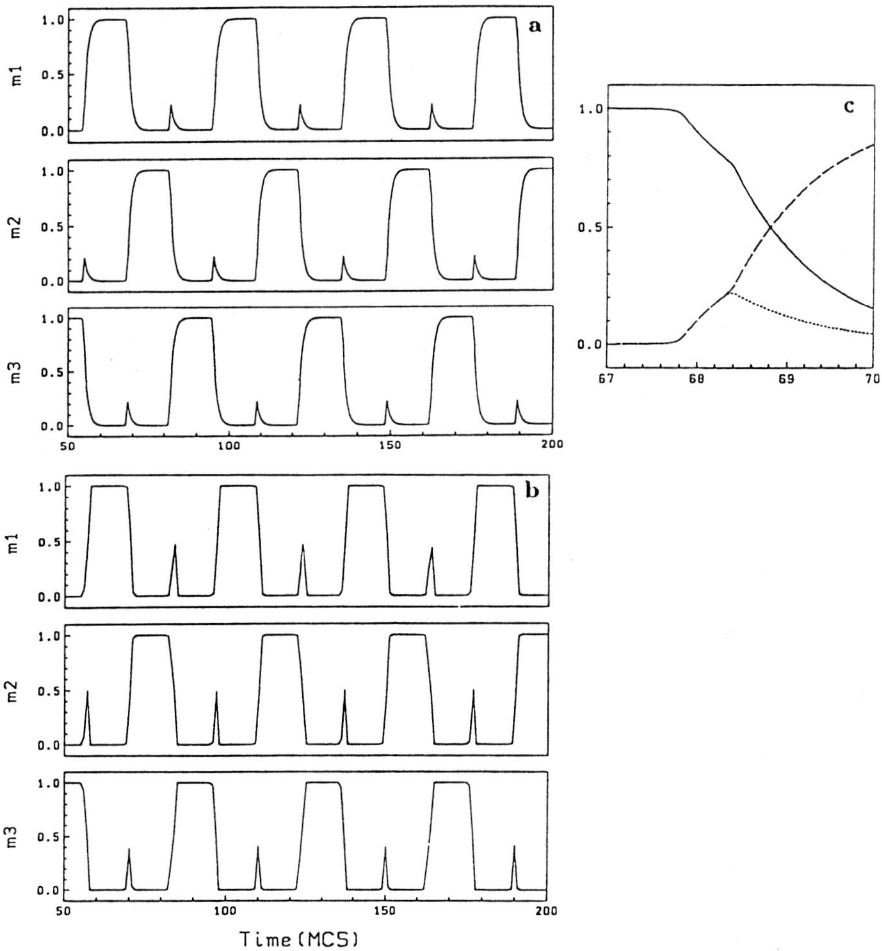

Fig. 1.6a–c. Comparison between sequential (**a**) and parallel (**b**) dynamics in a cycle consisting of three patterns (q = 3). Each track represents the overlap $m_\mu(t)$, $1 \le \mu \le 3$, with one of the patterns of the cycle as a function of time. The inset on the right (**c**) shows that during a transition in **a** the sublattices do not flip simultaneously. On the other hand, the transitions in **b**, where we have parallel dynamics, are relatively sharp and retrieval produces fairly perfect block pulses. In both a and b there is an exponential delay with τ = 10 while $\beta = 30, \varepsilon = 1.7$, and Γ = 1 (in a). The figure also suggests that it usually suffices to plot the overlaps with one of the patterns, say, the first one of the cycle

there is only a single τ and that it is prewired. Whatever Δ, during retrieval $\Delta_{\rm R}$ is a function of ε, β, and τ – and, of course, the delay mechanism. We also assume that the cycle really runs, which need not be the case.

For δ-function delay, *only a single realization is possible*, $\Delta_{\rm R} \approx \tau$, which is nearly independent of β and ε once the cycle is there. Let us take, as a rather harmless example, $\beta = \infty, \varepsilon > 1$, suppose that a 3-cycle (1,2,3) has been stored, and present as an initial condition pattern 1 during $(-\tau, -3\tau/4)$ and pattern 2 during $(-3\tau/4, 0)$. From (6) we then see that we will retrieve pattern 2 during $(0, \tau/4)$, pattern 3 during $(\tau/4, \tau)$, and so on. The correct sequence (1,2,3) will not be restored.

For exponential delay, one can obtain an appropriate $\Delta_R \approx \Delta$ by carefully tuning ε and β [1.92, 93]. How such an ε can be learnt by the network is as yet unclear. Furthermore, long exponential delays ($\gtrsim$ 20 ms) are hard to find in nature and a distribution of them leads to destabilization [1.81]. And, worse, if the patterns just have a varying duration Δ_μ (with $\Delta_\mu \neq \Delta_\nu$), then an appropriate retrieval is no longer possible. On the other hand, the Hebb rule (1) in conjunction with a broad distribution of delays can easily cope with all these situations which cannot be handled by (3–6). Referring the reader to elsewhere in this book [1.93] for an extensive discussion of temporal association, we now turn to the problem of describing the behavior of a cycle in the presence of many other data and equipped with, indeed, a single τ.

Extensively Many Patterns. In the kind of system under consideration, noise is either thermal or it is produced by extensively many other patterns. Thermal noise is taken care of by the stochastic dynamics (Sect. 1.1.2). Suppose, then, that we have a finite cycle whose patterns are to be labeled by μ in the presence of *extensively* many (αN) patterns which are to be labeled by ν. For the sake of simplicity the patterns are taken to be stationary, but generalization of the argument below is straightforward. The question is: How do we incorporate the noise generated by the ν patterns in the dynamics of the cycle, i.e., the μ patterns?

The intuitive idea behind the solution [1.94] is that the system spends its time in that part of phase space which is strongly correlated with the μ patterns, most of its time even in a specific valley (the ergodic component). We make, therefore, a slight detour and return to the computation of the free energy for the model without any transition term ($\varepsilon = 0$); cf. Sects. 1.3.3 and 1.4.3. We single out the finitely many μ patterns which are involved in the cycle and average over the rest. Whatever the nonlinearity in $Q^{(1)}$, the expression for the free energy, (3.3.32) or (4.3.10), consists of three parts. The first refers to the μ patterns we concentrate on, the second depends only on the order parameter $\hat{q}$ which represents the noise generated by the other, extensively many patterns, and the last describes the interaction between the two groups of patterns in terms of $\hat{q}$ and the overlaps m_μ.

Fixing $\hat{q}$, one straightforwardly verifies that the first and third terms of the free energy can be derived from the effective Hamiltonian

$$H_{\text{eff}} = -\tfrac{1}{2} N \sum_{\mu} \Lambda_\mu \left(N^{-1} \sum_{i=1}^{N} \xi_i^\mu S_i \right)^2 - \sqrt{\hat{q} r} \sum_{i=1}^{N} z_i S_i \,, \tag{5.4.14}$$

where the z_i are independent Gaussians with mean zero and variance one, while $r = r(\hat{q})$ is a given function of $\hat{q}$, (3.3.33) or (4.3.11), or a generalization thereof, and Λ_μ is the embedding strength of pattern μ. So the extensively many patterns outside the cycle produce a Gaussian random field of variance $\hat{q}r(\hat{q})$. We then find [1.94], for instance, instead of (13),

$$\dot{m}_\mu = -\Gamma\left\{m_\mu - \left\langle\left\langle x_\mu \tanh\left[\beta\left(\sum_\nu x_\nu(m_\nu + \varepsilon\bar{m}_{\nu-1}) + \sqrt{\hat{q}r}z\right)\right]\right\rangle\right\rangle\right\}, \tag{5.4.15}$$

$$\dot{\hat{q}} = -\Gamma\left\{\hat{q} - \left\langle\left\langle \tanh^2\left[\beta\left(\sum_\nu x_\nu(m_\nu + \varepsilon\bar{m}_{\nu-1}) + \sqrt{\hat{q}r}z\right)\right]\right\rangle\right\rangle\right\}.$$

The double angular brackets denote an average not only over the $x_\mu = \pm 1$ but also over the Gaussian z. Compared to (13), the role of the extensively many patterns is to smooth the solution through the Gaussian z. In fact, their role is quite similar to that of temperature.

The above procedure has no claim of rigor, but as yet no exact solution to the problem of treating extensively many patterns in a fully connected network has been found[14]. Figure 1.7 shows, however, that our ansatz contains the essential physics. The agreement between the solution to (15) and the numerical simulation on a finite sample is excellent. The pulse forms agree exactly and the periods differ by only 2%.

Though (13) has been derived to describe exactly a temporal sequence with finitely many patterns and (15) has been conceived to treat approximately the dynamics of a cycle operating in the presence of extensively many other patterns, both equations can also be applied to the case without any delay, viz., the customary Hopfield model [1.103]. Just put $\varepsilon = 0$ in (13) and (15). Concentrating on a single pattern μ we may ask how m_μ develops in time if the system starts from a noisy input $m_\mu(0) < 1$ and in the presence of extensively many other patterns. Qualitively, the two cases of Fig. 1.8, $\alpha = 0.10 < \alpha_c = 0.14 < \alpha = 0.15$, are in excellent agreement with Monte Carlo simulations [1.104, 105]: for *small* times and as long as $\alpha < 2/\pi$ [1.135], $m_\mu(t)$ is *always increasing*. Below α_c there is

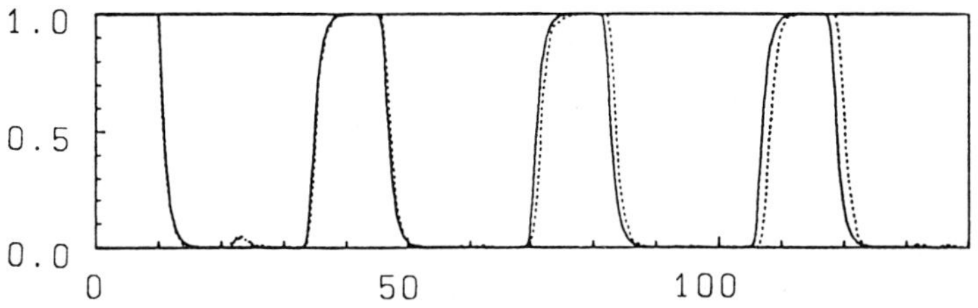

Fig. 1.7. Comparison between the analytic solution to (15) (*solid curve*) and numerical simulation on a finite sample of size $N = 1000$ (*dashed curve*), both with Glauber dynamics. Here we have $\beta = 10, \varepsilon = 1, \Gamma = 1$, and a δ-function delay with $\tau = 10$. A cycle of length three ($q = 3$) is running in the presence of 100 other patterns ($\alpha = 0.10$). Only the overlap with the first pattern of the cycle is shown

[14] In a *highly* diluted network with a connectivity of the order of $\ln N$ where N is the size of the system, dynamic correlations have been eliminated and an exact solution is possible [1.101, 102].

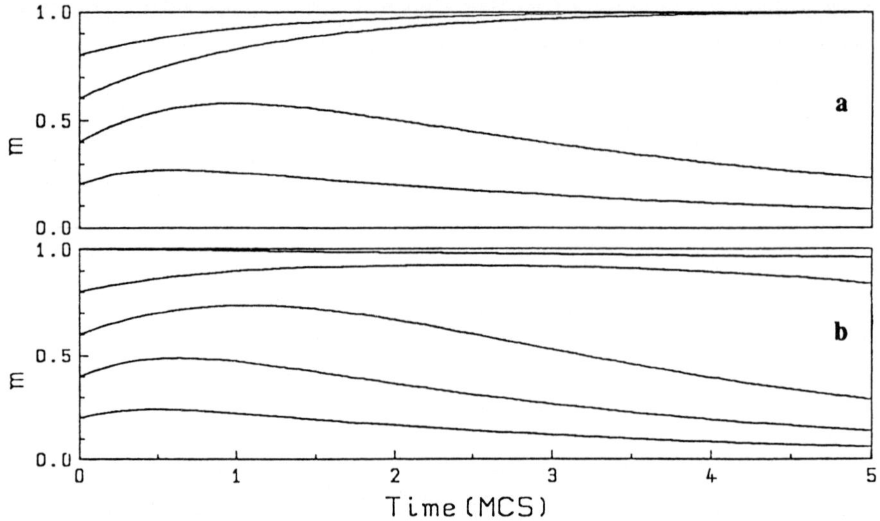

Fig. 1.8a, b. Time evolution of the overlap with a single pattern $m_\mu(t)$ in the Hopfield model with $\alpha = 0.10$ (**a**) and $\alpha = 0.15$ (**b**). For $\alpha < \alpha_c = 0.14$, there is an $m_c(\alpha) < 1$ such that $m_\mu(t)$ relaxes to a value close to 1 as $t \to \infty$ if $m_\mu(0) > m_c(\alpha)$ whereas it decays to 0 if $m_\mu(0) < m_c(\alpha)$. For $\alpha > \alpha_c$, we have $m_c(\alpha) = 1$ and the pattern is unstable. Note that $m_\mu(t)$ always starts by *in*creasing from $m_\mu(0)$

a critical value $m_c(\alpha)$ of $m_\mu(0)$ such that for $m_\mu(0) > m_c(\alpha)$ the overlap $m_\mu(t)$ converges to $m_\mu(\infty) \approx 1$. We find $m_c(0.10) \approx 0.50$, which is to be compared with $m_c(0.10) \approx 0.38$ found by Forrest [1.106]. One has to realize, though, that (15) has been conceived for a situation where the system is sitting at the bottom of a free-energy valley most of its time. Through a more intricate theory [1.104, 107] the agreement can be improved.

If $m_\mu(0) < m_c(\alpha)$, then $m_\mu(t)$ approaches zero as $t \to \infty$; at least, according to Fig. 1.8b. Simulations, however, show [1.104] that for $\alpha \gtrsim 0.08$ there is something like a "remanent magnetization", the system remembers where it came from $[m_\mu(0) > 0]$, and $m_\mu(\infty) > 0$. We have $m_\mu(\infty) \approx 0.36$ at $\alpha_c = 0.14$ and it decreases to 0.14, which is the value corresponding to the SK model [1.42, 137], as $\alpha \to \infty$. No theory explaining this remanent magnetization is yet known. All existing arguments [1.94, 104, 107] sooner or later use a Gaussian approximation, which then leads to a vanishing $m_\mu(\infty)$.

Internal Dynamics Coupled to an External Signal. The present modeling assumes a network that first "learns" a The present modeling a network that first "learns" a pattern by getting it imposed upon the net's constituents, the neurons. The given signal $S_i(t)$, $1 \le i \le N$, is inserted into (1) and encoded. As to Hebbian *learning*, this seems fair. However, one might complain that decoding during retrieval need not be as explicit as we have assumed it in that we *prescribe* initial conditions $S_i(t)$ during the time interval $(-\tau_{max}, 0)$. Usually there are receptors which feed their data into the network that has its own dynamics, i.e., its internal or eigendynamics. In addition, the incoming signal may be rather weak. For example, during a lecture (=learning) data are transmitted fairly loudly, during a test (=retrieval) they are whispered. All this can be modeled as follows [1.116].

We assume that the local field at i is

$$\begin{aligned} h_i(t) &= (1-\gamma)\sum_j J_{ij}\bar{S}_j(t) + \gamma h_i^{\mathrm{ext}}(t) \\ &= (1-\gamma)h_i^{\mathrm{int}}(t) + \gamma h_i^{\mathrm{ext}}(t)\ , \end{aligned} \tag{5.4.16}$$

where $h_i^{\mathrm{ext}}(t)$ is the signal that is fed into the network (via the receptors) and $h_i^{\mathrm{int}}(t)$ represents the time evolution of the network itself. Here we use (4) and assume $|h_i^{\mathrm{ext}}(t)| = 1$. Through γ we can vary the relative strength of h_i^{int} and h_i^{ext}, and the question is under what condition the network still operates as an associative memory, a "pattern" now being a *spatio-temporal* entity. There are three parameters in checking the performance of (16), viz., the inverse temperature β, γ, and the time during which h_i^{ext}, triggers the network. And, indeed, most of the time the associative capabilities are preserved well [1.116].

As a typical example we have taken the time-warp problem. Figure 1.9 shows that it is solved appropriately if, as always, the distortion is not too large. Note that the external signal may be rather weak (so whispering does work); e.g., in Fig. 1.9 the fraction $\gamma/(1-\gamma)$ of external signal to eigendynamics is $1/4$.

1.5.5 Intermezzo

Until now it was required that Prob $\{\xi_i^\mu = +1\} \equiv p = 0.5$. In other words, the *activity* $a \equiv \langle \xi^\mu \rangle = 2p - 1$, which is nothing but the magnetization associated with a pattern, had to be zero. As we have seen in (1.3), no extensive storage capacity is possible in a Hebbian network with $a \neq 0$. This *Intermezzo* is devoted to the problem how an extensive storage capacity can be restored. It can, but at the cost of losing the Hebbian representation (4.1). In the next section we treat *un*learning and show how pure Hebbian learning can function in spite of the activity being nonzero.

To handle low-activity patterns, we use the pseudoinverse technique of Sect. 1.5.2. The correlation matrix is easily calculated,

$$C_{\mu\nu} \equiv N^{-1}\sum_{i=1}^{N} \xi_i^\mu \xi_i^\nu = \delta_{\mu\nu}(1-a^2) + a^2 \tag{5.5.1}$$

plus terms of order $N^{-1/2}$ for $\mu \neq \nu$, which we neglect. The matrix C is of exactly the same form as the matrix Q in (3.3.26, 27), so its inverse is readily obtained,

$$\begin{aligned} (C^{-1})_{\mu\nu} &= (1-a^2)^{-1}[\delta_{\mu\nu} - a^2/(1-a^2+a^2 q)] \\ &\simeq (1-a^2)^{-1}\left[\delta_{\mu\nu} - \frac{1}{q} + \frac{1}{q}\left(\frac{1-a^2}{qa^2}\right)\right] . \end{aligned} \tag{5.5.2}$$

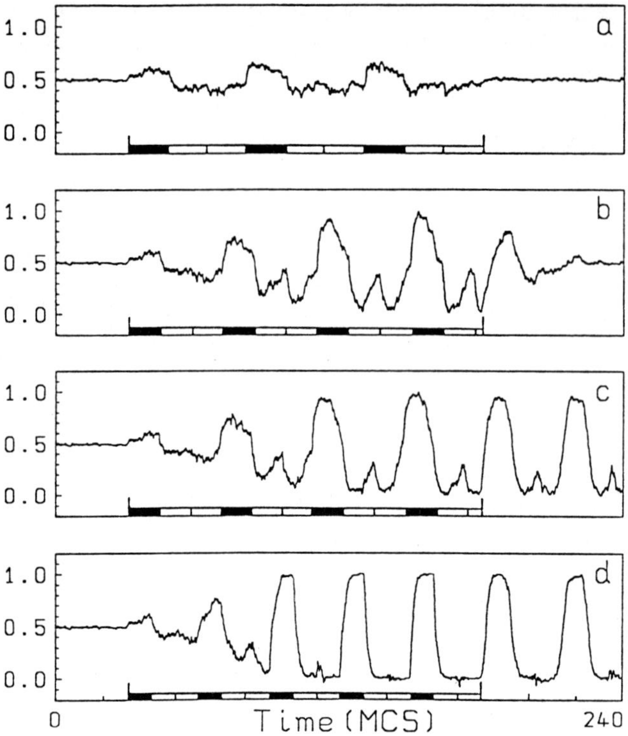

Fig. 1.9a–d. An external stimulus triggers the internal dynamics of a network with a broad, here uniform, *distribution* of delays between 0 and 30 Monte Carlo steps per spin (MCS). The network has been taught a cycle consisting of three patterns, each lasting Δ = 10 MCS; the learning rule used was (4.1). Both the 3-cycle and the 3-symmetric state ($m_1 = m_2 = m_3 = 0.5$) are stable. The system starts in the latter and is triggered during $30 \leq t \leq 180$ MCS by an external signal which is the original cycle *distorted in time*; see (4.16). The external signal is weak: $\gamma = 0.20$. From top to bottom (**a** – **d**) we have a frequency distortion by factors 0.6, 0.75, 0.77, and 1.0, compared to the original frequency, which is d. Each track represents the overlaps $m_1(t)$ with the first pattern of the cycle. If in a and b one presented the distorted cycle to the network within a period of time beyond $t = 180$, then the response would no longer change

Since C^{-1} is a $q \times q$ matrix and[15] $q \to \infty$, we have neglected terms of order q^{-3} and higher in (2). Inserting (2) into (2.11) we obtain

$$J_{ij} = N^{-1} \left\{ (1 - a^2)^{-1} \left[\sum_{\mu} \xi_i^{\mu} \xi_j^{\mu} - \frac{1}{q} \left(\sum_{\mu} \xi_i^{\mu} \right) \left(\sum_{\nu} \xi_j^{\nu} \right) \right] + 1 \right\} . \quad (5.5.3)$$

The last term in (3), viz. N^{-1}, comes from

$$\frac{1}{Na^2} \sum_{\mu,\nu} \frac{1}{q^2} \xi_i^{\mu} \xi_j^{\nu} = \frac{1}{Na^2} \left(\frac{1}{q} \sum_{\mu} \xi_i^{\mu} \right) \left(\frac{1}{q} \sum_{\nu} \xi_j^{\nu} \right) \simeq N^{-1} . \quad (5.5.4)$$

[15] For finite q, (5) below will do.

This ferromagnetic term is quite important. We will discuss its physics shortly. In the low-p limit, and with S_i replaced by $(S_i - a)$, the interaction (3) is optimal as it saturates the upper bound (3.11) [1.108].

Neglecting fluctuations of order $1/\sqrt{q}$ one can transform (3) into a form which looks, but is not, better,

$$J_{ij} = N^{-1}\left[\sum_{\mu} \frac{(\xi_i^\mu - a)(\xi_j^\mu - a)}{1 - a^2} + 1\right] . \tag{5.5.5}$$

The $(\xi_i^\mu - a)/\sqrt{1 - a^2}$ have mean zero and variance one as in the original Hopfield model; cf. (1.1.1). So apart from the ferromagnetic term N^{-1} (5) has quite a natural appearance. An alternative argument leading to (5) will be presented in Sect. 1.6. It is due to Bös [1.109] and is based on the notion of information hierarchy. Feigel'man and Ioffe [1.110] have found (5) independently. Whatever a, the model (5) *always* has the storage ratio $\alpha_c = 0.14$ of the Hopfield model, whereas (3) gives rise to a diverging $\alpha_c \sim -[(1+a)\ln(1+a)]^{-1}$ as $p \to 0$. The moral is that "fluctuating terms" may turn out to be quite important. From a neurophysiological point of view, the synaptic efficacy (5) has a drawback: the network has to preview the data and determine the activity a before it can store them. Unlearning can do without this.

Dropping the ferromagnetic term from (5) is bad [1.111] since without it α_c monotonically approaches zero as $p \to 0$. So what is the ferromagnetic term good for? This is seen most easily by performing a signal-to-noise-ratio analysis as in (3.5.7),

$$\begin{aligned} h_i^\mu &= \sum_{j(\neq i)} J_{ij}\xi_j^\mu \\ &= N^{-1}\left[(\xi_i^\mu - a)\sum_{j(\neq i)} \frac{(\xi_j^\mu - a)\xi_j^\mu}{1 - a^2} + \sum_{\substack{j(\neq i)\\ \nu(\neq\mu)}} \frac{(\xi_i^\nu - a)(\xi_j^\nu - a)\xi_j^\mu}{1 - a^2} + \sum_{j(\neq i)} \xi_j^\mu\right] \\ &= (\xi_i^\mu - a) + N^{-1}\sum_{\substack{j(\neq i)\\ \nu(\neq\mu)}} \frac{(\xi_i^\nu - a)(\xi_j^\nu - a)\xi_j^\mu}{1 - a^2} + a . \end{aligned} \tag{5.5.6}$$

The sum in the middle is the noise generated by the other patterns $(\nu \neq \mu)$. Whatever a, it always has the same variance, viz. α, as the noise term in (3.5.7). The first term on the right is the signal $(\xi_i^\mu - a)$, which, in view of the noise, becomes particularly bad as $p \to 0$, since in this limit $(\xi_i^\mu - a) \approx 0$ for most of the sites. However, the very last term a compensates for that and $(\xi_i^\mu - a) + a = \xi_i^\mu$ restores the original signal. Without a ferromagnetic term there is no compensation and, as we noted, the model behaves poorly. The fact that both the (restored) signal and the variance of the noise associated with (6) exactly agree with their counterparts of the Hopfield model makes it plausible that the storage capacity α_c of the two models is identical. A full-blown proof of this result can be found in Sect. 1.6. A signal-to-noise analysis for (3) proceeds along similar lines and is left to the reader.

For low-activity patterns there is also an excellent cycle generating mechanism [1.108]. According to (4.1), a cycle or temporal sequence is encoded as a τ-dependent quadratic form [1.85, 86]

$$\Delta J_{ij}(\tau) = N^{-1}\varepsilon_{ij}(\tau)\sum_{\mu,\nu} \xi_i^{\mu} Q_{\mu\nu}^{(\tau)} \xi_j^{\nu} \,. \tag{5.5.7}$$

To see why, just plug the cycle input $S_i(t) = \xi_i^{\nu(t)}$ for a given $\nu(t)$ into (4.1) and realize that $\nu(t)$ samples the labels of the cycle. If $a \neq 0$, then the only modification needed is to replace ξ_j^{ν} in (7) by $(\xi_j^{\nu} - a)/(1 - a^2)$. Note that this substitution is performed asymmetrically. For a single stationary pattern (and $\varepsilon_{ij} \equiv 1$) one then recovers (3) if one replaces a by $q^{-1}\sum_{\nu} \xi_j^{\nu}$, which seems fair as q becomes large, and adds the ferromagnetic N^{-1}.

1.5.6 Hebbian Unlearning

A model such as (5.6) can be criticized for at least two reasons. First, (5.6) is not really local since the synapse $j \to i$ has to know the overall (=global) activity a of the patterns. Second, in real-life situations the activity may vary from one pattern to the next. The rule (5.6) can handle this complication if one replaces a by a_{μ}, but in that case each pattern has to be previewed before it is handled – which seems implausible. Furthermore, the Hebb rule (4.1) does not "see" the a_{μ} either, in the case of stationary patterns the system ends up with a Hebbian coding *à la* (1.1.1), and an extensive storage capacity is illusory. However, as was suggested by Crick and Mitchison [1.112], nature may have taken care of this problem through unlearning.

Some time ago Crick and Mitchison hypothesized that the purpose of dream (REM = rapid eye movement) sleep [1.113] is to weaken certain undesirable modes in the network cells of the cerebral cortex. As a mechanism they proposed the "more or less random stimulation of the forebrain by the brain stem that will tend to excite the inappropriate modes" in conjunction with a reverse learning mechanism, *unlearning*, which "will modify the cortex (for example, by altering the strength of the individual synapses) in such a way that this particular activity is less likely in the future".

The above ideas were implemented by Hopfield et al. [1.114]. They loaded the Hopfield model with q stationary patterns $\{\xi_i^{\mu};\ 1 \le i \le N\}$, where N is the size of the system and the $1 \le \mu \le q$ label the patterns. The ξ_i^{μ} are independent identically distributed random variables which assume the value +1 with probability p and -1 with probability $(1-p)$. In the case of [1.114], $p = 0.5$. Before unlearning, the synaptic efficacies are

$$J_{ij} = N^{-1}\sum_{\mu=1}^{q} \xi_i^{\mu}\xi_j^{\mu} \,, \tag{5.6.1}$$

in agreement with (4.1). The dreaming procedure consists of three steps. (i) *Random shooting*: one generates a random initial configuration (there are a bunch of other procedures, though, which work about as well). (ii) *Relaxation*: one allows the system to relax (because $J_{ij} = J_{ji}$, it must relax) to a stationary configuration $\{\eta_i^d;\ 1 \le i \le N\}$ under a zero-temperature Monte Carlo or Glauber dynamics, i.e., a spin is flipped only if the energy $H_N = (-1/2)\sum_{i,j} J_{ij}S_iS_j$ is lowered. (iii) *Unlearning*: after dream d, all the $J_{ij}(i \neq j)$ are updated according to

$$J_{ij} \to J_{ij} - \frac{\varepsilon}{N}\eta_i^d\eta_j^d\,. \tag{5.6.2}$$

Note the minus sign in (2): the final state associated with dream d is weakened; it is "unlearned". The parameter ε is a small number: typically $\varepsilon \approx 0.01$. The index d labels the dreams and, if the procedure is repeated D times, $1 \le d \le D$. The unlearning (2) is *iterative* and purely Hebbian. In addition, it is *local* inasmuch as only data at i and j determine J_{ij}. In passing, we also note that the random shooting agrees extremely well with the neurophysiological picture of [1.112].

Hopfield et al. [1.114] published some positive results concerning unlearning in small samples ($N = 32$) with $p = 0.5$ and low loading $\alpha \approx 0.15$. The issue was revived recently, studied for a broad range of activities ($-0.5 \le a \le 0.5$) and system sizes ($100 \le N \le 400$), and shown to be surprisingly robust and faithful [1.115].

What, then, does unlearning achieve? First, it greatly improves the efficiency of the network. The number of parasitic or spurious states is drastically reduced and the domains of attraction of the individual patterns are expanded. Spurious states are mixture states at low $\alpha\,(\alpha \to 0)$ and spin-glass states related to the "remanent magnetization" of the patterns for $\alpha \gtrsim \alpha_c$. For $p \neq 0.5$, we even start with a single dominant state, the ferromagnet; cf. (1.3). Through unlearning, the storage capacity, i.e., the fraction $q_{\max}/N$, where $q_{\max}$ is the maximal number of patterns which can be retrieved reliably, is increased to a considerable extent. For example, it turns out that for unbiased patterns ($p = 0.5$) $q_{\max}$ increases from the Hopfield value $0.14N$ to $0.68\,^{+0.01}_{-0.03}N$, which is far beyond the $0.5N$ of the original pseudoinverse method (Sect. 1.2).

A second remarkable aspect of unlearning is that the optimal number of dreams D_{opt} does not depend on the activity of the neurons, or, more precisely, on the probability p of getting +1 in a random pattern. To determine D_{opt}, one needs a criterion. A very sensible one is to maximize the domains of attraction of the individual patterns. In principle, D_{opt} depends on four parameters, q, N, ε, and p. It then turns out that [1.115]

$$D_{\text{opt}}(q, N, p; \varepsilon) = \varepsilon^{-1}\left(c_0 + \frac{c_1}{N}\right)q \tag{5.6.3}$$

with $c_0 = 0.56 \pm .08$ and $c_1 = 22 \pm 10$, while ε has to be less than 0.05 (small enough). As we have already stressed, there is *no* dependence upon p. It also turns out that the performance of the network improves as $\varepsilon \to 0$.

A third aspect of unlearning, which is also worth noticing, is that using the simple Hebb rule (1) in conjunction with unlearning a network can store and retrieve patterns whose activity may vary from one to the next. This is consistent with D_{opt} being independent of p. Accordingly, the system need not predetermine the activity of the patterns, either during the learning or during the unlearning procedure. All this, including the random shooting, lends substantial support to the ideas of Crick and Mitchison [1.112].

As to the underlying mechanism, one can study [1.115] the J_{ij}, the local fields h_i, or the energy surface as D (=dreaming) proceeds. The change of the J_{ij} for $p \neq 0.5$ is quite explicit. As the network starts unlearning, the ferromagnetic state is dominant. After a while it has been deleted and new attractors, spurious states, appear, which are weeded out as well. The J_{ij} approach zero monotonically (sensible) and their variance also decreases; see Fig. 1.10a.

The zero-temperature dynamics is $S_i(t + \Delta t) = \text{sgn}[h_i(t)]$, where $h_i(t) = \sum_j J_{ij} S_j(t)$ is the local field at i. For a good retrieval we need $h_i^\mu \xi_i^\mu > 0$ for all μ; here h_i^μ has been defined in (5.7). When dreaming starts ($D = 0$) the h_i^μ are, as Fig. 1.10b shows, hopelessly inadequate since they cannot separate the states with $\xi_i = +1$ and the ones with $\xi_i = -1$. As D increases, a hole around $h = 0$ develops and, thus, autoassociative memory is restored.

Figure 1.11 displays the removal of the spurious states. Only some metastable states in the *direct* neighborhood of the individual patterns are left. Since the human eye (i.e., the human mind) is not very good in recognizing random patterns, it is more sensible to take the first eight letters A – H of our alphabet.

The letters in Fig. 1.12 are strongly correlated, as the reader can easily verify (they were not shifted with respect to each other), and their activity or, equiv-

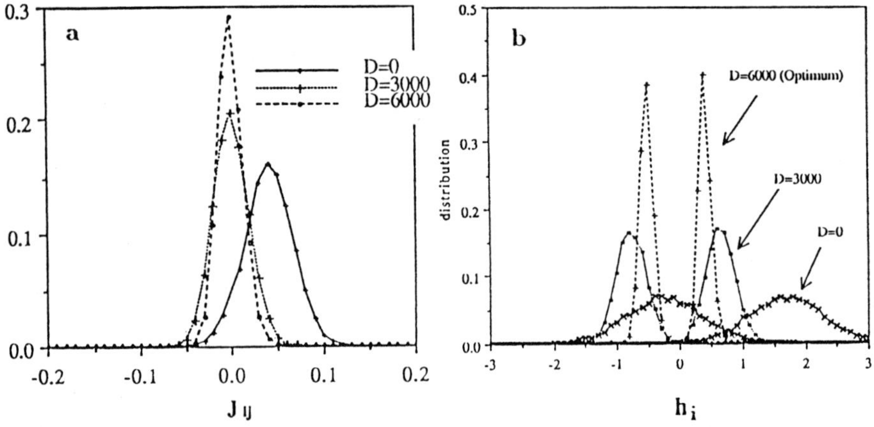

Fig. 1.10. (**a**) Distribution of the J_{ij} for different values of the total number of dreams D. Here $N = 400, q = 100, p = 0.7$, and $\varepsilon = 0.01$. The figure shows that the mean moves to zero (as it should) and the standard deviation decreases monotonically. $D_{\text{opt}} = 6000$. (**b**) Distribution of the local fields according to $\xi_i^\mu = +1$ (*right*) and $\xi_i^\mu = -1$ (*left*) for different values of D. Except for $p = 0.6$, the parameters are as in a

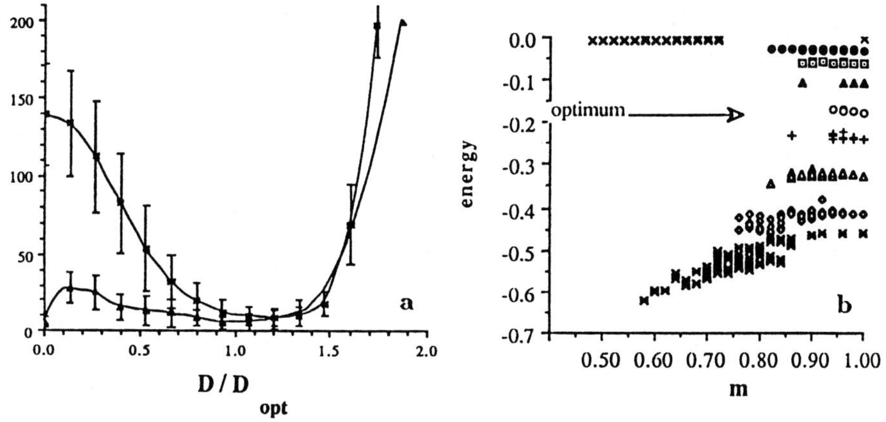

Fig. 1.11. (a) Number of parasitic states found in the vicinity of a pattern as a function of the total number of dreams D for N = 100 neurons and $\varepsilon = 0.01$. The two curves refer to $q = 50, p = 0.5$ and $q = 25, p = 0.7$. For $p \neq 0.5$, the ferromagnetic state (all spins up or all spins down) is eliminated first while at the same time the number of parasitic states increases. These are then weeded out as also happens in the case $p = 0.5$. Both curves have their minimum at D_{opt}. (b) Evolution of the energy landscape in the course of unlearning for $p = 0.5$. For various stages of the unlearning procedure, the energies of metastable states have been plotted as a function of their overlap m with a reference pattern. D increases from 0 (*bottom*) to 8000 (*top*) by steps of 1000; $D_{\text{opt}} \approx 4000$

Fig. 1.12. The first eight letters of the alphabet, A–H (*top*), have been stored in a network of N = 100 formal neurons. After unlearning, corrupted versions (*middle*) with about 12% noise are presented to the network, which then retrieves the original patterns (*bottom*). Note that the retrieved patterns closely resemble but need not coincide with their prototypes

alently, the fraction of black pixels varies (between 0.34 and 0.52). Two things are easily verified: (i) in spite of their strong correlation, the letters are retrieved well, and (ii) some additional attractors in their direct neighborhood are present; see in particular the letters F and G.

In a sense, D_{opt} is the (optimal) learning time. If one performs a numerical experiment on a serial computer, one has to realize that $N^2/2$ bonds have to be updated after each "dream" and that q is proportional to N. This gives N^3 as an estimate of the size dependence of the computer time needed – the very same estimate as for the perceptron algorithm (Sect. 1.5.3).

1.5.7 Forgetting

Why worry about forgetting? The simple answer is that the network should be able to "forget" so as to create space for new patterns and thereby avoid the blackout catastrophe that would occur if one stores more than $q = \alpha_c N$ patterns (Sects. 1.3.3 and 1.4.3). In this section we want to explore how forgetting can be modeled as an intrinsic property of the network. In psychology there are two viewpoints, *decay* and *interference*. Interference means that there are many data stored which interfere with each other. Forgetting then emerges as an intrinsic property of the network. The other viewpoint, decay, is that forgetting is just a matter of time, caused by a degrading of some chemical substance(s). If the patterns are labeled in such a way that the most recently stored pattern has number 1, then decay can be modeled simply by giving pattern n a weight $\varepsilon_n = \lambda^n$ with $\lambda < 1$. Weighted patterns have been treated in Sect. 1.3.3.

In modeling forgetting we would like to include (at least) three physiological requirements: (i) locality, preferentially *à la* Hebb; (ii) an encoding procedure which is *in*dependent of what memories have been stored previously; (iii) saturating synaptic efficacies, i.e., the $|J_{ij}|$ should remain bounded.

Both the decay and the interference viewpoint are included in the *iterative* prescription [1.57]

$$J_{ij}^{(\mu)} = \phi(\varepsilon_N \xi_i^\mu \xi_j^\mu + J_{ij}^{(\mu-1)}) , \tag{5.7.1}$$

where it is assumed that we store stationary, unbiased random patterns, $J_{ij}^{(\mu-1)}$ denotes the information up to and including pattern $\mu - 1$, and $\varepsilon_N \propto N^{-x}$ where $x \geq 0$ depends on the model, i.e., ϕ. The function ϕ is taken to be odd and concave for $x \geq 0$. We parenthetically note that (1) may be extended so as to include temporal sequences: if μ denotes a cycle, say, then $\xi_i^\mu \xi_j^\mu$ has to be replaced by $N\Delta J_{ij}(\tau)$ where $\Delta J_{ij}(\tau)$ is given by the Hebb rule (4.1).

We now list several examples. First, $\phi(x) = \lambda x$. If $\lambda = 1$, and $\varepsilon_N = \tilde{\varepsilon}/N$, then we recover the original Hopfield model (1.1.1). For $\lambda < 1$, we obtain the marginalist learning scheme of Toulouse et al. [1.54, 55]; see also Sect. 1.4.2. This is the decay viewpoint. Note that $\phi(x) = \lambda x$ does not saturate as $|x|$ becomes large. Second, we have the Hopfield–Parisi model with $\phi(x) = x$ for $|x| \leq 1$ and $\phi(x) = \text{sgn}(x)$ for $|x| \geq 1$ while $\varepsilon_N = (\tilde{\varepsilon}/N)^{1/2}$. The model has been proposed by Hopfield [1.56] and studied numerically by Parisi [1.57]. Third, $\phi(x) = \tanh(x)$ with $\varepsilon_N = \tilde{\varepsilon}/N$ is a typical representative of the family of ϕs which are odd and concave for $x \geq 0$. Note the different scaling of ε_N in these three families of examples.

The model (1) is solved as follows [1.58]. We rewrite (1) in terms of the general representation $J_{ij} = N^{-1} Q(\boldsymbol{\xi}_i; \boldsymbol{\xi}_j)$, where $Q(\boldsymbol{x}; \boldsymbol{y})$ is a synaptic kernel on $\mathbb{R}^q \times \mathbb{R}^q$ to be determined. We obtain that $J_{ij}^{(\mu)}$ follows from

$$Q(\boldsymbol{x};\boldsymbol{y}) = \phi(\varepsilon_N x_\mu y_\mu + \phi(\varepsilon_N x_{\mu-1} y_{\mu-1} + \ldots)) \,, \tag{5.7.2}$$

and note that Q satisfies the invariance condition (4.2.4) so that we can apply the general considerations of Sect. 1.4. In view of (4.3.10–12) we "only" have to determine the Λ_ϱ. The retrieval quality m_μ of pattern μ then follows from the fixed-point equation (4.3.12).

How do we determine the Λ_ϱ? This is done by mapping the original problem, finding the Λ_ϱ, onto determining the asymptotic behavior of a discrete dynamical system. The Λ_ϱ are given by (4.2.11),

$$\Lambda_\varrho = 2^{-q}\lambda_\varrho = 2^{-q}\sum_{\boldsymbol{x}} Q(\boldsymbol{e};\boldsymbol{x})v_\varrho(\boldsymbol{x}) \,, \quad v_\varrho(\boldsymbol{x}) = \prod_{i\in\varrho} x_i \,, \tag{5.7.3}$$

where $\boldsymbol{e} = (1, 1, \ldots)$. In view of (2), this suggests introducing the Markov chain

$$X_\mu = \phi(\varepsilon_N \eta_\mu + X_{\mu-1}) = \phi(\varepsilon_N \eta_\mu + \phi(\varepsilon \eta_{\mu-1} + X_{\mu-2})) = \ldots \,, \tag{5.7.4}$$

where the η_μ are independent, identically distributed random variables which assume the values ± 1 with equal probability. No doubt the η_μ in (4) are what the $\xi_i^\mu \xi_j^\mu$ with i and j fixed were in (1). If so, what is the initial condition in (1) when we start encoding the data? As stressed by Toulouse et al. [1.59], one should not take a *tabula rasa*, i.e., one should not start with $J_{ij} = 0$. In the present context this seems very natural: the organism has already gathered many impressions before the relevant data are stored. In technical terms, the Markov chain has already been iterated so long (q is so large) that it is in its *stationary state*; a random variable with the stationary distribution will be denoted by u. Colloquially, once the system is in the stationary state, it has forgotten its initial state completely.

Combining (3) and (4) we find for Λ_ϱ with $|\varrho| = 1$, which is an embedding strength of one of the patterns,

$$\Lambda_n = \langle \phi(\varepsilon_N \eta_n + \phi(\varepsilon_N \eta_{n-1} + \ldots \phi(\varepsilon_N \eta_0 + u)\ldots))\eta_0 \rangle_u \,. \tag{5.7.5}$$

We have now relabeled the ηs and exploited our assumption that the chain is in its stationary state so that $X_0 = u$. The labeling in (5) is such that $n = 1$ denotes the most recently stored pattern. The average over ξ_0 is easily performed, and we end up with

$$\Lambda_n = \tfrac{1}{2}\langle X_n(u+\varepsilon) - X_n(u-\varepsilon)\rangle_u \,, \tag{5.7.6}$$

where $X_n(y)$ is the state of the Markov chain which has started in y and performed n steps. The expression (6) makes clear that Λ_n is of the order of ε as $\varepsilon \to 0$. Since the Λ_n should remain of the order of one, we have to perform the rescaling $Q \to \varepsilon^{-1} Q$.

In the case of the Hopfield–Parisi model, the Markov chain (4) is an ordinary random walk on $[-1, 1]$ with step width ε_N and reflecting barriers at ± 1. The mathematics [1.117] is simplified by assuming $m\varepsilon_N = 1$ for some positive integer m. Then the stationary state gives the $(2m+1)$ accessible points equal weight, the average (6) can be performed, and [1.58]

$$\Lambda_n = \frac{8}{\pi^2}\varepsilon \sum_{s=1,3,5,\dots} s^{-2} \exp\left[-\left(\frac{\pi^2}{8}\varepsilon^2 n\right) s^2\right] . \tag{5.7.7}$$

Interestingly, Λ_n decreases *exponentially fast* in n as $n \to \infty$. Putting $t = \pi^2\varepsilon^2 n/8$, one can derive the estimate $\varepsilon^{-1}\Lambda_n \sim 1 - 4\sqrt{t}/\pi$ as $t \to 0$. The dependence upon ε in front of the sum makes explicit why one has to rescale Q so as to keep the Λ_n of order one as $N \to \infty$. Furthermore, the combination $\varepsilon^2 n$ appearing in the exponent in (7) suggests a scaling $\varepsilon_N = (\tilde{\varepsilon}/N)^{1/2}$ so as to store and retrieve a number of patterns proportional to N. In this way, therefore, Λ_n with $n = \alpha N$ remains *finite* as $N \to \infty$. It can be shown that the Λ_ϱ with $|\varrho| \neq 1$ are $o(\varepsilon)$ so that one can simplify $r(\hat{q})$ in close analogy to (4.3.14–16) and compute the storage capacity α_c and retrieval quality m_μ. Here the storage capacity is N^{-1} times the number of the most recently(!) stored patterns which can be retrieved satisfactorily. For a suitable choice of $\tilde{\varepsilon}$, the storage ratio can be as large as $\alpha_c = 0.04$ [1.58]; see Fig. 1.13. Since the patterns are not treated in a way which is permutation invariant, the retrieval quality m_μ is bound to depend on μ. From the point of view of theory, Parisi's numerics [1.57] was excellent since the theoretical predictions and the simulation results differed by no more than 2%.

The general case, represented by $\phi(x) = \tanh(x)$, is slightly more involved. It turns out, however, that Λ_n decreases exponentially fast in n as $n \to \infty$. The explanation [1.58] is that the Markov chain (4) is very strongly mixing (it is a K system) so that it "forgets" its past that fast.

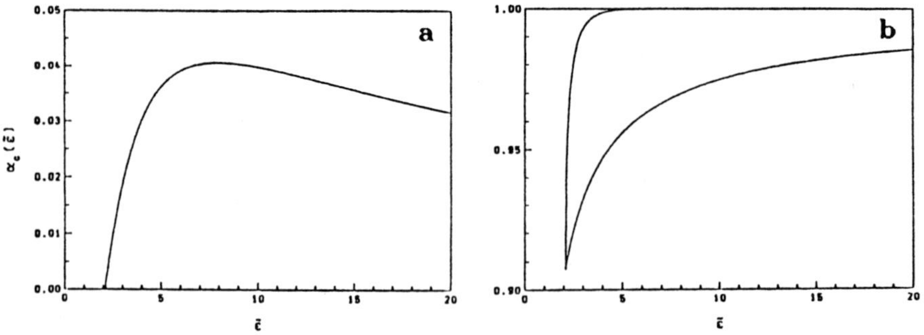

Fig. 1.13. (**a**) In the Hopfield–Parisi model, the last $\alpha_c N$ patterns can be remembered, whereas those beyond $\alpha_c N$ have been forgotten. Here α_c has been plotted as a function of $\tilde{\varepsilon}$ where, in (7.1), $\varepsilon_N = (\tilde{\varepsilon}/N)^{1/2}$. For $\tilde{\varepsilon} < \tilde{\varepsilon}_c$ no retrieval is possible. The retrieval quality of the most recently stored patterns (*upper curve*) and the ones on the verge of being forgotten (*lower curve*) have been plotted in (**b**). According to figure a (=theory), $\tilde{\varepsilon}_c = 2.09$, and a maximum $\alpha_c^{max} = 0.041$ is realized at $\tilde{\varepsilon}_{max} = 8.00$. These values are to be compared with Parisi's numerics [1.57]: $\tilde{\varepsilon}_c = 2.04$, $\alpha_c^{max} = 0.04$, and $\tilde{\varepsilon}_{max} = 8.16$. So theory and numerical simulation differ by no more than 2%

In all three cases, viz., $\phi(x) = \lambda x$ with $\lambda < 1$, the Hopfield–Parisi model, and $\phi(x) = \tanh(x)$ plus relatives, the upshot is that there exists a critical $\alpha_c \approx 0.04$ so that the most recently stored $\alpha_c N$ patterns can be retrieved at a low error rate whereas the ones beyond $\alpha_c N$ have been forgotten. Forgetting then means that the embedding strength Λ_n has become so small that the pattern is merged in the sea of noise generated by all the other patterns – a true interference phenomenon. The present theory also tells us something else. Namely, that decay, though a completely different mechanism from a psychological point of view, leads at the very end to the same mechanism of the memory's fading away by interference. The "interference" in the psychological sense is a truly intrinsic property of the network whereas decay is not. The performance of both is more or less identical [1.55, 58], though.

We finish this section with three remarks. First, one can also treat analytically a forgetful memory which is so highly diluted that dynamical correlations no longer occur [1.118]. Second, a *tabula rasa* assumption can be quite sensible if, for instance, one wants to model remembering a list of items. It is known from experimental psychology that the very first and the very last items of a list are remembered well whereas the ones in between fade away. One can use the ansatz (1) including a prefactor > 1 in front of $J_{ij}^{(\mu-1)}$ to reproduce this kind of behavior for *finite* N [1.119]. Second, one can also take care of Dale's law [1.120]: All synapses connected to a specific neuron are either excitatory or inhibitory. (There are exceptions to this rule, though.) Let η_j be +1 if j is taken to be excitatory and -1 if it is to be inhibitory. Then a simple model is defined by the iterative prescription

$$J_{ij}^{(\mu)} := J_{ij}^{(\mu)} \Theta(\eta_j J_{ij}^{(\mu)}) \,, \tag{5.7.8}$$

where the $J_{ij}^{(\mu)}$ on the right are given by (1) and $\Theta(x) = [\operatorname{sgn}(x) + 1]/2$. A slightly different model is obtained by defining $\phi(x) = 0$ for $x < 0$, $\phi(x) = x$ if $0 \le x \le 1$, and $\phi(x) = 1$ for $x > 1$, and putting

$$J_{ij}^{(\mu)} := \eta_j \phi\big(\eta_j [\varepsilon_N \xi_i^\mu \xi_j^\mu + J_{ij}^{(\mu-1)}]\big). \tag{5.7.9}$$

The Markov chain (4) induced by ϕ is an ordinary random walk on $[0, 1]$ with step size $\varepsilon_N = (\tilde{\varepsilon}/N)^{1/2}$ and reflecting barriers at 0 and 1. As before, changing ϕ means changing the Markov chain (4) in conjunction with ε_N. The dynamics of the models (8) and (9) is left to the reader as an interesting exercise. (Hint: exploit sublattices.) The prescription (9) is closely related to the "learning within bounds" algorithm of Toulouse et al. [1.54, 55, 59]; see in particular [1.54] for the numerical simulations. One might wish to invent a model with $J_{ij} = J_{ji}$ and solve it analytically. Do the theoretical predictions agree with the numerics of [1.54]? (Hint: you need not respect Dale's law literally.)

1.6 Hierarchically Structured Information

Most information presented to an organism has some internal, hierarchical structure. A somewhat academic but rather explicit example is a Linnean flora, which collects flowers that closely resemble each other in a family, gathers families into a class, classes into a category, and so on. The random patterns we have studied until now do not contain any internal structure except for their mutual overlaps $C_{\mu\nu} = N^{-1}\sum_i \xi_i^\mu \xi_i^\nu = a^2$ ($\mu \neq \nu$ and $N \to \infty$), where $a = \langle\xi\rangle$ is the activity. Since hierarchically structured information has attracted quite a bit of interest (see below) and is an intrinsically interesting notion, in this section we will analyze some basic aspects underlying it, paying specific attention to a version proposed by Parga and Virasoro.

In Sect. 1.6.1 we focus on the mathematical structures proper, in particular, on the notion of *martingale*, which is a key ingredient of most theories devoted to hierarchically structured information. A signal-to-noise-ratio analysis is presented in Sect. 1.6.2, the (thermo)dynamic stability problem is solved analytically and illustrated by some numerical simulations in Sect. 1.6.3, and the solution for a structure where the information layers have arbitrary weights is indicated in Sect. 1.6.4. In Sect. 1.6.5 we apply the previous considerations to networks with *non*vanishing activity and rederive (5.5.5). Finally, a discussion is presented in Sect. 1.6.6. For additional information the reader may consult Feigel'man and Ioffe [1.126].

1.6.1 Structured Information, Markov Chains, and Martingales

How can we model hierarchically structured information? To see how we can obtain a sensible answer to this question, we first analyze the essentials of the procedure which we have used in the Sects. 1.3,4.

The patterns to be stored in the couplings J_{ij} are N-bit words (ξ_i^μ; $1 \leq i \leq N$) which represent specific Ising-spin configurations. They are labeled by $1 \leq \mu \leq q$, with $q = \alpha N$, and one is usually interested in the case of extensively many patterns ($\alpha > 0$). In the Hopfield setup, the ξ_i^μ are taken to be *independent*, identically distributed random variables which assume the values ± 1 with equal probability. In other words, the ξ_i^μ are *unbiased.*

The local information available to neuron i is contained in the random vector $\boldsymbol{\xi}_i = (\xi_i^\mu;\ 1 \leq \mu \leq q)$. Following Hebb [1.15], we have required that J_{ij} be determined by $\boldsymbol{\xi}_i$ and $\boldsymbol{\xi}_j$ only so that $J_{ij} = N^{-1}Q(\boldsymbol{\xi}_i;\boldsymbol{\xi}_j)$ for some synaptic kernel Q on $\mathbb{R}^q \times \mathbb{R}^q$ (locality); cf. (1.3.1).

The requirement that the random variables ξ_i^μ be independent has been criticised by several authors [1.110, 122–124]. The main idea behind this critique is simple and rather convincing: the process of acquiring information is hierarchically structured and so are the errors. (An animal can confuse predator and prey only once in its life, whereas confusing two different kinds of prey is immaterial.)

One has to be careful, though. The independence of the random vectors $\boldsymbol{\xi}_i$, $1 \leq i \leq N$, can hardly be dropped. At best we could replace it by an ergodicity assumption [1.125], but we cannot go further than that as long as we want a well-defined free energy which does not depend on the specific (random) configuration we use. So we stick to independent, identically distributed $\boldsymbol{\xi}_i$, $1 \leq i \leq N$. The requirement, however, that the *components* ξ_i^μ, $1 \leq \mu \leq q$, be independent can be dropped. Phrased differently, the organization of the local data can be hierarchical.

In this chapter we study a modification of a "hierarchical" model proposed by Parga and Virasoro [1.124], which we think is by far the most interesting. The original motivation of these authors was rather eskatological, but we will introduce the hierarchy directly. For additional information, see Feigel'man and Ioffe [1.126]. In passing, we note that the data are stored in a *single* net, in contrast to proposals by Dotsenko [1.122] and Gutfreund [1.123].

We now turn to the definition of the local hierarchy [1.124], a Markov process. By the independence of the $\boldsymbol{\xi}_i$, it suffices to define the process at a single site, whose index will be dropped. See Fig. 1.14 for a genealogical tree of the hierarchy. Each generation is determined by its ancestors only. This is the Markov property. We start with a "godfather" ($k = 0$), who does nothing – as behooves a godfather. He will change his role and become "active" in Sect. 1.6.5, where we generalize the setup. The first generation ($k = 1$) consists of M_1 members and each of them has M_2 descendants, the second generation ($k = 2$). In the kth generation each member has M_{k+1} descendants, and so on until $k = K$. The assumption that all the members of a generation have the same number of descendents is convenient, not necessary. A member of the kth generation is represented by $\xi^k := \xi^{\mu_1,\ldots,\mu_j,\ldots,\mu_k}$, where for $1 \leq j \leq k$ the μ_j range between 1 and M_j. The pattern at level K and all their ancestors are ordered lexicographically; cf. Fig. 1.14.

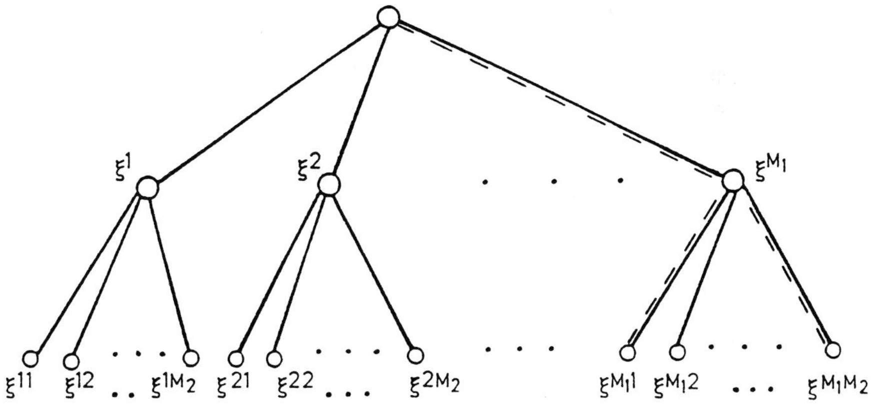

Fig. 1.14. Genealogical tree consisting of a godfather and two generations ($K = 2$). The states $\xi^{\mu\nu}$, the second generation (with $1 \leq \nu \leq M_2$), are descendants of ξ^μ, the first generation (with $1 \leq \mu \leq M_1$). The *dashed lines* refer to a branch of patterns and ancestors which are kept fixed in the replica method as explained in Sect. 1.6.3

Parga and Virasoro (PV) suggested the following distribution of the random variables ξ^k. They introduced numbers r_k such that $0 = r_0 < r_1 < \ldots < r_k < \ldots < r_K = 1$, allowed the values $\pm r_k$ for ξ^k, and required

$$\text{Prob}\,\{\xi^k|\xi^{k-1}\} = \tfrac{1}{2}[1 + \text{sgn}\,(\xi^k\xi^{k-1})r_{k-1}/r_k]\,. \tag{6.1.1}$$

$\text{Prob}\,\{\xi^k|\xi^{k-1}\}$ is the probability of getting ξ^k *given* ξ^{k-1}. Furthermore, given ξ^{k-1}, its descendants are sampled independently of each other. The Ising-spin configurations to be stored and retrieved are the $\xi_i^K = \pm 1$, where K stands for a specific $(\mu_1, \ldots, \mu_K)$ and $1 \le i \le N$.

Note that (1) is a *conditional* probability [1.125]. We condition with respect to the ancestor ξ^{k-1}. One easily verifies

$$\mathsf{E}\{\xi^k|\xi^{k-1}\} = \xi^{k-1}\,, \tag{6.1.2}$$

and since $(\xi^k)^2 = r_k^2$ we trivially get

$$\mathsf{E}\{(\xi^k)^2\} = r_k^2\,. \tag{6.1.3}$$

Here $\mathsf{E}\{\ldots\}$ denotes a mathematical expectation. The relation (2) says that the process we have defined is a *martingale* [1.127, 128]. To understand this valuable property we must make a small detour.

Martingale theory stems from stochastic considerations concerning gambling. Its relevance to probability theory was realized by Doob in the 1930s. Let $X_1, X_2, \ldots$ be a sequence of random variables describing a succession of games. We may think of X_k as the fortune of a gambler at time k. The game the gambler is playing is fair if

$$\mathsf{E}\{X_{k+1}|X_1, \ldots, X_k\} = X_k\,. \tag{6.1.4}$$

The point is that the expected gain on trial $k+1$ given the information up to and including trial k is $\mathsf{E}\{X_{k+1} - X_k|X_1, \ldots, X_k\} = X_k - X_k = 0$. This is what we intuitively expect from a fair game. For a Markov process, the conditional expectation simplifies and we get $\mathsf{E}\{X_{k+1}|X_1, \ldots, X_k\} = \mathsf{E}\{X_{k+1}|X_k\}$, i.e., X_{k+1} is determined by X_k only. Two aspects of the above example are worth noticing. As k proceeds we have an *in*creasing set of data, $(X_1, \ldots, X_k)$, and for a Markov process the martingale property is then expressed by (2).

Given the above martingale structure we can proceed to define the coupling constants,[16]

$$NJ_{ij} = \sum_{\mu\nu\varrho} \frac{(\xi_i^{\mu\nu\varrho} - \xi_i^{\mu\nu})(\xi_j^{\mu\nu\varrho} - \xi_j^{\mu\nu})}{1 - q_2} + \sum_{\mu\nu} \frac{(\xi_i^{\mu\nu} - \xi_j^{\mu})(\xi_j^{\mu\nu} - \xi_j^{\mu})}{q_2 - q_1} + \sum_{\mu} \frac{\xi_i^{\mu}\xi_j^{\mu}}{q_1}\,. \tag{6.1.5}$$

---

[16] Given (2) and (3), the pseudoinverse method (Sect. 1.5.2) also leads to (5); see Cortes et al. [1.129].

Here we took $K = 3$. The generalization to arbitrary K is evident. The q_k are overlaps of spin configurations, say μ and ν, in the lowest level K which have a *common ancestor* in the kth generation. So

$$
\begin{aligned}
\mu &= (\mu_1, \ldots, \mu_k, \mu_{k+1}, \ldots, \mu_K) \quad \text{and} \\
\nu &= (\mu_1, \ldots, \mu_k, \nu_{k+1}, \ldots, \nu_K) \quad \text{while} \\
q_k &= \lim_{N\to\infty} N^{-1} \sum_{i=1}^{N} \xi_i^\mu \xi_i^\nu = \mathsf{E}\{\xi^\mu \xi^\nu\}
\end{aligned}
\tag{6.1.6}
$$

with probability one [1.22]. In evaluating $\mathsf{E}\{\xi^\mu\xi^\nu\}$ one has to realize that averaging over ξ^μ and ξ^ν one first has to fix their ancestors $\mu' = (\mu_1, \ldots, \mu_{K-1})$ and $\nu' = (\mu_1, \ldots, \nu_{K-1})$ in layer $(K-1)$. By construction we then obtain, using the martingale property (2),

$$
\begin{aligned}
\mathsf{E}\{\xi^\mu\xi^\nu\} &= \mathsf{E}\{\mathsf{E}\{\xi^\mu\xi^\nu | \xi^{\mu'}, \xi^{\nu'}\}\} \\
&= \mathsf{E}\{\mathsf{E}\{\xi^\mu | \xi^{\mu'}\}\mathsf{E}\{\xi^\nu | \xi^{\nu'}\}\} \\
&= \mathsf{E}\{\xi^{\mu_1,\ldots,\mu_{K-1}} \xi^{\mu_1,\ldots,\nu_{K-1}}\} \, .
\end{aligned}
\tag{6.1.7}
$$

Continuing this process until we reach the common ancestor $\xi^{\mu_1,\ldots,\mu_k}$, we find by (3)

$$
q_k = \mathsf{E}\{\xi^\mu\xi^\nu\} = \mathsf{E}\{(\xi^{\mu_1,\ldots,\mu_k})^2\} = r_k^2 \, . \tag{6.1.8}
$$

In a similar vein one can show that the random variables $(\xi_i^{k+1} - \xi_i^k)/\sqrt{q_{k+1} - q_k}$ which occur in (5) all have mean zero and variance one, and that they are orthogonal. This is at the heart of the argument in Sect. 1.6.3 where the equivalence of the model (5) and the Hopfield model is established.

As it stands, (5) was not proposed by Parga and Virasoro [1.124]. They had substracted the empirical means $M_k^{-1}\sum_{\mu_k} \xi^{\mu_1,\ldots,\mu_{k-1}\mu_k}$ instead of the conditional expectation $\xi^{\mu_1,\ldots,\mu_{k-1}}$. As we will see shortly, precisely the latter leads, in conjunction with the martingale property, to a surprisingly simple mathematical structure. Furthermore, the conditional expectation is very natural since ξ^{k+1} is determined by ξ^k. We do not see any reason we should sample *all* the descendants of ξ^k so as to get the empirical mean. Family relations need not be that tight. Finally we note that as long as we work with Ising spins in the final layer, and thus with $\xi^K = \pm 1$, and use a prescription which intrinsically does not depend on k, we are bound to take the other ξ^k as dichotomic as well. This more or less singles out (1).

There is some freedom, though, in generating a genealogical tree. A more general procedure is the following [1.130]. Let $\mathcal{N}(a, b)$ denote the set of Gaussians with mean a and variance b. We start with sampling the ξ^1 from $\mathcal{N}(0, 1)$. Then, picking a specific ξ^1, we sample ξ^2 from $\mathcal{N}(\xi^1, 1)$. Next we pick a specific ξ^2 and sample ξ^3 from $\mathcal{N}(\xi^2, 1)$. And so on. By construction, each generation has the martingale property. The bonds J_{ij} are constructed as in (5) and the patterns to be stored are the $\operatorname{sgn}(\xi^k)$. Plainly, it is not necessary to take Gaussians. Any other (even) distribution will do as long as Prob $\{\xi^K = 0\} = 0$. In spite of that, we will stick to (1) throughout what follows.

In Sect. 1.6.3 we present an analytic solution to the free energy of the system (1) with extensively many patterns, i.e., with $q = \alpha N$ patterns where $\alpha > 0$. We show that at zero temperature [1.109, 110, 131]

$$\alpha_c^{\mathrm{PV}} = \alpha_c^{\mathrm{Hopfield}} = 0.14 \,. \tag{6.1.9}$$

More precisely, the fixed-point equations of the model (5) can be mapped exactly onto those of the Hopfield model with extensively many patterns, viz. (3.3.34), whatever the temperature. It is to be noted, though, that $q = \alpha N$ has to include not only the Ising spin configurations of the Kth layer but also *all the ancestors*, i.e., sgn (ξ_i^k), $1 \le k \le K$.

As a preparation for (9) we present a signal-to-noise ratio analysis in Sect. 1.6.2. The equivalence to the Hopfield model is shown in Sect. 1.6.3. There we derive the free energy, study the fixed-point equations which determine the order parameters, and establish the correspondence with the Hopfield model. Our results are illustrated and confirmed by some numerical simulation results which are presented at the end of Sect. 1.6.3.

1.6.2 Signal-to-Noise-Ratio Analysis

The minimal requirement for a functional memory is that the patterns be invariant under the dynamics, which is usually of the heat-bath type. A pattern is expected to have optimal stability at zero temperature. In this section we present a stability analysis at $T = 0$ under the simplifying assumption that the local field is a Gaussian random variable. In doing so we neglect intersite correlations.

A zero-temperature dynamics is deterministic. It is fully determined by the local field h_i experienced by the spin at site i,

$$S_i := \mathrm{sgn}\,(h_i)\,, \quad h_i = \sum_j J_{ij} S_j \,. \tag{6.2.1}$$

To simplify the notation we take $K = 2$. Then we have

$$N J_{ij} = \sum_{\mu,\nu} \frac{(\xi_i^{\mu\nu} - \xi_i^{\mu})(\xi_j^{\mu\nu} - \xi_j^{\mu})}{(1 - q_1)} + \sum_{\mu} \frac{(\xi_i^{\mu}\xi_j^{\mu})}{q_1} \,. \tag{6.2.2}$$

The question is: How many patterns can be stored in the J_{ij} so that a specific pattern, say $S_j = \xi_j^{\bar\mu\bar\nu}$, remains stable under the dynamics (1)? To answer this question we substitute (2) in (1), single out the terms $(\bar\mu, \bar\nu)$ and $\bar\mu$, and collect the rest in the primed sums below,

$$h_i = \xi_i^{\bar\mu\bar\nu} + N^{-1} \sum_{j \ne i} \times \left[\sum_{\mu,\nu}{}' \frac{(\xi_i^{\mu\nu} - \xi_i^{\mu})(\xi_j^{\mu\nu} - \xi_j^{\mu})}{(1 - q_1)} \xi_j^{\bar\mu\bar\nu} + \sum_{\mu}{}' \frac{(\xi_i^{\mu}\xi_j^{\mu})}{q_1} \xi_j^{\bar\mu\bar\nu} \right] , \tag{6.2.3}$$

which we rewrite as $h_i = \xi_i^{\bar{\mu}\bar{\nu}}(1 + \delta_i)$. The δ_i is a sum of independent random variables with mean zero, so it has a zero mean itself. Furthermore, δ_i is approximately Gaussian and all we have to do is calculate the average $\mathsf{E}\{\delta_i^2\}$ so as to find the variance. After some algebra we obtain a surprisingly simple result,

$$D \equiv \mathsf{E}\{\delta_i^2\} = N^{-1}[M_1 M_2 + M_1 - 2] \,. \tag{6.2.4}$$

If we put $M_1 M_2 = O(N)$ and allow both M_1 and M_2 to increase as a power of N, then (4) gives $N^{-1}(M_1 M_2) \equiv \alpha$, which is directly comparable to the Hopfield case.

It is even more surprising that the ancestor states $S_j = \xi_j^{\bar{\mu}}/\sqrt{q_1} = \xi_j^{\bar{\mu}}/r_1$ give rise to the very same result (4). A closer examination readily reveals what our strategy should have been. We have $M_1 M_2$ patterns in the second generation and M_1 ancestor patterns in the first, so altogether $q = M_1 M_2 + M_1$. If $q = \alpha N$, then the right-hand side of (4) leaves us with $D = \alpha$.

The proportionality factor α becomes "too large" if the event $\delta_i < -1$ occurs "too frequently". Its probability is

$$\text{Prob}\,\{\delta_i < -1\} = \frac{2}{\sqrt{\pi}} \int_{1/\sqrt{2D}}^{\infty} dy \exp(-y^2) = \text{erfc}\,(1/\sqrt{2D}) \,, \tag{6.2.5}$$

where erfc = 1 − erf is the complementary error function. We now discern two cases, neglecting intersite correlations.

First, the requirement that for a given pattern the probability of finding an error somewhere be less than one. That is, $N\text{Prob}\,\{\delta_i < -1\} < 1$ since there are N sites. As $N \to \infty$, the argument of the complementary error function in (5) has to become large. For large

$$x = \frac{1}{\sqrt{2D}}$$

we can replace erfc (x) by

$$\frac{1}{x\sqrt{\pi}} \exp(-x^2) \,.$$

Hence we find

$$\frac{1}{\sqrt{2\pi D}} \exp\left(-\frac{1}{2D}\right) < N^{-1} \tag{6.2.6}$$

and thus

$$D < [2 \ln N - \ln(2\pi D)]^{-1} \quad \Longrightarrow \quad \alpha < (2 \ln N)^{-1} \,. \tag{6.2.7}$$

Second, the requirement that for *all* patterns the probability of finding a single error be less than one. This gives $qN\text{Prob}\,\{\delta_i < -1\} < 1$ and thus using (6) once again we get

$$D < [4 \ln N + 2 \ln(\alpha) - \ln(2\pi D)]^{-1} \quad \Longrightarrow \quad \alpha < (4 \ln N)^{-1} \,. \tag{6.2.8}$$

Both (7) and (8) are identical with the requirements for α in the Hopfield case [1.13, 132], which suggests that also the extensive storage capacity of the two models agrees. In the next section it is shown that this is indeed the case.

1.6.3 Equivalence to the Hopfield Model

In this section we compute the free energy for a system with K generations of information (K arbitrary but finite) and extensively many patterns, and show how the fixed-point equations and the free energy can be reduced to their Hopfield equivalents. In the first subsection we compute the free energy itself, in the second we perform the reduction to the Hopfield model, and in the final subsection 6.3.c we show that also the stability criteria agree.

Free Energy. To compute the free energy $f(\beta)$ at inverse temperature β, we single out finitely many patterns and average over the remaining, extensively many patterns (plus their ancestors!) in the context of the replica method; cf. Sect. 1.2.5. The procedure starts with the relation

$$-\beta f(\beta) = \lim_{n\to 0} n^{-1}\phi(n)\,. \tag{6.3.1}$$

The function $\phi(n)$ is a real variable extension of

$$\phi(n) = \lim_{N\to\infty} N^{-1}\ln\langle Z_N^n\rangle\,, \quad n\in\mathbb{N}\,, \tag{6.3.2}$$

which is defined on the positive integers $\mathbb{N}$, with $Z_N = \mathrm{tr}\{\exp(-\beta H_N)\} = 2^{-N}\mathrm{Tr}\{\exp(-\beta H_N)\}$. Here and throughout what follows we often use $\langle\ldots\rangle$ instead of $\mathbb{E}\{\ldots\}$ to simplify the notation. The argument below mainly follows [1.109].

Before proceeding we fix some notation. First, we need a suitable representation of the J_{ij} in (5) for general K. Let us put

$$\tilde{\xi}^k = \frac{\xi^{\mu_1,\ldots,\mu_{k-1},\mu_k} - \xi^{\mu_1,\ldots,\mu_{k-1}}}{\sqrt{q_k - q_{k-1}}}\,, \tag{6.3.3}$$

where, by (1.8), $q_k = r_k^2$ and k stands for a multiindex $(\mu_1,\ldots,\mu_{k-1},\mu_k)$; cf. Fig. 1.14. If we write formally $1\le k\le K$, then this means that we sample

$$M_K\ldots M_2M_1 + M_{K-1}\ldots M_2M_1 + \ldots + M_2M_1 + M_1 = \alpha N \tag{6.3.4}$$

multiindices altogether. Because of (1.8) and the martingale property (1.2), the $\tilde{\xi}^k$ are normalized to one. Furthermore, if $k = (\mu_1,\ldots,\mu_k)$ and $l = (\nu_1,\ldots,\nu_l)$, then

$$\langle\tilde{\xi}^k\tilde{\xi}^l\rangle = \delta_{(\mu_1,\ldots,\mu_k),(\nu_1,\ldots,\nu_l)} \equiv \delta_{k,l}\,. \tag{6.3.5}$$

That is, the $\tilde{\xi}_k$ are orthonormal. This valuable property is a key to the argument below.

Using (3) and (4) we can write (1.5) as

$$NJ_{ij} = \sum_{1 \le k \le K} \tilde{\xi}_i^k \tilde{\xi}_j^k , \tag{6.3.6}$$

a form which already reminds us of the Hopfield model. The analogy would be perfect, *if the* $\tilde{\xi}^k$ *were independent.* By construction, however, they are not.

We now fix some patterns μ_K in the very last generation K. If we do, we also have to fix their ancestors, which are unambiguously defined. See again Fig. 1.14 and follow the dashed lines upwards. The patterns μ_K and their ancestors are all denoted by μ. There are *finitely* many of them, say s. The remaining, extensively many nodes of the genealogical tree are denoted by ν. In the context of the replica method, these have to be averaged. There are $\alpha N - s$ of them where αN is given by (4). If $\alpha > 0$, they will give rise to noise.

By (6), the Hamiltonian $H_N = (-1/2) \sum_{i,j} J_{ij} S_i S_j$ reappears as

$$H_N = -\frac{1}{2} \sum_{1 \le k \le K} \left(\frac{1}{\sqrt{N}} \sum_{i=1}^{N} \tilde{\xi}_i^k S_i \right)^2 , \tag{6.3.7}$$

and, as in Sects. 1.3.3 and 1.4.3, we split H_N up into $H_N^{(1)} + H_N^{(2)}$, where $H_N^{(1)}$ contains a sum over μ only and $H_N^{(2)}$ refers to the rest.

Instead of studying $\langle Z_N^n \rangle$ we note that we need only average the replicated $\exp(-\beta H_N^{(2)})$. The terms which then remain will turn out to be self-averaging. We have, from (7),

$$\left\langle \exp\left[-\beta \sum_{\sigma=1}^{n} H_N^{(2)}(\sigma) \right] \right\rangle = \left\langle \exp\left[\frac{\beta}{2} \sum_{\nu,\sigma} X_\sigma^2(\nu) \right] \right\rangle , \tag{6.3.8}$$

where, in close analogy to (3.3.5) and (4.3.3),

$$X_\sigma(\nu) = \frac{1}{\sqrt{N}} \sum_{i=1}^{N} \tilde{\xi}_i^\nu S_\sigma(i) \tag{6.3.9}$$

and the $1 \le \sigma \le n$ label the n replicas. The spin at site i is denoted by $S(i)$ instead of S_i so as to create room for the replica index σ.

As in Sects. 1.3.3 and 1.4.3, it is assumed that the $S_\sigma(i)$ live in μ-space and so are correlated with the μ-patterns, which are not averaged. We therefore take the $X_\sigma(\nu)$ as *Gaussian* random variables with mean zero and covariance matrix

$$\begin{aligned} C_{\nu\sigma;\nu'\sigma'} &= \langle X_\sigma(\nu) X_{\sigma'}(\nu') \rangle \\ &= N^{-1} \sum_{i=1}^{N} S_\sigma(i) S_{\sigma'}(i) \langle \tilde{\xi}_i^\nu \tilde{\xi}_i^{\nu'} \rangle \\ &= \delta_{\nu,\nu'} N^{-1} \sum_{i=1}^{N} S_\sigma(i) S_{\sigma'}(i) \equiv \delta_{\nu,\nu'} q_{\sigma\sigma'} . \end{aligned} \tag{6.3.10}$$

The nontrivial equality is the third one. The point is that owing to the martingale property (1.2) and the Markov structure the ξ^ν are orthonormal and (5) still holds, *despite the* ξ^μ*'s being kept fixed.*

The expression (10) tells us that the covariance matrix (3.3.6) of the Hopfield model and that of the hierarchical model (6) agree. The rest is plain sailing. Using (3.3.7) we obtain

$$\langle Z_N^n \rangle = \left\langle \mathrm{tr}_{S_\sigma} \left\{ \exp\left[-\beta \sum_\sigma H_N^{(1)}(\sigma) - \frac{1}{2}\alpha N \mathrm{Tr}\,\{\ln Q\} \right] \right\} \right\rangle , \tag{6.3.11}$$

where Q is defined by (3.3.10) with $\varepsilon_\nu \equiv 1$ and the noise term[17] $\mathrm{Tr}\,\{\ln Q\}$ follows from (3.3.11). The relevant order parameters are $q_{\sigma\sigma'}(\sigma \neq \sigma')$ and

$$m_{\mu\sigma} = N^{-1} \sum_{i=1}^{N} \xi_i^\mu S_\sigma(i) , \tag{6.3.12}$$

which generalizes the first equation of (3.3.13). Proceeding as in Sect. 1.3.3 we make the replica-symmetric ansatz $m_{\mu\sigma} = m_\mu$ and $q_{\sigma\sigma'} = \hat{q}$, and find in the limit $n \to 0$

$$\begin{aligned} -\beta f(\beta) = &-\frac{\beta}{2} \sum_\mu m_\mu^2 \\ &- \alpha\{\ln[1-\beta(1-\hat{q})] + \beta\hat{q}[1-\beta(1-\hat{q})]\} - \frac{1}{2}\beta^2\hat{q}r(1-\hat{q}) \\ &+ \left\langle \int \frac{\mathrm{d}z}{\sqrt{2\pi}} \mathrm{e}^{-z^2/2} \ln\{\cosh[\beta(x+\sqrt{\hat{q}r}z)]\} \right\rangle , \end{aligned} \tag{6.3.13}$$

where $r = r(\hat{q})$ is given by (3.3.33) and

$$x = \sum_\mu m_\mu \xi^\mu . \tag{6.3.14}$$

In addition, one has to choose those solutions of the fixed-point equations

$$m_\mu = \langle\langle \xi^\mu \tanh[\beta(x+\sqrt{\hat{q}r}z)]\rangle\rangle , \tag{6.3.15a}$$

$$\hat{q} = \langle\langle \tanh^2[\beta(x+\sqrt{\hat{q}r}z)]\rangle\rangle \tag{6.3.15b}$$

which maximize the free-energy functional (13). The double angular brackets denote an average both over the Gaussian z and the finitely many ξ^μ. We parenthetically note that we could have varied the branching ratio from site to site *without* altering the final result (13–15), provided αN in (11) includes both the patterns in the very last generation K and all their ancestors.

---

[17] To be precise, its prefactor should not be αN but $\alpha N - s$. Since s is finite, it is negligible compared to αN with $N \to \infty$.

Reduction to the Hopfield Model. We are interested in the thermodynamic stability of a *single* pattern, which may be ξ^K in the lowest level or $\mathrm{sgn}(\xi^k) = \xi^k/r^k$ in the upper levels ($k < K$). Here we will use repeatedly $r_k = \sqrt{q_k}$, which is (1.8). As in the previous subsection, a single index k stands for a multiindex $(\mu_1, \ldots, \mu_k)$. To get a consistent calculation of the free energy in a genealogical tree of K generations, we have to fix a *chain* $\{\xi^1, \ldots, \xi^k, \ldots, \xi^K\}$ which consists of K elements ($K < \infty$) and ranges between the godfather and the last generation.

For a chain we have to perform the substitution

$$-\frac{\beta}{2}\sum_{\mu} m_\mu^2 \quad \rightarrow \quad -\frac{\beta}{2}\sum_{k=1}^{K} m_k^2 \tag{6.3.16}$$

in (13) while

$$x = \sum_{k=1}^{K} m_k \frac{\xi^k - \xi^{k-1}}{\sqrt{q_k - q_{k-1}}} \,. \tag{6.3.17}$$

The K order parameters m_k, $1 \le k \le K$, satisfy the fixed-point equation

$$m_k = \left\langle\left\langle \frac{\xi^k - \xi^{k-1}}{\sqrt{q_k - q_{k-1}}} \tanh[\beta(x + \sqrt{\hat{q}r})z] \right\rangle\right\rangle \,. \tag{6.3.18}$$

Equations (17) and (18) replace (14) and (15a).

One now may ask: What solution to the fixed-point equations (18) characterizes a retrieval state? To answer this question we fix an l between 1 and K, assume $T = 0$, and put $S_i = \xi_i^l/r_l$. If our hypothesis holds that both the patterns and their ancestors are retrieval states, this is expected to be an excellent approximation to a ground state. The order parameters m_k correspond to

$$m_k = N^{-1}\sum_{i=1}^{N} \frac{\xi_i^k - \xi_i^{k-1}}{\sqrt{q_k - q_{k-1}}} S_i \,, \quad 1 \le k \le K \tag{6.3.19}$$

with N large ($N \rightarrow \infty$). To evaluate (19) with $S_i = \xi_i^l/\sqrt{q_l}$ we have to discern two cases: $k \le l$ and $k > l$. Using the martingale property (1.2) and the Markov structure we obtain with probability one [1.22]

$$\begin{aligned} m_k &= \sqrt{\left(\frac{q_k - q_{k-1}}{q_l}\right)} \,, \quad k \le l \\ m_k &= 0 \,, \quad k > l \,. \end{aligned} \tag{6.3.20}$$

Equation (20), which is only an approximation, suggests that the appropriate ansatz is

$$\begin{aligned} m_k &= \sqrt{\left(\frac{q_k - q_{k-1}}{q_l}\right)} m \,, \quad k \le l \\ m_k &= 0, \quad k > l \,, \end{aligned} \tag{6.3.21}$$

where m is a parameter which still has to be determined. By the martingale property, the ansatz (21) is consistent with m_k vanishing in (18) for $k > l$.

Inserting (21) into (16) we get a telescoping sum,

$$-\frac{\beta}{2}\sum_{k=1}^{l} m_k^2 = -\frac{\beta}{2}m^2 \sum_{k=1}^{l} \frac{q_k - q_{k-1}}{q_l} = -\frac{\beta}{2}m^2 \, . \tag{6.3.22}$$

Through the same principle, $x = \xi^l/\sqrt{q_l}m$, where $\xi^l/\sqrt{q_l} = \pm 1$ and m is determined by ($1 \le k \le l$)

$$\begin{aligned}
m &= \sqrt{\frac{q_l}{q_k - q_{k-1}}}\left\langle\left\langle \frac{(\xi^k - \xi^{k-1})}{\sqrt{q_k - q_{k-1}}} \tanh\left[\beta\left(\frac{\xi^l}{\sqrt{q_l}}m + \sqrt{\hat{q}r}z\right)\right]\right\rangle\right\rangle \\
&= (q_k - q_{k-1})^{-1}\left\langle\left\langle (\xi^k - \xi^{k-1})\xi^l \tanh\left[\beta\left(m + \frac{\xi^l}{\sqrt{q_l}}\sqrt{\hat{q}r}z\right)\right]\right\rangle\right\rangle \\
&= (q_k - q_{k-1})^{-1}\langle \xi^l(\xi^k - \xi^{k-1})\rangle \int \frac{\mathrm{d}z}{\sqrt{2\pi}} \exp\left(-\frac{z^2}{2}\right) \tanh[\beta(m + \sqrt{\hat{q}r}z)] \\
&= \int \frac{\mathrm{d}z}{\sqrt{2\pi}} \exp\left(-\frac{z^2}{2}\right) \tanh[\beta(m + \sqrt{\hat{q}r}z)] \, .
\end{aligned} \tag{6.3.23}$$

See also (3.4.1) for a single μ. A moment's reflection then suffices to realize that the ansatz (21) directly leads to the free energy of the Hopfield model, viz., (3.3.32) with $\varepsilon_\mu = \varepsilon_\nu = 1$.

Stability and Numerical Simulation. The proof is finished by showing that (21) gives rise to at least a local *maximum* of the free-energy functional (13). This means that its second derivative, a matrix, has to be negative-definite. Let us call this matrix D. Its $(\hat{q}, \hat{q})$ element $D_{\hat{q}\hat{q}}$ is as in the Hopfield case and need not be given explicitly. Since r is a given function of $\hat{q}$ let us put $\sqrt{\hat{q}r} = g(\hat{q})$. For a state characterized by (21) the mixed $(m, \hat{q})$ elements are

$$\begin{aligned}
&\beta^2 g'(\hat{q})\left\langle\left\langle z \cosh^{-2}\left[\beta\left(\frac{\xi^l}{\sqrt{q_l}}m + \sqrt{\hat{q}r}z\right)\right] \frac{\xi^k - \xi^{k-1}}{\sqrt{q_k - q_{k-1}}}\right\rangle\right\rangle \\
&= \beta^2 g'(\hat{q}) \int \frac{\mathrm{d}z}{\sqrt{2\pi}} \exp\left(-\frac{z^2}{2}\right) \\
&\quad \times z \cosh^{-2}[\beta(m + \sqrt{\hat{q}r}z)] \left\langle \frac{\xi^l}{\sqrt{q_l}} \frac{\xi^k - \xi^{k-1}}{\sqrt{q_k - q_{k-1}}}\right\rangle
\end{aligned} \tag{6.3.24}$$

Here l is fixed and refers to the ansatz (21). We rewrite (24) as $A_k = F(\beta, \hat{q})a_k$, where

$$a_k = \left\langle \frac{\xi^l}{\sqrt{q_l}} \frac{\xi^k - \xi^{k-1}}{\sqrt{q_k - q_{k-1}}}\right\rangle , \quad 1 \le k \le K \, , \tag{6.3.25}$$

is given by (20). Finally, the (m, m) matrix elements $D_{kk'}$ vanish if $k \ne k'$. The diagonal elements ($k = k'$) all equal

$$A = -\beta\left\{1 - \beta\int \frac{dz}{\sqrt{2\pi}} \exp\left(\frac{-z^2}{2}\right)\cosh^{-2}[\beta(m + \sqrt{\hat{q}r}z)]\right\} < 0\,. \quad (6.3.26)$$

Collecting terms we obtain

$$D = \left(\begin{array}{ccc|c} A & & 0 & A_1 \\ & \ddots & & \vdots \\ 0 & & A & A_K \\ \hline A_1 & \cdots & A_K & D_{qq} \end{array}\right). \quad (6.3.27)$$

In the Hopfield case with K patterns we would have obtained the same matrix *except* that $a_k = \delta_{kl}$ for a fixed l which refers to the pattern under consideration. Through a direct calculation one verifies that the characteristic equation for D is

$$(A - \lambda)^{K-1}[(D_{\hat{q}\hat{q}} - \lambda)(A - \lambda) - \left(\sum_k a_k^2\right) F^2(\beta, \hat{q})] = 0\,. \quad (6.3.28)$$

Since $\sum_k a_k^2 = 1$, this is *precisely* the characteristic equation for the Hopfield case. As for the retrieval states, the equivalence to the Hopfield model has been established.

The above theoretical predictions have been confirmed by numerical simulations on a two-layered structure [1.109]. Theory predicts that $\alpha N = M_1 M_2 + M_1$ patterns can be stored as long as $\alpha \leq \alpha_c = 0.14$. The retrieval quality m decreases from $m = 1$ at $\alpha = 0$ to $m = 0.97$ at $\alpha = \alpha_c$.

Figure 1.15a is a plot of the theoretical retrieval quality. Given (M_1, M_2), we assign to this point the number $M_1 M_2 + M_1$ as long as it is less than or equal to $\alpha_c N$ for some N. In the numerical work N was taken to be 100. Connecting points where $M_1 M_2 + M_1$ is constant we get a contour line. If this constant exceeds $\alpha_c N$, then the retrieval quality drops to zero and no retrieval is possible. Zero contour lines have not been indicated.

Real-life data, such as Fig. 1.15b, look slightly different. N is finite ($N = 100$) and beyond $\alpha_c N$ there is a remanent magnetization, so that we find a plateau beyond which the retrieval quality decreases again, though not as steeply as one might expect theoretically. The variance of the empirical data at the plateau level is about two, so the agreement between the two figures is quite satisfying. It fully confirms the fact that both the patterns *and their ancestors* are stored.

1.6.4 Weighted Hierarchies

One of the ideas behind the concept of information (software) hierarchy was that categorization of data is an important aspect of memory acquisition and that, if the original pattern cannot be perceived, at least the ancestor can be recognized (=retrieved) – a hierarchy of errors [1.124]. However, (3.6) tells us that all generations have equal weight, and (3.2) and (3.28) show that certain sums are telescoping and that they all reduce to expressions which describe a

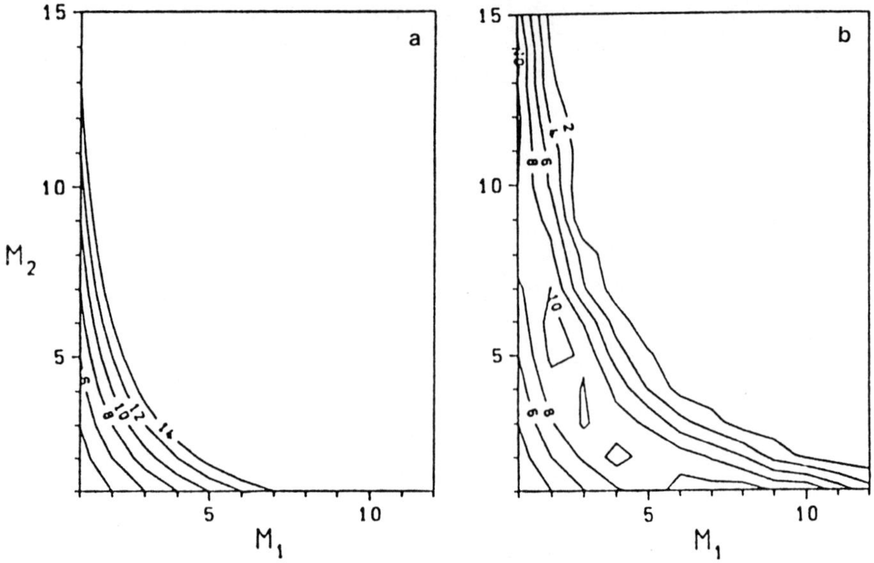

Fig. 1.15a, b. Contour lines indicating the number of patterns which are safely retrieved in a system of size $N = 100$. The information hierarchy consists of two generations (cf. Fig. 1.13). (**a**) Theoretical prediction: beyond $M_1M_2 + M_1 = 0.14N = 14$ no pattern can be retrieved. (**b**) Empirical retrieval quality for the same system: at each data point (M_1, M_2) the mean number of retrieved states, an average over 20 samples, has been indicated and points with the same value have been connected by a contour line

single pattern in the Hopfield model amidst αN other ones. Both the patterns and their ancestors *all* become stable or unstable at the very same moment so that a hierarchy of errors cannot occur as long as we treat patterns and ancestors on an equal footing. To get different behavior for different layers $1 \le k \le K$ we therefore have to give each layer or even each node k its own weight ε_k.

What changes occur in the considerations of the previous section if instead of (3.6) we put

$$NJ_{ij} = \sum_{1 \le k \le K} \varepsilon_k \tilde{\xi}_i^k \tilde{\xi}_j^k \tag{6.4.1}$$

for k a multiindex as in (3.3)? The answer is: hardly any. We can directly use the algebra of Sect. 1.3.3 that was devised for the generalized Hopfield model with weighted patterns. The noise is given by (3.3.9–11), and the free energy and the fixed-point equations follow from (3.13–15) provided we replace x in (3.14) by

$$x = \sum_{\mu} \varepsilon_\mu m_\mu \tilde{\xi}^\mu \, . \tag{6.4.2}$$

We are done. As, for instance, [1.109] and the discussion in [1.126: Sect. 4] show, one can give each generation a suitable weight so that, when the temperature is raised from zero, first the original patterns and then their ancestors, i.e., the classes, are recognized. Thus a recognition hierarchy can be realized. If "temperature" stands for general activity of the network, as it is induced, e.g., by fright, then this kind of behavior is even quite realistic.

1.6.5 Low-Activity Patterns

We have seen in Sect. 1.5.1 that for patterns with a nonvanishing activity $a = \langle\xi\rangle = 2p - 1$ the Hebb rule (1.1.1) does not allow an extensive storage capacity. We are now well-prepared to (re)derive a set of couplings, viz. (5.5.5), which generalize (1.1.1) to nonzero a and give a storage capacity which is identical to $\alpha_c^{\text{Hopfield}} = 0.14$. The argument below is due to Bös [1.109].

Until now the "godfather" in Fig. 1.14 did what it behooved a godfather to do, namely, nothing. Accordingly, both the patterns and their ancestors had *on the average* a vanishing activity,

$$\mathbb{E}\{\xi^k\} = \mathbb{E}\{\xi^k|\xi^{k-1}\} = \mathbb{E}\{\xi^{k-1}\} = \cdots = \mathbb{E}\{\xi^1\} = 0 \,. \tag{6.5.1}$$

Here we have used (1.1), the martingale property (1.2), and the fact that $r_0 = 0$. For the first generation, (1.1) then gives $\text{Prob}\{\xi^1 = \pm r_1\} = (1/2)[1 + \text{sgn}(\xi^1)r_0/r_1] = 1/2$ and, hence, $\mathbb{E}\{\xi^1\} = 0$. Note, however, that in a *specific* chain there is a preference to keep the spins parallel. That is, we have for $k > 1$

$$\begin{aligned} \text{Prob}\{+|+\} = \text{Prob}\{-|-\} &= \frac{1}{2}\left(1 + \frac{r_{k-1}}{r_k}\right) > \frac{1}{2}\,, \\ \text{Prob}\{+|-\} = \text{Prob}\{-|+\} &= \frac{1}{2}\left(1 - \frac{r_{k-1}}{r_k}\right) < \frac{1}{2}\,. \end{aligned} \tag{6.5.2}$$

For biased patterns, $\mathbb{E}\{\xi^1\} = a \neq 0$. This suggests that the godfather should become active. Reminding the reader that $\xi^k = \pm r_k$ we put $\xi^0 = r_0 = a$. This is the bias since ξ^0 is fixed. If we just want to study low-activity patterns, we take $K = 1, r_0 = a$, and $r_K = r_1 = 1$. So we only have the (god)father in layer 0 and the patterns $1 \le \mu \le q$ in the first generation. Then, from (1.1), we get

$$\text{Prob}\{\xi_i^\mu\} = \text{Prob}\{\xi_i^\mu|\xi_i^0\} = \frac{1}{2}\left(1 + \xi_i^\mu \frac{a}{r_k}\right) = \frac{1}{2}(1 + \xi_i^\mu a)\,, \tag{6.5.3}$$

which is the probability distribution for biased patterns with $\langle\xi\rangle = a$. Furthermore, since by (3.3)

$$\tilde{\xi}_i^0 = \frac{\xi_i^0}{a} = 1\,, \quad \tilde{\xi}_i^\mu = \frac{\xi_i^\mu - \xi_i^0}{\sqrt{1-a^2}} = \frac{\xi_i^\mu - a}{\sqrt{1-a^2}}\,, \tag{6.5.4}$$

(3.6) gives

$$NJ_{ij} = \sum_\mu \frac{(\xi_i^\mu - a)(\xi_j^\mu - a)}{1-a^2} + 1\,, \tag{6.5.5}$$

which is (5.5.5), as announced. We can repeat the arguments of Sect. 1.6.3. The only difference is that we now have an extra pattern, ($\xi_i^0 = 1$; $1 \le i \le N$), which is the ferromagnetic ground state belonging to (5), and an extra order parameter ($N \to \infty$)

$$m_0 = N^{-1} \sum_{i=1}^{N} \xi_i^0 S_i = N^{-1} \sum_{i=1}^{N} S_i \,, \tag{6.5.6}$$

which is the magnetization; cf. (3.12). Since it represents the (god)father, it always occurs in (3.14), which now reads

$$x = m_0 + \sum_{\mu} m_\mu \xi^\mu \,. \tag{6.5.7}$$

Incorporating m_0 in (3.13) and (3.15) we are done. From (3.4), the storage capacity is

$$\alpha_c^{\text{bias}} = \frac{q+1}{N} = \alpha_c^{\text{Hopfield}} = 0.14 \,, \tag{6.5.8}$$

whatever a *is*. This result has been confirmed by numerical simulations [1.109]. In passing we note that the physical meaning of the godfather in his new role has already been eludicated by the signal-to-noise-ratio analysis (5.5.6). A similar analysis shows that dropping the ferromagnetic 1 from (5) leads to

$$\alpha_c(a) \approx (1 - a^2)\alpha_c^{\text{Hopfield}} \le \alpha_c^{\text{Hopfield}} \,, \tag{6.5.9}$$

the inequality being strict for $a \neq 0$. To compensate for that, Amit et al. [1.112] introduced soft and rigid constraints. The alternative (5) seems simpler and also more physical since it precisely makes up for the $-a^2$ in $(1 - a^2)$ of (9).

1.6.6 Discussion

Do the considerations of the present chapter contribute to our understanding of certain phenomena as they occur in nature? The answer seems to be yes. There is, for instance, the syndrome *prosopagnosia*, which is related to the visual cortex and has been interpreted recently as a malfunctioning in the categorization process [1.133]. It can be defined as an impairment in identifying one individual among those that are visually similar to it and therefore can be said to belong to the same class. The patient can, say, recognize a car as such but cannot determine it any further. So the patient can determine the category to which a stimulus belongs but is unable to recognize the individual. (Imagine how frustrating it is to confuse a VW Beetle and a Porsche.) A very interesting explanation of this syndrome has been put forward by Virasoro [1.134].

Through (5.2) we have seen that going from one generation to the next the (+|+) and (−|−) combinations which "confirm the class" (C) are more probable than the (+|−) and (−|+) combinations, which can be said "to confirm the individual" (I). Virasoro has shown that there is also an asymmetry in the distribution of the local fields h_i. The fields on the I sites are more concentrated at small values of h and hence more sensitive to a destruction of synapses, e.g. by lesion, than those at C sites, which are more stable. A severe destruction of the synapses leads to *agnosia* (nothing can be identified) whereas a moderate destruction induces *prosopagnosia* in that the local I fields, which should confirm the individual, have been shifted towards a C value whereas those at the C sites still have the correct sign. Hence the class can be recognized but the individual cannot.

Dotsenko [1.122] and Gutfreund [1.123] work with hardware instead of software hierarchies. Bearing in mind Sect. 1.6.5, the Gutfreund idea can be explained fairly easily. The notion of biased pattern is generalized to a layered structure through the prescription

$$\text{Prob}\,\{\xi^k|\xi^{k-1}\} = \tfrac{1}{2}(1 + \xi^k\xi^{k-1}a_{k-1})\,, \tag{6.6.1}$$

where $0 < a_{k-1} < 1$ for $1 \le k \le K$ and all the ξs are ± 1. One directly verifies that $\mathsf{E}\{\xi^k|\xi^{k-1}\} = a_{k-1}\xi^{k-1}$. The couplings for layer k are therefore taken to be $\tilde{\xi}_k = (\xi^k - a_{k-1}\xi^{k-1})/(1 - a_{k-1}^2)^{1/2}$. However, (1) does not represent a martingale. In addition, there is no ferromagnetic term either so that the dynamics has to be constrained to fix the activity and hence keep the storage capacity at a reasonable value; otherwise the poor performance (5.9) would occur. Since the activity is bound to vary from generation to generation, Gutfreund had to store each generation *in its own net*, tune the constraint for each net *separately*, and assume a ferromagnetic feedforward coupling from site i in net $k-1$ to site i in net k for all $1 \le i \le N$ and all $1 \le k \le K$. This gives rise to another drawback: to guarantee retrieval in the final generation, the first one is hardly loaded because (see Fig. 1.14) there are always more progeny than ancestors. Otherwise the notion of hierarchy is without meaning.

A different hardware approach to hierarchically structured information has been proposed by Dotsenko [1.122]. In his work, the hierarchy is implemented in a purely geometrical way which is reminiscent of a multidimensional version of Dyson's hierarchical model [1.136]. The lowest level consists of blocks of spins, each block constituting a Hopfield model by itself. For each block we get a block spin. At the next level, the block spins are grouped together so as to form another set of blocks, each of them being a Hopfield model by itself – and so on. According to a signal-to-noise-ratio analysis as quoted by Gutfreund [1.123], the associative capabilities of this model seem to be somewhat restricted. We think, though, that further work is needed.

1.7 Outlook

Stepping back for an overview, we see that considerable insight has been gained in understanding the dynamical behavior, both storage and retrieval, of a *single* net. If one couples different nets so as to form a feedforward structure, then each network is a bottleneck by itself. So one should understand its dynamics *per se* before embarking on a multilayer structure [1.138].

One of the exciting aspects of a real neural network (in some circles, a neuronal network) is its massively parallel operation. The sequential and parallel dynamics which we have studied so extensively, useful as they are to gain new theoretical insight, do not seem to be realistic: the sequential dynamics because it is too slow, the parallel dynamics because a simultaneous updating of all the neurons is out of question. A more realistic dynamics consists in an independent (sequential) updating of different, disjoint groups of neurons which communicate more or less randomly. It can be shown that this kind of dynamics can combine convergence to a fixed point with fast, because massively parallel, operation [1.139].

The level of activity in the brain is certainly much lower than what was assumed in the original Hopfield model. Furthermore, it seems reasonable to respect Dale's law to some extent [1.140] and assume that only excitatory synapses can learn [1.141, 142]. All this alters the setup but it does not change the overall philosophy nor does it require a drastic modification of the techniques, ideas, and notions expounded in the previous sections.

Summarizing, through the "physical" input, i.e., most of the references [1.1–144], the *collective* aspects of neural networks, though long overlooked, have finally got the attention they deserved. In this way new insight has been gained into an old problem, viz., understanding at least some formal aspects of the brain – a problem that will continue to fascinate us for some time to come.

Acknowledgments. The present manuscript is an extended version of lectures which one of the authors (J.L. v.H.) delivered at the Sektion Physik der Ludwig-Maximilians-Universität München during the spring term of 1989. It is a great pleasure to thank Professor H. Wagner for his hospitality and the stimulating atmosphere at his department. We also would like to thank our collaborators S. Bös, A. Herz, B. Sulzer, and M. Vaas for their help and advice. Part of this work has been supported by the Deutsche Forschungsgemeinschaft (via the SFB 123).

References

1.1 N. Wiener, *Cybernetics* (Wiley, New York, and Hermann, Paris, 1948)
1.2 J. J. Hopfield, Proc. Natl. Acad. Sci. USA **79**, 2554–2558 (1982)
1.3 T. E. Posch, USCEE report **290** (1968)
1.4 M. Minsky and S. Papert, *Perceptrons: An Introduction to Computational Geometry* (MIT Press, Cambridge, Mass., 1969) An expanded 2nd edition appeared in 1988. This book is a gold mine of insight.
1.5 J. J. Hopfield, Proc. Natl. Acad. Sci. USA **81**, 3088–3092 (1984)
1.6 W. S. McCulloch and W. Pitts, Bull. Math. Biophys. **5**, 115–133 (1943)

1.7 K. Huang, *Statistical Mechanics* (Wiley, New York, 1963); a 2nd edition appeared in 1987
1.8 P. Peretto, Biol. Cybern. **50**, 51–62 (1984)
1.9 K. Binder, in *Monte Carlo Methods in Statistical Physics*, edited by K. Binder (Springer, Berlin, Heidelberg, 1979) pp. 1–45
1.10 R. J. Glauber, J. Math. Phys. **4**, 294–307 (1963)
1.11 N. Metropolis, A. W. Rosenbluth, M. N. Rosenbluth, A. H. Teller, and E. Teller, J. Chem. Phys. **21**, 1087–1092 (1953)
1.12 H. E. Stanley, *Introduction to Phase Transitions and Critical Phenomena* (Oxford University Press, New York, 1971)
1.13 D. J. Amit, H. Gutfreund, and H. Sompolinsky, (a) Phys. Rev. A **32**, 1007–1018 (1985); (b) Phys. Rev. Lett. **55**, 1530–1533 (1985); (c) Ann. Phys. (N.Y.) **173**, 30–67 (1987)
1.14 J. J. Hopfield and D. W. Tank, Science **233**, 625–633 (1986)
1.15 D. O. Hebb, *The Organization of Behavior* (Wiley, New York, 1949) p. 62
1.16 J. L. van Hemmen and R. Kühn, Phys. Rev. Lett. **57**, 913–916 (1986)
1.17 J. L. van Hemmen, D. Grensing, A. Huber, and R. Kühn, J. Stat. Phys. **50**, 231–257 and 259–293 (1988)
1.18 S. R. S. Varadhan, *Large Deviations and Applications* (Society for Industrial and Applied Mathematics, Philadelphia, PA, 1984). This work has become a classic.
1.19 O. E. Lanford, "Entropy and equilibrium states in classical statistical mechanics", in *Statistical Mechanics and Mathematical Problems*, edited by A. Lenard, Lecture Notes in Physics, Vol. **20** (Springer, New York, Berlin, Heidelberg, 1973) pp. 1–113. This elegant paper was seminal to, e.g., Refs. [1.20] and [1.21].
1.20 R. S. Ellis, *Entropy, Large Deviations, and Statistical Mechanics* (Springer, New York, Berlin, Heidelberg, 1985); Ann. Prob. **12**, (1984)
1.21 J. L. van Hemmen, "Equilibrium theory of spin glasses: Mean field theory and beyond", in *Heidelberg Colloquium on Spin Glasses*, edited by J. L. van Hemmen and I. Morgenstern, Lecture Notes in Physics, Vol. **192** (Springer, New York, Berlin, Heidelberg 1983), in particular, the Appendix; "The theory of large deviation and its applications in statistical mechanics", in *Mark Kac Seminar on Probability and Physi3cs, Syllabus 1985–1987*, edited by F. den Hollander and R. Maassen, CWI Syllabus Series No. 17 (CWI, Amsterdam, 1988) pp. 41–47
1.22 J. Lamperti, *Probability* (Benjamin, New York, 1966)
1.23 A. C. D. van Enter and J. L. van Hemmen, Phys. Rev. A **29**, 355–365 (1984)
1.24 See, for instance, Ref. [1.12, Sect. 1.6.5]
1.25 N. G. de Bruyn, *Asymptotic Methods in Analysis*, 2nd Edition (North-Holland, Amsterdam, 1961) Sect. 1.4.2; a Dover edition has been published recently.
1.26 J. L. van Hemmen, Phys. Rev. Lett. **49**, 409–412 (1982); J. L. van Hemmen, A. C. D. van Enter, and J. Canisius, Z. Phys. B **50**, 311–336 (1983)
1.27 A. W. Roberts and D. E. Varberg, *Convex Functions* (Academic, New York, 1973)
1.28 J. L. van Hemmen, Phys. Rev. A **34**, 3435–3445 (1986)
1.29 J. L. van Hemmen, D. Grensing, A. Huber, and R. Kühn, Z. Phys. B **65**, 53–63 (1986)
1.30 J. L. van Hemmen and R. G. Palmer, J. Phys. A: Math. Gen. **12**, 3881–3890 (1986)
1.31 D. Grensing and R. Kühn, J. Phys. A: Math. Gen. **19**, L1153–L1157 (1986)
1.32 J. L. van Hemmen and R. G. Palmer, J. Phys. A: Math. Gen. **12**, 563–580 (1979)
1.33 In fact, as is also discussed at length in Ref. [1.32], in practical work the extension is not unique.
1.34 C. M. Newman, Neural Networks **1**, 223–238 (1988)
1.35 J. L. van Hemmen and V. A. Zagrebnov, J. Phys. A: Math. Gen. **20**, 3989–3999 (1987)
1.36 J. L. van Hemmen would like to thank M. Bouten (LUC, Diepenbeek) for his insistence on physical transparence
1.37 L. Breiman, *Probability* (Addison-Wesley, Reading, Mass., 1968) Sects. 11.3 and 11.4, including problem 11.6
1.38 D. Grensing, R. Kühn, and J. L. van Hemmen, J. Phys. A: Math. Gen. **20**, 2935–2947 (1987)
1.39 J. L. van Hemmen, Phys. Rev. A **36**, 1959–1962 (1987)
1.40 J. Marcinkiewicz, Sur une propriété de la loi de Gauss, Math. Z. **44**, 612–618 (1939). The theorem has been rediscovered several times. For a textbook presentation, see: H. Richter, *Wahrscheinlichkeitstheorie*, 2nd Edition (Springer, Berlin, Heidelberg, 1966) pp. 213–214
1.41 *Heidelberg Colloquium on Spin Glasses*, edited by J. L. van Hemmen and I. Morgenstern, Lecture Notes in Physics, Vol. **192** (Springer, Berlin, Heidelberg, 1983)

1.42 D. Sherrington and S. Kirkpatrick, Phys. Rev. Lett. **35**, 1792–1796 (1975). The SK model is expected to describe a spin glass in sufficiently high dimensions ($d > 8$).
See also: M. Mézard, G. Parisi, and M. A. Virasoro, *Spin Glass Theory and Beyond* (World Scientific, Singapore, 1987). It is fair to say that this book is devoted almost exclusively to the SK model

1.43 W. A. Little, Math. Biosci. **19**, 101–120 (1974); W. A. Little and G. L. Shaw, Math. Biosci. **39**,281–290 (1978)

1.44 A. Crisanti, D. J. Amit, and H. Gutfreund, Europhys. Lett. **2**, 337–341 (1986)

1.45 J. A. Hertz, G. Grinstein, and S. A. Solla, in *Heidelberg Colloquium on Glassy Dynamics*, edited by J. L. van Hemmen and I. Morgenstern, Lecture Notes in Physics, Vol. **275** (Springer, New York, Berlin, Heidelberg, 1987) pp. 538–546

1.46 D. J. Amit, in: Ref. [1.45: pp. 466–471]; A. Treves and D. J. Amit, J. Phys. A: Math. Gen. **21**,3155–3169 (1988)

1.47 A. Crisanti and H. Sompolinsky, Phys. Rev. A **36**, 4922–4939 (1987) and **37**, 4865–4874 (1988)

1.48 M. V. Feigel'man and L. B. Ioffe, Intern. J. Mod. Phys. B **1**, 51–68 (1987)

1.49 A simple and elegant proof can be found in: J. Lamperti, *Stochastic Processes* (Springer, New York, 1977) pp. 107–112. See also R. Kindermann and J. L. Snell, *Markov Random Fields and their Applications*, Contemporary Mathematics Vol. 1 (American Mathematical Society, Providence, Rhode Island, 1980) pp. 52–61

1.50 J. F. Fontanari and R. Köberle, Phys. Rev. A **36**, 2475–2477 (1987)

1.51 S. Grossberg, Neural Networks **1**, 17–61 (1988), in particular, Sect. 1.9

1.52 R. Kühn and J. L. van Hemmen, Graded-Response Neurons (Heidelberg, 1987, unpublished); R. Kühn, S. Bös, and J.L. van Hemmen: Phys. Rev. A **43**, RC (1991)

1.53 Equations (2.2.8) and (2.2.9) were discovered independently by J. Jędrzejewski and A. Komoda, Z. Phys. B **63**, 247–257 (1986)

1.54 J.-P. Nadal, G. Toulouse, J.-P. Changeux, and S. Dehaene, Europhys. Lett. **1** (1986) 535–542 and **2**, 343 (E) (1986)

1.55 M. Mézard, J.-P. Nadal, and G. Toulouse, J. Phys. (Paris) **47**, 1457–1462 (1986)

1.56 J. J. Hopfield, in *Modelling in Analysis and Biomedicine*, edited by C. Nicolini (World Scientific, Singapore, 1984) pp. 369–389, especially p. 381

1.57 G. Parisi, J. Phys. A: Math. Gen. **19** , L617–L620 (1986)

1.58 J. L. van Hemmen, G. Keller, and R. Kühn, Europhys. Lett. **5**, 663–668 (1988)

1.59 G. Toulouse, S. Dehaene, and J.-P. Changeux, Proc. Natl. Acad. Sci. USA **83**, 1695–1698 (1986)

1.60 See Table I on p. 271 of Ref. [1.17]

1.61 H. Sompolinsky, Phys. Rev. **34**, 2571–2574 (1986)

1.62 H. Sompolinsky, in *Heidelberg Colloquium on Glassy Dynamics*, edited by J. L. van Hemmen and I. Morgenstern, Lecture Notes in Physics, Vol. **275** (Springer, New York, Berlin, Heidelberg, 1987) pp. 485–527

1.63 There is the physiological rule "low-efficacy synapses degenerate." See: J.-P. Changeux, T. Heidmann, and P. Patte, in *The Biology of Learning*, edited by P. Marler and H. Terrace (Springer, New York, Berlin, Heidelberg 1984) pp. 115–133

1.64 J. L. van Hemmen and K. Rzążewski, J. Phys. A: Math. Gen. **20**, 6553–6560 (1987)

1.65 G. Toulouse, in: Ref. [1.62: pp. 569–576]. Toulouse considers a slightly different model (learning within bounds) with $\alpha_c \approx 0.015$. Estimating the connectivity Z of neurons involved in short-term memory to be of the order $Z \approx 500$, he finds that *at most* $\alpha_c Z \approx 7$ items can be stored. It is known from experimental psychology that the short-term memory capacity of humans is 7 ± 2 items (a rather famous number). If more items have to be stored, none of them can be retrieved, i.e., they are all forgotten. If the Hopfield model is overloaded, no retrieval is possible either.

1.66 R. Penrose, Proc. Cambridge Philos. Soc. **51**, 406–413 (1955) and **52**, 17–19 (1956); these papers are strongly recommended reading. The mathematics and numerics of the pseudoinverse is discussed at length in: T. N. E. Greville, SIAM Review **2**, 15–43 (1960), and A. Albert, *Regression and the Moore–Penrose Pseudoinverse* (Academic, New York, 1972)

1.67 T. Kohonen, IEEE Trans. Comput. **C-23**, 444–445 (1974); see also Kohonen's book, *Associative Memory* (Springer, New York, Berlin, Heidelberg, 1977)

1.68 L. Personnaz, I. Guyon, and G. Dreyfus, J. Phys. (Paris) Lett. **46**, L359–L365 (1985). These authors rediscovered the pseudoinverse in the nonlinear context (5.2.1), which they reduced to (5.2.2). A slightly more general, also linear, problem had been solved previously by Kohonen, see Ref. [1.67].

1.69 F. R. Gantmacher, *The Theory of Matrices*, Vol. I (Chelsea, New York, 1977) Sects. IX. 3 and 4.

1.70 S. Diederich and M Opper, Phys. Rev. Lett. **58**, 949–952 (1987)

1.71 I. Kanter and H. Sompolinsky, Phys. Rev. A **35**, 380–392 (1987)

1.72 A. M. Odlyzko, J. Combin. Theory Ser. A **47**, 124–133 (1988)

1.73 L. Personnaz, I. Guyon, and G. Dreyfus, Phys. Rev. A **34**, 4217–4228 (1986). The authors use a parallel dynamics and show, for instance, that cycles cannot occur.

1.74 F. Rosenblatt, *Principles of Neurodynamics* (Spartan Books, Washington, DC, 1961)

1.75 Ref. [1.4: Chap. 11] gives a lucid discussion

1.76 E. Gardner, J. Phys. A: Math. Gen. **21**, 257–270 (1988)

1.77 See also Chap. 3 by Forrest and Wallace. It contains a nice appendix, which supplements well the arguments presented here; in particular, the case $\kappa > 0$ in (5.3.9)

1.78 T. M. Cover, IEEE Trans Electron. Comput. **EC–14**, 326–334 (1965); P. Baldi and S. Venkatesh, Phys. Rev. Lett. **58**, 913–916 (1987)

1.79 W. Krauth and M. Mézard, J. Phys. A: Math. Gen. **20**, L745–L752 (1987)

1.80 C. F. Stevens, Nature **338**, 460–461 (1989) and references quoted therein; Nature **347**, 16 (1990)

1.81 The signal may also be "smeared out" by the capacitance of the dendritic tree. This gives rise to an exponential delay with an RC time τ'. Since the time window associated with τ' is rather narrow (a few milliseconds), certainly when compared with the axonal delay τ, it will be neglected here. See, however, A. Herz, B. Sulzer, R. Kühn, and J. L. van Hemmen, in *Neural Networks: From Models to Applications*, edited by L. Personnaz and G. Dreyfus (I.D.S.E.T., Paris, 1989) pp. 307–315

1.82 V. Braitenberg, in *Brain Theory*, edited by G. Palm and A. Aertsen (Springer, New York, Berlin, Heidelberg, 1986) pp. 81–96

1.83 R. Miller, Psychobiology **15**, 241–247 (1987)

1.84 K. H. Lee, K. Chung, J. M. Chung, and R. E. Coggeshall, Comp. Neurol. **243**, 335–346 (1986)

1.85 A.V.M. Herz, B. Sulzer, R. Kühn, and J. L. van Hemmen, Europhys. Lett. **7**, 663–669 (1988)

1.86 A.V.M. Herz, B. Sulzer, R. Kühn, and J. L. van Hemmen, Biol. Cybern. **60**, 457–467 (1989)

1.87 S. R. Kelso, A. H. Ganong, and T. H. Brown, Proc. Natl. Acad. Sci. USA **83**, 5326–5330 (1986)

1.88 R. Malinow and J. P. Miller, Nature **320**, 529–530 (1986)

1.89 The fact that Prob $\{\xi_i^\mu = +1\} \equiv p = 0.5$ also allows numerical simulations at a reasonable system size N; cf. Refs. [1.85] and [1.86]. For small p, numerical simulation is out of the question since either N is so small that the statistics is no good or N is so large that even most supercomputers have memory problems

1.90 L. N. Cooper, in *Nobel Symposia*, Vol. **24**, edited by B. and S. Lundqvist (Academic, New York, 1973) pp. 252–264

1.91 D. Kleinfeld, Proc. Natl. Acad. Sci. USA **83**, 9469–9473 (1986)

1.92 H. Sompolinsky and I. Kanter, Phys. Rev. Lett. **57**, 2861–2864 (1986)

1.93 R. Kühn and J. L. van Hemmen, this volume, Chap. 7

1.94 U. Riedel, R. Kühn, and J. L. van Hemmen, Phys. Rev. A **38**, 1105–1108 (1988); U. Riedel, diploma thesis (Heidelberg, February 1988)

1.95 J. Hale, *Theory of Functional Differential Equations* (Springer, New York, Berlin, Heidelberg, 1977)

1.96 R. Bellman and K. L. Cooke, *Differential Difference Equations* (Academic, New York, 1963)

1.97 N. D. Hayes, J. London Math. Soc. **25**, 226–232 (1950)

1.98 L. S. Pontryagin, Amer. Math. Soc. Transl. series 2, **1**, 95–110 (1955)

1.99 P. Peretto, Neural Networks **1**, 309–321 (1988)

1.100 L. F. Abbott and T. B. Kepler, J. Phys. A: Math. Gen. **22**, L711–L717 (1989)

1.101 B. Derrida, E. Gardner, and A. Zippelius, Europhys. Lett. **4**, 167–173 (1987)

1.102 R. Kree and A. Zippelius, this volume, Chap. 6

1.103 The equations for the Hopfield case with finitely many patterns, i.e., (13) with $\varepsilon = 0$, have been rediscovered by A. C. C. Coolen and Th. W. Ruijgrok, Phys. Rev. **38** (1988) 4253–4255 and M. Shiino, H. Nishimori, and M. Ono, J. Phys. Soc. Jpn. **58** (1989) 763–766. Here too the notion of sublattice is instrumental.
1.104 S. Amari, Neural Networks **1**, 63–73 (1988)
1.105 An illustration that should not be taken too seriously can be found on p. 561 in: Ref. [1.62]
1.106 B. Forrest, J. Phys. A: Math. Gen. **21**, 245–255 (1988)
1.107 H. Horner, D. Bormann, M. Frick, H. Kinzelbach, and A. Schmidt, Z. Phys. B **76**, 381–398 (1989)
1.108 W. Gerstner, J. L. van Hemmen, and A.V.M. Herz, manuscript in preparation; J.L. van Hemmen, W. Gerstner, A.V.M. Herz, R. Kühn, and M. Vaas, in *Konnektionismus in Artificial Intelligence und Kognitionsforschung*, edited by G. Dorffner (Springer, Berlin, Heidelberg, 1990) pp. 153–162
1.109 S. Bös, R. Kühn, and J. L. van Hemmen, Z. Phys. B **71**, 261–271 (1988); S. Bös, diploma thesis (Heidelberg, August 1988)
1.110 M. V. Feigel'man and L. B. Ioffe, Int. J. Mod. Phys. B **1**, 51–68 (1987)
1.111 D. J. Amit, H. Gutfreund, and H. Sompolinsky, Phys. Rev. A **35**, 2293–2303 (1987)
1.112 F. Crick and G. Mitchison, Nature **304**, 111–114 (1983)
1.113 E. R. Kandel and J. H. Schwartz, *Principles of Neural Science*, 2nd Edition (Elsevier, New York, 1985) Chap. 49
1.114 J. J. Hopfield, D. I. Feinstein, and R. G. Palmer, Nature **304**, 158–159 (1983)
1.115 J. L. van Hemmen, L. B. Ioffe, R. Kühn, and M. Vaas, Physica A **163**, 386–392 (1990); M. Vaas, diploma thesis (Heidelberg, October 1989); J.L. van Hemmen, in *Neural Networks and Spin Glasses*, edited by W.K. Theumann and R. Köberle (World Scientific, Singapore 1990), pp. 91–114
1.116 A.V.M. Herz, in *Connectionism in Perspective*, edited by R. Pfeifer, Z. Schreter, F. Fogelman–Soulié, and L. Steels (North-Holland, Amsterdam, 1989), Ph.D. thesis (Heidelberg, September 1990), and work in preparation
1.117 W. Feller, *An Introduction to Probability Theory and Its Applications*, Vol. I, 3rd Edition (Wiley, New York, 1970) Sect. 1.XVI.3
1.118 B. Derrida and J.-P. Nadal, J. Stat. Phys. **49**, 993–1009 (1987)
1.119 N. Burgess, M. A. Moore, and J. L. Shapiro, in *Neural Networks and Spin Glasses*, edited by W. K. Theumann and R. Köberle (World Scientific, Singapore, 1990) pp. 291–307
1.120 J. C. Eccles, *The Understanding of the Brain*, 2nd Edition (McGraw-Hill, New York, 1977)
1.121 C. Meunier, D. Hansel, and A. Varga, J. Stat. Phys. **55**, 859–901 (1989)
1.122 V. S. Dotsenko, J. Phys. C **18**, L1017–L1022; Physica A **140**, 410–415 (1986)
1.123 H. Gutfreund, Phys. Rev. A **37**, 570–577 (1988)
1.124 N. Parga and M. A. Virasaro, J. Phys. (Paris) **47**, 1857–1864 (1986)
1.125 J. Lamperti, *Stochastic Processes* (Springer, New York, Berlin, Heidelberg 1977); for the mathematically minded there is a neat summary of conditioning in Appendix 2.
1.126 M. V. Feigel'man and L. B. Ioffe, this volume, Chap. 5
1.127 J. Doob, Am. Math. Month. **78**, 451–463 (1971)
1.128 K. L. Chung, *A Course in Probability Theory*, 2nd Edition (Academic, New York, 1974) Chap. 9
1.129 C. Cortes, A. Krogh, and J. A. Hertz, J. Phys. A: Math. Gen. **20**, 4449–4455 (1987)
1.130 N. Parga, private communication
1.131 A. Krogh and J. A. Hertz, J. Phys. A: Math. Gen. **21**, 2211–2224 (1988)
1.132 R. J. McEliece, E. C. Posner, E. R. Rodemich, and S. S. Venkatesh, IEEE Trans. Inf. Theory **IT-33**, 461–492 (1987)
1.133 A. R. Damasio, H. Damasio, and G. W. van Hessen, Neurology (NY) **32**, 331–341 (1982)
1.134 M. A. Virasoro, Europhys. Lett. **7**, 293–298 (1988)
1.135 B. Derrida, E. Gardner, and P. Mottishaw, J. Phys. (Paris) **48**, 741–755 (1987)
1.136 F. J. Dyson, Commun. Math. Phys. **12**, 91–107 (1969) and 212–215. For a back-of-the-envelope discussion, see C. J. Thompson in *Nonlinear Problems in the Physical Sciences and Biology*, edited by I. Stakgold, D. D. Joseph, and D. H. Sattinger, Lecture Notes in Mathematics, Vol. **322** (Springer, Berlin, Heidelberg, 1973) pp. 308–342, in particular, pp. 329–330
1.137 W. Kinzel. Phys. Rev. B **33**, 5086–5088 (1986)
1.138 E. Domany and R. Meir, this volume, Chap. 9

1.139 R. Kühn, J. Lindenberg, G. Sawitzki, and J. L. van Hemmen, manuscript in preparation
1.140 D. J. Amit, K. Y. M. Wong, and C. Campbell, J. Phys. A: Math. Gen. **22**, 2039–2043 (1989)
1.141 A. Treves and D. J. Amit, J. Phys. A: Math. Gen. **22**, 2205–2226 (1989); H. Sompolinsky, Physics Today **41** / 12, 70–80 (1988)
1.142 J. Buhmann, preprint (USC, 1989)
1.143 A. Frumkin and E. Moses, Phys. Rev. A **34**, 714–716 (1986); E. Goles and G. Y. Vichniac in *Neural Networks for Computing*, edited by J. S. Denker, AIP Conf. Proc. **151** (American Institute of Physics, New York, 1986) pp. 165–181
1.144 G.A. Kohring: J. Stat. Phys. **59**, 1077–1086 (1990)
1.145 H. Steffan and R. Kühn, Z. Phys. B **95**, 249–260 (1994)
1.146 D.J. Amit and M.V. Tsodyks, Network **2**, 259–274 (1991)
1.147 R. Kühn, Habilitationsschrift (Heidelberg, April 1991)
1.148 W. Gerstner and J.L. van Hemmen, in: *Models of Neural Networks II*, edited by E. Domany, J.L. van Hemmen and K. Schulten (Springer, New York, 1994) Ch. 1
1.149 J.J. Hopfield and D.W. Tank, Biol. Cybern. **52**, 141 (1985); D.W. Tank and J.J. Hopfield, IEEE Trans. on Circuits and Systems **33**, 533 (1985)
1.150 R. Kühn, in: *Statistical Mechanics of Neural Networks*, edited by L. Garrido, Springer Lecture Notes in Physics **398** (Springer, Heidelberg, 1990) pp 19–32; see also Ref. [1.52]
1.151 R. Kühn and S. Bös, J. Phys. A: Math. Gen. **26**, 831-857 (1993)
1.152 C.M. Marcus and R.M. Westervelt, Phys. Rev. A **40**, 501–504 (1989)
1.153 J.R.L. de Almeida and D.J. Thouless, J. Phys. A: Math. Gen. **11**, 983–990 (1978)

2. Information from Structure: A Sketch of Neuroanatomy

Valentino Braitenberg

With 6 Figures

Synopsis. After the discovery of electrical phenomena in brains, which suggested recording with electrodes as the obvious approach to the problem of cerebral function, the field of neuroanatomy underwent a long period of stagnation. It is now once more an important source of information, which model makers are tapping abundantly (those who are not would be well advised to consider doing so).

I will sketch a short overview of the main trends in neuroanatomy, past and present, and then try to establish something like a method for us to follow when translating the information on the structure of a brain into statements about its operation.

2.1 Development of the Brain

Structure remains dead if it is not seen in relation to some dynamic process. The neuroanatomist is bound to generate ideas when he steeps his own brain in microscopic imagery, and he can hardly communicate his findings without first weaving them into a story. A very exciting process that has illuminated anatomical research is that of development and with it the two aspects of Darwinian evolution, on the one hand, and individual development from the egg to the finished product, on the other. In some of the earlier literature these two processes were terminologically confused and sometimes even identified as one under the supposition that individual development is bound to repeat the steps of the development of the species. The two aspects of development are, however, quite separate in neuroanatomy, where they have left their mark on two distinct fields, both of which continue to flourish, of comparative anatomy and of embryology (nowadays often called "differentiation").

Comparative anatomy had its heyday at the beginning of this century, when, with the zoological stock-taking nearing its completion, thanks to improved methods of histology (satisfactory fixation of the tissue and a variety of routine staining procedures for cells and fibers) an illustrated catalog of the brains of a large sample of animal species was realized. The implicit hope was that the variety of animal brains *per se* should spell out a story in terms of which was descendant from which and, more interesting still, what differences in size of a certain part of the brain in different animal species could be related to particular habits, talents, and instincts of these animals. But the endeavor of comparative anatomy was, on the whole, disappointing. The dynamics which seem to incite the brains to ever greater perfection remained obscure, and the correlations between particular bulges of the brain in different species with their specific idiosyncrasies in behavior remained anecdotal. The general tendency today is that we first have to learn to relate neuroanatomical structure with function in some special case (which we may stumble on with luck), before going on to interpret the variety apparent in the zoological world.

Ontogenetic development, on the other hand, is a hot topic at present. It is the favorite playground of molecular biologists who are now satisfied with their knowledge of the basic carriers of information in the genetic channel and want to extend genetics to the question of how the genetic information is expressed in the complete organism. This kind of neuroanatomy is charming because it offers ample opportunity for straightforward experimentation, the possibilities of interference with the processes of development being many, well defined, and easy to observe in their effects. But it is a branch of biology concerned with an information channel quite unrelated to the flow of information between sense organs and motor output in the adult animal which is what neurophysiology is all about. In the problems dealt with in developmental neuroanatomy, the nervous system could well be replaced by some other part of the organism, the bones, the liver, or the skin, without any change in the basic strategy of the experimentation.

2.2 Neuroanatomy Related to Information Handling in the Brain

Here again we can distinguish between different approaches, one dominant in the past and one wich may become important in the future.

In the past 100 years the leading theme was that of "localization of function". This was a program aspiring to assign special mental faculties already bearing traditional names in everyday language, or such as emerged from a more and more refined psychology, to particular sites in the brain. In its extreme philosophical position, upheld for instance by C. and O. Vogt, the task was entirely fulfilled by a parcellation of mental faculties and a parallel parcellation of the brain with a mapping of the pieces of one classification into the pieces of the other. The question of the details in the neuroanatomy, which made a particular region of the brain responsible for "language" and another responsible for "attention", was not only neglected but declared further to be hopelessly beyond our powers of comprehension.

The other neuroanatomy starts with the advent of electronic computing devices. The objection of those brain scientists who held the inextricable complexity of the nervous system as a matter of principle could be countered with the argument that electronic computers were inextricable to them too, but not at all mysterious in their plan. It is this new neuroanatomy, which would see fiber patterns in the brain in terms of abstract schemes of computation, that I want to describe now.

2.3 The Idea of Electronic Circuitry

As soon as the idea of a circuit emerged as a basic unit of information handling in telephone and radio engineering, the search for circuits in the brain became a possibility. The resemblance of the diagrams describing fiber connections in the brain with the circuit diagrams of electronics was striking (Fig. 2.1) and is indeed not accidental. One common denominator between electronic devices and brains is their fibrous structure, indicating the common functional principle of transmission of signals from and to specified locations rather than by diffuse chemical or electrical interactions. Once this was recognized the search began for applications of the idea of circuits to particular pieces of the nervous system where function and behavior were presumed known. There were two main trends. Some scientists hoped to find examples of electronic-like circuitry in small animals whose behavior was presumed to be more schematic than that of the vertebrates. Others searched within more complex brains for regions of the

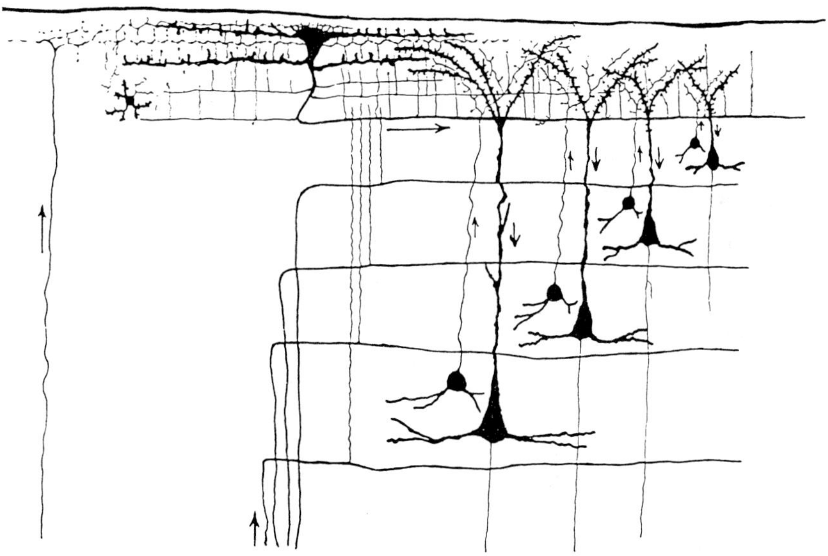

Fig. 2.1. Anatomical diagrams of real neuronal networks are often inspired by the circuit diagrams of electronics. The illustration shows Marin-Padilla's [2.2] idealization of the network in the upper layers of the mammalian cerebral cortex. Typically these "neuronal wiring diagrams" the number of neurons is by at least a factor of 1000 smaller than the number of neurons contained in the same space of the real brain. Also, the diagram suggests very specific relations between individual neurons, when in reality the connections are quite diffuse

gray substance with a particularly high degree of order, carrying the promise of easy intuitive interpretation. As an example of the first kind of study I will mention a network in the visual system of the fly.

2.4 The Projection from the Compound Eye onto the First Ganglion (Lamina) of the Fly

The network of Fig. 2.2 [2.1] is striking for two reasons. It is built with absolute precision, and we can easily interpret the simple rule according to which the matrix of the connections is built. The compound eye of the fly (like that of most other arthropods) is a collection of almost identical units, the so-called ommatidia. Each ommatidium has its own lens, its photosensitive elements positioned in the focal plane of the lens, and its own separate connections to the brain. There are 3000 ommatidia on each side. The optical axes of the ommatidia pierce a celestial sphere surrounding the animal in a fairly regular hexagonal array of points covering almost the whole sphere. The lines of sight defined by the seven photosensitive elements of each ommatidium are so arranged that one of them looks in the direction of the axis of that ommatidium and six look in a direction parallel to the axes of six neighboring ommatidia. A consequence

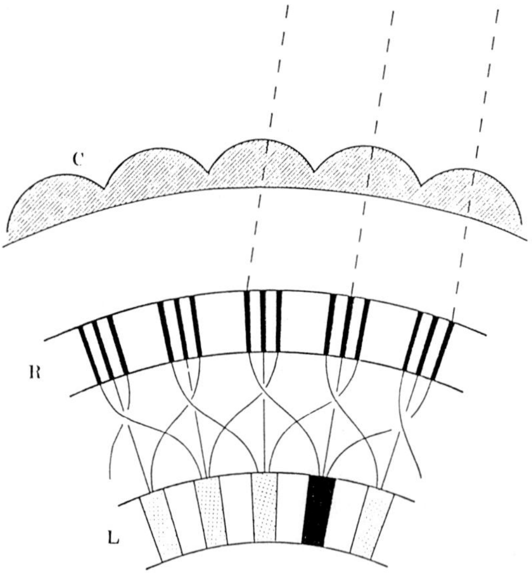

Fig. 2.2. The optical and neuronal projection of a distant point of visual environment onto the visual brain of the fly. C is the cornea of the compound eye with separate lenses having slightly divergent optical axes. R is the retina composed of three photosensitive elements in the focal plane of each lens (in reality there are seven to each lens). L is the first visual ganglion of the brain (the lamina) composed of as many compartments as there are lenses in the cornea. The fibers between R and L are so arranged as to make visual input from the same point seen by different lenses (parallel dashed lines) converge onto the same compartment of the brain (black) (From [2.12])

of this is that each point of the visual field of the compound eye is "seen" by seven photosensitive elements housed in seven different ommatidia. The rule that defines the wiring between the eye and the brain is simply this: all the fibers emanating from light-sensitive elements whose lines of sight pierce the celestial sphere at the same point come together in one compartment of the brain. The brain has as many compartments as there are ommatidia in the eye, arranged in a similar order. Each compartment collects fibers from seven ommatidia and each ommatidium distributes fibers into seven different compartments of the brain. The result is not striking, since it simply corresponds to an orderly mapping of visual space onto the brain. What is striking is the precision of the mechanisms of growth responsible for the production of this wiring (experience is not involved: the full set of fibers is already preformed in the pupa before the insect ever used its compound eye), as well as the nature of the forces guiding the fibers to their correct destination. But this is a problem for the embryologists. In terms of informational neuroanatomy the striking proposition is that this network embodies knowledge of geometrical optics: the geometry of the lens, its index of refraction, its focal length, the fact that a convex lens projects a sharp picture of distant objects in the focal plane, and the fact that this picture is rotated by 180°. All of this would have to be known precisely by an engineer if he were to design the "wiring", the network of fibers between the fly's eye and its brain. Thus, at least in this one example, we can say that we understand a network fully and that brains embody knowledge of the physical world. There are other examples of networks at successive levels of the fly's visual system which are built with the same precision, even if their functional interpretation is, as yet, only tentative.

2.5 Statistical Wiring

The neuroanatomist, with his eye trained on the beautiful simplicity of some insect nerve-nets, experiences disappointment when he looks at the nerve tissue of the human or other vertebrate brain. The impression in most cases is that of disorderly connections constrained only by statistical rules. These may be expressed in terms of the average size of neuronal elements, the density of the spread of their fibrous appendages, the statistics of the directions in which the fibers are arranged, and the mixture of neuronal elements of different kinds. It is of course possible, theoretically, that such apparent disorder actually hides a very complicated scheme which we have not yet been able to grasp. But this is not very likely. The main argument against this supposition of more order than meets the eye is that it would require a vast amount of genetic specification — more than the genetic code would allow.

The very element of the vertebrate nervous system, the neuron with its (generally) widespread dendritic and axonal ramification, embodies a statistical principle (Fig. 2.3). The dendritic tree, the receiving part of the neuron, is best de-

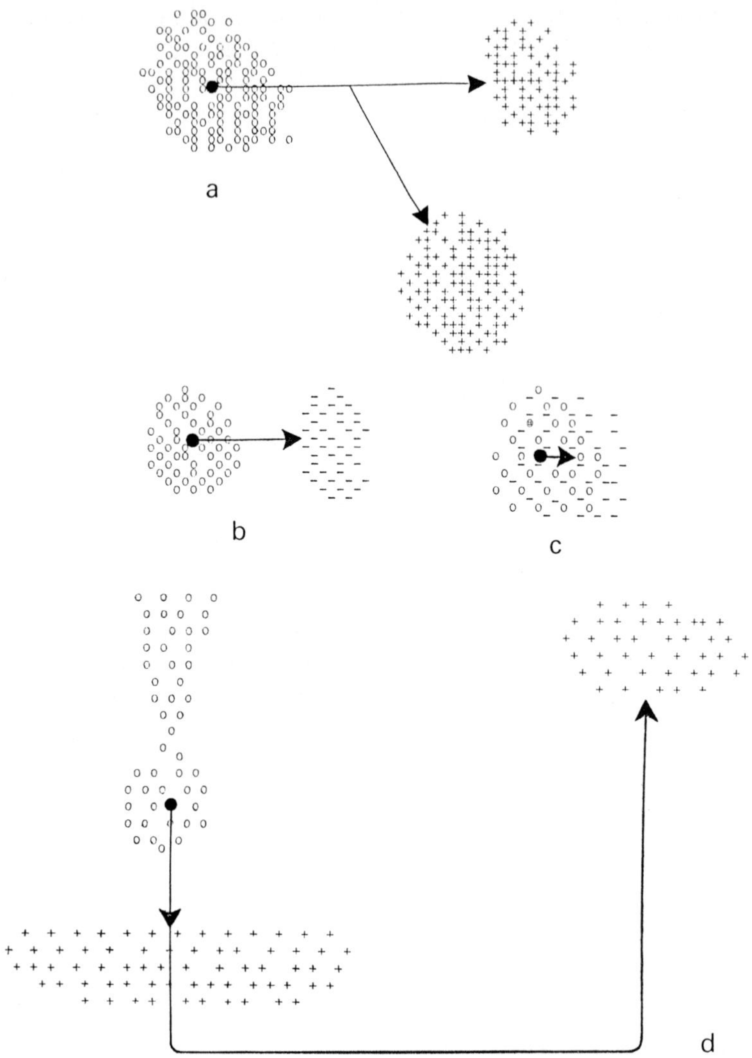

Fig. 2.3a–d. Neurons as devices operating on the distribution of activity in the nerve tissue. Each neuron consists of a cloud of input points (*small circles*) connected to one or more clouds of excitatory (*plusses*) or inhibitory (*minusses*) output points. Between the input and the output there is a cell body (*black circle*) in which all the input of that neuron is added, and a fiber (*arrow*) representing the threshold element. When the activity in the input cloud is above threshold, positive activity or negative activity (meaning prevention of activity) is produced in the output clouds. This in turn affects the input clouds of other neurons (not shown) which share the same space. All these clouds abundantly permeate each other, the distance between the neurons in the tissue being much smaller (by a factor of 10 or so) than the diameter of the clouds. (**a**) neurons with two separate excitatory output clouds. (**b**) neuron with inhibitory output cloud. (**c**) the inhibitory output cloud is often almost coincident with the input cloud of the same neuron, which thus changes the sign of the activity without changing the geometry of its distribution. (**d**) the pyramidal cell, the typical neuron of the cerebral cortex, with its peculiar input cloud and two excitatory output clouds: one in the vicinity and one in a distant region of the cortex

scribed as a cloud of several thousand "input points" (= synapses) distributed randomly around the location of the neural cell body, and the axonal tree, with its ramifications, as an equally random cloud of output points. Since the length of the branches of each neuron generally exceeds (by a factor of 5 or even considerably more) the separation of neighboring neurons, the clouds of input points of different neurons are vastly intermingled, and the same is true for the clouds of output points. In addition, of course, the input and the output clouds of different neurons share the same space and we may infer the coupling between two neurons from the product of the density of input points of one with the density of the output points of the other. If there were no local constraints to this picture, the description of couplings in the nervous system would be possible in terms of a continuum of superimposed fields describing the probability of the presence of various kinds of neuron with their presynaptic and postsynaptic components. It would be sufficient to add a rule describing the synaptic transmission and the non-linear mechanism of spike generation in the axon to obtain a picture of the gray substance quite close to some present-day models.

In reality we find many instances in which such a diffuse homogeneous and isotropic scheme does not apply. Even a rough examination of the statistics of neuronal connections in different parts of the brain reveals anisotropies, striking differences in the range and density of neuronal connections, and quite a variety of shapes of dendritic and axonal trees. This does not necessarily contradict the statistical view of connectivity, but sets interesting limits to it which are also subject to interpretation in terms of information-handling schemes.

2.6 Symmetry of Neural Nets

It is quite obvious that general statements about the geometry of interneuronal connections correspond to general statements about the type of computation performed there. A system with only short-range connections will perform primarily narrow-range operations; a system with connections completely independent of distance will perform a computation which is quite free of geometric contraints. A 3-dimensional nerve net containing only randomly oriented fibers is not likely to transform its input in terms of a 2-dimensional or 1-dimensional geometry, etc. (Fig. 2.4).

There are some interesting symmetries to be observed in brains. The bilateral symmetry, valid for all vertebrate brains (with some minor exceptions) and for most invertebrate brains, is fundamentally related to orienting behavior controlled by a motor apparatus with the same kind of symmetry. The up–down asymmetry is of course ultimately related to the direction of gravity and the front-to-back asymmetry to the predominant direction of locomotion.

Within the gray substance it is also possible to define types of symmetry [2.4] by stretching the abstract definition somewhat and replacing the geometric concept of identity by that of practical indistinguishability by a histologist: the

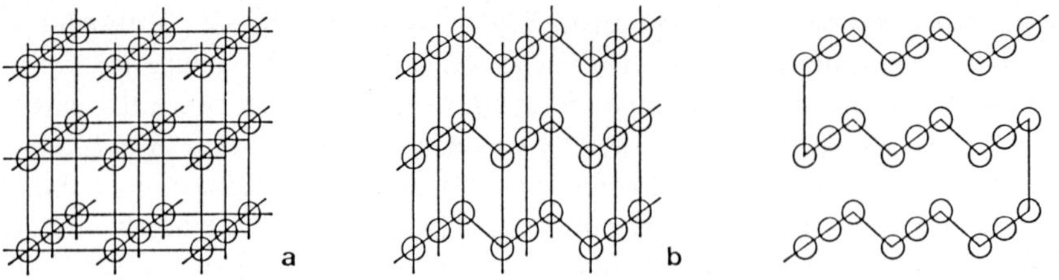

Fig. 2.4a–c. The macroscopic layout of neurons in the neural tissue is compatible with different kinds of connectivity, as in the examples: (**a**) three dimensional, (**b**) two dimensional, (**c**) one dimensional. Two-dimensional patterns are fairly common, the so-called cortices

symmetry of a piece of nerve tissue is defined by all the movements – transpositions, inversions, rotations – of a block of tissue in the microtome that will result in indistinguishable histological sections. When my fellow neuroanatomist puts a piece of cerebral or cerebellar cortex into his microtome in order to section it, he will not notice if I replace his piece by a neighboring one. But he would notice the replacement if his block contained the interpeduncular nucleus or had been taken from the hypothalamus or from any other part of the brain containing small regions with distinct, well-defined structures. The cerebral and the cerebellar cortex show translational symmetry; the hypothalamus does not. If the block of tissue were round, I could rotate it by any angle without his noticing it in the case of cerebral cortex, but I could only rotate it by 180° in the case of the cerebellar cortex, which has a definite anisotropy.

We are thus led to the inescapable conclusion that the different kinds of elaboration of signals attributed to the various pieces of nerve tissue must follow the same symmetry. A similar conclusion is possible in the cases where we observe gradients of size and density of the elements in one or more directions of the nerve tissue. It is very tempting to relate them to gradients in the sensory field that particular piece of brain deals with, or to gradients in the intensity of the reaction to stimuli depending on their position. Such considerations may not give us all the information we require for a reconstruction of the function from the structure, but they provide us with strong clues. I intend to show an example of an anisotropic cortex, the cerebellar cortex of vertebrates, and then an example of a gradient, in one of the structural parameters interpretable in terms of the behavior of a flying insect.

2.7 The Cerebellum

As I have already stated, the cerebellum [2.5] is a cortex in the sense of a sheet of gray substance with a basic connectivity repeated throughout its extent but lacking any appreciable local variation. The input fibers arrive with a uniform density everywhere and the output emanates from every small part of the sheet. The peculiar thing about the cerebellar cortex is the arrangement of two distinct populations of fibers at right angles to each other. One set, the so-called parallel fibers, runs in the direction which, if the cortex were stretched out so as to undo the folding, would be arranged in a transverse direction. The parallel fibers form a thick felt in which the other elements are embedded. Their synaptic connections with the output elements, the Purkinje cells, are excitatory. The other population of fibers, less numerous than the parallel fibers, is arranged at right angles to them and also connected with the Purkinje cells. These synapses, however, are inhibitory. There are also two sets of input fibers. First, the so-called mossy fibers connect with the cells from which the parallel fibers arise. Second, the so-called climbing fibers connect one to one (with some exceptions which I disregard here) directly to the Purkinje cells. To complete the synaptic arrangement of the cerebellar cortex, the parallel fibers make excitatory connections with the cells from which the inhibitory fibers arise. (The only element we leave out in this picture is a local inhibitory element, the so-called Golgi cell, which, since its axon terminates in the immediate neighborhood of its dendrites, does not contribute to the geometric interplay of signals in the cerebellar cortex.)

This remarkably skeletal arrangement leads to the following description: a signal entering through a mossy fiber is spread sideways, symmetrically to the left and to the right through the parallel fibers, reaching and probably exciting a row of 100 or so Purkinje cells (Fig. 2.5). At the same time it also reaches a

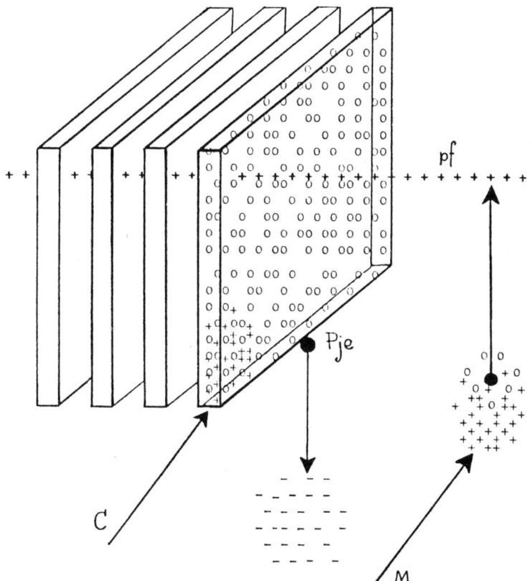

Fig. 2.5. The elementary mesh of the neuronal network in the cerebellar cortex. The graphic conventions are the same as in Fig. 2.3. The output element Pje (Purkinje cell) has an input cloud confined to a flat volume, which does not overlap with that of its neighbors. The output is inhibitory. The input fibers are of two kinds: C (the "climbing fibers") connected one to one with the output elements and M (the "mossy fibers") connected to the output via an excitatory interneuron, whose linear output cloud pf ("parallel fiber") passes through more than one hundred input clouds of Purkinje cells. This arrangement has suggested the importance of timing in the cerebellum. Some inhibitory interneurons which are also present there are not shown in the diagram

row of inhibitory interneurons whose axons spread, again symmetrically, forward and backward to inhibit the neighboring rows of Purkinje cells (not shown in Fig. 2.5). Thus the excited row will stand out on a background of muted activity. And so the basic unit of operation is a one-dimensional one. This led very early on to the interpretation that the structure of the cerebellar cortex has something to do with time. This explanation is all the more convincing given that the shape of the elements in the parallel-fiber–Purkinje-cell system do indeed seem to be specialized for an accurate definition of the arrival times of signals at the Purkinje cells of different locations. In fact, bending and stretching, as is commonly observed in those cortices where only the neighborhood relations between the elements seem to matter and not their distance, is definitely avoided, or rather is limited to one direction only. When the cerebellar cortex is folded, which is sometimes the case (especially in larger animals), the folds all run in the direction of the parallel fibers. Folds at right angles to this do not occur, so that parallel fibers take a straight course. Also, the dendritic trees in a row of Purkinje cells contacted by a bundle of parallel fibers are flattened so as to define more accurately a slice in time, and they do not overlap. If this is a device for measuring times, its resolution of about a tenth of a millisecond can be deduced from the spacing of the Purkinje cells and from the velocity of signals in the parallel fibers. Such a resolution may prove useful in many calculations connected with motor activity, a task in which the cerebellum seems to be involved. The simple-minded interpretation that signals in parallel fibers scan a motor program by reaching certain Purkinje cells at pre-established intervals of time did not lead very far because such programs could not have a duration of more than about 10 milliseconds, an upper limit set by the length of the individual parallel fibers. Parallel fibers are, however, staggered to form long rows, the length of which, considered as channels in time, correspond to a few tenths of a second in larger brains. Such a length of time makes more sense, matching that of many elementary motor acts [2.7]. It is also the duration of what in speech production is called syllable. A more recent interpretation of the space–time elaboration in the cerebellum [2.6] is that of a velocity detector in the tissue of parallel fibers all arranged in the same direction and all having approximately the same conduction velocity. There is one speed at which the input signals may move across the cerebellar cortex and elicit the maximal activity there, each new input signal being added to a wave of internal propagation along parallel fibers to form something like a tidal wave. Stationary input, or input moving more slowly or more rapidly, will have a lesser effect. Considering the inhibitory fibers, it is interesting to notice that when such a tidal wave is established it will also maximally inhibit the neighboring "time channels" keeping the activity as it were in a groove, as if to eliminate the competing actions that may be represented there. This translation of structural facts into a functional picture is inescapable here, even if not enough is known about the connection of the cerebellum with the rest of the brain to be able to make real use of it.

What is known, however, complies rather well with the model derived from anatomy. Interference with the normal functioning of the cerebellum, for instance when blocking the output channels by cooling or by surgical measures, makes it clear that the contribution of this organ to normal function is in the realm of sequencing muscular contractions in motor acts. In particular, what is lost when the cerebellum is damaged is that part of the muscular coordination which deals with the inertial forces generated by the movement. When the cerebellum is imparied, movements overshoot because the momentum generated by the go-signal is not annulled by a well-timed braking signal. Pendular movements become irregular because the angular momentum transmitted from one segment of the limb to another is not balanced by appropriate muscular tensions at the correct phase, etc. Experiments with monkeys show that these refinements of movement are learned, and only what is learned, apparently in the cerebellar circuits, breaks down after cerebellar lesions.

It is always fascinating to find some basic aspects of physics embodied in the anatomy. Inertial forces in movement are stronger with fast movements. Fast movements in the cerebellum correspond to shorter grooves. Shorter grooves activate fewer output (Purkinje) cells. It is noteworthy that these influence the muscle tonus (the DC component of muscular contraction) in a negative way, since their synapses onto the system controlling the tonus are inhibitory. Thus the tonus will be stronger (= less inhibited) with fast movements and more relaxed with slower movements, as it should be.

2.8 Variations in Size of the Elements

Just as symmetries in the structure of nerve nets correspond to symmetries in the abstract space of information handling, an anatomical gradient, meaning a systematic variation in the size of some elements throughout a nerve-net, can also be related to a variation of the performance in different parts of the corresponding sensory field. Again the visual system of the fly provides an instructive example (Fig. 2.6).

As I have described in a previous section of this article, the visual system of the fly is represented in an orderly manner as an array of discrete points on the surface of the (first) visual ganglion. Throughout the whole ganglion, at each input point, signals are relayed to a group of neurons of the second order which are identical throughout the array, except that they vary in size. Leaving aside some of the elements, the most striking fact is that the synapses leading from each input fiber to two second-order elements L_1 and L_2 are exactly the same, thereby constituting a puzzle: would it not be simpler to let one element receive the information, allowing for it to branch if more copies of the same signals are needed? The answer lies in the difference in size of the two elements to which each input fiber connects. Apparently they act as different electronic filters, the bigger one providing a larger membrane surface and therefore a larger capacitance in the conduction of electrical disturbances produced by the synapses. This anatomical observation is in good agreement with a model proposed by

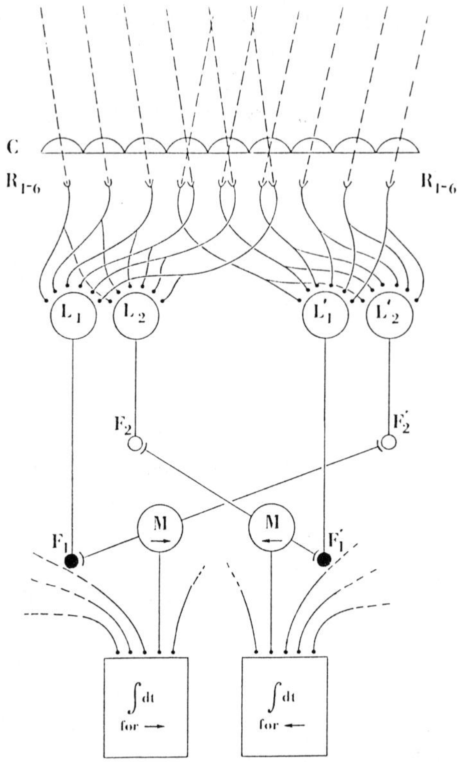

Fig. 2.6. Information flow in the visual system of the fly, from the visual input (*dashed lines*) to the two elements which signify movement in two opposite directions (*boxes below*). Each of the compartments of the first visual ganglion (L in Fig. 2.2) contains two elements L_1 and L_2 which receive exactly the same input. This makes sense if their different sizes can be interpreted as two different electronic filters (F_1 and F_2) through which the signals are relayed to the next level. There the movement detectors (M) are fed by one input point through one kind of filter (F_1) and by the neighboring point through the other kind (F_2). This makes the movement detectors asymmetrical and therefore directional. The (relative and absolute) sizes of L_1 and L_2 and therefore presumably the properties of the filters F_1 and F_2 vary in a systematic way over the surface of the eye, and so does the intensity of the reaction elicited by movement in the visual field (from [2.12])

Reichardt [2.8] for the neural interaction underlying movement perception. This model is based on a multiplying unit receiving input from neighboring input lines (M in Fig. 2.6), but through filters with different time-constants (F_1 and F_2). Since there is such a movement detector interposed between any two neighboring input points, it is of course important that each of the inputs is available in two versions, passed through the two different filters. The agreement is even more remarkable considering the intensity of a fly's reaction to movement perceived, which varies in different locations of the visual field [2.9]. The difference between the two fibers L_1 and L_2 varies in a similar way [2.10], with the maximum of the anatomical difference corresponding to the position of the maximal reaction.

2.9 The Cerebral Cortex

The cerebral cortex in mammals, from mice to men, is the largest piece of gray substance, about half of the total within the skull. On good, though vague, grounds it is often associated with higher mental functions. The anatomy of the cortex has been variously interpreted: the results range from total defeatism in the face of a totally specified, irreducibly complex network, to that of a random structure easily amenable to statistical treatment. The divergence continues to be fed by experimental evidence pointing on one hand to a striking homogeneity of the cortical tissue and, on the other to an ever more detailed local specialization of its tasks. Our own anatomical studies tend to stress the statistical view. The pattern of the connections within the cortex seems to be dictated by a rule requiring the maximal dispersion of signals from any element to as large as possible a number of elements. To a certain extent the connections within the cortex abolish neighborhood relations so that the probability of a connection between two elements is not any simple function of their distance. In this view the large majority of neuronal elements of the cortex are of one type, the pyramidal cells, and therefore the vast majority of connections within the cortex are between neurons of the same sort. This leads directly to the interpretation of the cortex as an associative matrix, all the more so since the synapses between pyramidal cells are of a special kind which we now believe to be involved in learning (synapses residing on dendritic spines). This view had to face the challenge of very specialized functions, such as those described by Hubel and Wiesel [2.11] and their followers for the visual cortex. We were able to demonstrate how many of the effects which at first sight seem to require a very specific wiring can be explained by invoking only very slight statistical variations in the distribution of the inhibitory neurons, which are interspersed as a minority within the pyramidal cells [2.12].

2.10 Inborn Knowledge

The examples given so far have mainly concerned mechanisms of input transformation, of the detection of such elementary things as movement in the visual field, or of motor sequences. We know that the brain also incorporates knowledge in the form of inborn programs such as occur in instinct behavior: feeding, fighting, mating, singing. Although the locations in the brain involved in the storage of this kind of behavioral information are known with fair precision in some cases, we have no detailed knowledge of the connections between the neurons mediating such behavior. This indicates the direction in which anatomy will have to go once the question of neuronal mechanisms is solved in principle.

References

2.1 V. Braitenberg: Patterns of projection in the visual system of the fly I. Retina-lamina projections. Exp. Brain Res. **3**, 271–298 (1967); V. Braitenberg: *On the Texture of Brains. Neuroanatomy for the Cybernetically Minded* (Springer, Berlin, Heidelberg 1977)

2.2 M. Marin-Padilla: "Neurons of layer I. A developmental analysis", in *Cerebral Cortex*, Vol. I, ed. by A. Peters and E.G. Jones (Plenum, New York 1984) pp. 447–478

2.3 V. Braitenberg: *On the Texture of Brains. Neuroanatomy for the Cybernetically Minded* (Springer Berlin, Heidelberg 1977)

2.4 V. Braitenberg: The concept of symmetry in neuroanatomy, Ann. New York Acad. Sci. **299**, 186–196 (1977)

2.5 V. Braitenberg and R.P. Atwood: Morphological observations on the cerebellar cortex, J. Comp. Neur. **109**, 1–33 (1958); V. Braitenberg: "Is the cerebellar cortex a biological clock in the millisecond range?", in *Progress in Brain Research. The Cerebellum*, Vol. 25, ed. by C.A. Fox and R.S. Snider (Elsevier, Amsterdam 1967) pp. 334–346; V. Braitenberg: "The cerebellum and the physics of movement: some speculations", in *Cerebellum and Neuronal Plasticity*, ed. by M. Glickstein, Ch. Yeo, J. Stein, (Plenum, New York 1987) pp. 193–207

2.6 V. Braitenberg: The cerebellum revisited, J. Theoret. Neurobiol. **2**, 237–241 (1983)

2.7 V.B. Brooks and W.T. Thach: "Cerebellar control of posture and movement", in *Handbook of Physiology*, Section 1: *The Nervous System*, ed. by V.B. Brooks (American Physiological Society, Bethesda, Maryland, 1981)

2.8 W. Reichardt: The insect eye as a model for analysis of uptake, transduction and processing of optical data in the nervous system, *34. Physikertagung Salzburg* (Teubner, Stuttgart 1970)

2.9 W. Reichardt: Musterinduzierte Flugorientierung, Naturwissenschaften **60**, 122–138 (1973); C. Wehrhahn and K. Hansen: How is tracking and fixation accomplished in the nervous system of the fly?, Biol. Cybernetics **38**, 179–186 (1980)

2.10 V. Braitenberg and H. Hauser-Holschuh: Patterns of projection in the visual system of the fly II. Quantitative aspects of second order neurons in relation to models of movement perception, Exp. Brain Res. **16**, 184–209 (1972)

2.11 D.H. Hubel and T.N. Wiesel: Functional architecture of macaque monkey visual cortex, Proc. Roy. Soc. Lond. Ser. B **198**, 1–59 (1977)

2.12 V. Braitenberg: Charting the visual cortex, in *Cerebral Cortex*, Vol. 3, ed. by A. Peters and E.G. Jones (Plenum, New York 1985) pp. 379–414; V. Braitenberg: "Two views of the cerebral cortex", in *Brain Theory*, ed. by G. Palm and A. Aertsen (Springer, Berlin, Heidelberg 1986), pp. 81–96; V. Braitenberg, A. Schüz: Cortex: Hohe Ordnung oder größtmögliches Durcheinander?, Spektrum der Wissenschaft **5**, 74–86 (1989), and: *Anatomy of the Cortex* (Springer Berlin, Heidelberg 1991)

3. Storage Capacity and Learning in Ising-Spin Neural Networks

Bruce M. Forrest and David J. Wallace

With 5 Figures

Synopsis. A review of recent results on the storage capacity and content-addressability of Ising-like models of neural networks is presented, including those employing perceptron-type learning algorithms.

3.1 Introduction

One particular class of neural-network models which has been the subject of enormous interest in recent years, particularly amongst the physics community, may be termed the *Ising-spin neural network.* In this review we shall be concerned with its role as a model of a memory for the storage and retrieval of information in a content-adressable manner. We shall begin by introducing the model and describing the close analogy [3.1] with a system of Ising spins, and then proceed to describe the Hopfield model [3.2], which was largely responsible for stimulating the interests of physicists in this class of model. After reviewing its properties, we shall then consider improvements to the model, in particular iterative "learning" algorithms designed to enhance the network's content-addressability.

3.1.1 The Model

The model is essentially a collection of interacting McCulloch–Pitts [3.3] neurons. The behavior of each neuron is simplified to such an extent that it may only exist in one of two states of activity: fully firing or quiescent. In anticipation of the analogy of the network of neurons with a system of Ising spins, we shall denote the state of the ith neuron at time t by $S_i(t) = +1$ if it is firing, and $S_i(t) = -1$ if it is not.

The dynamics of each of the N neurons – the way in which they respond to their environment – is determined solely through the local potential

$$h_i(t) \equiv \sum_{j=1}^{N} T_{ij} S_j(t) - U_i \tag{3.1}$$

that neuron i experiences at time t. U_i is the *threshold* of the neuron, the amount which the total incoming signal $\sum_{j=1}^{N} T_{ij} S_j(t)$ into neuron i must exceed in order that the neuron may fire. The connection strength T_{ij} represents the *synapse* from neuron j into neuron i and determines how strongly the firing activity of neuron j influences that of neuron i. T_{ij} may be either positive (excitatory), negative (inhibitory), or zero (no connection present).

The simple dynamical law governing the time evolution of the neural activities is such that they switch on or off according to whether their momentary local potential (3.1) is positive or negative, i.e.,

$$S_i(t+\tau) = \text{sgn}(h_i(t)) \ , \tag{3.2}$$

where τ represents the response time of the neurons.

In the Ising-spin analogy, we see that each "spin" S_i aligns with its local "field" h_i and from (3.1) we see that the synaptic connections T_{ij} are just like the exchange or coupling constants between the spins.

It need hardly be said that this is an extreme simplification of the behavior of a real network of interacting neurons, such as in the human brain. There the number of neurons are some 10^{10}, each a complex entity in its own right and each of which interacts with anything up to the order of 10^4 other neurons. The total population of synapses then must be of the order of 10^{14}. Nonetheless, the underlying philosophy of most classes of neural-network models – and this one is certainly no exception – is that it may be possible to account for some of the powerful processing capabilities of their real (biological) counterparts by attempting to reduce the complex details at the ("microscopic") level of the individual neuron to a few essential mechanisms, and to concentrate on the general, collective behavior of the ("macroscopic") system as a whole. Such a philosophy is by no means foreign to physics, and certainly not to statistical physics, where, as here, we are dealing with a system comprising of a huge number of interacting subsystems. In many cases we can explain the macroscopic (or bulk) behavior of the system by keeping only the barest necessary details at the microscopic (subsystem) level and looking for their collective phenomena.

3.1.2 Content-addressable Memory

The primary motivation in studying neural-network models comes from the tremendous processing power at the disposal of real neural systems. One of the most important of the capabilities of the human brain is that of content-adressable memory (CAM). In such a memory the entries which are stored can each be evoked by a cue (input) which sufficiently resembles the desired entry: i.e., each piece of stored information is retrieved on the basis of its content. This ability gives rise to fault-tolerance with respect to input errors: the memory is able to cope with distorted input stimuli. It can also process the information (access the entries) very rapidly: consider, for example, that the recognition of words or faces can normally be done in fractions of a second and yet typically involve a choice from amongst an enormous database containing perhaps hundreds of thousands of alternatives. Furthermore, this speed is achieved despite the fact that the timescales on which neurons respond are merely of the order of milliseconds. Consequently such recognition tasks cannot involve merely a sequential search through all of the possible candidates.

This type of memory is, of course, in sharp contrast to that of a conventional computer which stores data in specific locations, each designated an address – usually a sequence of bits. It must be supplied with the exact bit-address or else, even if only one bit is wrong, a different memory location will be accessed and, in principle, a completely unrelated piece of data will be evoked.

We shall see that, despite the simplicity of the model introduced above, it can exhibit rich behavior and can account for many of these desirable features, such as content-adressable memory, fast processing, and fault-tolerance to both input and structural damage.

In the model we can regard the state of the network of N neurons at time t as an N-component binary vector $\boldsymbol{S}(t) \equiv \{S_1(t), \ldots, S_N(t)\}$, or, equivalently a "spin configuration".

The network is prepared in an initial state corresponding to the firing pattern $\boldsymbol{S}(0)$ and is updated according to the dynamics (3.2). Several choices are available. One possibility is a "lockstep-parallel" dynamics, in which the new state of every neuron is determined at each update from the state of the network at the previous time-step. Alternatively, the nodes may be updated serially, either in random or sequential order, using the new value of the updated neuron in the calculation of the new state of the next neuron. The behavior of the network depends on which particular type of dynamics is specified and may be extremely complicated. The simplest interpretation of memory in the model is that a pattern $\boldsymbol{S}^*$ is "stored" if it corresponds to a persistent firing pattern, i.e., if $\boldsymbol{S}(t) = \boldsymbol{S}^*$ for all t exceeding some $t' > 0$. From (3.2) we see that this can only happen if the state $\boldsymbol{S}^*$ corresponds to spin configuration in which every spin is aligned with its local field:

$$S_i^*(t) h_i(t) > 0 \,, \quad i = 1, \ldots, N \,. \tag{3.3}$$

"Learning" or "training" the network then corresponds to adjusting the connections T_{ij} such that each member of a given ("nominal") set of vectors is indeed a fixed point of the dynamics.

3.1.3 The Hopfield Model

One of the best-known models which falls into this category is the Hopfield model [3.2]. Although most of the ideas incorporated in the model were not really new (some can be traced back to McCulloch and Pitts [3.3] and Hebb [3.4]), Hopfield did bring together many important ideas into one model. Before we describe the model, let us first consider an important qualitative feature identified by him.

If each diagonal element of the connection matrix is zero:

$$T_{ii} = 0 \tag{3.4a}$$

and if the connections are symmetric:

$$T_{ij} = T_{ji} \,, \tag{3.4b}$$

then it is straightforward to show that an energy function exists,

$$E = -\tfrac{1}{2} \sum_{i=1}^{N} \sum_{j=1}^{N} T_{ij} S_i S_j \,, \tag{3.5}$$

which is monotonically decreasing under *sequential* updating of the neurons. Neither of conditions (3.4) is plausible in real neural systems: the first implies that the state of a neuron is unaffected by its own state at the previous time-constant, while the second would imply that neuron i has the same influence on neuron j as j has on i. (In reality synapses are very much unidirectional effects.) However, the existence of an energy function (3.5) provides a simpler picture of the nature of memory in the model and, as we shall see shortly, has allowed extensive analysis of the model using the powerful analytical tools of equilibrium statistical mechanics. Since a spin-flip (neuron state change) will only occur if it reduces the energy (3.5), every local minimum of E must be stable against any single spin-flip and so must correspond to a fixed point of the dynamics (3.2). Hence the states stored in the memory are the local minima of E.

We can now also envisage the content-adressable nature of the memory: a stored item will be recalled by specifying enough of its content to ensure that the initial state of the network is in the basin of attraction of the desired item. If this is the case then the network dynamics will evolve the initial configuration to the required local minimum.

Also in this picture, we can see another desirable property of the network model: robustness to "structural damage" as well as to input noise. The "death" of some fraction of the synaptic connections may produce only small changes in the form of the energy surface, so that the stored states in memory are likely to suffer only small perturbations. This property arises owing to the *distributed* nature of the memory: the fact that a given state is stored in the memory depends in general on the values of all of the connections and so the stored state is not localized in the same sense as the stored information in a conventional computer memory.

The particular choice of connection strengths associated with the Hopfield model is based on the suggestion by Hebb [3.4] that the synaptic connection between two neurons tends to be enhanced if the two neurons are simultaneously active. A simple way of implementing this in the context of the model here is to use the prescription proposed by Cooper et al. [3.5]. Given that we wish to store a set of p nominated patterns $\{\xi_i^\mu; i = 1, \ldots, N; \mu = 1, \ldots, p\}$, where each ξ_i^μ is an independent random variable assuming the values +1 or -1 with equal probablility, then the connections are constructed according to:

$$T_{ij} = \frac{1}{N} \sum_{\mu=1}^{p} \xi_i^\mu \xi_j^\mu \,, \quad T_{ii} = 0 \tag{3.6}$$

(and the thresholds U_i are set to zero).

The first thing we should note is that this choice obeys the conditions (3.4) and so an energy function exists. We can easily appreciate the rationale for this choice. Firstly, it is a local "training" procedure in that the change in T_{ij} due to the introduction of an additional pattern only depends on the activities of the two neurons which it connects. Secondly, when the network is in state $\boldsymbol{\xi}^{\mu}$, substituting (3.6) into (3.1), we see that the local field experienced by the ith neuron will be

$$h_i = \xi_i^{\mu} + \frac{1}{N} \sum_{\nu \neq \mu} \sum_{j \neq i} \xi_i^{\nu} \xi_j^{\nu} \xi_j^{\mu} \,. \tag{3.7}$$

The first term on the right-hand side will reinforce the stability of the spin ξ_i^{μ} and so is like a "signal" term. The other one, however, is essentially the sum of $(N-1)(p-1)$ random variables, each ± 1, thus giving rise to a "noise" term roughly of a Gaussian nature with mean zero and standard deviation $\simeq \sqrt{p/N}$. So, as long as the "storage ratio" p/N (traditionally denoted by α) is not too large (compared with 1), we would expect (3.6) to be a successful prescription. If we were to treat this "signal-plus-noise" analysis exactly, we would have to properly take into account the correlations which appear between the noise terms appearing for each spin i and each pattern μ. (This is done, e.g., in [3.6]).

In the above we have used a "spin variable" $(1,-1)$ representation, but it is also possible to work with (1,0) variables defined by $V_i \equiv (S_i + 1)/2$ (we shall refer to this as the "V-model"). For a given choice of T_{ij}, the dynamics of the two models can be made equivalent by setting the (otherwise arbitrary) thresholds in (3.1) to be $U_i = -\sum_j T_{ij}$. If the thresholds in both models were chosen to be zero, then we would have two distinct models.

Another thing we should note about the model for any choice of symmetric connections is that if a pattern is stored, then its complement (obtained by flipping all the spins) will also be stored, assuming zero thresholds. This is not the case for the V-model with zero thresholds.

3.1.4 The Spin-glass Analogy

The strong analogy of the model as a system of interacting Ising spins has already been pointed out above. This becomes even stronger for the case of symmetric T_{ij} when we realize that the energy function (3.5) which governs the single spin-flip dynamics is just like the Hamiltonian for the Ising model of a uniaxial magnet. One main difference, however, is that in Ising-spin models of uniaxial magnets, these exchange constants are typically restricted to nearest- or next-nearest-neighboring spins only. In general, in these neural-network models, the connections can be fully connected in that each neuron may influence and be influenced by any other.

In the Hopfield model if just one pattern is stored then it is equivalent to the Mattis model [3.7], which only has two energy minima: the stored pattern itself and its complement. It is just a fully connected ferromagnetic model subjected to a gauge transformation (locally redefining "up"/"down" to be "aligned"/"flipped" with respect to the spin in the stored pattern). At the other extreme, where the number of stored patterns becomes very large ($p \to \infty$), each T_{ij} tends to a Gaussian variable of zero mean (as a consequence of the central-limit theorem). This is very similar to the long-range Sherrington–Kirkpatrick (SK) model [3.8] of a spin-glass in which each bond is typically drawn from a Gaussian distribution. (However, in the model here there are still correlations amongst the T_{ij} as they are each built from a common set of random bit patterns.) It is known that such a model exhibits very complex behavior as a result of the disorder in the system and of the frustration (the competing ferromagnetic/antiferromagnetic interactions) that is present. In fact, one of the consequences is that the number of local energy minima increases exponentially with the system size [3.9].

In some sense, then, we would expect the case of finite p to be an interpolation between the Mattis and the SK spin-glass models.

3.1.5 Finite Temperature

Noise can be introduced into the deterministic dynamics (3.2) in a parameterized fashion. For example (as in the Little [3.10] model), the state $S_i(t+\tau)$ at time $t+\tau$ is no longer certain to align with its local field, but instead adopts the value S' with probability

$$P(S_i(t+\tau) = S') = \frac{1}{1+\exp(-2\beta h_i S')} . \tag{3.8}$$

β measures the amount of noise present (with $\beta \to \infty$ recovering the determinstic dynamics (3.2)) and is analogous to an inverse "temperature" parameter. In the symmetric T_{ij} case this analogy becomes clear when we realize that, since (3.8) under sequential dynamics obeys detailed balance, the network will relax into a stationary (equilibrium) state, these being distributed according to the Boltzmann factor $\mathrm{e}^{-\beta E}$, where E is the configurational energy (3.5) of the state into which the system relaxes. So, just as in equilibrium statistical mechanics, if we are interested in the properties of these stationary states, we have to deal with the free energy of the system

$$F = -T \ln \sum_{\boldsymbol{S}} \mathrm{e}^{-\beta E(\boldsymbol{S})} . \tag{3.9}$$

One property which is of prime importance is the extent to which a nominal pattern resembles any of the equilibrium (i.e., stored) states. This can be measured by determining the *overlap* m^μ between a nominal pattern $\boldsymbol{\xi}^\mu$ and a stationary state $\boldsymbol{S}$ of the system:

$$m^\mu \equiv \frac{1}{N} \sum_{i=1}^{N} S_i \xi_i^\mu . \tag{3.10}$$

We can see that $m^\mu = 1$ only if $\boldsymbol{S}$ and $\boldsymbol{\xi}^\mu$ are identical patterns, $m^\mu = -1$ if they are each other's complement, and $m^\mu = O(1/\sqrt{N})$ if they are uncorrelated with each other. If $m^\mu = O(1)$, we say that it is a *macroscopic* overlap, while if it is $O(1/\sqrt{N})$ or less, it is a *microscopic* overlap. Note that m^μ is related to the "hamming distance" d^μ between the patterns (the fraction of spins which differ) by $d^\mu = (1 - m^\mu)/2$.

Amit et al. [3.11] have explored in some depth the properties of the Hopfield model storing p random uncorrelated patterns on N neurons. Their mean-field analysis, based on the replica methods [3.12, 13] developed for spin-glass models, is exact in the thermodynamic limit, i.e., for $N \to \infty$. (However, we should be aware that the replica method employed for averaging over the disorder in the system due to the different possible realizations of nominal patterns, is only strictly valid for temperatures $T > T_R$, the AT temperature [3.14]. In particular, the $T = 0$ (deterministic) case violates this criterion. It has nonetheless been shown [3.15] that the corrections due to "replica-symmetry breaking" are small.)

We shall present here a summary of the results for the zero-temperature model in terms of the storage ratio $\alpha \equiv p/N$.

- For $\alpha \ll 1/4 \ln N$ the nominal patterns are stored perfectly: no spins are wrongly aligned.
- For $\alpha \leq 1/4 \ln N$ errors begin to appear: minima of the energy differ from the nominal patterns at a few spin sites.
- For $p \geq 3$ "spurious" minima are created in the energy surface: these are linear combinations (admixtures) of nominal states. All of the nominal states are still themselves successfully stored, albeit with a few errors, as each one lies very close to an energy minimum. Owing to their high correlation with a nominal state, these nearby stable states are designated *ferromagnetic* (FM) states.
- For finite α ($p = O(N)$) another class of spurious states appear: the *spin-glass* (SG) states. Unlike the mixture states, these are uncorrelated with any of the embedded nominal states. The number of SG states is exponential in N.
- As α is increased beyond a first critical value, α_1^c, a phase transition occurs in the sense that the energy of the spurious states becomes lower than that of the FM ("memory") states. The network still performs as an associative memory for the nominal states as the FM states are still energy minima (with a depth of $O(N)$).
- There is a second phase transition at a higher value α_2^c, at which the FM memory states associated with the nominal states cease to exist. The only remaining stable states are SG states and so the storage prescription (3.6) suffers complete loss of memory capacity.

Amit et al. calculated the two critical storage ratios as

$$\alpha_1^c \simeq 0.05\,, \quad \alpha_2^c \simeq 0.14$$

within a replica-symmetric mean-field approximation.

Although we might have anticipated it from a simple "signal-plus-noise" analysis (*à la* (3.7)), it is in fact nontrivial to show that the corresponding critical storage ratios α_1^c and α_2^c have roughly half of these values when we use the V-model. This is shown in [3.6].

The themodynamics of the *projector* model [3.16, 17] has also been solved [3.18]. In this model, the choice of connections is

$$T_{ij} = \frac{1}{N} \sum_{\nu,\mu=1}^{N} \xi_i^\mu C_{\nu\mu}^{-1} \xi_j^\nu \,, \tag{3.11}$$

where the correlation matrix is

$$C_{\mu\nu} \equiv \frac{1}{N} \sum_{k=1}^{N} \xi_k^\nu \xi_k^\mu \,.$$

T_{ij} is a *projection* matrix onto the subspace spanned by the nominal patterns. The C_{ij} matrix can also be regarded as a *pseudo-inverse* of the pattern matrix $X_{i\nu} \equiv \xi_i^\nu$; $T = X(X^{\mathrm{t}}X)^{-1}X^{\mathrm{t}}$.

For a finite number of nominal patterns ($\alpha = 0, N \to \infty$) (3.6) and (3.11) are identical as the patterns of the Hopfield model become completely uncorrelated, $C_{\mu\nu} = \delta_{\mu\nu}$. When $\alpha > 0$ the projector model will store all of the patterns perfectly up to $\alpha = 1$, provided that C^{-1} exists. (This is guaranteed to be the case if the nominal patterns are linearly independent.) It turns out [3.18] that there is also a first-order phase transition as α is raised beyond $\alpha_c = 1$. Unlike the Hebb rule (3.6), the general form of (3.11) doesn't provide a simple, local modification to each T_{ij} on presentation of an additional nominal pattern.

3.2 Content-addressability: A Dynamics Problem

The powerful tools of equilibrium statistical mechanics allow us then to determine the accuracy of the memory (given the connections T_{ij}): we can find out how close the dynamically stable states are to the desired nominal states. This is, of course, an essential prerequisite for a good content-adressable memory, but it is not alone sufficient.

Given that a certain choice of $\{T_{ij}\}$ stores our set of nominal patterns within some desirable accuracy, we must also find out how good the resultant content-addressability of the memory is. This is directly related to the size of the regions of attraction around each stored state. The memory would be of little use if we were to find that, although it stored every nominal pattern exactly, the regions of attraction around them were negligible – in such a case the memory would have to be presented with almost the complete pattern in order to retrieve it. Ideally we would like the basins of attraction around each pattern to be as large as possible. We might consider the optimal behavior to be

$$S(\infty) = \xi^{\mu} \quad \text{where} \quad \mu = \max_{\nu} \frac{1}{N} \sum_{i=1}^{N} S_i(0)\xi_i^{\nu} \,,$$

i.e., the network produces the nominal pattern μ which most resembles the input pattern $S(0)$. (Of course, one way to guarantee finding the desired nominal pattern μ would be to compute the overlap of the input pattern with each of the p nominal patterns and choose the one with the greatest overlap. This would require $\mathrm{O}(pN)$ direct bit comparisons then $\mathrm{O}(\ln p)$ steps to find the best match. Assuming that the network has the correct behavior and produces the desired output pattern, as it evolves it is simultaneously "considering" all of the nominal patterns and requires $\mathrm{O}(N^2)$ integer operations per sweep.)

Determining the sizes of the basins of attraction is a dynamics problem. We must look at the configurational flow of the network: given that the network is initialized in some state which has been obtained by adding noise (randomly flipping the spins) in a nominal configuration, how does the state of the network subsequently evolve? Does it flow towards and then stabilize sufficiently close to the desired nominal state? The answer to these questions can also depend on the order in which the neurons are updated.

As is so often the case in statistical mechanics, the dynamics problem is more difficult to solve than the statics one (of storage accuracy). Consider, for example, that the SK model is well understood in terms of the static properties of the Parisi solution, but dynamical questions such as relaxation times and remanence remain unsolved.

There have been some recent attempts at tackling this problem analytically, but it has not yet proved possible to do so without making approximations. The main problem is in the strong correlations which arise. The local field at any spin during a given updating sweep of the network depends on what happened to all the spins during the previous sweep. That in turn depends on the preceding sweep to that one, and so on back in time.

If the network is updated completely in parallel, and if we consider its time evolution only in terms of its overlap with one of the patterns (ξ^{μ}, say) and ignore higher correlations, then Kinzel [3.19] has shown that

$$m^{\mu}(t+\tau) = \int \mathrm{e}^{-z^2/2} \tanh\left[\beta(m^{\mu}(t) + \Delta(t)z)\right] \, \mathrm{d}z \,, \tag{3.12}$$

where

$$\Delta^2 = \left[1 - (m^{\mu}(t))^2\right] \left\langle \sum_j J_{ij}^2 \right\rangle + m^{\mu}(t)^2 \delta \,,$$

with $\delta = \alpha$ ($\delta = 0$) for the Hopfield (projector) model. The angular brackets indicate an average over the possible realisations of the nominal patterns ξ^{μ}. (This is in fact an exact result for an extremely diluted asymmetric model [3.20].) $m^{\mu}(\infty)$ would tell whether the μth pattern has been successfully retrieved or not.

Horner [3.21] has also been able to study the basins of attraction in the Hopfield model within a mean-field approximation. There one finds that at values of $\alpha < 0.138$ the initial overlap $m^{\mu}(0)$ must be greater than a critical minimum overlap to retrieve the nominal pattern. For example, at $\alpha = 0.1$, this minimum overlap is around 0.45.

Gardner et al. [3.22] were able to obtain exact solutions for $m^{\mu}(t)$ with parallel updating, but their analysis was limited to the first two updating sweeps of the network only.

Meir and Domany [3.23] have shown that it is possible to solve the dynamics exactly when we deal with a *layered feed-forward* network. In such networks the neurons are arranged in successive layers with connections only between layers and only in one (the "forward") direction. The state of each layer only depends on the state of the preceding layer, thus eliminating the troublesome correlations due to feedback. They found that the network displays critical behavior in that in order to retrieve a stored pattern μ the overlap m^{μ} of the input pattern with the desired pattern must exceed a critical value $m^{c}(\alpha)$, where $N\alpha$ patterns are stored. Below we shall see numerical evidence that the Hopfield model also seems to produce this dynamical critical behavior.

3.2.1 Numerical Tests

We can always resort to numerical simulations to test the content-adressability of the model. Figure 3.1 shows the results of one such test [3.24] applied to the Hopfield model storing $N\alpha$ patterns on N neurons. Serial (single spin-flip) dynamics were used in these simulations. A pattern is deemed to have been successfully recalled if the final state of the network differs from it in no more than

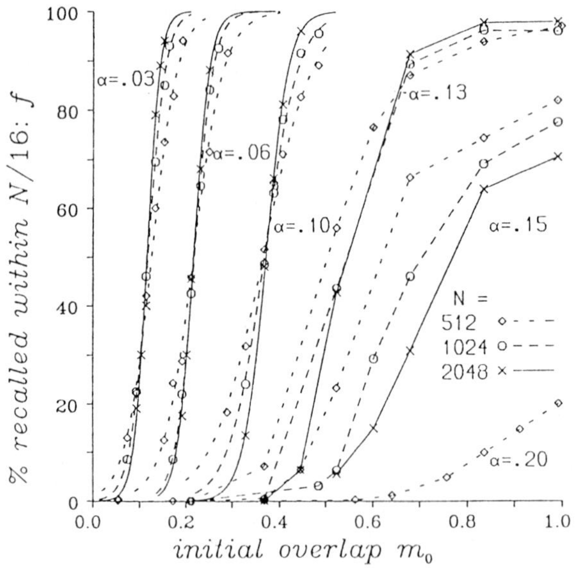

Fig. 3.1. The fraction of nominal states in the Hopfield model recalled with less than $N/16$ errors from initial states having overlap m_0

$N/16$ spins. This margin of recall error allows for the possibility of satellite (or "ferromagnetic") states highly correlated with the nominal states being stored instead of the nominal states themselves [3.11]. Shown in Fig. 3.1 for three different system sizes N is the fraction $f(m_0)$ of nominal states that are successfully recalled from initial noisy patterns having a specific overlap m_0 with the nominal state.

We can see that the behavior for the three lower values of α (and also, to a lesser extent, for $\alpha = 0.13$) is in marked contrast to that of $\alpha = 0.15$. While for increasing N the performance for $\alpha = 0.15$ shows a gradual worsening, the behavior of the lower values suggests an approach towards a discontinuous jump from no recall ($f = 0$) to certain recall ($f = 1$). This is borne out by fitting the finite-size scaling forms

$$f(m_0) = \frac{1}{1 + C\exp[N(m_c - m_0)]}\,, \tag{3.13}$$

to the curves for the three lower values of α, where C and a are positive constants independent of N (but possibly dependent upon α). This form is equivalent to expressing the ratio of probability of recall to that of nonrecall in the following scaling form:

$$\frac{f(m_0)}{1 - f(m_0)} = C\exp[aN(m_0 - m_c)]\,. \tag{3.14}$$

In the thermodynamic limit ($N \to \infty$), we would imply from this that, in addition to its critical behavior with respect to storage capacity, the Hopfield model also exhibits a first order dynamical phase transition. The "control parameter" here is m_0, the initial overlap of the network with a nominal pattern. For low values of m_0 the system is in a dynamically "disordered" phase and the nominal pattern is not recalled ($f(m_0) = 0$). As m_0 is increased past a critical overlap m_c, there is a discontinuous change in the probability of recall (which is like an "order parameter" here) and the desired pattern is definitely retrieved ($f(m_0) = 1$). Estimates for the crtical overlaps can be obtained [3.24] by extrapolating the results to $N^{-1} \to 0$ (see Table 3.1).

Table 3.1. Numerical estimates of C, a, and m_c

α	C	a	m_c
0.03	0.755	0.036	0.111(10)
0.06	1.086	0.028	0.218(13)
0.10	1.070	0.020	0.373(17)

We would expect $m_c \to 1$ as $\alpha \to \alpha_2^c$, since higher values of α implies the nonexistence of any minima in the vicinity of the nominal patterns and, of course, of any basins of attraction around them.

3.3 Learning

We have seen that the Hopfield model using the Hebb prescription (3.6) is somewhat limited in its ability to store random uncorrelated bit patterns: it fails to store more than 0.15 of them and for a general choice of connections it should be possible to store up to $2N$ such patterns [3.25, 26]. Moreover, even for $\alpha < \alpha_2^c$, this storage prescription cannot ensure perfect recall of the patterns.

In an attempt to find a better solution to this storage problem, we can turn to "learning algorithms": iterative error-correcting techniques which are applied to the synaptic connections in an attempt to endow the network with the desired storage capabilities. We may view this as a dynamical process that is reciprocal to that of recall. In the latter process, fixed synaptic connections govern the dynamical response of the neurons, while in the former, the activity of the neurons determine the evolution of the synapses.

3.3.1 Learning Perfect Storage

The first such learning algorithm [3.27, 28] we shall consider is conceptually simple in nature. At each learning iteration, every nominal pattern is tested for stability and in every one which fails, those unstable spins (ξ_i^μ) are reinforced by adding a Hebbian (3.6) term to all the synaptic connections impinging on those spins:

$$T_{ij} \to T_{ij} + \frac{1}{N}\xi_i^\mu \xi_j^\mu \; ; \quad j = 1, \ldots, N \, . \tag{3.15}$$

In other words, we construct an error mask ε_i^μ at each spin i for each nominal pattern μ:

$$\varepsilon_i^\mu \equiv \tfrac{1}{2}\left\{1 - \operatorname{sgn}\left(\xi_i^\mu \sum_{j=1}^{N} T_{ij}\xi_j^\mu\right)\right\} \tag{3.16a}$$

so that ε_i^μ is O(1) if the ith spin is (is not) correctly aligned with its local field. The connections T_{ij} are then modified accordingly by

$$T_{ij} \to T_{ij} + \frac{1}{N}\varepsilon_i^\mu \xi_i^\mu \xi_j^\mu \, . \tag{3.16b}$$

The symmetry of the connections can be preserved if we instead employ

$$T_{ij} \to T_{ij} + \frac{1}{N}(\varepsilon_i^\mu + \varepsilon_j^\mu)\xi_i^\mu \xi_j^\mu .$$

We repeatedly test all the nominal states and when (or if!) all of the error masks are zero, the learning is complete. This is perceptron learning.

The algorithm can indeed ensure perfect storage of all the nominal patterns; not just for $\alpha < \alpha_2^c$, but even up to at least $\alpha = 1$. However, the promising behavior for the higher values of α is short-lived since upon closer scrutiny we

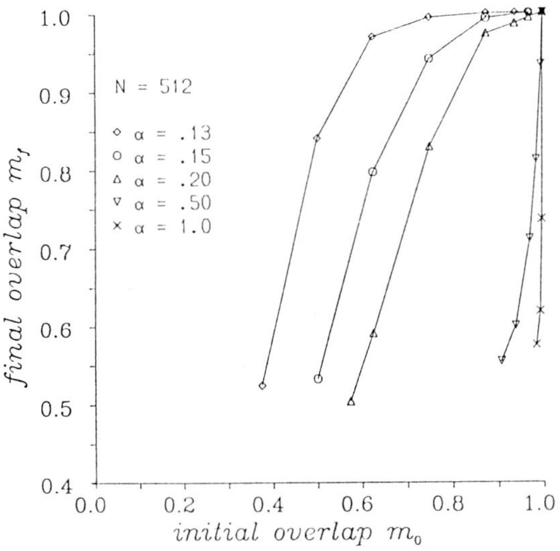

Fig. 3.2. Mean final overlap after iteration of noisy nominal states stored using the learning algorithm (3.16)

find that the resultant basins of attraction around the $N\alpha$ patterns for $\alpha = 0.5$ or 1 are negligibly small (Fig. 3.2).

Perhaps this should not have been all that surprising when we realize that all we are demanding of the patterns is that they are dynamically stable: the alignment of a spin and its local field need not necessarily be large, it need only be positive. The small alignments created for these larger values of α are easily disturbed (turned negative) by flipping only a few of the spins in a nominal configuration. This can then in turn flip other properly aligned spins and the state of the system flows further away from the nominal state and so the network fails to recall the pattern from just a small amount of noise.

3.3.2 Enforcing Content-addressability

So, in order not only to store (stabilize) the patterns, but also to ensure that they have appreciable content-addressability, we must attempt to "excavate" basins of attraction around them. One such approach is that of the algorithms of the following type [3.24, 29–32]. We attempt to ensure that the alignment of spin and local field is not just positive but greater than some minimum (positive) bound:

$$\xi_i^\mu h_i \equiv \xi_i^\mu \sum_j T_{ij} \xi_j^\mu > B > 0 \,, \tag{3.17}$$

This we could do by simply modifying the error masks in (3.16) to

$$\varepsilon_i^\mu \equiv \tfrac{1}{2} \left\{ 1 - \operatorname{sgn}\left(\xi_i^\mu \sum_{j=1}^{N} T_{ij} \xi_j^\mu - B \right) \right\} , \tag{3.18a}$$

and then change the T_{ij} as before:

$$T_{ij} \to T_{ij} + \frac{1}{N}\varepsilon_i^\mu \xi_i^\mu \xi_j^\mu \quad \text{or} \quad T_{ij} + \frac{1}{N}(\varepsilon_i^\mu + \varepsilon_j^\mu)\xi_i^\mu \xi_j^\mu . \tag{3.18b}$$

In this way we would hope that basins of attraction would be implicitly created around the patterns.

However, if we are to adopt this approach, we must be wary of the trap of simply scaling up the sizes of the connections. Suppose we were to increase all of the T_{ij} by a constant multiplicative factor, $k(> 0)$, say. Then first of all we note that the dynamics of the network would stay the same:

$$S_i(t+\tau) = \text{sgn}\left\{\sum_j (kT_{ij})S_j(t)\right\} = \text{sgn}\left\{\sum_j T_{ij}S_j(t)\right\} . \tag{3.19}$$

Now suppose we wished to find $\{T_{ij}\}$ such that all of the spin to local field alignments were at least $B > 0$, i.e., satisfying (3.17), but that we had only succeeded in finding a set $\{t_{ij}\}$ that satisfied

$$\xi_i^\mu \sum_j t_{ij}\xi_j^\mu > b \,, \quad (b < B) \,.$$

Then all we have to do to solve our original problem (3.17) is to scale up all of the t_{ij} by an amount $k \equiv B/b$:

$$T_{ij} \equiv \frac{B}{b}t_{ij} \,,$$

for then

$$\xi_i^\mu \sum_j T_{ij}\xi_j^\mu \equiv \xi_i^\mu \sum_j (kt_{ij})\xi_j^\mu > kb \quad (\equiv B) \,.$$

Hence to avoid this happening, we must ensure that the bound B in (3.17) is of the same order of magnitude as $\xi_i^\mu h_i$. If the T_{ij} are $O(T)$ then $h_i(\boldsymbol{\xi}^\mu)$ will be $O(T\sqrt{N})$ since it consists of N terms, each $O(T)$ and each positive or negative with equal probability. Hence, this means that B must also be $O(T\sqrt{N})$. One possible choice [3.24] is

$$B = M\langle |T_{ij}|\rangle \sqrt{N} \,, \tag{3.20}$$

where $\langle\ \rangle$ denote the average with respect to all of the connections in row i of the T_{ij} matrix:

$$\langle |T_{ij}|\rangle \equiv \frac{1}{N}\sum_{j=1}^{N} |T_{ij}| \,.$$

We now have that M is the proper bound to be varied, as it is $O(1)$ and independent of N and of any scaling of the T_{ij}.

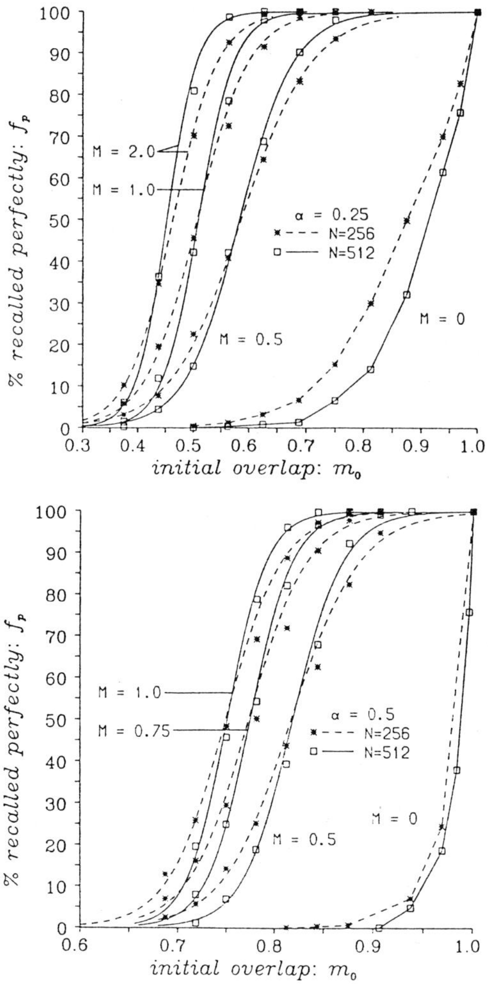

Fig. 3.3. After learning with (3.18): the fraction of perfectly recalled nominal states from various initial overlaps at $\alpha = 0.25$ (*top*) and $\alpha = 0.5$ (*bottom*)

Note that the first algorithm that we considered (3.16) just corresponds to $M = 0$ here. Implementing the algorithm (3.18–20) on networks of $N = 256$ and 512 neurons [3.24], we find that, in addition to perfect storage, content-addressability is indeed also achieved for the nonzero values of M at $\alpha = 0.25$ and 0.5 (Fig. 3.3). Scaling forms similar to those in Fig. 3.1 are fitted and we can see from the trend from $N = 256$ to 512 that appreciable basins of attraction will probably be created as $N \to \infty$.

As with the Hopfield model the existence of a critical minimum overlap m_c is suggested (this time it depends on M as well as on α). Estimates for m_c at $\alpha = 0.25$ are

$$m_c(M = 0.5) = 0.58(3)\ , \quad m_c(M = 1.0) = 0.51(2)\ ,$$
$$m_c(M = 2.0) = 0.44(2)\ ,$$

and at $\alpha = 0.5$:

$$m_c(M = 0.5) = 0.82(4)\,, \quad m_c(M = 0.75) = 0.77(3)\,,$$
$$m_c(M = 1.0) = 0.75(3)\,.$$

This behavior contrasts with that of $M = 0$ where the suggested trend is to negligible regions of attraction and hence to no content-addressability.

3.3.3 Optimal Learning

In the approach adopted by Gardner [3.30, 32], the connections J_{ij} are normalized to

$$\frac{1}{N}\sum_{j\neq i} J_{ij}^2 = 1\,. \tag{3.21a}$$

This, of course, is an equally valid precaution against arbitrary rescaling of the connections, as it enforces $O(J_{ij}) = 1$. The stringent condition – analogous to (3.17) – which is then imposed on the spin to local field alignments is

$$\frac{1}{\sqrt{N}}\xi_i^\mu \sum_{j\neq i} J_{ij}\xi_j^\mu > K\,. \tag{3.21b}$$

Since $O(J_{ij}) = 1$ and so $O(\sum_j J_{ij}\xi_j^\mu) = \sqrt{N}$, we see that K is O(1) and is the analog of M in (3.20). (We shall see below that K and M are in fact related merely by a multiplicative constant despite arising from rather different formalisms.)

Given a set of nominal random bit-patterns, is it at all possible to choose an appropriate set of interactions $\{J_{ij}\}$ subject to the normalization (3.21a) such that all of the patterns will have the required stability (3.21b)? Gardner [3.30, 32] has provided the answer to this question by calculating the fractional number of connection matrices $\{J_{ij}\}$ which satisfy (3.21b) out of all those obeying (3.21a).

We can regard a given choice of connection matrix $\{J_{ij}\}$ as occupying one point in the $N(N-1)$-dimensional phase-space spanned by the $N(N-1)$ independent elements J_{ij} $(i \neq j)$ of the matrix (assuming a general asymmetric matrix). All matrices obeying (3.21a) will occupy some volume of this space and contained within it will be all those (if any) satisfying both (3.21a) and (3.21b). The ratio of the second volume to the first will give the fractional number of solutions. If this yields a nonzero value, then there must be at least one way of choosing an appropriate set of connections.

The required ratio is

$$V = \frac{\int \prod_{i\neq j} \mathrm{d}J_{ij} \prod_\mu \theta\left(\xi_i^\mu \sum_{j\neq i} J_{ij}\xi_j^\mu - K\sqrt{N}\right)\delta\left(\sum_{j\neq i} J_{ij}^2 - N\right)}{\int \prod_{j\neq i} \mathrm{d}J_{ij}\delta\left(\sum_{j\neq i} J_{ij}^2 - N\right)} \tag{3.22}$$

The δ-functions enforce (3.21a), whilst the θ (threshold) functions ensure compliance with (3.21b).

This type of calculation is a kind of statistical mechanics in the space of couplings between spins: here the spins are quenched at the values ξ_i^μ and the couplings J_{ij} are allowed to vary. In problems such as spin-glasses [3.8, 12] or the storage capacity of a network using a particular choice of connections, the *inverse* problem is solved: the couplings are quenched and it is the spins which are varied.

The particular value of V will depend on the particular realization of nominal patterns chosen, so to obtain the typical value of V, the dependence of (3.22) on the choice of nominal patterns must be eliminated by averaging the result over all possible realizations of the nominal patterns (which can be done, once again, with the help of the replica trick). It turns out that for $N\alpha$ patterns, there exists a maximum possible K for which at least one solution exists, its value K^* being given by [3.30, 32]

$$\int_{-K^*(\alpha)}^{\infty} \frac{\exp(-t^2/2)}{\sqrt{2\pi}} \left(t + (K^*(\alpha))^2\right) \mathrm{d}t = \frac{1}{\alpha} . \tag{3.23}$$

This has the behavior $K^* \to 0$ as $\alpha \to 2$, consistent with the result that $\alpha = 2$ is the maximum storage capacity for random uncorrelated bit patterns [3.25, 26].

In the light of the above calculation we can now obtain a prediction for the optimal value of the bound M in (3.20). First of all we must expose the relationship between M and K. To do this we shall scale the connections T_{ij} to conform with the normalization (3.21a). We define new connections J_{ij} by

$$J_{ij} \equiv \frac{T_{ij}}{\sqrt{\left(\sum_{j=1}^{N} T_{ij}^2\right)/N}} \tag{3.24}$$

(recall that we are free to do so since the neural dynamics will remain unaffected). Now (3.17) is

$$\xi_i^\mu \sum_{j=1}^{N} T_{ij}\xi_j^\mu > M\sqrt{N}\langle|T_{ij}|\rangle .$$

Dividing this through by $\sqrt{\sum_{j=1}^{N} T_{ij}^2}$ then gives

$$\frac{1}{\sqrt{N}}\xi_i^\mu \sum_{j=1}^{N} J_{ij}\xi_j^\mu > M\frac{\langle|T_{ij}|\rangle}{\sqrt{\langle T_{ij}^2\rangle}} ,$$

which, by comparison with (3.21b), tells us that

$$M \equiv K\frac{\sqrt{\langle T_{ij}^2\rangle}}{\langle|T_{ij}|\rangle} . \tag{3.25}$$

Finally, we note that the distribution of connections $\{T_{ij}\}$ remains Gaussian of zero mean [3.24] after completion of the learning, so that the ratio multiplying

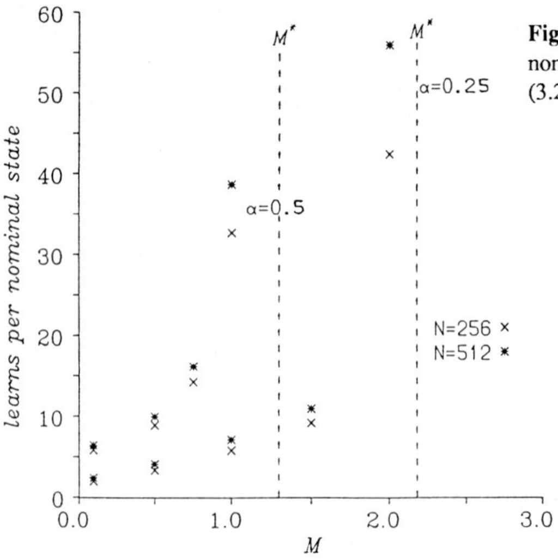

Fig. 3.4. The number of learns required per nominal state using (3.18) with B given by (3.20)

K in (3.25) is nothing more than the width of a Gaussian variable (of mean zero) divided by its mean absolute value, i.e., we have that $M = K\sqrt{\pi/2}$.

From this and (3.23) we find that the optimal values $M^*(\alpha)$ for Fig. 3.3 should be $M^*(0.25) \simeq 2.18$ and $M^*(0.5) \simeq 1.30$. The largest values attempted were 2.0 and 1.0 respectively. Figure 3.4 gives an indication of the total effort needed to learn to completion as M is increased.

In our procedure, one "learn" consists of a learning sweep through any nominal state which has not yet been perfectly stored. The optimal values of M^* are indicated: learning becomes rapidly more difficult as M approaches M^*, in accord with the analytical results of Gardner.

We should also note here that the numerical simulations used a symmetric version of the learning algorithms, but that the analytical results for K^* assume general asymmetric connections. Thus it might not necessarily be possible to attain the optimal M using symmetric connections. Nonetheless, the numerical results show that values approaching M^* can be achieved, and that imposing stronger alignments between spin and local field in the nominal patterns does indeed implicitly create appreciable content-addressability.

Even though the calculation of (3.22) provides us with the answer to whether a solution for the choice of connections exists, it may seem rather surprising that an algorithm such as (3.18) manages to find a solution: it is, after all, a deterministic search through a very high $(N(N-1))$-dimensional solution space of quasi-continuous variables. (They become continuous in the thermodynamic limit of $N \to \infty$.) We may initially have thought it necessary to resort to more powerful, complicated schemes such as those involving a stochastic search (for example, a type of simulated annealing). However, in addition to telling us whether any solution exists, Gardner has also demonstrated [3.30, 32] a convergence theorem for the algorithm which proves that it will converge to a solution provided one exists (which should be the case if $K < K^*$). The proof of the theorem (see the Appendix) is essentially an extension of the perceptron convergence theorem

[3.33] to the case of many threshold functions. The algorithm must be executed in series through the nominal patterns for the theorem to hold. The symmetric version can also be shown to converge to a solution, but there is a further requirement: the error masks must be calculated *in parallel* over the sites if we are to be assured of convergence.

As further evidence that these learning schemes can install finite content-addressability, more recently Kepler and Abbot [3.34] have studied the recall properties of networks trained near saturation (i.e., with values of K close to K^* in (3.21b)). They have observed that, under parallel updating, a stored pattern $\boldsymbol{\xi}^\mu$ will be retrieved from an initial state having overlap m_0 with it, only if, after one parallel sweep, the state has traveled at least half of the distance to $\boldsymbol{\xi}^\mu$. That is, the overlap m_1 after one sweep must satisfy

$$\frac{m_1 - m_0}{1 - m_0} \geq \frac{1}{2} \, . \tag{3.26}$$

This observation[1] then also implies the existence of a critical minimum overlap m_c, where

$$\frac{m_1 - m_c}{1 - m_c} = \frac{1}{2} \, , \tag{3.27}$$

such that all states having initial overlap $m_0 > m_c$ with a nominal pattern will be in that pattern's domain of attraction and so will be guaranteed to retrieve it. m_c turns out to be function only of K^*.

For the learning algorithm of Krauth and Mézard [3.31], Opper [3.35] has recently been able to apply a replica approach to find the mean number of learning steps required to obtain optimal stability of $N\alpha$ random patterns in the thermodynamic limit, with α remaining finite.

The algorithm seeks to impose a minimum stability on the spins and does so by determining at each learning step which patterns (if any) do not have a minimum stability (3.17) at spin site i. Out of those which do not, the one ξ_i^ν with the *lowest* stability $\xi_i^\nu h_i^\nu$ is found and the connection T_{ij} is adjusted:

$$T_{ij} \to T_{ij} + \frac{1}{N} \xi_i^{\nu(i)} \xi_j^{\nu(i)} \quad (j \neq i) \, , \tag{3.28}$$

where $\nu(i)$ denotes the pattern which has the worst stability at site i. The algorithm is finished when all patterns obey (3.17).

For optimal stability, the mean number of learning cycles (through all the spins) is Δ^{-2} where [3.35]

$$\int_{-\Delta}^{\infty} \frac{\exp -t^2/2}{\sqrt{2\pi}} \left(t + (\Delta)^2\right) \mathrm{d}t = \frac{1}{\alpha} \, , \tag{3.29}$$

i.e., an identical form to (3.23). Δ diverges as $(2 - \alpha)^{-2}$ for $\alpha \to 2$, which once again agrees with the maximum capacity of $2N$ random patterns [3.25, 26].

---

[1] Since the time of writing, the criterion advocated by Kepler and Abbot has been called into question by J. Krätzschmar and G.A. Kohring (J. Phys. France **51**, 223 (1990).

3.3.4 Training with Noise

An alternative approach to the type of learning algorithm described above is to attempt to explicitly create domains of attraction around each nominal pattern. This involves training the network to associate noisy versions of each nominal pattern with the desired pattern.

Specifically, suppose we have a state S^μ which is a noisy version of one of the nominal states, $\boldsymbol{\xi}^\mu$, say, such that they differ in Nf spin sites for some fraction f $(0 \leq f < 1/2)$. Then to train the network to associate S^μ with $\boldsymbol{\xi}^\mu$ we can iterate the state S^μ once and then define error masks

$$\varepsilon_i^\mu(S^\mu) \equiv \tfrac{1}{2}\left\{1 - \operatorname{sgn}\left(\xi_i^\mu \sum_{j=1}^{N} T_{ij} S_j^\mu\right)\right\} \tag{3.30a}$$

which are 0 (1) according to whether site i of pattern μ is (is not) correctly retrieved in one sweep from the noisy state S^μ. We then update the connections as before:

$$\Delta T_{ij} = \frac{1}{N}\varepsilon_i^\mu \xi_i^\mu S_j^\mu \quad (i \neq j)\,, \tag{3.30b}$$

or

$$\Delta T_{ij} = \frac{1}{N}(\varepsilon_i^\mu \xi_i^\mu S_j^\mu + \varepsilon_j^\mu \xi_j^\mu S_i^\mu) \quad (i \neq j)\,,$$

if we wish to maintain symmetric connections.

A convergence theorem (see the Appendix) can also be proved [3.36] for this algorithm, which once again parallels the perceptron convergence theorem [3.33].

We shall present an application of this algorithm in the following section.

3.3.5 Storing Correlated Patterns

As we noted earlier, there is an intrinsic limitation of $2N$ on the number of random uncorrelated patterns which can be stored by a network with pairwise connections. However, what would happen if we tried to store patterns which contained some correlations with each other?

One simple way of imposing correlations amongst the patterns is to choose all of them with the same "magnetization" m:

$$\langle \frac{1}{N} \sum_{i=1}^{N} \xi_i^\mu \rangle = m\,, \tag{3.31}$$

meaning that in each pattern on average there are $(1+m)N/2$ spins up $(-1 \leq m \leq 1)$. If we were to do this, we would find that the storage ability of the network would improve in that, as m is increased from 0 (the uncorrelated case), the optimal value of K in (3.21b) would increase and the maximum number of patterns stored by the network would increase beyond $2N$ [3.32].

As a further example of storing correlated patterns and as an example of the performance of (3.30), we will now consider the storage of English words containing a given number, L, of letters. The number of such words is in principle 26^L, but in reality the correlations between letters reduce the number of recognisably "English" words to much less than this. (Other similar examples of this nature would be phonetically spelled words or syntactically allowed sentences composed of syntactic elements.)

We shall store a given L-letter word on a network of $N = nL$ neurons by assigning n neurons to each letter, representing a particular letter by some n-bit pattern: these could be random patterns, for example, or n-pixel representations of the letter. We might anticipate that the network could successfully tackle such a problem when we recall that the prototype spurious states generated by the Hebb prescription (3.6) are admixtures of three or more of the prescribed patterns. These are useful generealizations here: for example, with $L = 3$, "SAT" is a $(2/3, 2/3, 2/3)$ mixture of "SIT', "CAT" and "SAD". To the extent that the language is determined by such pair correlations, these specific mixture states may improve the storage capacity compared to storing random words.

Figure 3.5 shows the training success of prescription (3.30) for the case of storing 64 four-letter words on 256 neurons [3.36] (i.e., $L = 4$, $n = 64$, $p = 64$). Each training cycle consisted of up to one entire sweep through all sites and all patterns for one noisy initial vector per pattern, using the symmetric version of (3.30) in parallel on the sites and in series on the patterns. The training schedule consisted of up to 20 cycles of zero noise – in order to learn the patterns themselves – followed by a multiple of 16 cycles with noise of 4, 8, 9, 10, 11, or 12 spin-flips per letter, finishing off with a further session of up to 20 cycles with zero noise to ensure that the patterns themselves remain learnt. The percentage of letters which are exactly recovered starting from noisy vectors with the same amount of noise is plotted against the number of training cycles.

Figure 3.5a shows the results for completely random patterns; Fig. 3.5c for a dictionary of 64 real words, and Fig. 3.5b for the same dictionary, but with the letters jumbled at random amongst the "words" (thus preserving the letter frequency). The learning in Fig. 3.5b is faster than Fig. 3.5a owing to correlations coming from different words containing letters which are the same, and Fig. 3.5c is even faster owing to the pair correlations between letters occuring in real words. Finally we note that successful learning was also obtained with 15% noise on a dictionary of 1760 real words – considerably larger than the system size $N = 256$.

Hence, correlations amongst the patterns which are being stored can considerably enhance the performance of the network, with respect to both learning and retrieval.

What are the relative merits and failings of the two types of algorithm which we have considered here? To answer this quantitatively would require direct comparison by numerical simulation, for only then could we see which one learned fastest, which one could cope with the most input noise for a given storage ratio, etc. We can, nevertheless, compare and contrast them qualitatively.

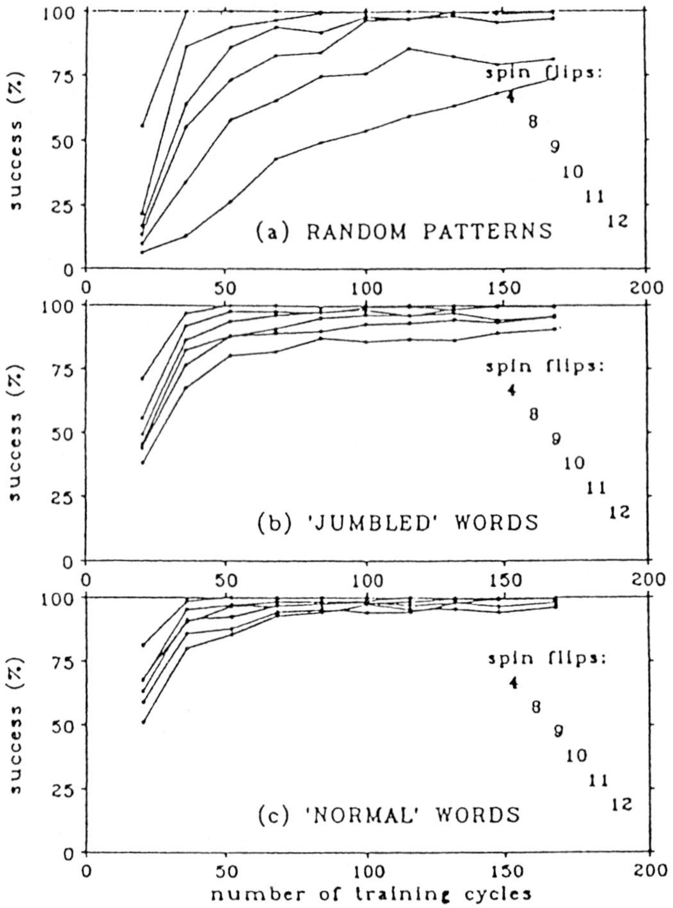

Fig. 3.5a–c. Learning curves for (**a**) random patterns, (**b**) random letters, and (**c**) words

Both will converge to a solution, *provided one exists*: for the particular form (3.21) of the first type (3.18), we know a priori whether a solution does indeed exist (for random patterns or for patterns with the particular correlation given by (3.31), but not for (3.30). Algorithms such as (3.18) only act on the nominal patterns themselves and so only involve one iteration sweep per nominal state during learning. The "training with noise" algorithms test more than one noisy pattern per nominal state: the number of possible states differing from a nominal state in a given amount of sites (i.e., for a given amount of noise) increases exponentially in the number of sites which differ. However, the number of states required to train on in order to achieve finite content-addressability may not necessarily suffer this exponential growth. Further, it can provide the opportunity of creating anisotropic basins of attraction by choosing an appropriate set of noisy patterns to train on. By its very nature, it also demands retrieval of nominal patterns from noisy versions in one sweep.

Although both algorithms are optimal in the sense that they can generate the maximum content-addressability for a given number of patterns, they are not optimal with respect to minimizing the number of learning iterations required. For example, faster convergence may be achieved by taking into account the magnitude of the error at each learning cycle [3.37, 38].

The algorithm (3.30) could also be generalized to train the network to recover patterns after two iterations. For example, (3.30b) could be changed to $\Delta T_{ij} = \varepsilon_i^\mu \xi_i^\mu S_j^{\mu\prime}$, where $S^{\mu\prime}$ is the configuration after one iteration from S^μ and ε_i^μ is the error mask after two iterations.

3.4 Discussion

Perceptron-type learning algorithms can greatly improve upon the storage limitations of the Hopfield model, which fails to store more than $0.15N$ random patterns, falling short of the theoretical limit of $2N$. Nevertheless, it does function as a reasonable content-addressable memory in its retrieval phase ($\alpha < \alpha_2^c$), with the associated basins of attraction exhibiting behavior consistent with a first-order dynamical phase transition.

Learning algorithms such as (3.16) can only ensure the accuracy of the stored nominal states and don't seem to be able to endow the patterns with appreciable regions of attraction. This must be done either implicitly, as in (3.18) – by enforcing a minimum stability on the spins – or explicitly, as in (3.30) – by ensuring noisy versions of a nominal pattern will correctly associate to it.

The performance seems to be enhanced when the nominal patterns contain correlations: the storage capacity and the optimal stability increases for (3.21), while the rate of learning and subsequent retrieval performance for (3.30) are improved.

These perceptron-type algorithms are conceptually simple in nature and despite being deterministic, are guaranteed to find a solution, given that one exists (and providing we execute them serially through the patterns). They also have the appealing quality of being *local* algorithms as the changes to each synapse only depend on the activity of the two neurons which it connects.

There are, of course, a great many modifications and variations of these Ising-like models which we haven't touched upon in this review. For example, we have throughout only considered zero thresholds U_i in (3.1), but the retrieval may be improved by choosing U_i appropriately: the retrieval of the nominal patterns in the Hopfield model is enhanced if the U_i are correlated with the input pattern [3.39]. It is also possible to incorporate higher than pairwise interactions into these models, generalizing (3.1) to

$$h_i(t) = \sum_{j_1, j_2, \dots, j_m} T_{ij_1 j_2 \dots j_m} S_{j_1} S_{j_2} \dots S_{j_m} .$$

Under conditions analogous to (3.4) an energy function exists and the Hebbian prescription (3.6) can be generalized in an obvious way. These "multi-connected" nets also undergo phase transitions and their absolute storage capacity per weight is not significantly improved over that of the pairwise connected net; exact results can be obtained in the limit $m \to \infty$ without resorting to replica methods. Details can be found in Gardner [3.40] and references therein. We have also only considered fully connected models, but the learning algorithms here could easily be tailored to deal with partially connected networks [3.41], or with networks in which the connections are restricted in their magnitude: "clipped" [3.42] or even restricted to ± 1 only [3.43, 44]. These latter two variants – connections restricted by connectivity and magnitude – may prove crucial for practical applications.

Acklowledgements. We would like to acknowledge the great scientific debt which we owe to Elizabeth Gardner. We would also like to thank J.L. van Hemmen for helpful comments.

Appendix

Here we provide a proof of the convergence theorem for the learning algorithm

$$\varepsilon_i^\mu \equiv \theta\left(K\sqrt{\sum_{j=1}^{N} T_{ij}^2} - \xi_i^\mu \sum_{j=1}^{N} T_{ij}\xi_j^\mu\right), \tag{3.32a}$$

$$T_{ij} \to T_{ij} + \frac{1}{N}\varepsilon_i^\mu \xi_i^\mu \xi_j^\mu , \tag{3.32b}$$

where $\theta(x) = 1$ if $x \geq 0$ and 0 if $x < 0$.

The theorem shall hold for a general *asymmetric* connection matrix T_{ij}, which allows the learning (3.32) to be performed independently for each neuron i. That is, we may update all of the incoming connections T_{ij}, $j = 1, 2, \ldots, N$, for each i independently.

First of all, let us introduce the scalar product of two interaction matrices T and U at site i,

$$(T \cdot U)_i \equiv \sum_{j=1}^{N} T_{ij}U_{ij} , \tag{3.33}$$

and the norm of T at site i,

$$\|T\|_i \equiv \sqrt{(T \cdot T)_i} . \tag{3.34}$$

The convergence of the algorithm can be proved *provided a solution for the* T_{ij} *exists*. Suppose, then, that there exists a solution T_{ij}^* such that

$$\xi_i^\mu \sum_{j=1}^{N} T_{ij}^* \xi_j^\mu > (K + \delta)\|T^*\|_i , \tag{3.35}$$

where δ is some positive number for each pattern μ and each site i.

The proof of convergence then follows by considering the convergence of the following quantity as learning proceeds,

$$X_i^{(n)} \equiv \frac{(T^{(n)} \cdot T^*)_i}{\|T^{(n)}\|_i \|T^*\|_i} , \tag{3.36}$$

where $T^{(n)}$ is the connection matrix after the nth application of (3.32).

We shall now show that if learning were to proceed indefinitely, the quantity $X_i^{(n)}$ would eventually increase above 1. Since $X_i^{(n)}$ is bounded above by 1, by the Schwarz inequality, this is impossible and hence the algorithm must terminate.

The change in the numerator of (3.36) at time step n is

$$\Delta(T^{(n)} \cdot T^*)_i = \frac{1}{N}\varepsilon_i^\mu \sum_{j=1}^{N} \xi_i^\mu \xi_j^\mu T_{ij}^* > \frac{1}{N}\|T^*\|_i (K+\delta) \tag{3.37}$$

by (3.35), and so at time step n the numerator of (3.36) is bounded below,

$$(T^{(n)} \cdot T^*)_i > \frac{1}{N}\|T^*\|_i (K+\delta) n , \tag{3.38}$$

where, for convenience, we have assumed the starting point of the learning process to be $T^{(0)} = 0$. We are free to do so since the only a priori assumption of the theorem is the existence of at least one solution T^*. Note that, from the work of Gardner, [3.30, 32], we know whether a solution J_{ij} for (3.21) exists, and such a solution will also solve (3.32).

The change in the denominator of (3.36) at time step n comes from the change in the norm of $T^{(n)}$:

$$\begin{aligned}\Delta(T^{(n)}T^{(n)})_i &= \sum_{j=1}^{N}\left(T_{ij}^{(n)} + \frac{1}{N}\varepsilon_i^\mu \xi_i^\mu \xi_j^\mu\right)^2 - \sum_{j=1}^{N}\left(T_{ij}^{(n)}\right)^2 \\ &= 2\frac{1}{N}\varepsilon_i^\mu \sum_{j=1}^{N} T_{ij}^{(n)} \xi_i^\mu \xi_j^\mu + \frac{1}{N}\varepsilon_i^\mu .\end{aligned} \tag{3.39}$$

But either $\varepsilon_i^\mu = 0$ or $\varepsilon_i^\mu = 1$, in which case $\sum_j T_{ij}^{(n)} \xi_i^\mu \xi_j^\mu < K\|T^{(n)}\|_i$, and hence

$$\Delta(T^{(n)}T^{(n)})_i < \frac{1}{N}\varepsilon_i^\mu \left(2K\|T^{(n)}\|_i + 1\right) . \tag{3.40}$$

Now

$$\Delta(T^{(n)}T^{(n)})_i = \sum_{j=1}^{N}\left(T_{ij}^{(n)} + \delta T_{ij}^{(n)}\right)^2 - \sum_{j=1}^{N}\left(T_{ij}^{(n)}\right)^2 \simeq 2\sum_{j=1}^{N} T_{ij}^{(n)} \delta T_{ij}^{(n)} ,$$

since $\delta T_{ij}^{(n)} = \varepsilon_i^\mu \xi_i^\mu \xi_j^\mu / N$, and so $\sum_{j=1}^{N} (\delta T_{ij}^{(n)})^2 = \mathrm{O}(1/N)$. Also

$$\Delta\|T^{(n)}\|_i$$

$$= \sqrt{\sum_{j=1}^{N}\left(T_{ij}^{(n)} + \delta T_{ij}^{(n)}\right)^2} - \sqrt{\sum_{j=1}^{N}\left(T_{ij}^{(n)}\right)^2} \simeq \frac{\sum_{j=1}^{N} T_{ij}^{(n)}\delta T_{ij}^{(n)}}{\sqrt{\sum_{j=1}^{N}\left(T_{ij}^{(n)}\right)^2}} \,.$$

Hence we have that

$$\Delta\|T^{(n)}\|_i \simeq \frac{\Delta(T^{(n)}T^{(n)})_i}{2\|T^{(n)}\|_i} < \frac{1}{N}\left(K + \frac{1}{2\|T^{(n)}\|_i}\right) \tag{3.41}$$

for $\varepsilon = 1$ using (3.40).

Suppose that the algorithm has been iterated n times (i.e., $\varepsilon_i^\mu = 1$ has occurred n times) and has not terminated. $X_i^{(m)} < 1$ must hold at each step $m \leq n$, i.e.,

$$\|T^{(m)}\|_i > \frac{(T^{(m)}T^*)_i}{\|T^*\|_i}$$

and so, by (3.38),

$$\|T^{(m)}\|_i > \frac{1}{N}m(K + \delta) \,. \tag{3.42}$$

Then, since (3.41) holds at every $m \leq n$, we have

$$\|T^{(n)}\|_i < \frac{1}{N}Kn + \frac{1}{2N}\sum_{m=1}^{n-1}\frac{N}{m(K+\delta)} \,. \tag{3.43}$$

Now,

$$\sum_{m=1}^{n-1}\frac{1}{m} < 1 + \int_1^{n-1} x^{-1}\mathrm{d}x \,,$$

and thus

$$\|T^{(n)}\|_i < \frac{1}{N}Kn + \frac{1}{2(K+\delta)}\,[1 + \ln(n-1)] \,.$$

Combining this and (3.38) yields

$$X_i^{(n)} > \frac{K + \delta}{K + (\ln(n-1) + 1)N/2(K+\delta)n} \,. \tag{3.44}$$

Therefore $X_i^{(n)}$ becomes larger than one for sufficiently large n, contradicting the hypothesis that the algorithm does not terminate.

Convergence of the symmetric version of (3.32), where (3.32b) is replaced by

$$T_{ij} \to T_{ij} + \frac{1}{N}(\varepsilon_i^\mu + \varepsilon_j^\mu)\xi_i^\mu\xi_j^\mu \,,$$

can be proved in a similar fashion. The symmetry imposed upon the the connections $T_{ij} = T_{ji}$ introduces a dependence among the sites, and the algorithm must be executed *in parallel* over the sites. The proof of convergence then follows by defining the scalar product between matrices T and U at site i to be

$$(T \cdot U) \equiv \sum_{\substack{i,j=1 \\ i \neq j}}^{N} T_{ij} U_{ij}$$

instead of (3.33).

The proof of the convergence of the "training with noise" algorithm (3.30) follows in a similar fashion. Instead of (3.35), one assumes that there exists a solution for the connections T^*_{ij} such that

$$\xi_i^\mu \sum_{j=1}^{N} T^*_{ij} S_j^\mu > \|T^*\|_i \delta$$

for some positive δ and for each possible noisy initial vector S_i^μ. The error masks ε_i^μ are defined by (3.30). The change in the numerator of (3.36) at the nth learning step is

$$\Delta(T^{(n)} T^*)_i = \frac{1}{N} \varepsilon_i^\mu \xi_i^\mu \sum_{j=1}^{N} T^*_{ij} S_j^\mu > \frac{1}{N} \|T^*\|_i \delta \,, \tag{3.45}$$

while the change in the denominator comes from

$$\Delta(T^{(n)} T^{(n)})_i = 2 \frac{1}{N} \varepsilon_i^\mu \xi_i^\mu \sum_{j=1}^{N} T^*_{ij} S_j^\mu + \frac{1}{N} \varepsilon_i^\mu < \frac{1}{N} \varepsilon_i^\mu \,, \tag{3.46}$$

since $\varepsilon_i^\mu = 1$ only if $\xi_i^\mu \sum_{j=1}^N T^*_{ij} S_j^\mu < 0$, otherwise $\varepsilon_i^\mu = 0$. As before, the two inequalities (3.45) and (3.46) can be combined to produce

$$X_i^{(n)} > \delta \sqrt{\frac{n}{N}} \,. \tag{3.47}$$

Once again, this would exceed 1 for large enough n, so the algorithm must converge at a finite number of steps.

References

3.1 Cragg and H.N.V. Temperley: EEG Clin. Neurophysiol. **6**, 85 (1954)
3.2 J.J. Hopfield: *Proc. Natl. Acad. Sci. USA* **79**, 2554 (1982)
3.3 W.S. McCulloch and W.A. Pitts: Bull. Math. Biophys. **5**, 115 (1969)
3.4 D.O. Hebb: *The Organisation of Behaviour* (Wiley, New York 1949)
3.5 L.N. Cooper, F. Liberman, and E. Oja: Biol. Cybern. **33**, 9 (1979)
3.6 A.D. Bruce, E.J. Gardner and D.J. Wallace: J.Phys A **20**, 2909 (1987)
3.7 D.C. Mattis: Phys. Lett. **56A**, 421 (1976)

3.8 D. Sherrington and S. Kirkpatrick: Phys. Rev. Lett. **35**, 1792 (1975)
3.9 M.A. Moore: in *Statistical and Particle Physics: Common Problems and Techniques*, edited by K.C. Bowler and R.D. Kenway, (SUSSP Publ., Univ. of Edinburgh 1984) pp. 303–357
3.10 W.A. Little: Math. Biosc. **19**, 101 (1974)
3.11 D.J. Amit, H. Gutfreund, and H. Sompolinsky: Phys. Rev. Lett. **55**, 1530 (1985); Phys. Rev. A **32**, 1007 (1985); Ann. Phys. **173**, 30 (1987); Phys. Rev. A **32**, 2293 (1987)
3.12 S.F. Edwards, and P.W. Anderson: J. Phys. F **5**, 965 (1975)
3.13 G. Parisi, J. Phys. A **13**, 1887 (1980)
3.14 J.R.L. de Almeida and D.J. Thouless: J. Phys. A **11**, 983 (1978)
3.15 A. Crisanti, D.J. Amit, and H. Gutfreund: Europhys. Lett. **2**, 337 (1987)
3.16 T. Kohonen: *Self-Organisation and Associative Memory*, (Springer, Berlin, Heidelberg 1984)
3.17 L. Personnaz, I. Guyon, and G. Dreyfus: Phys. Rev. A **34**, 4217 (1987)
3.18 I. Kanter and H. Sompolinsky: Phys. Rev. A **35**, 380 (1987)
3.19 W. Kinzel: Z. Phys. B **60**, 205 (1985); **62**, 267 (1986)
3.20 B. Derrida, E. Gardner, and A. Zippelius: Europhys. Lett. **4**, 167 (1987)
3.21 H. Horner: Phys. Blätter **44**, 29 (1988)
3.22 E. Gardner, B. Derrida, and P. Mottishaw: J. Physique **48**, 741 (1987)
3.23 R. Meir and E. Domany: Phys. Rev. Lett. **59**, 359 (1987); Europhys. Lett. **4**, 645 (1987); Phys. Rev. A **37**, 608 (1988)
3.24 B.M. Forrest: J. Phys. A **21**, 245 (1988)
3.25 T.M. Cover: IEEE Trans. Electron. Comp. **EC-14**, 326 (1965)
3.26 S. Venkatesh: in *Proc. Conf. on Neural Networks for Computing, Snowbird, Utah*, (AIP Conf. Proc. 151), edited by J.S. Denker, (American Institute of Physics, New York 1986) p. 440; PhD thesis, Cal. Inst. Tech. (1986)
3.27 D.J. Wallace: in *Lattice Gauge Theory – a Challenge to Large Scale Computing*, edited by B. Bunk, K.H. Mutter, and K. Schilling, (Plenum, New York 1986) pp. 313-330
3.28 A.D. Bruce, A. Canning, B. Forrest, E. Gardner, and D.J. Wallace: in *Proc. Conf. on Neural Networks for Computing, Snowbird, Utah*, AIP Conf. Proc. 151, edited by J.S. Denker, (American Institute of Physics, New York 1986) pp. 65-70
3.29 S. Diederich and M. Opper: Phys. Rev. Lett. **58**, 949 (1987)
3.30 E. Gardner: Europhys. Lett. **4**, 481 (1987)
3.31 W. Krauth and M. Mézard: J. Phys. A **20**, L745 (1987)
3.32 E. Gardner: J. Phys. A **21**, 257 (1988)
3.33 M. Minsky and S. Papert: in *Perceptrons*, (MIT Press, Cambridge, Mass. 1969)
3.34 T.B. Kepler and L.F. Abbott: J. Physique **49**, 1657 (1988)
3.35 M. Opper: Phys. Rev. A **38**, 3824 (1988)
3.36 E. Gardner, N. Stroud, and D.J. Wallace: in *Neural Computers: From Computational Neuroscience to Computer Science*, edited by Eckmiller, (Springer, Berlin, Heidelberg 1988) pp. 251-260; J. Phys. A: Math. Gen. **22**, 2019 (1989)
3.37 P. Peretto: Neural Networks **1**, 309 (1988)
3.38 L.F. Abbott and T.B. Kepler: J. Phys. A **22**, L711 (1989)
3.39 A. Engel, H. Englisch, and A. Schütte: Europhys. Lett. **8**, 393 (1989)
3.40 E. Gardner: J. Phys. A **20**, 3453 (1987)
3.41 A. Canning and E. Gardner: J. Phys. A **15**, 3275 (1988)
3.42 G. Parisi: J. Phys. A **19**, L617 (1986)
3.43 H. Sompolinsky: Phys. Rev. A **34**, 2571 (1986)
3.44 J.L. van Hemmen: Phys. Rev. A **36**, 1959 (1987)

4. Dynamics of Learning

Wolfgang Kinzel and Manfred Opper

With 8 Figures

Synopsis. Supervised learning in attractor networks which perform as an associative memory is investigated. Two learning algorithms, the PERCEPTRON of optimal stability and the ADALINE, are derived from optimization problems and exact results for their dynamics are obtained. The ADALINE is extended to networks with binary synapses and is studied numerically. The basins of attraction during the learning process are calculated using a Gaussian approximation. Finally analytical results for forgetting in the ADALINE network are presented.

4.1 Introduction

One of the remarkable properties of neural networks is their ability to learn specific tasks. According to Hebb [4.1] a network learns by synaptic plasticity; the connection strength between two neurons slowly adjusts to correlations between their neural activity. In models of neural networks synaptic plasticity can easily be formulated: the change of the synaptic strength between two neurons is a function of their activity and their local potentials. It has been demonstrated that by such a mechanism simple networks of formal neurons can learn fairly complicated computational tasks [4.2].

In the simplest case a neural network consists of one input layer of neurons, a set of independent output neurons and one layer of synaptic connections between input and output (PERCEPTRON [4.3]). In the learning state, input and output neurons are fixed according to some presented examples and the synaptic weights are adjusted by some simple rules (supervised learning). Two such rules are well known, the ADALINE (adaptive linear network) of Widrow and Hoff [4.4] and the PERCEPTRON of Rosenblatt [4.5]. Both algorithms have been analyzed [4.6] in the context of the Hopfield model [4.7] of neural networks; they are the subject of the present article.

However, a simple perceptron can only perform a rather limited class of tasks [4.3]. But a network with one additional layer of "hidden units" in addition to input and output neurons can already represent any function. For such networks algorithms of supervised learning have also been studied. If the synapses are changed by the gradient of the quadratic deviation of the output neurons from the desired state, this algorithm is called "backpropagation" [4.8]. In this case the hidden neurons are fixed by the synaptic weights.

If the neurons are modeled by binary elements, back propagation cannot be applied. But recently several learning algorithms have been suggested for this case, too. One method considers the internal representation of the hidden units as the essential degrees of freedom for learning [4.9]. These internal states modify the synapses, which in turn change the internal states. A second method finds the synaptic weights by a heuristic optimization procedure [4.10]. In a third approach, hidden units as well as additional layers are added only if the system cannot solve its task [4.11]. This method is guaranteed to converge but the number of hidden units may be very large. In all of these cases the network has learned nontrivial tasks such as assigning the parity of the input sequence to an output neuron. Interesting applications have been demonstrated, too; examples are text-to-phoneme conversion [4.12] and backgammon [4.13].

Supervised learning means that the networks learn by examples; i.e. for fixed input/output neurons the synapses and hidden units adjust according to given rules. Unsupervised learning has also been studied [4.14–17]. In this case only the input is given and the network finds the representations of all of the following layers of neurons by synaptic plasticity according to some Hebb rule as well as by rearrangement of the neurons. By this kind of self-organization the network learns to classify input patterns.

In the simplest models of neural networks a neuron is just a two-state threshold unit [4.18]. One may extend these models taking more physiological details into account [4.19–21]. In particular a delay of the signal from the activity of the connected neurons is necessary to learn a time sequence of patterns [4.22–24].

Learning in simple models of neural networks has also been considered in the context of biological functions. A completely disordered network can learn just by regression of synapses [4.25, 26]. On the basis of observations on the acquisition of song by birds, a model has been proposed which learns temporal sequences by selection [4.27].

A system of formal neurons may be regarded as a cellular automaton [4.28]. However, the rule of the dynamics of the neurons is rather restricted by the biologically motivated fact that a neuron receives the superposition of the activity signals of all of the connected neurons weighted by the corresponding synaptic efficiencies. Each neuron reacts to the sum of all weighted ativities only. Hence, if a neuron receives a signal from c other neurons, only c numbers define the dynamic evolution. On the other hand, in a general cellular automaton of binary elements a table of 2^c numbers defines the dynamics, and this table can easily be realized in standard computer hardware. Hence, such networks may be well suited for future realizations in machines; and learning has been studied for several types of such cellular automata [4.29–31].

In this paper we concentrate on recent results on learning in the Hopfield model [4.7] and its extensions. This means that we restrict ourselves to a system of totally connected binary units which perform as an associative memory. With respect to supervised learning such a system is essentially a perceptron [4.3–5] (single layer of synapses) or a multilayer network with fixed given internal representations of the stored patterns [4.32]. The reason for our restriction to the simplest models is the fact that in addition to numerical simulations [4.33–36] many analytical results have been obtained recently [4.5, 6, 14, 37–41].

In Sect. 4.2 several algorithms of supervised learning are introduced and the properties of the networks after learning are discussed. Sections 4.3 and 4.4 present analytical results on learning times for the ADALINE and PERCEPTRON, respectively. Learning with binary synapses which may be important for hardware realizations is defined and investigated in Sect. 4.5. The basins of attraction of the network during and after learning are studied on the basis of a mean-field approximation [4.42, 43] in Sect. 4.6. Forgetting during the learning process is calculated analytically in Sect. 4.7, and finally Sect. 4.8 presents an outlook.

4.2 Definition of Supervised Learning

Consider a network of N formal neurons S_i which are either active ($S_i = +1$) or quiescent ($S_i = -1$). Each neuron S_i receives an input field

$$h_i = \sum_j J_{ij} S_j \tag{4.1}$$

from all other neurons S_j, where J_{ij} is a real number modeling the efficiency of the synapse from neuron j to neuron i. A positive (negative) J_{ij} describes an excitatory (inhibitory) synapse. The time evolution of neuron S_i is modeled by

$$S_i(t+1) = \operatorname{sgn} h_i(t) \ , \tag{4.2}$$

where t labels a time step and the neurons may be updated either in a random sequence or in parallel.

Equation (4.2) describes the dynamics of the neurons S_i for a fixed matrix J_{ij} of synaptic weights. Learning occurs on a much slower timescale by a change of the weights J_{ij}. Here we consider supervised learning only. This means we consider a set of p patterns $\xi_i^\nu \in \{+1, -1\}$ $(i = 1, \dots N; \quad \nu = 1, \dots, p)$ and fix the neuron state S_i to one of the patterns, $S_i = \xi_i^\nu$. Then the coupling J_{ij} is changed according to some function f

$$J_{ij}(t+1) - J_{ij}(t) = f(\xi_i^\nu, \xi_j^\nu, h_i^\nu(t)) \ , \tag{4.3}$$

with

$$h_i^\nu(t) = \sum_j J_{ij}(t) \xi_j^\nu \ . \tag{4.4}$$

Equation (4.3) is a local learning rule, i.e. the change of the synapse depends only on the information of the two adjacent neurons. (In principle one could add a dependence on h_j^ν, but we do not expect any advantage.) The locality of synaptic plasticity may be plausible biologically, although a change due to a global signal cannot be ruled out, to our knowledge. But for hardware realizations a local mechanism is definitely of great advantage. The patterns ξ_i^ν may be presented randomly, sequentially, or in parallel; in the latter case f depends on the complete set $\{h_i^\nu(t)\}$.

The problem remains to find a good function f, i.e. an algorithm which yields a network which performs the desired task well. Here we consider an associative memory. This means that (4.3) should yield a matrix J_{ij} by which firstly all patterns ξ_i^ν are stable states of the neural dynamics (4.2) and secondly each pattern ξ_i^ν is an attractor of the dynamics $S_i(t)$ with a large basin of attraction. The first condition is obviously given by

$$\xi_i^\nu h_i^\nu > 0 \,, \tag{4.5}$$

while it is still not precisely known which properties of J_{ij} give a large basin of attraction. But it seems plausible that a large "stability parameter"

$$\Delta_{\nu,i} = \frac{\xi_i^\nu h_i^\nu}{\left(\sum_j J_{ij}^2\right)^{1/2}} \tag{4.6}$$

should give a large basin of attraction [4.37, 44].

For random patterns ξ_i^ν the space of all possible matrices $\{J_{ij}\}$ has been investigated by Gardner [4.44] using methods of the statistical mechanics of spin glasses [4.45, 46]. The maximal possible stability

$$\Delta = \max_{\{J_{ij}\}} \min_{\nu,i} \Delta_{\nu,i} \tag{4.7}$$

has been calculated as a function of the number p of stored patterns. Gardner obtains with $\alpha = p/N$ and for $N \to \infty$ [4.44]

$$\alpha \int_{-\Delta}^{\infty} \frac{\mathrm{d}z}{\sqrt{2\pi}} \mathrm{e}^{-z^2/2} (z + \Delta)^2 = 1 \,. \tag{4.8}$$

Hence it is possible to store the maximal number, $p = 2N$, of random patterns in a network of N neurons ($\alpha_c = 2$ for $\Delta = 0$).

According to (4.5) the learning rule (4.3) should increase the fields. In the simplest case the change of the synapses is proportional to the gradient of those fields:

$$J_{ij}(t+1) - J_{ij}(t) = \frac{\gamma}{N} \frac{\partial}{\partial J_{ij}} \xi_i^\nu h_i^\nu = \frac{\gamma}{N} \xi_i^\nu \xi_j^\nu \,. \tag{4.9}$$

If the initial matrix is zero and if all patterns ν are presented once one obtains the couplings

$$J_{ij} = \frac{\gamma}{N} \sum_\nu \xi_i^\nu \xi_j^\nu \,. \tag{4.10}$$

This is the Hopfield model [4.7], whose attractors have been calculated analytically using spin-glass theory [4.47]. It has a maximal storage capacity of $\alpha_c = 0.145$ which is far below the maximal possible capacity of $\alpha_c = 2$. Furthermore (4.5) is not obeyed for the patterns ξ_i^ν but for states very close to them; it is not the patterns but the states close to them that are the attractors of the network (although this is not a serious drawback).

An improvement of the sequential learning procedure is obtained if one applies (4.8) only to a pattern which does not obey the stability equation (4.5). Formally this reads

$$J_{ij}(t+1) = J_{ij}(t) + \frac{\gamma}{N}\Theta(c - \xi_i^\nu h_i^\nu)\xi_i^\nu \xi_j^\nu \,, \tag{4.11}$$

where $\Theta(x)$ is the step function and an additional threshold $c > 0$ has been introduced. This is the famous PERCEPTRON algorithm, for which a convergence theorem exists [4.5]: if a matrix J_{ij} exists which obeys $\xi_i^\nu h_i^\nu > c$ then the algorithm (4.11) converges to it.

However, (4.11) does not necessarily lead to a matrix of maximal stability Δ, (4.8). But a modification of this algorithm has been proven to give maximal stability [4.37]: (4.10) has to be applied only to the worst pattern; i.e. to the pattern ν with minimal stability. In the limit of $c \to \infty$ one obtains the desired matrix with optimal stability given by (4.8).

A different version of a learning algorithm is obtained from the request that all fields $\xi_i^\nu h_i^\nu$ are not larger but equal to a given threshold c. For $c = 1$ this condition reads

$$\xi_i^\nu h_i^\nu = 1 \tag{4.12}$$

for all patterns ν and sites i. The corresponding synaptic change is obtained by minimizing the quadratic deviation from (4.12)

$$E_i^\nu = \frac{\gamma}{N}(1 - \xi_i^\nu h_i^\nu)^2 \,. \tag{4.13}$$

If the change of J_{ij} is chosen proportional to the negative gradient of E_i^ν, one obtains (with $c = 1$)

$$J_{ij}(t+1) - J_{ij}(t) = \frac{\gamma}{N}[1 - \xi_i^\nu h_i^\nu(t)]\xi_i^\nu \xi_j^\nu \,. \tag{4.14}$$

This rule is similar to (4.9), but now the prefactor is weighted according to the condition (4.12). The algorithm is equivalent to the ADALINE rule [4.4]. If all patterns are presented in parallel, it reads

$$J_{ij}(t+1) = J_{ij}(t) + \frac{\gamma}{N}\sum_\nu [1 - \xi_i^\nu h_i^\nu(t)]\xi_i^\nu \xi_j^\nu \,. \tag{4.15}$$

In order to maximize the stability, (4.6) one may simultaneously minimize the norm $\sum_j J_{ij}^2(t)$; then the rule (4.14) is modified to

$$J_{ij}(t+1) = \left(1 - \frac{\mu}{N}\right) J_{ij}(t) + \frac{\gamma}{N}[1 - \xi_i^\nu h_i^\nu(t)]\xi_i^\nu \xi_j^\nu , \tag{4.16}$$

where μ is an additional parameter.

A matrix J_{ij} which obeys (4.12) can be obtained directly from this equation. Obviously the identity matrix

$$J_{ij} = \delta_{ij} \tag{4.17}$$

fulfills condition (4.12). But this network stabilizes all possible states $\{S_i\}$; in particular it is useless for an associative memory. A better matrix is the projector onto the linear space spanned by the p patterns $\{\xi_i^\nu\}$ [4.48]. Such a projector can be constructed by (pseudo)inverting (4.12), one obtains [4.48–50]

$$J_{ij} = \frac{1}{N} \sum_{\nu,\mu} \xi_i^\nu (C^{-1})_{\nu\mu} \xi_j^\mu , \tag{4.18}$$

where C is the correlation matrix

$$C_{\nu\mu} = \frac{1}{N} \sum_i \xi_i^\nu \xi_i^\mu . \tag{4.19}$$

C^{-1} exists up to $p = N$ *linear independent* patterns, and hence one has $\alpha_c < 1$. In fact an analytical calculation of the attractors of the network (4.18) gives $\alpha_c = 1$ if the self-coupling J_{ii} is set to zero. If J_{ii} is included in (4.18) then the stored patterns have a nonzero basin of attraction for $\alpha_c < 0.5$ only [4.50].

The dynamics of the PERCEPTRON (4.10) and the ADALINE (4.13) and (4.14) are investigated in Sects. 4.3 and 4.4, respectively. If the synaptic weights are restricted to two values $J_{ij} \in \{1/\sqrt{N}, -1/\sqrt{N}\}$ then similar algorithms can be derived from (4.5) and (4.13). This case is studied in Sect. 4.5.

4.3 Adaline Learning

In this section we study the dynamics of the ADALINE rule (4.15) for the storage of $p = \alpha N$ random patterns in the limit $N \to \infty$. It is shown that the learning process can also be cast into a set of network equations, where now the patterns interact via their correlations. Finally we solve the dynamics using mean-field methods of statistical mechanics.

A similar model in continuous time under external noise has been independently studied by J. Hertz et al. [4.14]. To characterize *typical* features of the algorithms' performance we shall consider the total quadratic error of the learning process at time t:

$$E(t) = (Np)^{-1} \sum_{i,\nu} (1 - \xi_i^\nu h_i^\nu(t))^2 = (Np)^{-1} \sum_{i,\nu} (E_i^\nu(t))^2 , \tag{4.20}$$

which for random patterns becomes self-averaging, i.e. independent of the special realizations of the patterns (only on their probability distribution) for $N, p \to \infty$.

Using (4.14) we easily derive the linear recursion

$$E_i^\nu(t+1) = E_i^\nu(t) - \gamma \sum_\nu B_{\mu\nu}(i) E_i^\nu(t) , \tag{4.21}$$

where

$$B_{\mu\nu}(i) = N^{-1} \xi_i^\mu \xi_i^\nu \sum_{j, j \neq i} \xi_j^\mu \xi_j^\nu \quad \text{and} \quad E_i^\mu(0) = 1 ,$$

valid for initially empty ($J_{ij}(0) = 0$) networks without self-couplings ($J_{ii}(t) = 0$). This type of network equation describes the interference of the patterns owing to their correlations during the learning process. Although for random patterns a single element $B_{\mu\nu}$ is typically of the order of $1/\sqrt{N}$ ($\mu \neq \nu$), the net effect of these correlations cannot be neglected when $p \propto N$.

We shall calculate $E(t)$ exactly for the simplest ensemble of patterns, where ξ_i^μ is set equal to ± 1 with probability $1/2$ independently for each i and μ. Replacing the sum over i by an average over the random patterns and solving the linear system (4.21) we find

$$E(t) = p^{-1} \overline{\sum_\nu (E^\nu(t))^2} = p^{-1} \sum_{\mu\nu} E^\mu(0) \overline{\left[(\mathbb{1} - \gamma B)^{2t} \right]_{\mu\nu}} E^\nu(0) , \tag{4.22}$$

where now the index i has been omitted. From the form of B it can be seen that the averages over the off-diagonal elements ($\mu \neq \nu$) vanish for unbiased ($\overline{\xi_i^\mu} = 0$) patterns. Thus we are left with

$$\begin{aligned} E(t) &= p^{-1} \mathrm{Tr} \left\{ \overline{\left[(\mathbb{1} - \gamma B)^{2t} \right]} \right\} = p^{-1} \overline{\sum_\alpha (1 - \gamma \lambda_\alpha)^{2t}} \\ &= \int \mathrm{d}\lambda (1 - \gamma\lambda)^{2t} \varrho(\lambda) , \end{aligned} \tag{4.23}$$

where Tr denotes the trace and λ_α the eigenvalues of B, and

$$\varrho(\lambda) = p^{-1} \overline{\sum_\alpha \delta(\lambda - \lambda_\alpha)} \tag{4.24}$$

is the average density of eigenvalues of the matrix B. The calculation of the average density of states for large matrices with random elements is a well-known task in statistical physics. We shall briefly describe the method. Using a standard Gaussian transformation one has

$$p^{-1} \sum_\alpha \delta(\lambda - \lambda_\alpha) = \frac{2}{\pi p} \lim_{\varepsilon \to 0} \mathrm{Im} \left\{ \frac{\partial}{\partial \lambda} \ln Z(\lambda + \mathrm{i}\varepsilon) \right\} \tag{4.25}$$

with

$$Z(\lambda) = \int \prod_\nu \mathrm{d}y^\nu \exp \left[\frac{\mathrm{i}}{2} \sum_{\mu\nu} y^\mu (\lambda \delta_{\mu\nu} - B_{\mu\nu}) y^\nu \right] .$$

Z describes an auxiliary Gaussian model with random interactions at an "imaginary" temperature.

Usually the average of a logarithm would require the application of the famous "replica trick" [4.45,46]. For the present model, however, it can be shown that logarithm and average commute, so one can also average over Z directly.[1]

Because of the correlations of the $B_{\mu\nu}$s, (4.25) is slightly more complicated then the standard Gaussian model. To perform the average, the quadratic form

$$\sum_{\mu\nu} y^\mu B_{\mu\nu} y^\nu = \sum_{j,j\neq i} \left(\sum_\mu \frac{y^\mu}{\sqrt{N}} \xi_i^\mu \xi_j^\mu \right)^2 \tag{4.26}$$

must be decoupled with the help of an auxiliary variable m_j. One obtains

$$\overline{Z}(\lambda) = K \int \prod_\nu \mathrm{d}y^\nu \int \prod_j \mathrm{d}m_j \times \exp\left[\frac{\mathrm{i}}{2}\lambda \sum_\mu (y^\mu)^2 + \frac{\mathrm{i}}{2} \sum_j m_j^2 - \frac{1}{2N} \sum_j m_j^2 \sum_\mu (y^\mu)^2 \right] , \tag{4.27}$$

where we have dropped terms of higher order in N^{-1}, and K is a constant independent of λ. The remaining integrals can be calculated in the mean-field limit $p, N \to \infty$ by saddle-point methods. Introducing the order parameter $q(\lambda) = p^{-1} \sum_\mu (y^\mu)^2$ we finally get for the density of eigenvalues

$$\begin{aligned} \varrho(\lambda) &= \pi^{-1} \lim_{\varepsilon \to 0} \mathrm{Re}\,\{q(\lambda + \mathrm{i}\varepsilon)\} \\ &= (2\pi\alpha\lambda)^{-1} \sqrt{(\lambda_2 - \lambda)(\lambda - \lambda_1)} + (\alpha - 1)/\alpha\ \Theta(\alpha - 1)\delta(\lambda) , \end{aligned} \tag{4.28}$$

with

$$\lambda_{1,2} = (1 \mp \sqrt{\alpha})^2 .$$

This differs from the standard semicircular law, which would be valid for statistically *independent* matrix elements $B_{\mu\nu}$. The resulting expression for the total error of learning

$$E(t) = (2\pi\alpha)^{-1} \int_{\lambda_1}^{\lambda_2} \lambda^{-1} (1 - \gamma\lambda)^{2t} \sqrt{(\lambda_2 - \lambda)(\lambda - \lambda_1)} \mathrm{d}\lambda + \frac{\alpha - 1}{\alpha} \Theta(\alpha - 1) \tag{4.29}$$

shows that for $\alpha = p/N < 1$ one can always find a value for γ, such that $\overline{E}(t)$ will decay to zero for $t \to \infty$, i.e. the condition (4.12) will be fulfilled exactly after learning. For $\alpha > 1$ one has $E(\infty) > 0$ and the algorithm is unable to

[1] The corresponding Edwards–Anderson parameter $p^{-1} \sum_\nu \langle y_a^\nu \rangle \langle y_b^\nu \rangle$ vanishes (brackets mean average over the "imaginary temperature" ensemble, a and b are replicas).

satisfy (4.12). This is a feature independent of the ensemble of patterns, because there are $p > N$ conditions for only $N - 1$ couplings J_{ij}, $j = 1 \dots N, j \neq i$.

Nevertheless, even for $\alpha > 1$ a least-square solution is achieved with a finite width for the distribution of internal fields. Thus a finite fraction of bits are wrongly stored for each pattern. This is in contrast to the corresponding network with *nonzero* diagonal couplings [4.38], where the matrix of couplings converges to the unit matrix for $\alpha \geq 1$, which trivially stabilizes arbitrary patterns (with a zero basin of attraction).

From the exponential decay of (4.29) for $\gamma < (1 + \sqrt{\alpha})^2/2$ we can define a learning time τ for the ADALINE algorithm. Different definitions are possible, depending on whether the bulk of patterns fulfills (4.12) or in the second case even the last pattern is properly stored. The first possibility has been studied in the continuum-time model, showing a divergence [4.14]

$$\tau \sim (1 - \alpha)^{-1} \quad \text{for} \quad \alpha \to 1 \, .$$

In the second case we use the fact that the behavior of $E(t)$ at very long times is governed by the eigenvalue of B with a maximum distance from γ^{-1}. For the optimal value

$$\gamma = (1 + \alpha)^{-1}$$

the corresponding learning time is minimal:

$$\tau = \left[\ln \left(\frac{1 + \alpha}{2\sqrt{\alpha}} \right) \right]^{-1} , \tag{4.30}$$

which for $\alpha \to 1$ diverges like

$$\tau \sim (1 - \alpha)^{-2} \, .$$

4.4 Perceptron Learning

Condition (4.12) leading to the ADALINE algorithm does not yield the maximal results for stability and storage capacities. In the following we shall show that the capacity can be improved by storing only an appropriate subset of patterns (differing from neuron to neuron) from which the network is able to generalize to the remaining patterns of the original learning set. An algorithm which realizes this concept is the PERCEPTRON algorithm with optimal stability [4.37, 39].

Looking back at the ADALINE rule (4.14) at threshold c one finds for the resulting coupling matrix the Hebbian-type expression

$$J_{ij}(\infty) = c \sum_{\nu} \frac{x_\nu(i)}{\sqrt{N}} \xi_i^\nu \xi_j^\nu \, , \tag{4.31}$$

where the x_νs satisfy

$$\sum_{\nu} B_{\mu\nu}(i) x_\nu(i) = 1 , \tag{4.32}$$

and $B_{\mu\nu}$ is the correlation matrix introduced in (4.21). $x_\nu(i)$ can be regarded as a measure of how strongly pattern ν has been enforced during the learning process at neuron i. For random patterns it can be shown that (for each i) the x_νs are Gaussian distributed with some positive mean and nonzero variance. This implies that a finite fraction of them must be *negative*. This strange result means that these patterns have been in fact *unlearnt*, rather than *learnt*. Setting their strength x_ν equal to zero (and shifting the remaining x_νs somewhat to maintain (4.32) would even increase the absolute value of their fields $\xi_i^\nu h_i^\nu$.

Now we shall prove that the requirement of optimal stability automatically leads to such picture. We consider a coupling matrix (4.31) derived from a Hebbian type of learning rule. If we require that $\xi_i^\nu h_i^\nu \geq c$ for all patterns ν and that in addition the stability (at fixed i) (4.6)

$$\Delta = \min_{\nu} \frac{\xi_i^\nu h_i^\nu}{\sqrt{\sum_j J_{ij}^2}} = \frac{c}{\sqrt{\sum_j J_{ij}^2}} \tag{4.33}$$

is maximized, we get

$$f_\mu \equiv \sum_{\nu} B_{\mu\nu} x_\nu \geq 1 \tag{4.34}$$

and for the optimal stability Δ

$$N\Delta^{-2} = \sum_{\mu\nu} x_\mu B_{\mu\nu} x_\nu = \sum_{\mu\nu} f_\mu (B^{-1})_{\mu\nu} f_\nu = \text{minimum} , \tag{4.35}$$

where we have assumed for simplicity that B is invertible. To find the minimum of this quadratic form with linear constraint one must distinguish between two types of pattern (for fixed neuron i): a class of patterns for which $f_\mu = 1$, i.e. which satisfy the condition (4.12) of ADALINE learning, and a second class for which $f_\mu > 1$. For the latter patterns the necessary condition for an extremum of H is

$$\frac{\partial H}{\partial f_\mu} = \sum_{\nu} (B^{-1})_{\mu\nu} f_\nu = x_\mu = 0 . \tag{4.36}$$

Thus optimal stability automatically leads to a division of the training set of patterns into a group which has to be enforced by learning and a second group onto which the network can generalize. The case of noninvertable Bs can be treated properly by introducing Lagrangian multipliers for the constraints (4.34). This will not alter our result if one additionally requires $x_\mu \geq 0$.

Krauth and Mezard have recently introduced a variant of the PERCEPTRON learning algorithm which yields optimal stability (4.33) in the limit, where the threshold c goes to infinity [37]. Numerical simulations show however that even the standard PERCEPTRON rule (4.11) achieves nearly optimal stability at a large threshold. From (4.11) (with $\gamma = 1$) we see that for such learning rules the quantity $cx_\mu(i)$ represents the number of times, pattern μ had to be enforced at neuron i. The total number $c\sum_\mu x_\mu(i)$ is therefore a measure for the learning time of the algorithm. Our result (4.36) means in this context that a subgroup of patterns has to perform only a small number of steps (relative to the threshold). Being driven by the other patterns they are learnt very fast.

Quantitative results for the distribution of x_μs in the optimal stability limit can be given for random patterns, when $p, N \to \infty$. Using an intuitive approach similar to the "cavity" method [4.46] we rederive a result which has been previously derived by one of us [4.39] by help of the replica method. The basic idea consists in adding a new pattern ξ_j^0, $j = 1, \ldots, N$ to the p patterns of the training set, thereby assuming that the probability distribution of x_0 (in the $p+1$ pattern system) coincides with the distribution for an arbitrary x_μ in the p pattern system for $p \to \infty$.

If the field cf. (4.34)

$$u_0 = \sum_{\nu>0} B_{0\nu} x_\nu \tag{4.37}$$

exceeds the value 1, the new pattern need not be stored explicitly, thus $x_0 = 0$. Since the ξ_j^0 are uncorrelated to the x_νs for $\nu > 0$, u_0 (the "cavity field") is a Gaussian random variable with $\overline{u_0} = 0$ and $\overline{u_0^2} = \Delta^{-2}$ (see (4.35)). Thus the probability that $x_0 = 0$ is given by

$$P_0 = \int_\Delta^\infty Dz\,; \quad Dz = \frac{1}{\sqrt{2\pi}} \mathrm{e}^{-z^2/2} \mathrm{d}z\,. \tag{4.38}$$

If on the other hand $u_0 < 1$, the pattern has to be learnt and x_0 must satisfy after learning the condition (4.32), which now reads

$$\sum_\nu B_{0\nu} \tilde{x}_\nu = x_0 + u_0 + g x_0 = 1\,, \tag{4.39}$$

where

$$g = \sum_{\nu>0} B_{0\nu} \frac{\delta x_\nu}{\delta x_0}$$

represents a reaction $x_\nu \to \tilde{x}_\nu = x_\nu + \delta x_\nu$, $\delta x_\nu = \mathrm{O}(1/\sqrt{N})$ of the previously stored patterns owing to the small disturbance generated by storing pattern 0. In the mean-field limit it is assumed that g is nonfluctuating (self-averaging). The $p(1 - P_0) = \alpha_{\text{eff}} N$ disturbed patterns with nonzero x_ν satisfy the condition for ADALINE learning (4.32). We make the crucial assumption that for the calculation of g these patterns (though being "selected" by the PERCEPTRON algorithm) can be treated as uncorrelated. Thus, if we had to calculate the distribution of x_0 in the pure ADALINE case (for $\alpha_{\text{eff}} N$ patterns) we would also arrive at (4.39) (now valid for all u_0) with the same value of $g = g(\alpha_{\text{eff}})$. We use this idea to find g from the results of our ADALINE calculations. Averaging

(4.39) over u_0 (which is Gaussian with zero mean in the ADALINE case) gives

$$\begin{aligned}(1+g)^{-1} &= \overline{x_0}(ADA) = \sum_\nu \overline{(B^{-1})_{\mu\nu}} = \int \varrho(\lambda)\lambda^{-1}\mathrm{d}\lambda \\ &= (1-\alpha_{\text{eff}})^{-1} ,\end{aligned} \tag{4.40}$$

where we have used the density for the eigenvalues of B, $\varrho(\lambda)$, from (4.28), for $\alpha = \alpha_{\text{eff}}$ and the fact that $\overline{B^{-1}_{\mu\nu}} = 0$ for $\mu \neq \nu$. Thus (4.38) to (4.40) can be summarized in one equation:

$$w(x_0) = P_0\delta(x_0) + \frac{\Theta(x)}{\sqrt{2\pi\sigma^2}} \exp\left(\frac{(x_0-m)^2}{2\sigma^2}\right) , \tag{4.41}$$

with

$$m = (1-\alpha_{\text{eff}})^{-1} , \quad \sigma^2 = \Delta^{-2}(1-\alpha_{\text{eff}})^{-2} ,$$

$$\alpha_{\text{eff}} = \alpha(1-P_0) = \alpha \int_{-\Delta}^{\infty} Dz .$$

$w(x)\mathrm{d}x$ gives the fraction of patterns with a number of learning steps between cx and $c(x+\mathrm{d}x)$ in the limit of large c. This holds exactly for the PERCEPTRON algorithm with optimal stability and at least approximately for the standard one. Finally we have to find an equation which determines the optimal stability Δ. From (4.35, 36) we have

$$\Delta^{-2} = N^{-1}\sum_\mu x_\mu = \alpha \int x w(x)\mathrm{d}x = \alpha\sigma \int_{-\Delta}^{\infty} Dz(z+\Delta) , \tag{4.42}$$

which can be finally cast into the form

$$\alpha \int_{-\Delta}^{\infty} Dz(z+\Delta)^2 = 1 . \tag{4.43}$$

This is E. Gardner's famous result for optimal stabilities [4.44]. Equations (4.38, 43) show that one can store up to $p = 2N$ random pattern by learning only up to $p' = N$ of them explicitly. As in the ADALINE case this takes a "learning time" $T = c\sum_\nu x_\nu$ diverging quadratically $\propto (2-\alpha)^{-2}$ for the critical capacity.

We have compared the analytical result (4.41, 43) with simulations of the standard algorithm at $c = 10$. Figure 4.1 shows the distribution $w(x)$ averaged over many samples of the patterns for $\alpha = 0.5$. Figure 4.2 gives the average number of updates $T = c\sum_\nu x_\nu$ for $N = 100$.

We expect that our analysis can be used to design new and faster learning rules with optimal stability. The linear growth of learning times with the threshold c is a drawback of the PERCEPTRON learning algorithms. Our results suggest that this problem could be circumvented if by some preprocessing at each neuron the patterns μ with $x_\mu = 0$ could be cast away from the training set. The remaining patterns could then be learnt much faster by an ADALINE

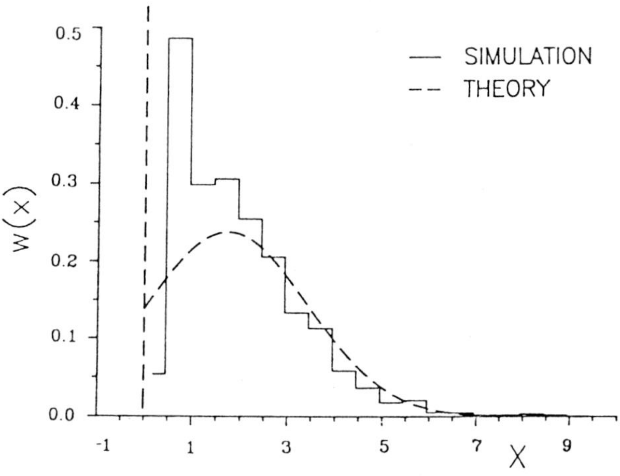

Fig. 4.1. Distribution of learning times $t = xc$ for a single pattern at storage capacity $\alpha = 0.5$ (PERCEPTRON algorithm). The δ-contribution at $x = 0$ has a weight $P_0 \simeq 0.15$

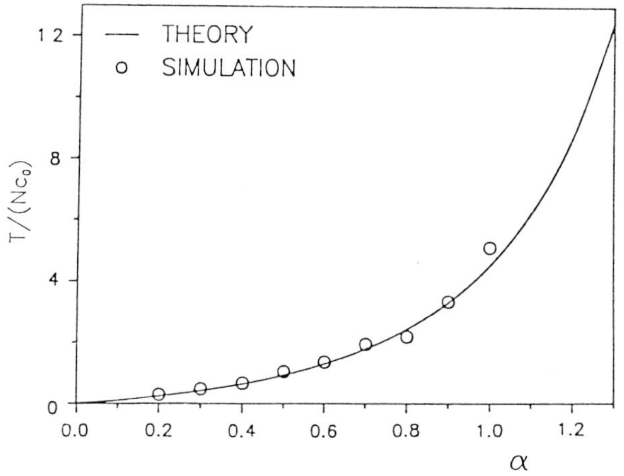

Fig. 4.2. Scaled average number of learning steps. The simulations are for $N = 100$ and $c = 10$, averaged over 50 sets of patterns

procedure. A second type of fast learning has been derived by applying standard optimization techniques directly to the minimization problem (4.34, 35) [4.51]. A learning rule using a Darwinian approach to optimization is at present under consideration [4.52].

4.5 Binary Synapses

In the previous sections the synapses are described by real number J_{ij}. For technical applications it may be useful to restrict the strength of the connections to binary values. In this case the learning algorithms of Sects. 4.3 and 4.4 cannot be applied.

Of course, one could first construct continuous couplings and then define binary couplings by the sign of the continuous ones. For the Hopfield model and the corresponding layered system this has been done and the model can still be solved analytically [4.53–55]. The synapses are defined as

$$J_{ij} = \frac{1}{\sqrt{N}} \mathrm{sgn} \left\{ \sum_{\nu} \xi_i^{\nu} \xi_j^{\nu} \right\} . \tag{4.44}$$

It turns out that this network is equivalent to the model with continuous synapses (4.2), but with additional static Gaussian noise. The maximal storage capacity α_c decreases by about 20–30%, but otherwise the properties of the associative memory are almost not affected by the restriction to binary synapses.

An attempt has been made to calculate the maximal possible storage capacity using the same methods as for (4.7) [4.56]. However, in this case the replica-symmetric solution of the phase space of binary synapses obviously gives wrong results. An upper bound is given by $\alpha_c \leq 1$, and a recent calculation indicates that $\alpha_c \approx 0.83$ (for random patterns [4.57]). A learning algorithm has been developed and investigated [4.36, 58] which follows the derivation of the ADALINE from (4.12). The quadratic form

$$E_i = \frac{1}{N} \sum_{\nu} (\lambda - \xi_i^{\nu} h_i^{\nu})^2 \tag{4.45}$$

is minimized with respect to the synapses

$$J_{ij} = \frac{\sigma_{ij}}{\sqrt{N}} \quad ; \quad \sigma_{ij} \in \{+1, -1\} . \tag{4.46}$$

The parameter λ should be optimized such that the network has a large storage capacity. Note that again different rows i of the matrix J_{ij} do not interact. Equation (4.45) may be written in the form (index i is omitted)

$$E = -\sum_{j,k} K_{jk} \sigma_j \sigma_k - \sum_j h_j \sigma_j - \alpha \lambda^2 , \tag{4.47}$$

with

$$K_{jk} = -\frac{1}{N} \sum_{\nu} \xi_j^{\nu} \xi_k^{\nu} , \tag{4.48}$$

$$h_j = \frac{2\lambda}{N} \sum_{\nu} \xi_i^{\nu} \xi_j^{\nu} . \tag{4.49}$$

Hence the optimization problem (4.45) with synaptic variables σ_j is equivalent to finding the ground state of a Hopfield model of neurons σ_j with a negative coupling, (4.49) and an additional external field h_j. The fields h_j obviously favor the clipped synapses (4.44) while the couplings K_{jk} try to prevent a condensation of the synapses $\{\sigma_j\}$ into one of the patterns $\{\xi_i^\nu \xi_j^\nu\}$. The network works as an associative memory only if the synapses have correlations of the order of $1/\sqrt{N}$ into all of the patterns. Note that the clipped Hopfield model (4.44) is related to the p-symmetric mixture state discussed by Amit et al. [4.47].

The minimum of E_i (4.45) has been estimated by replica theory [4.36, 58]. The replica-symmetric solution gives a phase transition at

$$\lambda_c = \sqrt{\frac{2}{\pi\alpha} - 1}\,. \tag{4.50}$$

For $\alpha < \alpha(\lambda_c)$ the "cost function" E_i is zero, i.e. all patterns are stable with local fields $\xi_i^\nu h_i^\nu = \lambda$. Above $\alpha(\lambda_c)$ E_i increases with the storage capactiy α.

Numerically a (local) minimum of E_i has been obtained by sequential descent [4.36, 58]. Initially the variables σ_j are set to random values ± 1; then the sites j are visited sequentially and the variable σ_j is changed if E_i decreases by this change. This is iterated until E_i has reached a local minimum. Figure 4.3 shows the result as a function of α. The numerical values of E_i are larger than the analytic ones and do not show any transition at $\alpha(\lambda_c)$. An initial start with the clipped Hopfield model (4.44) does not improve the results of a random start. However, a start from the clipped PERCEPTRON (4.11) gives a lower cost function E_i, and simulated annealing [4.59] minimizes E_i further. This shows that this optimization problem has many local minima. But no indications could be found that the numerical results finally converge to the analytic ones. In

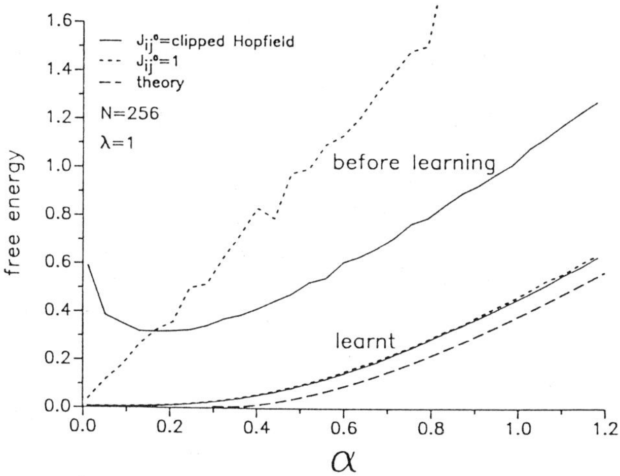

Fig. 4.3. Cost function E for $\lambda = 1$ as a function of storage capacity α for a network with binary synapses. The replica theory gives a transition at $\alpha_c = 1/\pi$. The numerical data are obtained from a decent algorithm. (From [4.36])

particular the phase transition at $\alpha(\lambda_c)$, (4.48) could not be found numerically, E_i is always larger than zero. This discepraney may be due the replica-symmetric approximation, similar to [4.56].

The cost function E_i (4.45) contains an additional parameter λ which can still be optimized. Unfortunately the value of E_i does not give much information about the properties of the corresponding associative memory. A low value of E_i means that most of the local fields $\xi_i^\nu h_i^\nu$ are close to the value λ, but this does not tell us much about attractors and basins of attraction of the corresponding network. Hence, the network obtained from minimization of E_i has been simulated and the basins of attraction have been calculated [4.36]. This will be discussed in the following section. It turns out that λ_c from (4.48) is not the optimal choice of the value of λ. A larger value

$$\lambda_m = \sqrt{\frac{2}{\pi\alpha}}, \tag{4.51}$$

which is an upper bound for $\langle \xi_i^\nu h_i^\nu \rangle$, gives a maximal storage capacity of $\alpha_c \approx 0.4$. This seems to be the optimal choice of λ for the sequential-descent algorithm. Note that for $\lambda \to \infty$ one obtains the clipped Hopfield model (4.44), which performs much more poorly ($\alpha_c \approx 0.11$) than the one obtained from learning with $\lambda = \lambda_m$. A learning algorithm which gives a matrix with maximal stability Δ (4.6) is still not known (anyway Δ is not known, either). An algorithm which is related to the PERCEPTRON would improve the worst pattern (lowest $\xi_i^\nu h_i^\nu$) by a change of a single synapse σ_{ij}. This is being investigated [4.60], but reliable results are still not availabe.

4.6 Basins of Attraction

An associative memory should have a large maximal storage capacity α_c, and each pattern ξ^ν should have an attractor which is close to ξ^ν (with respect to the Hamming distance) and has a large basin of attraction. However, it is not clear to what extent the learning rules of Sects. 4.3, 4.4, and 4.5 optimize these properties. For instance the PERCEPTRON (4.11) gives a matrix J_{ij} which has precisely the patterns ξ_i^ν as attractors, and a large stability Δ is expected to give large basin of attraction. But since the attractors may move away from the patterns and since the properties of the matrices which determine the size of the attractors are not known, yet, it cannot be ruled out that much better networks exist which may be obtained by learning algorithms very different from the ones discussed above.

Unfortunately basins of attractions still cannot be calculated exactly, although some progress has been made recently [4.61, 62]. Hence one has to rely on either numercial simulations [4.33–36] or on simple approximations [4.40, 43, 63].

Simulations have shown that PERCEPTRON learning improves the properties of the network with respect to the Hopfield model (in [4.35] a version of (4.11) was used which gives symmetric couplings $J_{ij} = J_{ji}$). Training the network with noisy patterns seems to give large basins of attraction, although it is very slow [4.33, 34].

Here we want to describe an analytic treatment of the dynamics of the neurons. In the simplest approximation, correlations are neglected and the distribution of local fields h_i is assumed to be Gaussian [4.26, 43]. For the Hopfield model these approximations become exact in the extremely diluted asymmetric model [4.42]. In this case the dynamics is described just by the overlap $m_\nu(t)$ to the patterns, with

$$m_\nu(t) = \frac{1}{N} \sum_i \xi_i^\nu \langle S_i(t) \rangle \, , \tag{4.52}$$

where $\langle \dots \rangle$ means an average over the initial state and the thermal noise. If the state $\{S_i(t)\}$ has nonzero overlap to only one pattern $\nu = 1$, the equation for $m_1(t)$ becomes at zero temperature [4.40, 43] (for parallel updating)

$$m_1(t+1) = \operatorname{erf}\left(\frac{m_1(t)}{\sqrt{2}\Delta(t)} \right) \, , \tag{4.53}$$

where the relative width $\Delta(t)$ of the field distribution is given by

$$\Delta^2(t) = [1 - m_1^2(t)]a + m_1^2(t)b \, , \tag{4.54}$$

with

$$a = \frac{\sum_j J_{ij}^2}{\overline{(\xi_i^1 h_i^1)}^2} \, , \quad b = \frac{\overline{(\xi_i^1 h_i^1)^2}}{\overline{(\xi_i^1 h_i^1)}^2} - 1 \, ;$$

$\overline{(\dots)}$ means an average over the patterns $\{\xi_i^\nu\}$. Hence, only two parameters of the matrix J_{ij} enter the retrieval dynamics, namely the norm of J_{ij} and the width of the field distribution of pattern $\nu = 1$, relative to the mean value of $\xi_i^1 h_i^1$. If the distribution of the fields $\xi_i^1 h_i^1$ is known it can be included in a generalized equation (4.53).

The basin of attraction can be calculated by iterating (4.53) numerically; it is given by the initial overlap $m_1^c(0)$, which separates the flow $m_1(t) \to m_1(\infty) \approx 1$ (retrieval) from the flow $m_1(t) \to 0$ (nonretrieval of pattern ξ_i^1).

Figure 4.4 shows the basins of attraction for the projector matrix, (4.19) and the matrix of largest stability (4.7), respectively [4.40]. In the first case the matrix may be learned by ADALINE, and one has

$$a = \frac{\alpha}{1-\alpha} \, , \quad b = 0 \, , \tag{4.55}$$

while in the second case, which is obtained from the PERCEPTRON, the exact distribution of local field h_i has been used (see also [4.64]). Although the PERCEPTRON gives a maximal storage capacity $\alpha_c = 2$ which is twice as large as that of the projector matrix $\alpha_c = 1$, the basins of attractions are similar for both cases.

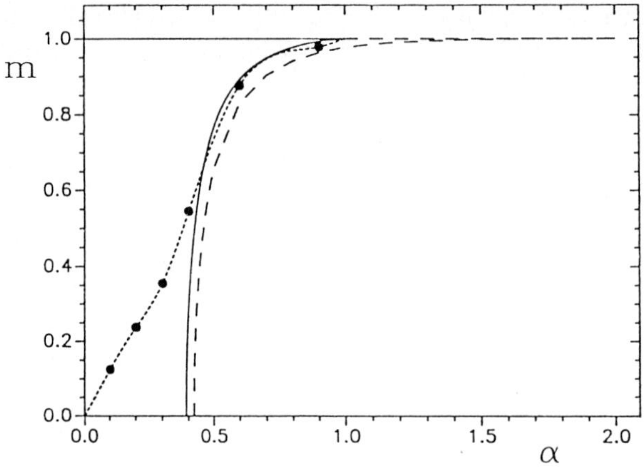

Fig. 4.4. Basins of attraction for the projector matrix (*solid line*) and the matrix of largest stability (*dashed line*). The points are numerical data of Kanter and Sompolinski [4.50]. m is the retrival overlap, and the lower branch separates the flow to m=1 from the flow to $m = 0$. (From [4.40])

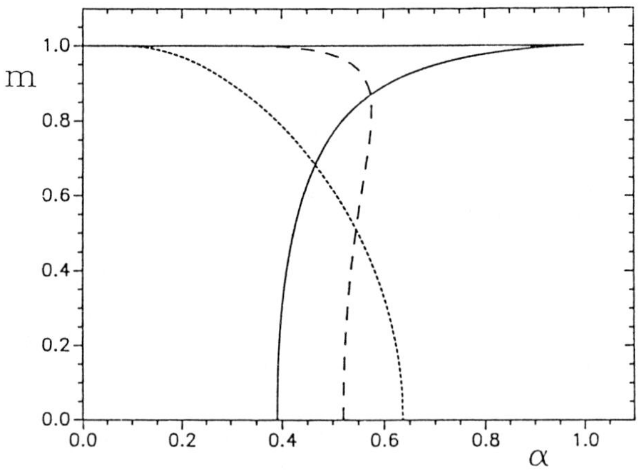

Fig. 4.5. Basins of attraction for a synaptic matrix which has been learned by parallel ADALINE. The *dotted line* is the first learning step, which gives the Hopfield model. The *dashed line* is from the matrix after the second step, and the *solid line* gives the basin after infinitely many steps (projector matrix). (From [4.40])

It is interesting to see how learning improves the properties of associative memories. This is shown in Fig. 4.5 for the ADALINE, (4.14), where the patterns are presented in parallel. Starting with zero couplings one obtains in the first learning step the Hopfield model with $a = b = \alpha$. In the second presentation of the patterns one finds [4.40].

$$a = \alpha \left(1 + \frac{\alpha}{(\alpha + 1)^2} \right) , \quad b = \frac{\alpha^2}{\alpha + 1} , \tag{4.56}$$

while for infinitely many learning steps one has the projector matrix with (4.55). Figure 4.5 shows the basins of attractions for the three cases.

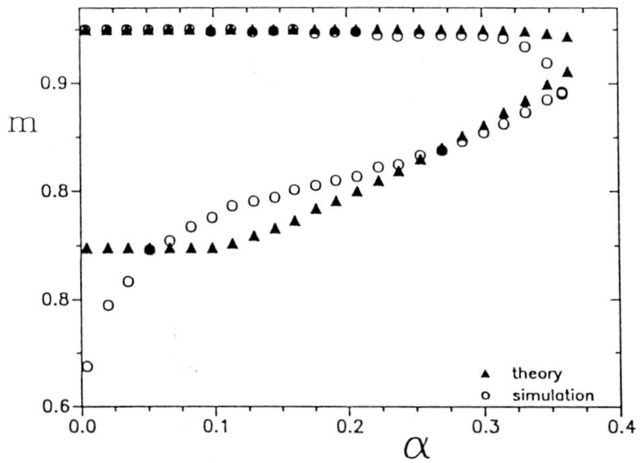

Fig. 4.6. Basin of attraction for bynary synapses which have been learned by optimizing the cost function $E(\lambda = 1)$ using a descent algorithm (compare with Fig. 4.3). The theory uses parameters taken from numerical data. (From [4.43])

Note that for $\alpha \approx 0.5$ the second learning step gives a larger basin of attraction than the matrix obtained after infinitely many steps. Supervised learning as discussed in this article tries to find perfect attractors, but this does not necessarily mean that the space of information which flows to those attractors is as large as possible.

For the networks with binary synapses obtained from optimizing (4.45) by descent, the basins of attraction have been calculated numerically [4.36]. But one may also take the two parameters a and b (4.55) from the final matrix J_{ij} and calculate the retrieval from the approximation (4.53) and (4.55). Figure 4.6 compares the results, where the parameter $\lambda = 1$ has been taken. One obtains a maximal storage capacity of $\alpha_c \approx 0.36$, although the basin of attraction is relatively small. The good agreement between simulation and approximation is due to a narrow field distribution close to the attractor $m \approx 1$ [4.40, 43]. A different initial matrix for the learning process, stochastic learning, or PERCEPTRON-like learning may still improve the associative memory with binary synapses [4.36, 60], but reliable results are not yet available.

4.7 Forgetting

When a neural network learns new patterns, the synaptic weights are changed. This may destabilize the previously learned patterns; hence the network forgets.

This is not the case for the Hopfield model (4.10), where all patterns are added to the connections J_{ij} with the same weight. Hence by adding new patterns all of the patterns are destabilized until above a critical capacity α_c all of the memory is suddenly erased, without prewarning.

Two mechanisms of gradual forgetting have been proposed for the Hopfield model: (i) the last pattern is stored with an increasing synaptic weight [4.25, 67], and (ii) the weights are replaced by a nonlinear function of the Hopfield couplings, e.g. the sign as in (4.44) [4.65–69]. In both cases the previously learned patterns are gradually erased.

If one improves the learning algorithm, for instance by the ADALINE or PERCEPTRON, the resulting matrix J_{ij} does not weight the patterns by the same amount. Furthermore, at least during the first learning steps, the synapses depend on the sequence on which the patterns are presented.

For the ADALINE rule with sequential updating, (4.9) these properties have been investigated analytically for the first learning cycle (sweep through the patterns) [4.41]. The algorithm is such that the last pattern ν is always perfectly stable, i.e. $\xi_i^\nu h_i^\nu = 1$ (for $\gamma = 1$). However, when a pattern is added it creates noise to the fields $\xi_i^\nu h_i^\nu$ of the previously learned patterns. If $p = \alpha N$ random patterns have been presented once by (4.14), the fields of the pattern number $p_0 = \alpha_0 N$ are distributed by a Gaussian with mean

$$\overline{\xi_i^{p_0} h_i^{p_0}} = \mathrm{e}^{-(\alpha-\alpha_0)}$$

and variance (4.57)

$$\overline{(\xi_i^{p_0} h_i^{p_0})^2} - (\overline{\xi_i^{p_0} h_i^{p_0}})^2 = \left(1 - \mathrm{e}^{-\alpha}\right) - \mathrm{e}^{-2(\alpha-\alpha_0)}\left(1 - \mathrm{e}^{-\alpha_0}\right) .$$

Furthermore, the norm of the matrix J_{ij} is given by

$$\sum_j J_{ij}^2 = 1 - \mathrm{e}^{-\alpha} . \tag{4.58}$$

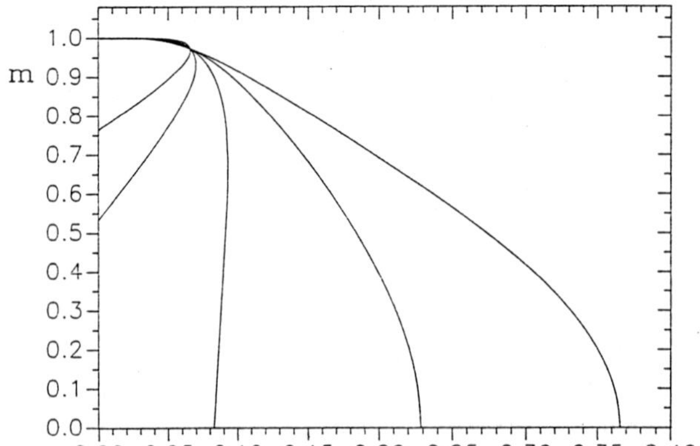

Fig. 4.7. Basins of attraction of pattern $p_0 = \alpha_0 N$ after learning $p = \alpha N$ patterns by applying sequential ADALINE in one sweep only. (From [4.41])

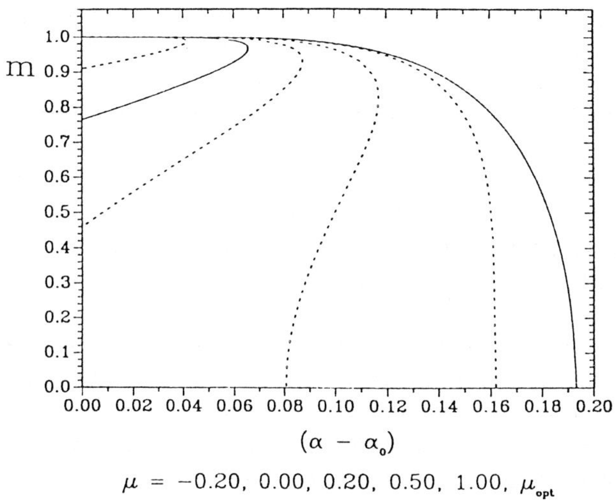

Fig. 4.8. Same as Fig. 4.7 for $\alpha \to \infty$, but with additional relaxation mechanism given by the parameter μ. For $\mu_{\text{opt}} \approx 2.09$ one obtains a maximal storage capacity. (From [4.41])

This gives both of the parameters a and b which determine the retrieval properties of the network from the approximation (4.53). Figure 4.7 shows the basins of attraction for different values of α_0 and α.

If the ADALINE is extended to (4.16) such that simultaneously the norm of the matrix is reduced by each larning step, one has an additional parameter μ which should be optimized. In this case one obtains for $\alpha, \alpha_0 \to \infty$

$$a = \frac{1}{1+2\mu}\, e^{2(1+\mu)(\alpha-\alpha_0)} , \quad b = \frac{1}{1+2\mu}\, (e^{2(1+\mu)(\alpha-\alpha_0)} - 1) . \tag{4.59}$$

At $m \approx 0$ the basin of attraction is bounded by $a = 2/\pi$; this gives $\alpha_c(\mu)$. Figure 4.8 shows the basins of attraction for different values of μ, after infinitely many patterns were presented ($\alpha \to \infty$). The maximal basin of attraction is given by $\mu_{\text{opt}} \approx 2.09$.

4.8 Outlook

The past few years have shown that theoretical physics can contribute useful results to the understanding of neural networks. The physicist's approach is somewhat different from those of his colleagues in biology or computer science. It concentrates on simple models and it searches for quantitative results on typical and infinitely large systems. Hence physics contributes to a general basic understanding of the cooperative behavior of neural networks which is definitely useful for research on more biological or applied problems.

In the beginning, physicists used methods of solid-state theory (the statistical mechanics of spin glasses) to calculate storage capacities and retrieval errors of neural networks performing as an associative memory. For fixed given synapses the attractors in the *dynamics of the neurons* could be calculated analytically. The second important step was to extend these methods to the *dynamics of the synapses*. The calculation of the space of all possible synaptic matrices gave the maximal storage capacity of attractor networks (without hidden units). Finally spin-glass methods were able to yield quantitative results on learning algorithms, as is shown in this contribution.

In the future this research may proceed in several directions. The models will be extended taking more biological details, such as synaptic delays or triple interactions, into account. Closer contact with microscopic mechanisms as well as with macroscopic functions is clearly desirable to understand biological networks. But models will also be extended to come closer to practical applications. Learning in multilayer networks with hidden units still cannot be studied analytically, even in the simplest cases. Fast and effective learning algorithms, for which our analytical studies might be helpful have yet to be developed. The structure of the network obviously depends on the task to be performed; this has to be studied in more detail.

The future will show to what extent physics will contribute to a solution of these problems, but the first few years of this research program have already been very successful.

Acknowledgement. This work has been supported by the Deutsche Forschungsgemeinschaft and the Höchstleistungsrechenzentrum Jülich. The authors thank H. Köhler, M. Biehl, and J. Kleinz for assistance.

References

4.1 D.O. Hebb: *Organization of Behavior* (Wiley, New York 1949)
4.2 D.E. Rumelhart and J.L. McClelland: *Parallel Distributed Processing: Explorations in the Microstructure of Cognition*, 2 Vols. (MIT Press, Cambridge, 1986)
4.3 M. Minsky and S. Papert: *Perceptrons*, (MIT Press, Cambridge 1988)
4.4 B. Widrow and M.E. Hoff: WESCON Convention, Report IV, 96 (1960)
4.5 F. Rosenblatt: Psychoanalytic Review **65**, 386 (1958)
4.6 S. Diederich and M. Opper: Phys. Rev. Lett. **58**, 949 (1987)
4.7 J.J. Hopfield: Proc. Natl. Acad. Sci USA **79**, 2554 (1982)
4.8 D.E. Rumelhart, G.E. Hinton, and R.J. Williams: Nature **323**, 533 (1986)
4.9 T. Grossman, R. Meir, and E. Domany: Complex Systems **2**, 255 (1989)
4.10 P. Rujan and M. Marchand: Complex Systems **3**, 229 (1989)
4.11 M. Mezard and J.P. Nadal: J. Phys. A: Math. Gen. **22**, 2191 (1989)
4.12 T.J. Sejnowski and C.R. Rosenberg: Complex Syst. **1**, 145 (1987)
4.13 G. Tesauro and T.J. Sejnowski: in *Neural Information Processing Systems*, edited by D.Z. Anderson (American Institute of Physics, New York, 1988) p.442
4.14 J. A. Hertz, A. Krogh, and G.I. Thorbergson: J. Phys. A: Math. Gen. **22**, 2133 (1989)
4.15 R. Meir and E. Domany: Phys. Rev. A **37**, 2660 (1988)
4.16 E. Domany, R. Meir and W. Kinzel: Europhys. Lett. **2**, 175 (1986)
4.17 R. Linsker: Proc. Natl. Acad. Sci USA **83**, 7508, 8390, 8779 (1986)
4.18 W.S. McCulloch and W.A. Pitts: Bull. Math. Biophys. **5**, 115 (1943)
4.19 J. Buhmann and K. Schulten: Europhys. Lett. **4**, 1205 (1987)
4.20 A.C.C. Coolen and C.C.A.M. Gielen: Europhys. Lett. **7**, 281 (1988)

4.21 M. Kerszberg and A. Zippelius: Phys. Scr. **T 33**, 54 (1990)
4.22 A. Herz, B. Sulzer, R. Kühn, and J.L. van Hemmen: Europhys. Lett. **7**, 663 (1988) and Biol. Cybern. **60**, 457 (1989)
4.23 H. Sompolinsky and I. Kanter: Phys. Rev. Lett. **57**, 2861 (1986)
4.24 P. Peretto and J.J. Niez: in *Disordered Systems and Biological Organization*, edited by E. Bienenstock, F. Fogelman-Soulié, and G. Weisbuch (Springer, Berlin, Heidelberg 1986) p. 171
4.25 G. Toulouse, S. Dehaene, and J.P. Changeux: Proc. Natl. Acad. Sci USA **83**, 1695 (1986)
4.26 W. Kinzel: Z. Phys. B **60**, 205 (1985)
4.27 S. Dehaene, J.P. Changeux, and J.P. Nadal: Proc. Nath. Acad. Sci USA **84**, 2727 (1987)
4.28 S. Wolfram: *Theory and Application of Cellular Automata* (World Scientific, Singapore 1986)
4.29 B.A. Huberman and T. Hogg: Phys. Rev. Lett. **52**, 1048 (1984)
4.30 P. Carnevali and S. Patarnello: Europhys. Lett. **4**, 1199 (1987)
4.31 K.Y.M. Wong and D. Sherrington: Europhys. Lett. **7**, 197 (1988)
4.32 R. Meir and E. Domany: Phys. Rev. A **37**, 608 (1988)
4.33 E. Gardner, N. Stroud, and D.J. Wallace: in *Neural Computers*, edited by R. Eckmiller and C. v. d. Malsburg, (Springer,Berlin, Heidelberg 1988) p. 251
4.34 G. Pöppel and U. Krey: Europhys. Lett. **4**, 979 (1987)
4.35 B.M. Forrest: J. Phys. A: Math. Gen. **21**, 245 (1988)
4.36 H. Köhler: Diplom thesis, Justus-Liebig-Universität Gießen (1989)
4.37 W. Krauth and M. Mezard: J. Phys. A: Math. Gen. **20**, L 745 (1987)
4.38 M. Opper: Europhys. Lett. **8**, 389 (1989)
4.39 M. Opper: Phys. Rev. A **38**, 3824 (1988)
4.40 J. Kleinz: Diplom thesis, Justus-Liebig-Universität Gießen (1989)
4.41 M. Biehl and W. Kinzel: unpublished
4.42 B. Derrida, E. Gardner, and A. Zippelius: Europhys. Lett. **4**, 167 (1987)
4.43 M. Opper, J. Kleinz, H. Köhler, and W. Kinzel: J. Phys. A: Math. Gen. **22**, L 407 (1989)
4.44 E. Gardner: J. Phys. A: Math. Gen. **21**, 257 (1988)
4.45 K. Binder and A.P. Young: Rev. Mod. Phys. **58**, 801 (1986)
4.46 M. Mezard, G. Parisi, and M.A. Virasoro: *Spin Glass Theory and Beyond* (World Scientific, Singapore 1987)
4.47 D.J. Amit, G. Gutfreund, and H. Sompolinsky: Ann. Phys. **173**, 30 (1987)
4.48 T. Kohonen: Selforganization and Associative Memory (Springer, Berlin, Heidelberg 1988)
4.49 L. Personnaz, I. Guyon, and G. Dreyfus: Phys. Rev. A **34**, 4217 (1986)
4.50 I. Kanter and H. Sompolinsky: Phys. Rev. A **35**, 380 (1987)
4.51 J.K. Anlauf and M. Biehl: Europhys. Lett. **10**, 687 (1989)
4.52 S. Diederich and M. Opper: Phys. Rev. A **39**, 4333 (1989) and unpublished
4.53 H. Sompolinsky: Phys. Rev. A **34**, 2571 (1986)
4.54 J.L. van Hemmen: Phys. Rev. A **36**, 1959 (1987)
4.55 E. Domany, W. Kinzel, and R. Meir: J. Phys. A: Math. Gen. **22**, 2081 (1989)
4.56 E. Gardner and B. Derrida: J. Phys. A: Math. Gen. **21**, 271 (1988)
4.57 W. Krauth and M. Opper: J. Phys. A: Math. Gen. **222**, L519 (1989)
4.58 H. Köhler, S. Diederich, M. Opper, and W. Kinzel: Z. Physik B **78**, 333 (1990)
4.59 S. Kirkpatrick, C.D. Gelatt, and M.P. Vecchi: Science 220, 671 (1983)
4.60 J. Anlauf: unpublished
4.61 H. Horner: Phys. Blätter **44**, 29 (1988)
4.62 R.D. Henkel and M. Opper: Europhys. Lett. **11**, 403 (1990)
4.63 T.B. Keppler and L.F. Abott: J. Phys. France **49**, 1657 (1988)
4.64 W. Krauth, M. Mezard and J.P. Nadal: Complex Systems **2**, 387 (1988)
4.65 G. Parisi: J. Phys. A **19**, L 617 (1986)
4.66 J.L.v. Hemmen, G. Keller, and R. Kühn: Europhys. Lett. **5**, 663 (1988)
4.67 B. Derrida and J.P. Nadal: Jour. Stat. Phys. **49**, 993 (1987)
4.68 T. Geszti and F. Pazmandi: J. Phys. A: Math. Gen. **20**, L 1299 (1987)
4.69 J.P. Nadal, G. Toulouse, J.P. Changeux, and S. Dehaene: Europhys. Lett. **1**, 535 (1986)

5. Hierarchical Organization of Memory

Michail V. Feigel'man and Lev B. Ioffe

With 2 Figures

Synopsis. The problems associated with storing and retrieving hierarchically organized patterns are analyzed, the performance of various prescriptions is discussed, and their efficiency is evaluated.

5.1 Introduction

It is very well known in cognitive psychology that human memory is organized hierarchically: it is far easier to memorize an object that resembles closely a known one than to store and recall a new and different one. One can easily assure oneself of this fact if one tries to memorize a picture consisting of a dozen lines which resembles nothing and a picture which resembles, even distantly, some well-known object. Evidently, we usually do not memorize the object but the details which differentiate it from other objects of the same class. On the other hand, we are more prone to mix the objects in the same class than to mix objects in different classes. Naturally, we would like to construct a model of a neural network which displays these features.

Keeping in mind this goal, we should first reformulate those intuitively evident requirements which have to be imposed on a model into rigorous mathematical statements. As usual, we suppose that all information concerning an object is presented to us in the form of an N-bit pattern $\xi_i = \pm 1$ ($i = 1, \ldots, N$). The similarity between two patterns ξ_i and ζ_i is measured by their overlap q defined by $q = N^{-1} \sum_i \xi_i \zeta_i$ where i runs from 1 to N. For similar patterns the overlap is close to unity whereas for uncorrelated patterns it is random variable with zero mean and small ($N^{-1/2}$) variance, since in this case every term in the sum $\sum_i \xi_i \zeta_i$ has zero mean and unit variance.

Then we would like to define the notion of *class* of patterns as the set of all patterns that have mutual overlaps *larger* than q. The intuitive idea behind this definition is that "similar" patterns have a large mutual overlap. Unfortunately, this definition is not internally consistent. For instance, consider the following example of three patterns with overlaps $q_{12} = q_{13} = 1/2$, $q_{23} = 0$, which can exist in a four-bit network ($\xi^1 = (1, 1, 1, 1)$, $\xi^2 = (1, 1, 1, -1)$, $\xi^3 = (1, 1, -1, 1)$). Obviously, if in this example $0 < q < 1/2$, then we get into trouble since we cannot decide which of the two patterns (ξ_2 or ξ_3) should be included in the same class with the pattern ξ^1. This type of categorization problem is absent only if the set of mutual overlaps between patterns obeys a severe rule: if overlap q_{12} between any two patterns is less than q and overlap q_{13} with some other pattern is also less than q then overlap q_{23} should also be less than q. If for a given set of patterns only one such q exists, then only one class can be defined, i.e., the class which contains patterns with mutual overlaps less than q. In general, a set of q_i ($i = 1, \ldots, k$) exists so that a ramified categorization scheme can be defined: each pattern ξ^α belongs to some class which, in turn, belongs to some category and so on. Thus, each pattern can be specified by a string $\{\mu_1, \ldots, \mu_k\}$ which specifies its category (μ_1), class (μ_2), and so on, and finally the pattern itself (μ_K) and, hence, $q_1 < q_2 < \ldots < q_K$.

In the limit $K \to \infty$, a categorization is possible for *any* q, provided for any three patterns we have $q_{23} \geq \min(q_{13}, q_{12})$, which is known as an *ultrametricity* condition. The set of patterns then forms a so-called ultrametric space [5.14] and the string $\{\mu_1, \ldots, \mu_K\}$ becomes infinite. The condition of ultrametricity imposes a severe restriction on the set of patterns. Nevertheless, it deserves a special consideration since it provides a deep analogy between the present problem and the physical problem of a spin glass, where the ultrametric set of patterns naturally appears.

5.2 Models: The Problem

Now we turn to the construction of models which allow the storage and successful retrieval of categorized information. In all models that we shall discuss each neuron is decribed by a bit variable $S_i = \pm 1$ and the information can be stored in the real symmetric matrix of synaptic junctions J_{ij} only. In the retrieval process the initial state of neurons $S_i(t=0)$ corresponds to a partial, deteriorated pattern. Then we apply a dynamics so that each variable S_i is aligned along the "internal field" $h_i = \sum_i J_{ij} S_j$ through the prescription $S_i = \mathrm{sgn}(h_i)$. The process stops when a final state is reached in which all variables S_i ("spins") are aligned along their fields. At every step of this process the energy function

$$E = -\tfrac{1}{2} \sum_{ij} J_{ij} S_i S_j \tag{5.1}$$

is lowered and the final state corresponds to a local minimum of this function. Since the energy function is bounded from below this local minimum always exists. The final state is the retrieved pattern. Our goal is to construct a matrix J_{ij} which allows the correct retrieval of a given categorized set of patterns. In other words, we should find a matrix J_{ij} so that $\xi_i^\alpha = \mathrm{sgn}(h_i^\alpha)$ with $h_i^\alpha = \sum_j J_{ij} \xi_j^\alpha$ for all patterns ξ_i^α. If the condition $\xi_i^\alpha = \mathrm{sgn}(h_i^\alpha)$ is fulfilled for all sites and all patterns then the exact retrieval is possible. However, we do not usually need such a severe requirement and can be satisfied if for each prototype pattern ξ_i^α some other pattern $\tilde{\xi}_i^\alpha$ exists which differs from the prototype only in a small number of sites and is the final state of the retrieval process: $\tilde{\xi}_i^\alpha = \mathrm{sgn}(\tilde{h}_i^\alpha)$.

Certainly, for a correct performance of the memory, the overlap $\tilde{q}$ between the prototype ξ^α and the real pattern $\tilde{\xi}^\alpha$ should be more than any overlap between different patterns.

For *un*correlated patterns $\{\xi^\alpha\}$ the matrix J_{ij} which solves the problem of correct retrieval was proposed quite a while ago by Cooper [5.1], who implemented some qualitative ideas of Hebb [5.2] which were made operational by Hopfield [5.3]

$$J_{ij} = \frac{1}{N} \sum_\alpha \xi_i^\alpha \xi_j^\alpha . \tag{5.2}$$

It can be proved that this prescription allows the correct retrieval of $K \sim N$ uncorrelated patterns, but only a few patterns of macroscopically large [$q \sim \mathrm{O}(1)$] overlap can be stored and retrieved. That is, even a single class of patterns cannot be retrieved successfully with storage prescription (5.2). To show this, let us consider the fields h_i acting in the presence of the state $S_i = \xi_i^\alpha$, $1 \leq i \leq N$. If in this state all fields h_i are aligned with ξ_i^α, then ξ_i^α is a final state of the retrieval process and prescription (5.2) works, and vice versa. The fields h_i can be represented in the form

$$h_i = \xi_i^\alpha + \sum_{\beta(\neq\alpha)} \xi_i^\beta q_{\beta\alpha} . \tag{5.3}$$

For *un*correlated patterns the second term in (5.3) is a random variable with zero mean and variance $\sqrt{K/N}$; thus for $K \ll N$ the h_i is aligned along ξ_i^α, whereas for correlated patterns q is of the order of one and the second term in (5.3) is large, which makes the correct retrieval of one class of a few correlated patterns impossible.

5.3 A Toy Problem: Patterns with Low Activity

Before we plunge into the details of construction of the matrix which allows the correct retrieval of a general categorized set of patterns, it is useful to study a toy problem of retrieval of a single class of correlated patterns. We can further simplify the problem if we assume that correlation between patterns is due to a low average level of activity. Specifically, we suppose that for each pattern neuron i fires ($\xi_i = -1$) with small probability p and is quiescent ($\xi_i = +1$) with probability $1 - p$. Apart from being instructive, this problem is also interesting in itself from a biological viewpoint since in a real brain only a small portion of all neurons is firing at the same moment.

As we have shown above, the Hebbian rule (5.2) allows the correct retrieval of only a few correlated patterns. A general algorithm which allows the correct retrieval of any number of patterns less than or equal to N was proposed by Personnaz et al. [5.4] (see also Kohonen [5.5]) and studied by Kanter and Sompolinsky [5.6]. To describe this algorithm, which we henceforth refer to as the *quasi-inverse rule*, we consider the generalization of the Hebbian rule (5.2):

$$J_{ij} = \frac{1}{N} \sum_{\alpha,\beta} \xi_i^\alpha \xi_j^\beta B_{\alpha\beta} \tag{5.4}$$

with some matrix $B_{\alpha\beta}$. The matrix B should be chosen so that the condition $\mathrm{sgn}(h_i^\alpha) = \xi_i^\alpha$ is satisfied for all patterns ξ_i^α. If all the patterns ξ_i^α, $1 \le \alpha \le K$, are linearly independent, then an even stronger condition $h_i^\alpha = \xi_i^\alpha$ can be satisfied. Indeed, if we choose

$$(B^{-1})_{\alpha\beta} = \frac{1}{N} \sum_j \xi_j^\alpha \xi_j^\beta , \tag{5.5}$$

where B^{-1} means matrix inversion, then, inserting (5.5) into definition (5.4) and evaluating the local fields, we get $h_i^\alpha = \xi_i^\alpha$. The matrix inversion implied by (5.5) is possible if the determinant of the right-hand side of this equation is not zero, i.e., if the ξ_i^α are linearly independent – as predicted. If the ξ_i^α, $1 \le \alpha \le K$, are linearly dependent, then (5.5) has to be reinterpreted as a *quasi*- or *pseudo*-inverse [5.4, 5]. Hence the name quasi-inverse rule.

The above algorithm has two drawbacks. (i) The procedure of matrix inversion is very unlocal: it requires the simultaneous knowledge of *all* patterns whereas the desirable algorithm should allow an iterative storage of patterns. (ii) The maximum number of stored patterns equals N since N is the maximum number of linearly independent vectors ξ_i^α. Therefore, the information stored decreases with p: $I = pN^2 \log_2 p$ bits. The desirable algorithm should allow us to store a larger number of correlated patterns which contain the same amount of information as a set of uncorrelated patterns. To overcome the first drawback we should look for other, less general, but simpler learning rules that allow the correct retrieval of correlated patterns. Fortunately, studying these simplified rules we shall find the algorithm which allows the retrieval of $N(p \ln p)^{-1}$ patterns, and thus overcome the second drawback of the rule (5.5). Moreover, this algorithm will provide useful hints to what is wrong with the quasi-inverse rule and how algorithms with maximal storage capacity can be constructed.

We start with the simplification of the learning rule (5.4, 5) for the set of patterns with low levels of activity p. Each of these patterns can be represented in the form $\xi_i^\alpha = \overline{\xi} + \delta\xi_i^\alpha$ where $\overline{\xi}$ is the mean level of activity: $\overline{\xi} = N^{-1} \sum_i \xi_i^\alpha = (1-2p)$. The variable $\delta\xi_i^\alpha$ characterizing individual patterns has zero mean by definition and the $\delta\xi_i^\alpha$ of different patterns are uncorrelated since in our simple problem there is no other correlation between patterns apart from their common low level of activity. Note that we do not impose the condition $\sum_i \delta\xi_i^\alpha = 0$ on each pattern but treat the variables $\delta\xi_i^\alpha$ as independent random variables: $\langle \delta\xi_i^\alpha \delta\xi_j^\beta \rangle = 4p(1-p) \times \delta_{ij}\delta_{\alpha\beta}$ where δ_{ij} and $\delta_{\alpha\beta}$ are Kronecker deltas. Accordingly, the mean activity of each pattern can differ from p by a small amount (proportional to $pN^{-1/2}$). Then the equation of the $B_{\alpha\beta}$-matrix (5.5) acquires a simple form,

$$(B^{-1})_{\alpha\beta} = (1-2p)^2 + \frac{1}{N}\sum_i \delta\xi_i^\alpha \delta\xi_i^\beta$$
$$= (1-2p)^2 + \delta^{\alpha\beta} 4p(1-p)\,, \tag{5.6}$$

where we neglect small ($N^{-1/2}$) Gaussian fluctuations of the sum $\sum_i \delta\xi_i^\alpha \delta\xi_i^\beta$. The matrix in the right-hand side of (5.6) can be inverted easily:

$$B_{\alpha\beta} = -\frac{(1-2p)^2}{4p(1-p)[K(1-2p)^2 + 4p(1-p)]} + \frac{\delta_{\alpha\beta}}{4p(1-p)}\,, \tag{5.7}$$

where K is the number of patterns. Since we are interested only in the case $K \gg 1$ we can neglect the second term in square brackets in (5.7). Inserting the resulting expression for the $B_{\alpha\beta}$-matrix into (5.4) we get:

$$J_{ij} = \frac{1}{N}\left\{1 + \frac{1}{4\Delta}\sum_\alpha \delta\xi_i^\alpha \delta\xi_j^\alpha\right\} \tag{5.8}$$

where

$$\Delta = \Delta_0 = p(1-p)\,.$$

We would like to emphasize that this form of the J_{ij}-matrix is nothing but a simplification of the general quasi-inverse rule for the special case of a large set of patterns with low level of activity. However, the number of patterns K should not be very large ($K \ll N$) because otherwise we cannot neglect the small fluctuations of $\sum_i \delta\xi_i^\alpha \delta\xi_i^\beta$ which we omitted in deriving (5.6). Thus, prescription (5.8) is equivalent to the quasi-inverse rule if $1 \ll K \ll N$. In this special case the rule (5.5) becomes local. We can generalize it slightly by regarding Δ as a free parameter. These generalized rules with $\Delta \neq \Delta_0$ will display new features that will be useful for more complex problems involving categorized sets of patterns.

To study the algorithm (5.8) for various choices of the parameter Δ and for an arbitrary number of patterns we evaluate the local fields h_i^α and perform a simple "signal-to-noise-ratio" analysis:

$$h_i^\alpha = \frac{1}{N}\sum_{j(\neq i)}\left[1 + \frac{1}{4\Delta}\sum_\beta \delta\xi_i^\beta \delta\xi_j^\beta\right]\xi_j^\alpha$$
$$= \xi_i^\alpha \frac{\Delta_0}{\Delta} + \left[1 - \frac{\Delta_0}{\Delta}\right](1-2p) + \frac{1}{4N\Delta}\sum_{\beta(\neq\alpha)} \delta\xi_i^\beta \sum_{j(\neq i)} \delta\xi_j^\beta \xi_j^\alpha\,. \tag{5.9}$$

The first term in the right-hand side of (5.9) represents the "signal", the second produces the uniform bias which tries to unify all patterns, and the third is the usual random Gaussian noise with standard deviation $\Delta_0/\Delta\sqrt{K/N}$. Certainly, the uniform bias should not overwhelm the signal: $|\Delta/\Delta_0 - 1|(1-2p) < 1$. If this inequality is fulfilled and the number of patterns K is not too large say, $K \lesssim N$, then the signal term in (5.9) dominates the other terms and correct retrieval is possible. Naively one would expect that, since at $\Delta = \Delta_0$ the bias term in

(5.9) is absent, $\Delta = \Delta_0$ is the best possible choice. However, a more careful analysis shows [5.7] that this choice is not the best but only the simplest because at $\Delta = \Delta_0$ the maximum storage capacity α (defined as the minimal value of K/N at which any correlation between the retrieved and the stored pattern is lost) is the same as the storage capacity of the Hopfield model with uncorrelated patterns: $\alpha_c \simeq 0.14$, but this value increases with Δ/Δ_0 and reaches the maximum $\alpha_c \simeq 0.20$ at $\Delta/\Delta_0 \simeq 1.18$. The nonzero positive bias $(1 - \Delta_0/\Delta) > 0$ favors positive local fields, i.e., quiescent neuron states ($\xi_i = +1$). This fact results in important consequences for the performance of an overloaded system since it means that for slight overloading the noise term [the third term in (5.9)] destroys the individual patterns but that any retrieved pattern still has a low activity. Only at a further increase of Δ/Δ_0 does the noise result in complete disorder. In other words, the intermediate noise mixes patterns in the same class *but leaves the class intact*. This property is extremely desirable for retrieval of a general categorized set of patterns.

Thus, we have eliminated the first drawback of the quasi-inverse algorithm and find that its simplification (5.8) is a local rule. However, the number of patterns that we can store using the algorithm (5.8) is even less than that obtained by the quasi-inverse rule. Therefore, our next problem is to modify the learning rule (5.8) in such a way that a larger number of patterns can be stored. To find this modification we consider the "signal-to-noise" analysis (5.9) once again and try to diminish the influence of the noise term. We note that it can be rewritten in the form

$$f_i^\alpha = \frac{1}{4N\Delta} \sum_{\beta(\neq\alpha)} \left\{ \delta\xi_i^\beta \sum_j \delta\xi_j^\beta + \delta\xi_i^\beta \sum_j \delta\xi_j^\beta (\xi_j^\alpha - 1) \right\} \tag{5.10}$$

and at low activity the main contribution to the noise comes from the first term inside the braces. This term depends weakly on α and, thus, can be eliminated easily. For instance, we can introduce thresholds h_{0i} that modify the energy functional (5.1):

$$E = -\tfrac{1}{2} \sum_{ij} S_i J_{ij} S_j + \sum_i h_{0i} S_i \,, \tag{5.11}$$

so that the stability condition becomes $\xi_i^\alpha = \mathrm{sgn}(\tilde{h}_i^\alpha)$ with modified local fields $\tilde{h}_i = h_i - h_{0i}$. If we choose

$$h_{0i} = \frac{1}{4N\Delta} \sum_\beta \delta\xi_i^\beta \sum_j \delta\xi_j^\beta \,, \tag{5.12}$$

then the first term in the noise (5.10) is cancelled and the amplitude of the noise is diminished dramatically. The same result can also be achieved by appropriate tuning of the J_{ij}-matrix [5.21], but here we discuss a model with modified thresholds since the prescription (5.8, 11, 12) then acquires a very simple and natural form in the so-called V-representation. In this representation each neuron is described by a variable $V_i = (1 - S_i)/2$ which is 0 or 1. In terms of these "occupation number" variables the energy functional (5.11) becomes

$$E = -\tfrac{1}{2}\sum_{ij} V_i T_{ij} V_j + \sum_i V_i \theta_i \,,$$
$$\theta_i = 2\Big[\sum_j J_{ij} - h_{0i}\Big] \,, \quad T_{ij} = 4J_{ij} \,. \tag{5.13}$$

We see that thresholds which we inserted by hand in the S-representation (5.12) disappear in the V-representation in which all thresholds are $\theta_i = \theta = 2$.

In the S-model with thresholds h_{0i} given by (5.12) or in the V-model with uniform thresholds, only the second term in (5.10) is left in the noise f_i^α which is by a factor of $\sqrt{p}$ smaller than the noise f_i^α. Thus, this noise is less than the signal if $K \ll N/p$, which means that we have achieved our goal of constructing a model in which the information content does not depend on the level of activity. A more sophisticated analysis of the V-model [5.8] confirms this conclusion and produces the exact value of the storage capacity (storage ratio):

$$\alpha_c = \frac{1}{2p|\ln p|}\left[1 - \frac{1}{\sqrt{|\ln p|}}\right] . \tag{5.14}$$

Thus, we see how starting from the quasi-inverse rule we were able to modify it and get the explicit form for the local learning rule (5.8, 11, 12) with large storage capacity.

Another approach to the problem was proposed by Gardner and coworkers [5.19], who proposed using an iterative procedure to construct the interaction matrix and prove that this iterative procedure always converges if there is at least one matrix J_{ij} which solves the problem of retrieval of a given set of patterns. At each iterative step of this algorithm some pattern ξ_i^λ and its noisy analog S_i^λ are considered. Here S_i^λ is the pattern ξ_i^λ distorted at sites chosen randomly with density c. First an *error mask* is defined,

$$\varepsilon_i^\lambda = \tfrac{1}{2}\left[1 - \mathrm{sgn}\left(\xi_i^\lambda \sum_j J_{ij} S_j^\lambda\right)\right] , \tag{5.15}$$

which takes the value 0 (or 1) if the site i is correctly (or incorrectly) retrieved, i.e., if the local field h_i generated by S_i^λ is aligned along (or opposite to) ξ_i^λ. Then the matrix J_{ij} is updated:

$$J_{ij}^{\mathrm{new}} = J_{ij}^{\mathrm{old}} + \varepsilon_i^\lambda \xi_i^\lambda S_j^\lambda + \varepsilon_j^\lambda \xi_j^\lambda S_i^\lambda \quad . \tag{5.16}$$

This iteration procedure converges to a matrix J_{ij}, which, if it exists, has a wonderful property: the dynamical evolution which starts from the noisy pattern S_i^λ leads, after one sweep over all sites, to the stored pattern ξ_i^λ. The serious problem of convergence, once the existence of such a matrix J_{ij} is guaranteed, was solved by Gardner et al. [5.19]. In particular, Gardner [5.9] has shown that the matrix J_{ij} for the problem of low-activity neurons exists if the number of patterns is less than $N/p|\ln p|$, in agreement with the result (5.14). For a full discussion of Gardner's work, see Chap. 3.

Before we turn to the discussion of the general problem let us summarize our main conclusions for the problem of low-activity patterns. For this problem we have shown that a matrix J_{ij} exists which can be expressed locally through patterns ξ_i^λ and which allows the correct retrieval of $N/p|\ln p|$ patterns. Furthermore, we have provided its form (5.8, 11, 12) explicitly.

5.4 Models with Hierarchically Structured Information

We now turn to the more complex problem of storing a hierarchy of patterns. The solution to the general problem of storing a given, arbitrary set of categorized patterns is not yet available but a special case which entails the *construction* of the patterns, has been investigated extensively by different groups [5.7, 10–13]. In its simplest form, all sites are taken to be independent and at a given site i all patterns of a single class are equivalent, i.e., *given* an ancestor ξ_i^ν they are all generated by the very same probability mechanism: $\xi_i^{\nu\mu} = \tilde{\xi}_i^\nu$ with probability $(1-p)$ and $\xi_i^{\nu\mu} = -\tilde{\xi}_i^\nu$ with probability p where $\tilde{\xi}_i^\nu$ is such that $(1-2p)\tilde{\xi}_i^\nu = \xi_i^\nu$. Then the mathematical expectation of $\xi_i^{\nu\mu}$ *given* ξ_i^ν is just ξ_i^ν itself. This is an important property of a procedure which we now want to analyze in some detail.

A general and rather ingenious scheme has been proposed by Parga and Virasoro [5.10]. Though these authors were strongly motivated by considerations stemming from the solution of the Sherrington–Kirkpatrick model [5.14, 15], which we will not dwell upon, one can introduce their scheme directly so as to exhibit a rather interesting mathematical structure [5.11].

For a hierarchy consisting of L levels, one chooses numbers r_l such that $0 \le r_0 < r_1 < \ldots < r_l < \ldots < r_L = 1$. The level zero consists of a "godfather", which may, but need not, produce descendants; cf. Fig. 5.1. One then allows the values $\pm r_k$ for $\xi^{\mu_1, \mu_2, \ldots, \mu_k}$ in the kth level and requires

$$\mathrm{Prob}\left\{\xi^{\mu_1, \ldots, \mu_{k-1}, \mu_k} | \xi^{\mu_1, \ldots, \mu_{k-1}}\right\}$$
$$= \frac{1}{2}\left[1 + \mathrm{sgn}\left(\xi^{\mu_1, \ldots, \mu_{k-1}, \mu_k} \xi^{\mu_1, \ldots, \mu_{k-1}}\right) r_{k-1}/r_k\right] .$$

Here $\mathrm{Prob}\left\{\xi^{\mu_1, \ldots, \mu_{k-1}, \mu_k} \mid \xi^{\mu_1, \ldots, \mu_{k-1}}\right\}$ is the probability of getting $\xi^{\mu_1, \ldots, \mu_{k-1}, \mu_k}$ *given* $\xi^{\mu_1, \ldots, \mu_{k-1}}$. It is a *conditional* probability, conditional with respect to the ancestor $\xi^{\mu_1, \ldots, \mu_{k-1}}$. The Ising spin configurations to be stored and retrieved are the $\xi_i^L = \pm 1$ (where L stands for a specific $\mu_1, \ldots, \mu_L$) and $1 \le i \le N$.

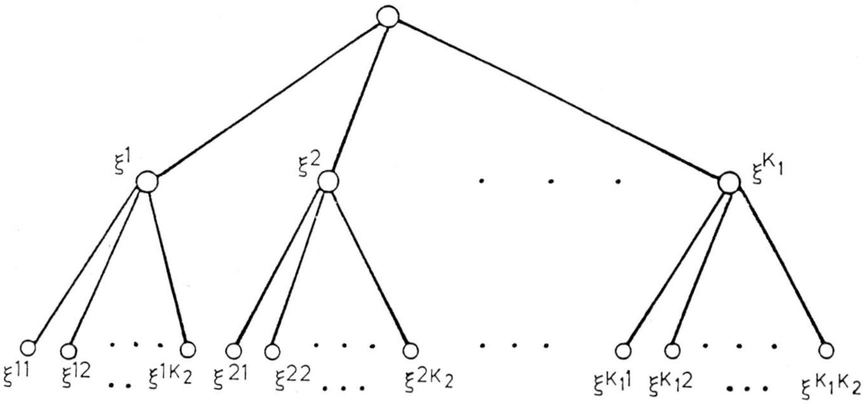

Fig. 5.1. A two-level hierarchy: A genealogical tree consisting of a "godfather" (zeroth level) and two generations ($L = 2$). The states $\xi^{\mu\nu}$ belong to the second generation and the ξ^{μ} to the first. The $\xi^{\mu\nu}$ are descendents of ξ^{μ}. For the sake of convenience, it was assumed that $1 \leq \mu \leq K_1$ while $1 \leq \nu \leq K_2$, whatever μ

The above probabilistic setup has a remarkable property. One easily verifies

$$\mathsf{E}\left\{\xi^{\mu_1,\ldots,\mu_{k-1},\mu_k} \mid \xi^{\mu_1,\ldots,\mu_{k-1}}\right\} = \xi^{\mu_1,\ldots,\mu_{k-1}} \tag{5.17}$$

where $\mathsf{E}\{\ldots\}$ is a mathematical expectation. In particular, here it is the conditional expectation of $\xi^{\mu_1,\ldots,\mu_{k-1},\mu_k}$ *given* its ancestor $\xi^{\mu_1,\ldots,\mu_{k-1}}$.

The relation (5.17) says that the Markov process which we have defined possesses an additional, valuable property: it is a *martingale* [5.11]. This notion stems from the theory of gambling. It allows, among other things, a straightforward evaluation of overlaps. Suppose that two patterns μ and ν have a common ancestor $(\mu_1, \ldots, \mu_l)$ in the lth level. That is,

$$\mu = \left(\mu_1, \ldots, \mu_l, \mu_{l+1}, \ldots, \mu_L\right) \quad \text{and} \quad \nu_1 = \left(\mu_1, \ldots, \mu_l, \nu_{l+1}, \ldots, \nu_L\right) .$$

We are interested in the overlap

$$q_l = \lim_{N\to\infty} \frac{1}{N} \sum_{i=1}^{N} \xi_i^{\mu} \xi_i^{\nu} = \mathsf{E}\{\xi^{\mu}\xi^{\nu}\} .$$

Owing to the martingale property, the right-hand side is evaluated easily. We have

$$\begin{aligned}\mathsf{E}\{\xi^{\mu}\xi^{\nu}\} &= \mathsf{E}\left\{\mathsf{E}\{\xi^{\mu}\xi^{\nu} \mid \xi^{L-1}\}\right\} \\ &= \mathsf{E}\left\{\xi^{\mu_1,\ldots,\mu_{L-1}}\xi^{\mu_1,\ldots,\nu_L}\right\} ,\end{aligned}$$

where ξ^{L-1} is the ancestor of ξ^{μ} in the $(L-1)$th level. Continuing this process we finally reach the common ancestor $\xi^{\mu_1,\ldots,\mu_l}$ and find

$$q_l = \mathsf{E}\{\xi^{\mu}\xi^{\nu}\} = \mathsf{E}\left\{\left(\xi^{\mu_1,\ldots,\mu_l}\right)^2\right\} = r_l^2$$

The coupling constants of Parga and Virasoro [5.10], though originally derived through an ultrametricity argument, agree with the ones which we will present shortly in (5.18). It turns out that the martingale property (5.17) is a mainstay for the evaluation of the corresponding free energy. The reader may consult Bös et al. [5.11] for details.

How do we choose the coupling constants J_{ij}? To guess a solution, we use again the line of thought that has led us to the result (5.8) for the low-activity patterns, i.e., we construct the matrix $B_{\alpha\beta}$ of the quasi-inverse algorithm, evaluate it by neglecting fluctuation terms, and get:

$$J_{ij}N = \sum_{\mu\nu\varrho} \frac{(\xi_i^{\mu\nu\varrho} - \xi_i^{\mu\nu})(\xi_j^{\mu\nu\varrho} - \xi_j^{\mu\nu})}{1 - q_2} + \sum_{\mu\nu} \frac{(\xi_i^{\mu\nu} - \xi_i^{\mu})(\xi_j^{\mu\nu} - \xi_j^{\mu})}{q_2 - q_1} + \sum_{\mu} \frac{\xi_i^{\mu}\xi_j^{\mu}}{q_1} . \tag{5.18}$$

Here we have taken the number of levels to be $L = 3$. The generalization to arbitrary L is self-evident. The q_k are overlaps of patterns (classes) which belong to the same class (category) in the next generation. For the case of a two-level hierarchy with only one class (ancestor) $\xi_i^1 = +1$, (5.18) simplifies to (5.8) with $\Delta = \Delta_0$, and we recover our low-activity patterns of the previous section.

Analogously to the low-activity case, we can generalize the prescription (5.18) slightly by introducing parameters δ_l,

$$J_{ij}N = \delta_2^{-1} \sum_{\mu\nu\varrho} \frac{(\xi_i^{\mu\nu\varrho} - \xi_i^{\mu\nu})(\xi_j^{\mu\nu\varrho} - \xi_j^{\mu\nu})}{1 - q_2} + \delta_1^{-1} \sum_{\mu\nu} \frac{(\xi_i^{\mu\nu} - \xi_i^{\mu})(\xi_j^{\mu\nu} - \xi_j^{\mu})}{q_2 - q_1} + \sum_{\mu} \frac{\xi_i^{\mu}\xi_j^{\mu}}{q_1} . \tag{5.19}$$

As for the low-activity case we can employ the signal-to-noise-ratio analysis to study the prescription (5.19) and show that the performance of the model is governed by the *total* number of patterns stored, $K_{\text{tot}} = K_1 \ldots K_L$, and that the relevant parameter is the ratio $\alpha = K_{\text{tot}}/N$. As before, the result $\alpha_{\max} \sim \mathrm{O}(1)$ means that the amount of information that can be stored in the network is small compared to the information capacity of the Hopfield prescription for uncorrelated patterns.

To obtain more quantitative results one should use the statistical-mechanical method which was developed for the spin-glass problem [5.14, 15] and applied to neural networks by Amit, Gutfreund, and Sompolinsky [5.16]. To keep the notation simple, we shall show how the method works for a two-level hierarchy.

In the statistical-mechanical approach, all properties of the network are expressed through the partition sum

$$\varrho(\chi) = \sum_{\{S\}} \exp[-\beta H(\boldsymbol{S})] , \quad H(\boldsymbol{S}) = E(\boldsymbol{S}) - \sum_i \chi_i S_i , \tag{5.20}$$

where the sum is over all spin configurations $\{\boldsymbol{S}\}$, $E(\boldsymbol{S})$ is the energy (5.1), the χ_i are auxiliary fields, and β is the inverse temperature which measures the intensity of noise. Differentiating $\ln \varrho$ with respect to $\beta\chi_i$ one can obtain any spin correlation function. Introducing the prescription (5.19) into the energy (5.1) we get the Hamiltonian $H(\boldsymbol{S})$ for the present model:

$$H(\boldsymbol{S}) = \frac{-1}{N}\left\{\sum_\mu \frac{1}{q_1}\left[\sum_i S_i\xi_i^\mu\right]^2 + \frac{1}{\delta_1(1-q_1)}\sum_{\mu\nu}\left[\sum_i S_i\delta\xi_i^{\mu\nu}\right]^2\right\} - \sum_i \chi_i S_i\,, \tag{5.21}$$

where $\delta\xi_i^{\mu\nu} = \xi_i^{\mu\nu} - \xi_i^\mu$. If one reads Δ/Δ_0 for δ_1 and Δ_0 for $1-q_1$, then one recovers the low-activity prescription (5.8). The factor $\exp[-\beta H(\boldsymbol{S})]$ acquires a simpler form if we introduce auxiliary Gaussian variables m_λ and $m_{\lambda\mu}$ to linearize the squares and rewrite the partition sum:

$$\varrho(\chi) = \sum_{\{\boldsymbol{S}\}}\int Dm_\lambda Dm_{\lambda\mu}\exp\left[-\beta\tilde{H}(\boldsymbol{S}, m_\lambda, m_{\lambda\mu})\right]\,,$$

$$\tilde{H}(\boldsymbol{S}, m_\lambda, m_{\lambda\mu}) = \sum_\mu\left\{\frac{Nq_1}{2}m_\mu^2 - m_\mu\sum_i S_i\xi_i^\mu\right\} \tag{5.22}$$

$$+\,\delta_1^{-1}\sum_{\mu\nu}\left\{\frac{N(1-q_1)}{2}m_{\mu\nu}^2 - m_{\mu\nu}\sum_i S_i\delta\xi_i^{\mu\nu}\right\} - \sum_i \chi_i S_i\,.$$

The meaning of the auxiliary variables m_μ and $m_{\mu\lambda}$ becomes more transparent in the mean-field approximation, in which we get

$$m_\mu = \frac{1}{Nq}\sum_i\langle S_i\rangle\xi_i^\mu\,, \quad m_{\mu\nu} = \frac{1}{N(1-q)}\sum_i\langle S_i\rangle\delta\xi_i^{\mu\nu}\,. \tag{5.23}$$

Thus we see that m_μ describes the projection of a state $\langle S_i\rangle$ onto a class ξ_i^μ whereas $m_{\mu\nu}$ describes how close the state $\langle S_i\rangle$ is to a specific pattern $\xi_i^{\lambda\mu}$. To study the retrieval properties of the J_{ij}-matrix we should consider spin configurations which are close to an individual pattern, say $\mu_0\nu_0$. For these configurations, m_μ and $m_{\mu\nu}$ are of the order of one for only a single pattern, specified by $\mu_0\nu_0$, and are small for any other $\mu\nu$. The mean deviations of the variables m_μ and $m_{\mu\nu}$ from the mean-field solution (5.23) are of the order of $1/N$, so we can neglect them for the variables m_{μ_0} and $m_{\mu_0\nu_0}$ which describe the overlap with a given pattern and its class, but we should take these fluctuations into account for *all other* variables. To do this we average (5.22) over the random variables ξ_i^μ, $\delta\xi_i^{\lambda\mu}$, keeping terms which contain $\xi_i^{\lambda_0}$ and $\delta\xi_i^{\lambda_0\mu_0}$ fixed, and obtain

$$\varrho(\chi) = \sum_{\{\boldsymbol{S}\}}\int Dm_\lambda^\sigma Dm_{\lambda\mu}^\sigma\exp\left[-\beta\tilde{H}\left(\boldsymbol{S}^\sigma, m_\lambda^\sigma, m_{\lambda\mu}^\sigma\right)\right]\,,$$

$$
\begin{aligned}
\tilde{H}\left(\boldsymbol{S}^\sigma, m_\lambda^\sigma, m_{\lambda\mu}^\sigma\right) &= \sum_\sigma \left\{ \left[\frac{q_1 N}{2} \left(m_{\mu_0}^\sigma\right)^2 - m_{\mu_0} \sum_i S_i^\sigma \xi_i^{\mu_0} \right] \right. \\
&+ \delta_1^{-1} \left[\frac{N(1-q_1)}{2} \left(m_{\mu_0\nu_0}^\sigma\right)^2 - m_{\mu_0\nu_0}^\sigma \sum_i S_i^\sigma \sigma \xi_i^{\mu_0\nu_0} \right] - \sum_i \chi_i^\sigma S_i^\sigma \\
&+ \left. \sum_\mu \frac{q_1 N}{2} \left(m_\mu^\sigma\right)^2 + \delta_1^{-1} \sum_{\mu\nu} \frac{(1-q_1)N}{2} \left(m_{\mu\nu}^\sigma\right)^2 \right\} \\
&- \frac{1}{2}\beta \sum_{\sigma\sigma'} \left[\sum_i S_i^\sigma S_i^{\sigma'} \right] \left[q_1 \sum_\mu{}' m_\mu^\sigma m_\mu^{\sigma'} + \delta_1^{-2}(1-q_1) \sum_{\mu\nu}{}' m_{\mu\nu}^\sigma m_{\mu\nu}^{\sigma'} \right] .
\end{aligned}
\tag{5.24}
$$

Here the notation $\sum'$ means that summation is restricted to $\mu \neq \mu_0$, $\mu\nu \neq \mu_0\nu_0$. In deriving (5.24) we have used the replica approach (see [5.14, 15]), where σ labels the replicas, and expanded the result of the averaging procedure with respect to m_μ and $m_{\mu\nu}$ up to second order. This is justified since the variation of each of these variables is small. To cope with the last term in (5.24) we note that it can be regarded as a factor of two sums consisting of a large number of independent terms. Before integrating over m_μ^σ and $m_{\mu\nu}^\sigma$ we can therefore replace $N^{-1} \sum_i S_i^\sigma S_i^{\sigma'}$ by its average:

$$
\frac{1}{N} \sum_i S_i^\sigma S_i^{\sigma'} \approx Q^{\sigma\sigma'} , \quad Q^{\sigma\sigma'} = \langle \langle S_i^\sigma S_i^{\sigma'} \rangle_T \rangle .
$$

In the replica-symmetric approximation

$$
Q^{\sigma\sigma'} = Q + (1-Q)\delta_{\sigma\sigma'} , \tag{5.25}
$$

where Q is the Edwards–Anderson order parameter: $Q = \langle \langle S \rangle_T^2 \rangle$. Here $\langle \ldots \rangle_T$ means average over thermal fluctuations and $\langle \ldots \rangle$ average over the randomness. Now we are able to integrate over the variables m_μ and $m_{\mu\nu}$. To simplify the notation we suppose that the number of classes $K_1 \ll N$ so that we need only keep track of the contribution of $m_{\mu\nu}^\sigma$ to the effective energy of the spin variables and neglect the contribution of m_μ^σ. We then find

$$
\begin{aligned}
H(\boldsymbol{S}) = &- \sum_\sigma \left[\frac{q_1 N}{2} m_{\mu_0}^2 - m_{\mu_0}^\sigma \sum_i S_i^\sigma \xi_i^{\mu_0} \right] \\
&+ \frac{1}{\delta_1} \sum_\sigma \left[\frac{N(1-q_1)}{2} \left(m_{\lambda_0\mu_0}^\sigma\right)^2 - m_{\lambda_0\mu_0}^\sigma \sum_i S_i^\sigma \delta \xi_i^{\lambda_0\mu_0} \right] \\
&- \sum_i \chi_i S_i - \beta\alpha r \sum_{\sigma,\sigma'} \sum_i S_i^\sigma S_i^{\sigma'} ,
\end{aligned}
$$

where

$$
r = \frac{\delta^{-2}}{[1-(1-Q)\beta\delta^{-1}]^2} \quad \text{and} \quad \alpha = K_{\text{tot}}/N .
$$

The part of the Hamiltonian $H(S)$ which depends on the variables S_i is a sum of independent terms for each site i. It can be regarded as a sum of energies associated with individual variables S_i in the field

$$h_i = \chi_i + m_{\mu_0}\xi_i^{\mu_0} + \frac{1}{\delta_1} m^{\sigma}_{\lambda_0\mu_0}\delta\xi_i^{\lambda_0\mu_0} + z\sqrt{\alpha r}\ , \tag{5.26}$$

where z is the Gaussian with variance $\langle z^2\rangle = 1$. The average $\langle S_i\rangle$ in field h_i is $\langle S_i\rangle = \tanh(\beta h_i)$, which together with the mean-field equations (5.23) for m_{λ_0} and $m_{\lambda_0\mu_0}$ yields a closed system of equations for m_{λ_0}, $m_{\lambda_0\mu_0}$, and Q:

$$\begin{aligned} m_{\lambda_0} &= \frac{1}{Nq}\sum \xi_i^{\lambda}\langle\tanh(\beta h_i)\rangle_z\ , & m_{\lambda_0\mu_0} &= \frac{1}{N}\frac{1}{1-q}\sum_i \delta\xi_i^{\lambda}\langle\tanh(\beta h_i)\rangle_z \\ r &= Q^2[\delta-(1-Q)\beta]^{-2}\ , & Q &= \left\langle \tanh^2(\beta h_i)\right\rangle_z\ . \end{aligned} \tag{5.27}$$

The h_i are given by (5.26) and the average over z should be performed with weight $\exp(-z^2/2)\mathrm{d}z/\sqrt{2\pi}$.

At $\delta = 1$, (5.26, 27) possess a set of solutions with $m_{\lambda\mu} = m_0(T)$, $m_\lambda = m_0(p)$ which correspond to the retrieval states $\xi_i^{\lambda\mu}$ in the lowest level.

Moreover, there are also solutions with $m_{\lambda\mu} = 0$, $m_\lambda = m_0(\beta)$ corresponding to the ancestor or basic pattern ξ_i^{λ}. All these solutions are degenerate in free energy, and the final equations for $m_0(\beta)$ and $q(\beta)$ coincide with those previously derived for the uncorrelated Hopfield model (for more details see [5.7, 11, 13]). Therefore, the "phase diagram" for this model at $\Delta = \Delta_0$ coincides with that obtained by Amit et al. [5.16]; in particular, retrieval states exist at $\beta^{-1} < T_{\mathrm{M}}(\alpha)$, where $T_{\mathrm{M}}(0) = 1$ and $\alpha_{\mathrm{c}}(\beta^{-1} = 0) \simeq 0.14$. Thus, it appears in this case that hierarchical organization of patterns does not lead to any improvement of the system behavior with respect to the case of uncorrelated random patterns. However, there are several ways to generalize this model in order to utilize the hierarchical organization of patterns.

The most obvious generalization is to consider the same model but with $\delta > 1$ [5.7]. Since there is no fear of confusion, we have replaced δ_1 by δ. For $\delta > 1$, the degeneracy between basic patterns ξ_i^{λ} and individual patterns $\xi_i^{\lambda\mu}$ is broken and the basic patterns have a lower free energy. Therefore, there are two distinct phase transition lines $T_1(\alpha,\delta)$ and $T_2(\alpha,\delta)$ in the (T,α) "phase diagram". Retrieval states correspond to the basic patterns existing at $T < T_1(\alpha,\delta)$, whereas individual patterns $\xi_i^{\lambda\mu}$ can be retrieved at $T < T_2(\alpha,\delta) < T_1(\alpha,\delta)$ only. The temperatures $T_{1,2}(\alpha,\delta)$ can be obtained numerically; here we present some analytical results. As $\alpha \to 0$ we obtain:

$$\begin{aligned} &T_1(\alpha,\,\delta) \to T_1(\delta) = 1\ , \\ &T_2(\alpha,\,\delta) \to T_2(\delta) \simeq 1 - 4(\delta-1) \quad \text{as} \quad (\delta-1)\to 0\ , \\ &T_2(\alpha,\,\delta) \to T_2(\delta) \simeq \frac{4(1-\delta/\delta_{\mathrm{m}})}{\ln(1-\delta/\delta_{\mathrm{m}})} \quad \text{as} \quad \left(1-\frac{\delta}{\delta_{\mathrm{m}}}\right) \to 0\ , \end{aligned}$$

where $\delta_m = 2(1-p)(1-2p)$ is the maximal δ below which there is a nonzero storage capacity. In the limit $\alpha \to 0$ the first transition is second order, as opposed to the second one, which is a first-order transition and becomes second order at $\delta > 1$ only. Note that the temperatures $T_{1,2}(\alpha, \delta)$ obtained above correspond to the appearance of metastable retrieval states (and thus are analogous to the line $T_M(\alpha)$ discussed above).

The storage capacity of the system at $T = 0$ is also determined by two characteristic values $\alpha_1(\delta)$ and $\alpha_2(\delta)$; cf. Fig. 5.2. At $\alpha < \alpha_1(\delta)$ the basic images can be retrieved and at $\alpha < \alpha_2(\delta) < \alpha_1(\delta)$ all the individual patterns are discernible. The value of $\alpha_1(\delta)$ is determined by the condition of solvability of the equation

$$t\left[\sqrt{2\alpha} + \frac{2}{\sqrt{\pi}} e^{-t^2}\right] = \delta \operatorname{erf}(t) . \tag{5.28}$$

The value of $\alpha_2(\delta)$ is determined by the solvability of the equations below, which are valid only for highly correlated individual patterns, i.e., when $p \ll 1$,

$$\begin{aligned} \frac{\sqrt{2\alpha}}{1-B} t_1 &= \delta \operatorname{erf}(t_1) , \\ \frac{\sqrt{2\alpha}}{1-B} t_2 &= \operatorname{erf}(t_2) - [\delta - 1] \operatorname{erf}(t_1) , \\ B &= \sqrt{\frac{2}{\pi\alpha}} \left[(1-\Delta_0) \exp\left(-t_1^2\right) + \Delta_0 \exp\left(-t_2^2\right)\right](1-B) , \end{aligned} \tag{5.29}$$

where $4\Delta_0 = 1 - q_1$. At $\delta = 1$, both (5.28) and (5.29) reduce to a single equation coinciding with the corresponding equation (4.8) in [5.16] so that $\alpha_1(\delta) = \alpha_2(1) = \alpha_c \simeq 0.14$. The dependence of $\alpha_1(\delta)$ and $\alpha_2(\delta)$ upon Δ has been obtained numerically and is shown in Fig. 5.2. If $\delta \le \delta' \simeq 1.16$, then α_2 equals α_1 and both of them increase with δ. For $\delta > \delta'$ we find that $\alpha_2 < \alpha_1$ and, at still larger δ, α_2 begins to decrease and tends to zero as $\delta \to \delta_m$. Its maximum value is

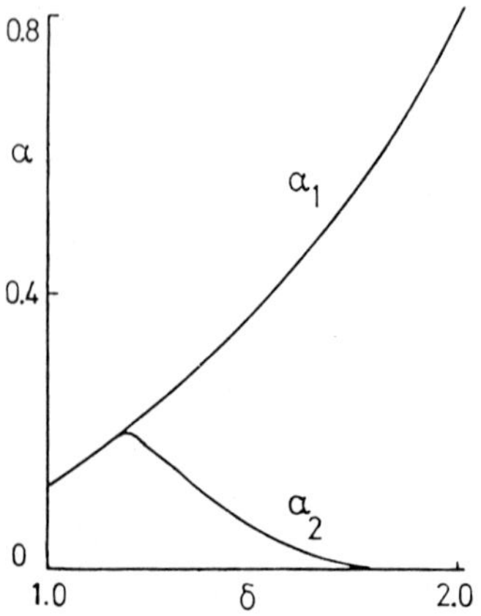

Fig. 5.2. Storage capacity α of a two-level hierarchy as a function of the dimensionless parameter δ (see text)

$$\alpha_{2\,\max} \simeq 0.20 \tag{5.30}$$

which is attained at $\delta \simeq 1.18$. See Fig. 5.2.

Thus, it is possible to increase the storage capacity slightly by an appropriate tuning of the parameter δ. More interesting, however, is the possibility of ordering the process of recognition in a hierarchical manner.

Indeed, we can at first keep the temperature in the interval $T_2(\alpha, \delta) < T < T_1(\alpha, \delta)$ and recognize rapidly the class (defined by the ancestor or basic image ξ_i^λ) to which the input pattern belongs. Then we have to decrease the temperature below $T_2(\alpha, \delta)$ so that the detailed identification of the original pattern can occur. Note here the interesting similarity between the above process and the method of optimization by simulated annealing [5.17].

A different way of producing similar results was proposed by Krogh and Hertz [5.13], who have considered a many-level hierarchy in a way similar to that of Parga and Virasoro [5.10]. Moreover, in their model the basic patterns ξ_i^λ are considered correlated (i.e., $K_1^{-1} \sum_\lambda \xi_i^\lambda = a_0 \neq 0$) so that an additional "zero-level" term should be added [cf. (5.8)] to the J_{ij}-matrix:

$$NJ_{ij} = J_0 + \frac{1}{q_1} \sum_{\lambda}^{K_1} \xi_i^\lambda \xi_i^\lambda + \frac{1}{q_2 - q_1} \sum_{\lambda\mu} \left(\xi_i^{\lambda\mu} - \xi_i^\lambda\right) \left(\xi_j^{\lambda\mu} - \xi_j^\lambda\right)$$
$$+ \frac{1}{1 - q_2} \sum_{\lambda\mu} \left(\xi_i^{\lambda\mu\nu} - \xi_i^{\lambda\mu}\right) \left(\xi_j^{\lambda\mu\nu} - \xi_i^{\lambda\mu}\right) . \tag{5.31}$$

Expression (5.31) is written for a three-level hierarchy; the coefficients in front of the second, third, etc. terms in J_{ij} are fixed and correspond to $\delta_K = 1$ at any hierarchy level higher than the zeroth, whereas the value of J_0 was varied in order to increase the storage capacity. Moreover, a finite threshold (or external field) was considered, i.e., the full Hamiltonian is

$$H = -\frac{1}{2} \sum_{ij} J_{ij} S_i S_j - h \sum_i S_i . \tag{5.32}$$

Krogh and Hertz [5.13] have studied the storage capacity as a function of a_0, J_0 and h. At $1 - a_0 \ll 1$ and the optimal relation between J_0 and h, their results are very similar to ours, e.g., $\alpha_c \simeq 0.19$. Moreover, the patterns at different hierarchy levels have different average "magnetizations" which lead (at finite h) to a splitting of their energies. Thus, an appropriate tuning of the h value opens the possibility of focusing on some chosen level of the hierarchy.

5.5 Extensions

As we have shown, there are various prescriptions which allow the correct retrieval of a categorized set of patterns. For instance, in the model that we have discussed in detail the retrieval properties depend on the parameter δ. This raises a general question: What are the generic properties of both the J_{ij}-matrix leading to correct retrieval and of the retrieval process itself? The answer to this question demands the use of a sophisticated technique developed by Gardner [5.9], but the effort is worthwhile.

The point is that Gardner [5.9] has indicated an iterative algorithm to construct the J_{ij}-matrix. Convergence is guaranteed once a solution exists. This was shown to be the case for a large class of models. The related procedure of learning with noise has been treated by Gardner et al. [5.19]. For details, see the review by Forrest and Wallace, Chap. 3.

Very interesting results have also been obtained by Virasoro [5.18], who studied the distribution of errors in the retrieval process. The errors were induced by random destruction of synapses. He has shown that for a general form of the J_{ij}-matrix the deterioration of synapses results at first in impairment of distinction between different patterns in the same class. This is an important observation because for a real brain it is known that deterioration of synapses results in a syndrom called prosopagnosia, which means inability to recognize individual patterns visually among those that are similar to it, such as faces of people. Virasoro's idea was to study the distribution of the local fields at sites which confirm the class, $\xi_i^{\lambda\mu} = \xi_i^{\lambda}$, and those which distinguish individual patterns, $\xi_i^{\lambda\mu} = -\xi_i^{\lambda}$. He has shown that the second distribution is nearer to zero so that noise generated by random deterioration of synapses changes the local fields more easily on the sites which distinguish individual patterns.

5.6 The Enhancement of Storage Capacity: Multineuron Interactions

All the models of hierarchical memory organization discussed above are characterized by the storage capacities $\alpha_c \sim O(1)$, i.e. the maximum total number of stored patterns K_{tot} is of the order of N. This limitation can be easily understood from the information–theoretical viewpoint. Indeed, the total amount of information that can be stored in $N(N-1)/2$ synaptic strengths J_{ij} is $I_{max} = N^2/2\log_2 M$ where M is the number of the discrete values allowed for each J_{ij}. The information contained in the two-level hierarchy of patterns is

$$I = NK\left\{p\log_2(p^{-1}) + (1-p)\log_2\left[(1-p)^{-1}\right]\right\} . \tag{5.33}$$

Here the "ancestors" have been forgotten (without loss of generality) and p is the probability of differing from the class ξ_i^{λ}; cf. (5.17). The estimate of K follows from the inequality $I < I_{max}$. In the Hopfield model, $M = 2K+1$, so we get $2K < N\log_2 N$ which is a correct but weak inequality. In the limit $p \to 0$, our storage prescription implies $M \approx K/p$ and we get

$$K < \frac{N}{2}\,\frac{\log_2(N/p^2)}{p\log_2(p)} . \tag{5.34}$$

This estimate correlates well with the result (5.14). In order to store, say, $K \sim O(N^2)$ patterns we would need a storage prescription that can use effectively $M \sim \exp(N)$ allowed values of each J_{ij}. Even if such a prescription existed, it is hard to believe that such a very fine tuning can be realized in any biological or technical system. So we will look for an alternative. In passing we note that the crucial role of the discreteness of the J_{ij} was discovered in recent studies of general bounds on the storage capacity [5.7, 9].

These simple considerations show that in order to enhance the storage capacity considerably one should use many more memory-containing elements. On the other hand, the way to introduce these elements is almost obvious if the storage prescription (5.19) is considered. Indeed, let us assume that pattern recognition proceeds in a hierarchical manner so that the class (λ) has already been recognized. Then all the information concerning the detailed structure of other $(\lambda' \neq \lambda)$ classes, i.e., the terms

$$\sum_{\mu,\lambda'(\neq\lambda)} \delta\xi_i^{\lambda'\mu}\delta\xi_j^{\lambda'\mu} ,$$

is not needed in the retrieval process and it would be natural to "turn off" this part of the stored information. This can be done by the following modification of the J_{ij}-matrix:

$$\tilde{J}_{ij} = \sum_{\lambda=1}^{K_1} \xi_i^\lambda \xi_j^\lambda + \frac{1}{4\Delta}\sum_{\lambda=1}^{K_1} p_\lambda \sum_{\mu=1}^{K_2} \delta\xi_i^{\lambda\mu}\delta\xi_j^{\lambda\mu} , \tag{5.35}$$

where the $p_\lambda = N^{-1}\sum_i \xi_i^{(\lambda)} S_i$ serve as projection operators which turn on the information about the detailed structure of class (λ) when the system state $\{S_i\}$ has a nonzero overlap with the basic pattern ξ_i^λ. Thus, the interaction matrix J_{ij} depends on the system state $\{S_i\}$; that means the appearance of "triple" interaction in the energy functional,

$$\begin{aligned} H &= -\frac{1}{2}\sum_{ij} \tilde{J}_{ij} S_i S_j \\ &= -\frac{1}{2N}\sum_{ij}\sum_{\lambda} \xi_i^\lambda \xi_j^\lambda S_i S_j - \frac{1}{4N^2\Delta}\sum_{i,j,m}\sum_{\lambda} \xi_m^\lambda \\ &\quad \times \sum_{\mu=1}^{K_\lambda} \delta\xi_i^{\lambda\mu}\delta\xi_j^{\lambda\mu} S_i S_j S_m . \end{aligned} \tag{5.36}$$

Note the *three*-spin interaction associated with $S_i S_j S_m$.

We will not dwell upon the interesting problem of the biological relevance of multineuron interactions (see e.g. [5.20]) and proceed to the discussion of the properties of model (5.36). The statistical mechanics of this model was studied [5.7] by methods similar to those of Amit et al. [5.16]. The main qualitative difference is that now the relation between "spin-glass" order parameter q and "noise parameter" r contains both the total number of patterns K_{tot} and $K_2^{(\lambda)}$:

$$r = \frac{\alpha^{(1)} q}{[1 - \beta(1-q)]^2} + q \left(\frac{\Delta_0}{\Delta} \right)^2 \sum_\lambda \frac{\alpha_\lambda^{(2)}}{\left[m_\lambda^{-1} - \beta(1-q)\Delta_0/\Delta \right]^2} \tag{5.37}$$

[cf. (5.26)]. Here $\alpha^{(1)} = K_{\text{tot}}/N$, $\alpha_\lambda^{(2)} = K_2^{(\lambda)}/N$, and $K_2^{(\lambda)}$ are the numbers of individual patterns in the class λ. Therefore, the retrieval quality of the patterns which belong to some chosen class λ_0 (so that $m_\lambda = 0$ for all $\lambda \neq \lambda_0$) depends on $\alpha_{\lambda_0}^{(2)}$ and $\alpha^{(1)}$ only. This means that all the values $\alpha^{(1)}$, $\alpha_\lambda^{(2)}$ can be of the order of one so that

$$K_{\text{tot}} \sim \mathrm{O}(N^2) \,. \tag{5.38}$$

Our construction can be generalized to the n-level hierarchy along the lines of preceeding subsection [cf. (5.33)]:

$$N J_{ij}^{(n)} = N J_{ij}^{(n-1)} + \frac{1}{4\Delta^n} \sum_{\lambda_1 \ldots \lambda_n} \hat{P}_{\lambda_1 \ldots \lambda_{n-1}} \left(\xi_i^{\lambda_1 \ldots \lambda_n} - \tilde{\xi}_i^{\lambda_1 \ldots \lambda_{n-1}} \right) \times \left(\xi_i^{\lambda_1 \ldots \lambda_n} - \tilde{\xi}_i^{\lambda_1 \ldots \lambda_{n-1}} \right) , \tag{5.39}$$

where

$$\hat{P}_{\lambda_1 \ldots \lambda_{n-1}} = \hat{P}_{\lambda_1 \ldots \lambda_{n-2}} N^{-1} \sum_i \sigma_i \left(\xi_i^{\lambda_1 \ldots \lambda_{n-1}} - \tilde{\xi}_i^{\lambda_1 \ldots \lambda_{n-2}} \right) .$$

Then we can store

$$K_{\text{tot}} \sim \mathrm{O}(N^n) \tag{5.40}$$

patterns at the cost of introducing $(n+1)$-neuron interactions. The estimate (5.40) can be explained through the following speculation. The amount of the information stored in the K_{tot} patterns is $I \sim \mathrm{O}(N K_{\text{tot}})$. The maximum information which can be contained in $\mathrm{O}(N^{n+1})$ couplings $J_{i_1 \ldots i_{n+1}}$ is $I_{\max} \sim N^{n+1} \log_2 M$, where M is the number of the allowed values of each $J_{i_1 \ldots i_{n+1}}$. The inequality $I < I_{\max}$ yields roughly the estimate (5.40).

The general form of the phase diagram for the two-level model resembles the phase diagram of the model (5.8) with $\Delta > \Delta_0$. If all $\alpha_\lambda^{(2)}$ are equal to $\alpha_\lambda^{(2)} = \alpha^{(2)}$, then there are two phase transitions. The high-temperature one is due to the appearance of the basic retrieval states corresponding to the classes of patterns. At somewhat lower temperatures the second transition occurs and the detailed form of the patterns can be discerned. The second transition is first-order even in the limit $\alpha^{(2)} \to 0$. Generally, at arbitrary $\alpha_\lambda^{(2)}$, these transitions split into two sequences of transitions at the temperatures $T_1^{(\lambda)} = T_1 \left(\alpha^{(1)}, \alpha_\lambda^{(2)} \right)$ and $T_2^{(\lambda)} = T_2 \left(\alpha^{(1)}, \alpha_\lambda^{(2)} \right)$.

The main drawback of the model considered here is that it needs a large number of complicated units providing *multiple* interactions, e.g., $\mathrm{O}(N^2)$ triple interactions for the two-level hierarchy. Probably, one can improve the situation using the projector operators $\hat{P}_\lambda$ defined on a randomly diluted subset of lattice sites:

$$\hat{P}_\lambda = \frac{1}{Nc} \sum_m w_m \xi_m^\lambda S_m \tag{5.41}$$

where $w_m = 1$ with probability $c \ll 1$ and $w_m = 0$ otherwise.

Summarizing, we have indicated in this section how the storage capacity of the network can be increased substantially at the cost of introducing a three-or-more-neuron interaction.

5.7 Conclusion

In this paper we have shown that, for the simplest set of categorized information in which patterns in each class obey the martingale condition (5.17), there exists a matrix J_{ij} which allows the retrieval of $K = \alpha N$ patterns. Its explicit form was given in (5.19). Two important problems remain unsolved: in this storage prescription the information capacity is low and it can hardly be used to store more general sets of categorized patterns which satisfy the more flexible condition that in each class μ the overlap between patterns is larger than q_μ. In the specific case of low-activity patterns, which can be regarded as a set of patterns all belonging to the same class, the storage capacity is increased by a factor $[p|\ln p|]^{-1}$ (p activity) in the V-representation of neurons or if we introduce thresholds in the usual model with S-representation. We have also shown that the storage capacity can be increased substantially at the cost of introduction of multineuron interactions. There exists an S-representation, though, with a simple but *non*local *two*-neuron interaction [5.21], which also saturates the Gardner bound [5.9].

Since hierarchically structured information is omnipresent, efficient methods for its storage and retrieval are highly desirable, but, though important advances have been made, a complete theory incorporating, e.g., spatio-temporal structures is not yet available.

Acknowledgments. We thank J.L. van Hemmen for his generous help in preparing the manuscript. One of us (L.B.I.) also would like to express his gratitude to the members of the neural-networks group in Heidelberg for the hospitality extended to him during his stay there, when this work was finished.

References

5.1 L.N. Cooper: In *Proceedings of the Nobel Symposium on Collective Properties of Physical Systems*, ed. by B. Lundqvist and S. Lundqvist (Academic, New York 1973) pp. 252–264

5.2 D.O. Hebb: *The Organization of Behavior* (Wiley, New York 1949)

5.3 J.J. Hopfield: Proc. Nat. Acad. Sci. USA **79**, 2554 (1982)

5.4 H. Personnaz, I. Guyon, G. Dreyfus: J. Phys. (Paris) Lett. **46**, L359 (1985)
5.5 T. Kohonen: IEEE Trans Comput. **23**, 444 (1974); see also T. Kohonen: *Associative Memory* (Springer, New York, Berlin, Heidelberg 1978)
5.6 I. Kanter, H. Sompolinsky: Phys. Rev. A **35**, 380 (1987)
5.7 M.V. Feigel'man, L.B. Ioffe: Int. J. Mod. Phys. B **1**, 51 (1987)
5.8 M.V. Feigel'man, M. Tsodyks: Europhys. Lett. **6**, 101 (1988)
5.9 E. Gardner: J. Phys. A: Math. Gen. **21**, 257 (1988)
5.10 N Parga, M.A. Virasoro: J. Phys. (Paris) **47**, 1857 (1986)
5.11 S. Bös, R. Kühn, J.L. van Hemmen: Z. Phys. B. **71**, 261 (1988)
5.12 C. Cortez, A. Krogh, J.A. Hertz: J. Phys. A: Math. Gen. **20**, 4449 (1987)
5.13 A. Krogh, J.A. Hertz: J. Phys. A: Math. Gen. **21**, 2211 (1988)
5.14 A review of the ultrametric structures which appear in the Sherrington–Kirkpatrick model of a spin glass can be found in M. Mézard, G. Parisi, M.A. Virasoro: *Spin-Glass Theory and Beyond* (World Scientific, Singapore 1987)
5.15 A comprehensive review of spin-glass theories and their relationships with experiment can be found in K. Binder, A.P. Young: Rev. Mod. Phys. **58**, 801 (1986). For additional information, see *Heidelberg Colloquium on Glassy Dynamics*, edited by J.L. van Hemmen and I. Morgenstern, Lecture Notes in Physics, Vol. **275** (Springer, New York, Berlin, Heidelberg 1987)
5.16 D. Amit, H. Gutfreund, and H. Sompolinsky: Ann. Phys. **173**, 30 (1987)
5.17 S. Kirkpatrick, C.D. Gelatt, and M.P. Vecchi: Science **220**, 671 (1983)
5.18 M.A. Virasoro: Europhys. Lett. **7**, 293 (1988)
5.19 E. Gardner, N. Stroud, D.J: Wallace: in *Neural Computers: From Computational Neuroscience to Computer Science*, ed. by R. Eckmiller (Springer, New York, Berlin, Heidelberg 1988), pp. 251–260
5.20 P. Peretto, I. Niez: Biol. Cybern. **54**, 53 (1986)
5.21 L.B. Ioffe, R. Kühn, J.L. van Hemmen: J. Phys. A: Math. Gen. **22**, L1037 (1989)

6. Asymmetrically Diluted Neural Networks

Reiner Kree and Annette Zippelius

Synopsis. We review the properties of an exactly solvable model for an asymmetric neural network which is strongly diluted. We discuss in detail why the model can be solved exactly and derive its explicit solution in the framework of a new formalism for the dynamics of Ising spins. Results are presented for the static retrieval properties, time-dependent correlations and the distribution of neural activities.

6.1 Introduction

Models of interconnected formal neurons have been proposed by Little [6.1] and Hopfield [6.2] as a framework for content-addressable, associative memory. In these models the state of neuron i ($i = 1, \ldots, N$) is characterized by a single variable s_i. Each neuron is connected to many other neurons via synaptic connections, which are characterized by their synaptic efficacies J_{ij}. In the simplest approximation the total postsynaptic potential is just the sum of all the neuron's inputs: $h_i(t) = \sum_{j(\neq i)} J_{ij} s_j(t)$. From a physiological point of view, h_i is a rather crude model of the postsynaptic potential, because axonal temporal delays are neglected as well as spatial and temporal correlations on the dendritic tree.

Systems of this type have been studied extensively [6.3], especially for neurons of the McCulloch–Pitts type [6.4], for which s_i is a two-state variable, distinguishing only between an "active" state ($s_i = 1$) and a "passive" state ($s_i = 0$). In the absence of noise a two-state neuron will be active for some interval of time Δ, after its postsynaptic potential has exceeded a threshold h_0

$$s_i(t + \Delta) = \Theta\left(h_i(t) - h_0\right) \ .$$

When noise is present, it will be active only with a probability $P_\beta(h_i) < 1$, which depends on the noise level β^{-1} and the local field h_i.

If all the synaptic couplings are assumed *symmetric* ($J_{ij} = J_{ji}$), the long-time properties of such a network can be studied as a problem in *equilibrium statistical mechanics*, provided only that the noise obeys the requirement of detailed balance [6.5]. As an example let us just mention the well-known single-spin-flip Glauber dynamics [6.6], which leads to a stationary distribution

$$P_{\text{st}} \sim \exp(-\beta H)$$

with

$$H = -\frac{1}{2}\sum_{i \neq j} J_{ij} s_i s_j + \sum_i h_0 s_i .$$

It was the major proposition of [6.1, 2] that a network can be used to store and associatively retrieve p patterns $\{\xi_i^\nu\}_{i=1,\ldots,N}^{\nu=1,\ldots,p}$ of neuronal activity by an appropriate choice of the couplings $J_{ij} = J_{ij}(\{\xi_i^\nu\})$. A physiologically motivated *a priori choice* is the Hebb learning rule $J_{ij} = J_0 \sum_{\nu=1}^{p} \xi_i^\nu \xi_j^\nu$, where the ξ_i^ν take values ± 1, corresponding to stored neuronal activities $(\xi_i^\nu + 1)/2 = \{0, 1\}$.

This obviously produces symmetric synaptic couplings *connecting each neuron with all the other neurons* of the network. The statistical thermodynamics of a Little–Hopfield network equipped with Hebb's learning rule leads to exact analytic expressions for equilibrium properties [6.7], which determine, for example, the retrieval quality in the presence of noise and the storage capacity of the memory. This analytic solution was a major success of the application of equilibrium statistical mechanics to neural-network models and inspired the study of many variants of these models using the same method (e.g. models with modified learning rules [6.3]).

But soon there also arose some interest in the effects of (a) asymmetric couplings [6.8–10] and (b) dilution of synapses [6.11–13]. There is a diversity of different motivations to consider asymmetric couplings, e.g.:

1) The assumption $J_{ij} = J_{ji}$ lacks neurophysiological plausibility, because single (chemical) synapses operate unidirectional [6.14].
2) The majority of neurons act on other neurons either mainly excitatorily $(J_{ij}) > 0$ or inhibitorily $(J_{ij}) < 0$. This is clearly in contradiction with symmetric couplings [6.14].
3) One may wonder whether the properties of associative memory actually *require* the symmetry of couplings or whether the results obtained are robust against asymmetry.
4) The symmetric neural network possesses lots of attractors which do not correspond to memorized states (e.g., metastable mixture states [6.15] and the spin-glass attractor [6.7]). One may speculate about a constructive role of asymmetry in a learning rule which turns unwanted fixed points (corresponding to confused memory states) into chaotic time dependent trajectories [6.8].
5) Associative recall of static patterns should not remain the only task a neural network can perform. Both neurobiology and technical applications require more complex situations, such as memorization of pattern sequences and of pattern cycles. Storage of time-dependent attractors, however, inevitably leads to asymmetric synaptic couplings [6.9].
6) From the point of view of statistical mechanics, asymmetric networks are of interest as a subject of nonlinear dynamics [6.16].

The motivation to study diluted networks is twofold:

1. In biological networks the connectivity is high but not complete.
2. Dilution offers the possibility of investigating structured networks.

It should not come as a surprise that a Little–Hopfield network with an arbitrary choice of asymmetric couplings cannot be solved analytically. Nevertheless in the limit of strong dilution an exact solution can be achieved, as first pointed out by Derrida et al. [6.12, 17]. We shall discuss the features of this model that make it solvable, and obtain its solution with the help of combinatorial arguments in Sect. 6.2. We shall then present an alternative approach, which is based on a generating functional for noisy dynamics of two-state variables, in Sect. 6.3. Previous work on the dynamics of Ising spins was either based on a Langevin equation for soft spins [6.18] or on a master equation for the probability distribution [6.19]. We suggest using instead a Langevin-type equation for two-state variables $\sigma_i = 2s_i - 1 = \{\pm 1\}$

$$\sigma_i(t+1) = \mathrm{sgn}(h_i(t) - h_o + \phi_i(t))$$

with a fluctuating part $\phi_i(t)$ of the local field. This allows for the derivation of a generating functional in close analogy to the Onsager–Machlup formalism [6.20].

Finally, in Sect. 6.4, we shall discuss some extensions of the model as well as related work.

6.2 Solvability and Retrieval Properties

Let us consider a Little–Hopfield network of N two-state neurons represented by Ising-variables $\sigma_i = \pm 1$. By definition, the synaptic couplings of the network are given by an asymmetrically diluted version of Hebb's rule,

$$J_{ij} = \frac{1}{K} c_{ij} \sum_{\mu=1}^{p} \xi_i^\mu \xi_j^\mu \, , \tag{6.1}$$

where the activities ξ_i^μ of neuron i in pattern μ are uncorrelated random variables with probabilities $P(\xi_i^\mu = 1) = 1/2$ for all $i = 1, ..., N$ and $\mu = 1, ..., p$. The $c_{ij} = \{0, 1\}$ are independent random variables for each pair (i, j) and thus represent both *dilution* and *asymmetry*. The probability $P(c_{ij} = 1)$ of finding a nonvanishing synaptic coupling J_{ij} is assumed to be

$$P(c_{ij}) = 1 = K/N \tag{6.2}$$

for all pairs (i, j), and we will be interested in systems with $K/N \ll 1$. Note that for this special type of asymmetry, synaptic couplings between a pair (i, j) of neurons are unidirectional ($J_{ij} \neq 0, J_{ji} = 0$) with probability $2(K/N)(1 - K/N)$ and symmetric with probability $(K/N)^2$.

The dynamics of the network is determined by the updating rule

$$\sigma_i(t+\Delta) = \begin{cases} +1 & \text{with probability } (1+e^{-2\beta h_i(t)})^{-1} \\ -1 & \text{with probability } (1+e^{2\beta h_i(t)})^{-1} \end{cases} , \tag{6.3}$$

where the local field is given by

$$h_i(t) = \sum_{j(\neq i)} J_{ij}\sigma_j(t) . \tag{6.4}$$

The updating can be done either simultaneously for all spins (in this case the local field is calculated once for all spins per time step) or for a randomly chosen single neuron (with a subsequent new calculation of the local field). We shall refer to these possibilities as *parallel* and *random sequential* dynamics respectively. In the latter case the time step Δ should be scaled with $1/N$ to make it comparable with the timescale of parallel dynamics.

A simple combinatorial argument indicates that the dynamics of this model is exactly solvable [6.17, 21] in the limit of strong dilution. The state of neuron i, i.e., $\sigma_i(t_n)$ depends on the states $\{\sigma_{ik}(t_{n-1})\}$ of the first-generation ancestor neurons which contribute to $h_i(t_{n-1})$. Each of the $\sigma_{ik}(t_{n-1})$ has its own ancestors which form a second generation. In this way we may recursively construct the complete set of ancestors of $\sigma_i(t_n)$, which connect it to the initial data at t_0. Assuming for the moment that the number of input neurons for every neuron is fixed and equal to $C > 1$, the list of ancestors (including $\sigma_i(t_n)$) has

$$A_n = 1 + C + C^2 + \ldots + C^n = \frac{(C^{n+1}-1)}{(C-1)} \tag{6.5}$$

entries, *amongst which there may be some referring to the same neuron.* It is precisely this subset of neurons which leads to the dynamical generation of correlations between ancestors. If the subset is empty (and no correlations are present between initial neuronal activities $\{\sigma_i(t_0)\}$), all the A_n neurons in the list of ancestors remain uncorrelated. If the C first-generation ancestors of every neuron are chosen at random, the probability P_d that all the $A_n - 1$ ancestors are different is simply given by

$$P_\mathrm{d} = \prod_{j=1}^{A_n-1}\left(1-\frac{j}{N}\right) = 1 - \frac{1}{2N}A_n(A_n-1) + \mathrm{O}\left(\frac{A_n{}^4}{N^2}\right) . \tag{6.6}$$

Thus $P_\mathrm{d} \to 1$ in the limit $N \to \infty$ if $C^n \ll N^{1/2}$. For *finite* C, the time domain of uncorrelated ancestors grows as $\ln N$ for $N \to \infty$. Correlations appear when the number of generations $n \simeq \mathrm{O}(\ln N/\ln C)$. Hence there are loops in the system which have a typical length of $\mathrm{O}(\ln N)$. On the other hand if $C \propto N^\alpha$ the time domain of uncorrelated ancestors is always bounded. We therefore define a *strongly diluted network* by the requirement that

$$\frac{\ln N}{\ln C} \to \infty \tag{6.7}$$

for $N \to \infty$, so that uncorrelated ancestors can be used for calculating dynamical properties in the limit of long times. Note that this does not exclude the possibility $C \to \infty$ (e.g. take $C \propto \ln N$).

It is straightforward to generalize the above argument to cases where the number of input neurons contributing to each h_i is fluctuating around an average value $\overline{C}$. Note that the prescription (6.2) of a fixed probability K/N for the presence of a directed synaptic coupling leads to a Poisson distribution for the number C of input neurons with $\overline{C} = K$, i.e.

$$P(C) = \mathrm{e}^{-K} \frac{K^C}{C!} \ . \tag{6.8}$$

Using the fact that ancestors are uncorrelated, it is relatively easy to obtain closed evolution equations for averaged quantities [6.12]. Consider, e.g., the overlap $m^\mu(t_n)$ of the neural activity with *one* of the learnt patterns which is defined by

$$m^\mu(t_n) = \frac{1}{N} \sum_{i=1}^{N} \langle\langle \xi_i^\mu \langle \sigma_i(t_n) \rangle_\beta \rangle_C \rangle_\xi \tag{6.9}$$

Here $\langle ... \rangle_\beta$ denotes the average over the noisy dynamics defined in (6.3). In the following we shall consider parallel dynamics in detail and comment only briefly on the modifications required for random sequential dynamics.

Let us put $\mu = 1$ (without loss of generality) and introduce the transformed Ising variables $\tau_i(t_n) = \xi_i^1 \sigma_i(t_n)$. Then the field $h_i(t_n)$ is given by

$$h_i(t_n) = \frac{1}{K} \sum_{r=1}^{K} \sum_{\mu=1}^{p} \xi_i^\mu \xi_{jr}^\mu \sigma_{jr}(t_n) = \frac{1}{K} \xi_i^1 \sum_{r=1}^{K} \tau_{jr}(t_n) + \Delta_i(t_n) \ , \tag{6.10}$$

where the sum over r runs over K randomly chosen neurons, which contribute to the input of neuron i, i.e. $\sum_r \tau_{jr} = \sum_j c_{ij} \tau_j$ and

$$\Delta_i = \frac{1}{K} \sum_{\mu=2}^{p} \sum_{r} \xi_i^\mu \xi_{jr}^\mu \xi_{jr}^1 \tau_{jr} \ . \tag{6.11}$$

Note that the first term on the right-hand side of (6.10) tends to stabilize configurations with $\sigma_j = \xi_j^1$, whereas Δ_i acts as a random destabilizing perturbation due to the other $(p-1)$ stored patterns with $\mu > 1$. This random noise can take on the values $\Delta_i = (p-1) - 2s/K$, s being an integer number with $0 \le s \le (p-1)K$. Using the fact that $P(\xi_i^\mu = 1) = 1/2$ holds for all $i = 1, ..., N$ and $\mu = 1, ..., p$, the probability distribution to find a given Δ_i for a fixed configuration $\{\tau_{jr}\}$ is simply binomial:

$$P(s) = \frac{1}{2^{K(p-1)}} \binom{K(p-1)}{s} \ . \tag{6.12}$$

The $\{\sigma_{jr}(t_n)\}$ are uncorrelated and so are the $\{\tau_{jr}(t_n)\}$. Hence their probability distribution factorizes $P(\{\tau_{jr}(t_n)\}) = \prod_r p(\tau_{jr})$ and $p(\tau_{jr}) = (1+m_{jr}\tau_{jr})/2$. Here $m_{jr} = \langle \tau_{jr}(t_n)\rangle_\beta$ is the overlap at site jr averaged over the thermal noise. It follows that $\tau_{jr}(t_n) = \pm 1$ with probability $(1 \pm m_{jr})/2$. The probability that the local field takes on the value

$$h_i(t) = \frac{1}{K}\xi_i^1 \sum_r \tau_{jr} + (p-1) - \frac{2s}{K} \tag{6.13}$$

is given by

$$P(\{\tau_{jr}\}, s) = \frac{1}{2^{K(p-1)}} \binom{K(p-1)}{s} \prod_{r=1}^{K} \left(\frac{1+m_{jr}\tau_{jr}}{2}\right) . \tag{6.14}$$

Hence we find

$$\begin{aligned}\langle\langle \tau_i(t+\Delta)\rangle_\beta\rangle_\xi &= \langle \xi_i^1 \mathrm{Tr}_{\{\tau_{jr}\}} \sum_{s=0}^{K(p-1)} P(\{\tau_{jr}\}, s) \tanh \beta h_i \rangle_{\xi^1} \\ &= \mathrm{Tr}_{\{\tau_{jr}\}} \sum_{s=0}^{K(p-1)} \frac{1}{2}\left[\tanh \beta \left(\frac{1}{K}\sum_r \tau_{jr} + (p-1) - \frac{2s}{K}\right)\right. \\ &\quad \left. + \tanh \beta \left(\frac{1}{K}\sum_r \tau_{jr} - (p-1) + \frac{2s}{K}\right)\right] P(\{\tau_{jr}\}, s) .\end{aligned} \tag{6.15}$$

The two terms in (6.15) are identical, as one can see by substituting $s \to K(p-1) - s$ in the last term. To perform the average over the connectivity c_{ij}, we note that the ancestor trees of all sites $\{jr\}$ are disjoint, so we can average over all the trees independently. After averaging over the patterns and the connectivity, m_i is the same for all sites. Furthermore, since $p(\tau_{jr})$ is a linear function of m_{jr}, the overlap $m(t+1)$ is a function of $m(t)$ only:

$$\begin{aligned}m(t+\Delta) &= \mathrm{Tr}_{\{\tau_{jr}\}} \sum_{s=0}^{K(p-1)} \binom{K(p-1)}{s} \frac{1}{2^{Kp}} \\ &\quad \times \prod_r (1+m\tau_{jr}) \left[\tanh \beta \left(\frac{1}{K}\sum_r \tau_{jr} + (p-1) - \frac{2s}{K}\right)\right] .\end{aligned} \tag{6.16}$$

The trace over $\{\tau_{jr}\}$ is easily performed if one notes that all terms with a fixed number n of negative τ_{jr} give the same contribution:

$$\begin{aligned}m(t+\Delta) &= \sum_{n=0}^{K} \sum_{s=0}^{K(p-1)} \binom{K(p-1)}{s} \binom{K}{n} \frac{1}{2^{Kp}} (1+m)^{K-n}(1-m)^n \\ &\quad \times \left(\tanh \frac{\beta}{K}(Kp - 2s - 2n)\right) \equiv f_K(m(t)) .\end{aligned} \tag{6.17}$$

This is our final result if all neurons are connected to a *fixed number* K of other neurons. If the connectivity is random with *average coordination number* K, then one has to average over the connectivity C with the result

$$m(t+\Delta) = \sum_{C=0}^{\infty} \frac{\mathrm{e}^{-K} K^C}{C!} f_C(m(t)) \equiv f(m(t)) \,. \tag{6.18}$$

For random sequential dynamics one finds

$$m(t+\Delta) = \left(1 - \frac{1}{N}\right) m(t) + \frac{1}{N} f(m(t)) \tag{6.19}$$

or, choosing $\Delta = 1/N$,

$$\lim_{\Delta \to 0} \frac{m(t+\Delta) - m(t)}{\Delta} = \frac{\mathrm{d}m(t)}{\mathrm{d}t} = -m(t) + f(m(t)) \,. \tag{6.20}$$

Of particular interest are the stationary states of the network which correspond to fixed points of the flow equations (6.18, 20). These can be calculated analytically in the limit of high connectivity, $K \to \infty$. We expect that the number of patterns p which can be stored in the network is proportional to the number of synapses per neuron. Choosing $p = \alpha K$, we show in Appendix A that in the limit $K \to \infty$, $p \to \infty$, and α finite, $f(m)$ simplifies to

$$f(m) = \int_{-\infty}^{\infty} \frac{\mathrm{d}y}{\sqrt{2\pi\alpha}} \, \mathrm{e}^{-y^2/2\alpha} \tanh \beta(m-y) \,. \tag{6.21}$$

This result could have been guessed from (6.12). In the limit $K(p-1) \to \infty$ the binomial distribution becomes a Gaussian distribution

$$P(s) = \sqrt{2\pi K(p-1)} \exp\left[-\frac{(s - K(p-1)/2)^2}{K(p-1)/2}\right] \,. \tag{6.22}$$

With the substitution $y = 2s/K - (p-1)$, we find

$$P(y) = P(s)\frac{\mathrm{d}s}{\mathrm{d}y} = \frac{1}{\sqrt{2\pi\alpha}} \exp\left(-\frac{y^2}{2\alpha}\right) \,. \tag{6.23}$$

We now look for solutions of the fixed-point equation $m^* = f(m^*)$. $m^* = 0$ is always a fixed point. It becomes unstable at a critical noise level β^*, when $f'(0) = 1$. For weak noise ($\beta^* < \beta$) a fixed point with $m^* \neq 0$ is stable. Its magnitude goes to zero as $\beta \to \beta^*$; the transition is second order, in contrast to fully connected models. At zero noise level the critical value of α is $\alpha_\mathrm{c} = 2/\pi$. For any finite α, $0 < \alpha < \alpha_\mathrm{c}$, the overlap is imperfect; it is exponentially close to one for small α and goes to zero as $\sqrt{\alpha_\mathrm{c} - \alpha}$ for α close to the maximal capacity of the network.

Several other quantities have been calculated for the strongly diluted network. We will present more detailed results in the next section.

If the connectivity is random, the magnetization is nonuniform. For low average connectivity K these fluctuations are strong. With the same type of argument as used above for the average overlap one can also calculate the local magnetization as a function of the number of incoming synapses [6.22]. Knowledge of the local structure of the network may be relevant for the location of errors.

6.3 Exact Solution with Dynamic Functionals

In this chapter we are going to discuss an alternative approach to the combinatorial arguments used in the previous section.

The stochastic process defined by the updating rule (6.3) can also be characterized by giving the probabilities $P[\sigma_0^T]$ of all the possible *histories* σ_0^T of the network, i.e. all sequences of neuronal activities $\sigma(t_n) = (\sigma_1(t_n)...\sigma_N(t_n))$ which start with $\sigma(t_0)$ at some initial value t_0 and pass through T prescribed configurations $\sigma(t_1), ..., \sigma(t_T)$ in T time steps. Thus $P[\sigma_0^T]$ is the joint, conditional probability

$$P[\sigma_0^T] \equiv P(\sigma(t_T), ..., \sigma(t_1) \mid \sigma(t_0)) . \tag{6.24}$$

As we shall see below, it is possible to calculate this quantity explicitly for strongly diluted networks. As one main advantage of this approach, the limit $K \to \infty$ can be taken in $P[\sigma_0^T]$ directly, and this leads to a considerable simplification of the algebra. Thus the combinatorial arguments, which become increasingly cumbersome for more complicated dynamic quantities (such as time delayed correlations) or more complicated learning rules, can be circumvented. For finite average coordination number, however, no such simplification can be expected.

As a further advantage a class of updating rules more general than (6.3) can easily be considered. This class is characterized by evolution equations

$$\sigma_i(t + \Delta) = \mathrm{sgn}\,(h_i(t) + \phi_i(t)) , \tag{6.25}$$

with $h_i(t)$ given by (6.4). $\phi_i(t)$ represents noise in the system, which is uncorrelated for different neurons and for different time steps. The random variable $\phi_i(t_n)$ is characterized by the probability density $w(\phi_i(t_n))$. We will assume that $w(\phi_i(t_n)) = w(-\phi_i(t_n))$ and consequently $\langle\phi_i(t_n)\rangle_\phi = 0$.

A parallel dynamics is completely defined by (6.25) and a choice for $w(\phi)$, whereas for random sequential dynamics one has to add the requirement that (6.25) only holds for one randomly chosen neuron per time step.

To recover the Glauber dynamics used in the previous section let us choose

$$w(\phi) = \frac{\beta}{2}[\cosh(\beta\phi)]^{-2} \tag{6.26}$$

Then $\langle \mathrm{sgn}(h_i(t) + \phi_i(t)\rangle_{\phi_i(t)}$ is easily calculated to be $\tanh(\beta h_i(t))$. As $\sigma_i = \{\pm 1\}$, its most general probability distribution is of the form $P(\sigma_i) = (1 + \langle\sigma_i\rangle\sigma_i)/2$, and thus we get $P(\sigma_i(t_{n+1}) = 1) = (1 + \tanh(\beta h_i(t)))/2$ in accordance with (6.3).

Another plausible choice for $w(\phi)$ is a Gaussian distribution

$$w(\phi) = \frac{\beta}{\sqrt{\pi}} \, \mathrm{e}^{-\beta^2\phi^2} , \tag{6.27}$$

which leads to $\langle \mathrm{sgn}\,(h+\phi)\rangle_\phi = \mathrm{erf}(\beta h)$. A parallel dynamics of this type has been considered by Peretto [6.23].

To find an expression for the probability of histories $P[\sigma_0^T]$ we represent it as the expectation value of a product of Kronecker deltas which only contribute if (6.25) is satisfied. Thus for parallel dynamics

$$P_\mathrm{p}[\sigma_0^T] = \left\langle\left\langle \prod_{n=0}^{T-1} \prod_{i=1}^{N} \langle \delta_{\sigma_i(t_{n+1}),\, \mathrm{sgn}(h_i(t_n)+\phi_i(t_n))} \rangle_\phi \right\rangle_C\right\rangle_\xi , \tag{6.28}$$

whereas for random sequential dynamics the Kronecker deltas should only contribute if one neuron is updated and all the others keep the activity value of the previous time step, i.e.

$$P_\mathrm{s}[\sigma_0^T] = \left\langle\left\langle \prod_{n=0}^{T-1} \frac{1}{N} \sum_{k=1}^{N} \left\{ \langle \delta_{\sigma_k(t_{n+1}),sgn(h_k(t_n)+\phi_k(t_n))} \rangle_\phi \right.\right.\right.$$
$$\left.\left.\left. \times \prod_{i(\neq k)} \delta_{\sigma_i(t_{n+1}),\sigma_i(t_n)} \right\} \right\rangle_C \right\rangle_\xi . \tag{6.29}$$

The sum $(N^{-1}\sum_k)$ is just the average over the randomly chosen updated neuron k.

Most of the further steps are very similar for parallel (p) and for random sequential (rs) dynamics. Therefore we will present them in the notation of p-dynamics and comment briefly on the necessary modifications for rs-dynamics later on.

To perform the average over the random noise $\phi_i(t_n)$ we use the identity $\delta_{\sigma,\tau} = (1+\sigma\tau)/2$, which holds for two-state variables $\sigma, \tau = \{\pm 1\}$. After the averaging $P_\mathrm{p}[\sigma_0^T]$ takes on the form

$$P_\mathrm{p}[\sigma_0^T] = \left\langle\left\langle \prod_{n=0}^{T-1} G_\mathrm{p}(\sigma(t_{n+1}) \mid \sigma(t_n)) \right\rangle_C\right\rangle_\xi ; \tag{6.30}$$

with

$$G_\mathrm{p}(\sigma(t_{n+1}) \mid \sigma(t_n)) = \prod_{i=1}^{N} \left\{ \frac{1}{2} \left[1 + \sigma_i(t_{n+1}) F\left(\sum_{j(\neq i)} J_{ij}\sigma_j(t_n) \right) \right] \right\} . \tag{6.31}$$

Here, $F(h) = \langle \mathrm{sgn}(h+\phi)\rangle_\phi$, and we have inserted the explicit form of the field $h_i(t_n) = \sum_{j(\neq i)} J_{ij}\sigma_j(t_n)$. To perform the remaining averages it is useful to introduce the probability density for histories of the field $h_0^T = (h(t_T)...h(t_0))$, i.e.,

$$p[h_0^T] = \left\langle \left\langle \prod_{n=0}^{T-1} \prod_{i=1}^{N} \delta(h_i(t_n) - \sum_{j(\neq i)} J_{ij}\sigma_j(t_n)) \right\rangle_C \right\rangle_\xi . \tag{6.32}$$

Equation (6.32) can be used to bring (6.30) into the form

$$P_p[\sigma_0^T] = \int_{-\infty}^{\infty} \left(\prod_{n=0}^{T-1} d^N h(t_n) \right) \prod_{n=0}^{T-1} G_p(\sigma(t_{n+1}) \mid h(t_n)) p[h_0^T] . \tag{6.33}$$

The average over the $\{c_{ij}\}$ can now be performed explicitly, if we represent the δ-functions in (6.32) as Fourier integrals. Making use of (6.2) we obtain

$$p[h_0^T] = \left\langle \int_{-i\infty}^{i\infty} \mathcal{D}(\tilde{h}) \left\{ \exp\left(\sum_{n,i} \tilde{h}_i(t_n) h_i(t_n) \right) \right\} \right.$$

$$\left. \times \prod_{i \neq j} \left\{ 1 - \frac{K}{N} + \frac{K}{N} \exp\left(-\frac{1}{K} \sum_{\nu=1}^{\alpha K} \xi_i^\nu \xi_j^\nu \sum_n \tilde{h}_i(t_n) \sigma_j(t_n) \right) \right\} \right\rangle_\xi , \tag{6.34}$$

with $\mathcal{D}(\tilde{h}) = \prod_{n=0}^{T-1} \prod_{i=1}^{N} (d\tilde{h}_i(t_n)/2\pi)$.

For a strongly diluted network with large average coordination number K, this expression contains two small parameters, which we use as expansion parameters, i.e., K/N and $1/K$. To lowest order in K/N we get

$$p[h_0^T] = \left\langle \int \mathcal{D}(\tilde{h}) \exp\left\{ \sum_{n,i} \tilde{h}_i(t_n) h_i(t_n) - L \right\} \right\rangle , \tag{6.35}$$

with

$$L = \frac{K}{N} \sum_{i \neq j}^{N} \left[1 - \exp\left\{ -\frac{1}{K} \sum_\nu \xi_i^\nu \xi_j^\nu \sum_n \tilde{h}_i(t_n) \sigma_j(t_n) \right\} \right] . \tag{6.36}$$

Now we expand the exponential in (6.36) in powers of $(1/K)$, keeping only the leading contributions to L in the limit $K \to \infty$. These are contained in the first- and second-order terms of this expansion. In Appendix B we show that it is sufficient to replace the term $K^{-2} \sum_\nu \sum_\mu \xi_i^\nu \xi_j^\nu \xi_i^\mu \xi_j^\mu$ which appears in second order by its average value α/K, so that L can be written in the following form:

$$L = \frac{1}{N} \sum_{i \neq j}^{N} \sum_{n=0}^{T-1} \sum_{\nu=1}^{p} \xi_i^\nu \xi_j^\nu \tilde{h}_i(t_n) \sigma_j(t_n)$$

$$- \frac{\alpha}{2N} \sum_{i \neq j}^{N} \sum_{n,m=0}^{T-1} \tilde{h}_i(t_n) \sigma_j(t_n) \tilde{h}_i(t_m) \sigma_j(t_m) . \tag{6.37}$$

Note that the expansion proceeds in powers of $K^{-1/2}$ (and K/N), because the K^{-1} in the exponential of (6.36) is accompanied by a factor $\sum_{\nu=1}^{\alpha K} \xi_i^\nu \xi_j^\nu \sim O(K^{1/2})$.

Using standard techniques it is shown in Appendix C that the quantities $N^{-1}\sum_i \xi_i^\nu \sigma_i(t_n)$ and $N^{-1}\sum_i \sigma_i(t_n)\sigma_i(t_m)$ which appear in (6.37) can be replaced by their averages, so that L becomes

$$L = \sum_i^N \sum_{n=0}^{T-1} \sum_{\nu=1}^{s} m^\nu(t_n)\xi_i^\nu \tilde{h}_i(t_n) - \frac{\alpha}{2}\sum_i^N \sum_{n,m=0}^{T-1} \tilde{h}_i(t_n) C(t_n,t_m) \tilde{h}_i(t_m) \,, \tag{6.38}$$

with

$$m^\nu(t_n) = \frac{1}{N}\sum_i \langle \xi_i^\nu \sigma_i(t_n)\rangle_{\phi,C,\xi} \,,$$
$$C(t_n,t_m) = \frac{1}{N}\sum_i \langle \sigma_i(t_n)\sigma_i(t_m)\rangle_{\phi,C,\xi} \,. \tag{6.39}$$

Note that in (6.38) the sum over patterns runs over a finite subset $\xi^1, \xi^2, ..., \xi^s$ of the $p = \alpha K$ stored patterns. It turns out that almost all of the p overlaps m^ν are $O(1/\sqrt{N})$ except for a finite number $m^1, m^2, ..., m^s$ which are $O(1)$. This has already been noted for the fully connected network where $p = \alpha N$ and the $O(N)$ contributions of the small overlaps $O(1/\sqrt{N})$ add up to a finite amount of static random noise which even disrupts the retrieval properties of the network if $\alpha > \alpha_c$ [6.7]. In a strongly diluted network, however, $p = \alpha K = o(N)$ and therefore contributions from random overlaps $m^\nu = O(1/\sqrt{N})$ always vanish as $K/N \to 0$.

After inserting (6.38) into (6.35) we may perform the integrations over the $\tilde{h}_i$. Then it becomes obvious that the distribution $p[h_0^T]$ is *Gaussian* and *uncorrelated* for different neurons, i.e. it can be written in the form

$$p[h_0^T] = \prod_{i=1}^N \left[\frac{1}{\sqrt{\det C}} \int \prod_n \frac{dy_i(t_n)}{\sqrt{2\pi}} \exp\left\{ -\frac{1}{2}\sum_{n,m} y_i(t_n) C^{-1}(t_n,t_m) y_i(t_m) \right\} \times \left\langle \delta\left(h_i(t_n) - \sum_{\mu=1}^{s} \xi_i^\mu m^\mu(t_n) + \sqrt{\alpha}\, y_i(t_n) \right) \right\rangle_\xi \right] . \tag{6.40}$$

Finally we insert (6.40) into (6.33) to obtain $P_p[\sigma_0^T]$. Note that $P_p[\sigma_0^T]$ factorizes for different neurons, i.e.

$$P_p[\sigma_0^T] = \prod_{i=1}^N P_{p,i}[\sigma_{i0}^T] \,. \tag{6.41}$$

Thus we have reduced the calculation of $P_p[\sigma_0^T]$ to the solution of a *self-consistent* dynamics of N uncoupled neurons with identical evolution equations

$$\sigma(t_{n+1}) = \mathrm{sgn}(h(t_n) + \phi(t_n)) \,, \tag{6.42}$$

where

$$h(t_n) = \sum_{\nu=1}^{s} m^\nu(t_n)\xi^\nu + \sqrt{\alpha}y(t_n) \tag{6.43}$$

and

$$\langle y(t_n)y(t_m)\rangle_{y,\xi} = C(t_n, t_m) \,. \tag{6.44}$$

To obtain a closed system of self-consistent equations we express the overlap m^ν and the correlation function C (6.39) in terms of the single neuron variable $\sigma(t_n)$:

$$m^\nu(t_n) = \langle \xi^\nu \langle \sigma(t_n)\rangle_\phi \rangle_{y,\xi} \,, \tag{6.45}$$

$$C(t_n, t_m) = \langle \sigma(t_n)\sigma(t_m)\rangle_{y,\phi,\xi} \,. \tag{6.46}$$

Let us now briefly comment on the modifications which appear for random sequential dynamics. Note that all the steps necessary to calculate $p[h_0^T]$ remain unaffected. Thus $P_s[\sigma_0^T]$ describes the dynamics of N uncoupled neurons in uncorrelated random fields h_i with one (randomly chosen) updated neuron per time step. Therefore (6.42–46) still hold, if supplemented by the condition that the updating probability per time step is just $\Delta = 1/N$. As a consequence, the evolution equation for $\langle\sigma(t_n)\rangle_\phi$ takes on the form

$$\langle\sigma(t_{n+1})\rangle_\phi = (1-\Delta)\langle\sigma(t_n)\rangle_\phi + \Delta F(h(t_n)) \tag{6.47}$$

for both p- and rs-dynamics with

$$\Delta = \begin{cases} 1/N & \text{for rs} \\ 1 & \text{for p} \end{cases} \,. \tag{6.48}$$

Note that the explicit solution of (6.47) for random sequential dynamics, i.e.,

$$\langle\sigma(t_{n+1})\rangle = \sum_{k=0}^{n} \frac{1}{N}(1-1/N)^{n-k} \tanh \beta h(t_k) + (1-1/N)^{n+1}\sigma(t_0) \tag{6.49}$$

converges to the result obtained from a continuous-time master equation for single-spin-flip Glauber dynamics [6.6, 13],

$$\langle\sigma(t_{n+1})\rangle_\phi = \sigma_0 \mathrm{e}^{-t} + \int_0^t \mathrm{d}t' \mathrm{e}^{-(t-t')} \tanh(\beta h(t')) \tag{6.50}$$

in the limit $N \to \infty$, $tN = n$.

Now we can easily construct evolution equations for averaged quantities. As a first example, consider the overlaps $m^\nu(t_n)$. Using (6.45) and (6.47) we get

$$m^\nu(t_{n+1}) = (1-\Delta)m^\nu(t_n) + f(m^1(t_n), \ldots, m^s(t_n)) \,, \tag{6.51}$$

with

$$f(m^1,...,m^s) = \left\langle \xi^\nu \left\langle F\left(\sum_{\mu=1}^{s} \xi^\mu m^\mu + \sqrt{\alpha} y\right)\right\rangle_y \right\rangle_\xi . \tag{6.52}$$

To perform the average over $y(t_n)$ in (6.52) we use the fact that $C(t_n, t_n) = 1$ which is implied by (6.46). As a consequence the average of any function $F(y(t_n))$ is just

$$\langle F(y(t_n))\rangle_y = \int_{-\infty}^{\infty} \frac{\mathrm{d}y}{\sqrt{2\pi}} \mathrm{e}^{-y^2/2} F(y) . \tag{6.53}$$

Thus we find

$$m^\nu(t_{n+1}) = \left\langle \xi^\nu \int_{-\infty}^{\infty} \frac{\mathrm{d}y}{\sqrt{2\pi}} \mathrm{e}^{-y^2/2} F\left[\sum_{\nu=1}^{s} m^\nu(t_n)\xi^\nu + \sqrt{\alpha} y\right]\right\rangle_\xi \tag{6.54}$$

for the case of p-dynamics. This result is a generalization of (6.18, 21) to the broader class of dynamics (6.25) and to states which have macroscopic overlaps with an arbitrary (but finite) number of stored patterns.

As an illustrative application of (6.54) (which has been discussed in [6.12]) consider configurations $\sigma(t)$ having macroscopic overlaps with two stored patterns ξ^1 and ξ^2 which are correlated such that $\langle \xi_i^1 \xi_j^2 \rangle = Q\delta_{ij}$ (i.e. $p(\xi_i^1 = \xi_i^2) = (1+Q)/2$ and $p(\xi_i^1 \neq \xi_i^2) = (1-Q)/2$). After performing the average over the patterns, the equations for m^1 and m^2 can be brought to the form

$$\begin{aligned} m^1(t_{n+1}) + \varepsilon m^2(t_{n+1}) = (1+\varepsilon Q) \int_{-\infty}^{\infty} \frac{\mathrm{d}y}{\sqrt{2\pi}} \mathrm{e}^{-y^2/2} F(m^1(t_n) \\ + \varepsilon m^2(t_n) + \sqrt{\alpha} y(t_n)) , \end{aligned} \tag{6.55}$$

with $\varepsilon = \pm 1$. Studying the fixed points of (6.55) one finds three qualitatively distinct types of behavior by varying the parameter α. For $\alpha > \alpha_c^{(1)} = (1+Q)^2 \alpha_c$ the only fixed point is $m_1^* = m_2^* = 0$, i.e. the system does not remember anything. For $\alpha_c^{(2)} = (1-Q)^2 \alpha_c < \alpha < \alpha_c^{(1)}$ one attractive fixed point appears with $m_1^* = m_2^* \neq 0$. The system remembers patterns 1 and 2 but cannot distinguish them. Finally, for $\alpha < \alpha_c^{(2)}$ two attractive fixed points ($0 \neq m_1^* \neq m_2^*$) appear which correspond to the two stored patterns ξ_1 and ξ_2. The value of α_c depends on the function F. For a deterministic updating rule ($F(x) = \mathrm{sgn}(x)$) $\alpha_c = 2/\pi$ (see Sect. 6.2).

The evolution equations of more complicated dynamical quantities can also be obtained from (6.42–46). As an example let us consider the time-delayed correlation function $C(t_n, t_m)$ for configurations which have a macroscopic overlap with one pattern, i.e. $m^1 = m \neq 0$, $m^2 = \ldots = m^s = 0$. As the dynamics is a stationary process, $C(t_n, t_m)$ becomes a function of $t_n - t_m$ if both t_n and t_m take on large values. From (6.42) and (6.46) we get

$$C(t_{n+1}, t_m) = (1-\Delta) C(t_n, t_m) + \Delta \langle \tanh \beta h(t_n) \langle \sigma(t_m)\rangle_\phi \rangle_{y,\xi} . \tag{6.56}$$

The average $\langle \sigma(t_n)\rangle_\phi$ can be inserted from (6.49) to obtain

$$C(t_{n+1},t_m) = (1-\Delta)C(t_n,t_m) + \sum_{k=0}^{m-1} \Delta^2(1-\Delta)^{m-1-k}$$
$$\times \langle \tanh \beta h(t_n) \tanh \beta h(t_k) \rangle_{y,\xi}$$
$$+ (1-\Delta)^m \langle \langle \sigma_0 \rangle_\phi \tanh \beta h(t_n) \rangle_{y,\xi} . \quad (6.57)$$

To carry out the remaining average over the $y(t)$, we use the generalization of (6.53) to functions $g(y(t_1), y(t_2), ..., y(t_n))$ of several variables. It is a straightforward exercise to show that the $y(t_1)...y(t_n)$ are Gaussian distributed with vanishing averages and a $(k \times k)$ variance matrix $C(t_i, t_j)$, $(i, j = 1, ..., k)$. Thus one gets, e.g.,

$$\langle g(y(t_1); y(t_2)) \rangle_y = (\det \hat{C})^{-1/2} \int \frac{dy_1 dy_2}{2\pi} g(y_1, y_2)$$
$$\times \exp \left\{ -\frac{1}{2} \begin{pmatrix} y1 \\ y_2 \end{pmatrix}^{\mathrm{T}} \hat{C}^{-1} \begin{pmatrix} y_1 \\ y_2 \end{pmatrix} \right\} , \quad (6.58)$$

with the (2×2) matrix

$$\hat{C} = \begin{pmatrix} 1 & C(t_1, t_2) \\ C(t_2, t_1) & 1 \end{pmatrix} . \quad (6.59)$$

Using the identity (6.58) and the property $F(x) = -F(-x)$ one obtains a closed evolution equation for $C(t_{m+n}, t_m)$. For large time arguments, $\lim_{m\to\infty} C(t_{n+m}, t_m) = C(t_{n+m} - t_m) = C(t_n)$, and this equation is of the form

$$C(t_{n+1}) = (1-\Delta)C(t_n) + \Delta \sum_{k=n+1}^{\infty} \Delta(1-\Delta)^{k-(n+1)} f(C(t_k), m) , \quad (6.60)$$

with

$$f(C,m) = \int_{-\infty}^{\infty} \frac{dz}{\sqrt{2\pi}} e^{-z^2/2}$$
$$\times \left\{ \int_{-\infty}^{\infty} \frac{d\eta}{\sqrt{2\pi}} e^{-\eta^2/2} F(m + \sqrt{\alpha(1-C)}\, \eta + \sqrt{\alpha C}\, z) \right\}^2 . \quad (6.61)$$

Here we have replaced $m(t_n)$ by its stationary value m and also omitted the term originating from the initial conditions. It can be seen from (6.49) that this term vanishes exponentially fast in the limit $t_m \to \infty$. In the limit of large time differences, i.e. $t_n \to \infty$, C reaches a fixed-point value $q = \lim_{n\to\infty} C(t_n)$ which is determined by

$$q = f(q,m) . \quad (6.62)$$

Let us discuss a few results which can be obtained from (6.60–62):

(a) For the case of deterministic updating (i.e. $F(x) = \mathrm{sgn}(x)$, $\phi = 0$), (6.62) has two solutions: $q = 1$ and $q = q_0(\alpha) < 1$, provided $\alpha < \alpha_c$. For small α, $q_0(\alpha)$ is easily calculated:

$$q_0(\alpha) \rightsquigarrow 1 - \frac{8}{\pi^2} \exp\left(-\frac{1}{\alpha}\right) . \tag{6.63}$$

If one linearizes (6.60) around these solutions one finds that $q = 1$ is locally unstable. As $q_0(\alpha) < 1$ is the stable solution, the neuronal activity pattern $\{\sigma(t)\}$ is not perfectly frozen in the absence of external time-dependent noise. In particular this implies that the retrieval phase is not charcaterized by fixed points of the neuronal activity pattern $\{\sigma_i\}$ as in a symmetric network [6.13].

(b) Another interesting question is the possibility of a spin-glass state, i.e. a stationary state with $m = 0$, $q \neq 0$. The solution $m = q = 0$ would become unstable if

$$\left(\frac{\partial f}{\partial q}\right)_{q=m=0} = 1 . \tag{6.64}$$

For the case of Glauber dynamics ($F(x) = \tanh(\beta x)$), $(\partial F/\partial q)_{m=q=0}$ is monotonically increasing with β and as $\beta \to \infty$ takes on the value $2/\pi < 1$, so there is no local instability of the $q = m = 0$ state [6.13].

(c) Note that for p-dynamics (6.60) takes on the simple form

$$C(t_{n+1}) = f(C(t_{n+1}), m) , \tag{6.65}$$

which implies that $C(t)$ reaches its asymptotic value q in one time step (starting from $C(0) = 1$) if the particular initial configuration σ_0 has been forgotten (i.e., $t_m \to \infty$).

The time-delayed correlation function $C(t)$ is a special case of the general multitime correlation function of neuronal activities $\tau(t_n) = \xi^1 \langle\sigma(t_n)\rangle_\phi$, i.e. $C(t_n, t_m, ..., t_k) = \langle\tau(t_n)\tau(t_m)...\tau(t_k)\rangle_\phi$. For large time arguments and parallel dynamics $C(t_n, t_m, ..., t_k) = \langle\langle\tau(t_n)\rangle_{\eta_n}\langle\tau(t_m)\rangle_{\eta_m}...\langle\tau(t_k)\rangle_{\eta_k}\rangle_z$. The same relation holds for sequential dynamics if all the time differences $(t_n - t_m, ...)$ are simultaneously considered large. In this case all the correlation functions can be obtained from the distribution $p(A)$ of neuronal activities given by

$$p(A) = \langle\delta(A - \langle\tau\rangle_\eta)\rangle_z = \int_{-\infty}^{\infty} \frac{\mathrm{d}z}{\sqrt{2\pi}} \mathrm{e}^{-z^2/2}$$
$$\times \delta\left\{A - \int_{-\infty}^{\infty} \frac{\mathrm{d}\eta}{\sqrt{2\pi}} \mathrm{e}^{-\eta^2/2} F(m + \sqrt{\alpha(1-q)}\eta + \sqrt{\alpha q}z)\right\} . \tag{6.66}$$

This expression was given in [6.13] for continuous-time Glauber dynamics. It has also been derived by combinatorial arguments [6.24] for $F(x) = \mathrm{sgn}(x)$. Note that $p(A)$ is completely determined by its first two moments m and q, which have to be calculated from (6.45) and (6.46). For $m = q = 0$, $p(A) = \delta(A)$, whereas $p(A)$ is continuous and does not contain δ-functions for $m \neq 0$, $q \neq 0$ even in the case of deterministic updating ($F(x) = \mathrm{sgn}(x)$).

6.4 Extensions and Related Work

The discussion of the strongly diluted model has been extended in several ways. It was noticed that strong dilution provides a limit in which many models can be solved analytically. This turned out to be very useful, in particular for a qualitative understanding of those models which cannot be handled analytically if they are fully connected. Derrida and Nadal discuss a network which accounts for the gradual deterioration of memory, i.e. forgetting, if new information is stored continuously [6.25]. Gutfreund and Mézard discuss the processing of temporal sequences in strongly diluted neural networks [6.26]. Gardner et al. consider an asymmetric learning rule which differentiates between pre- and postsynaptic neurons [6.27]; and Kerszberg and Zippelius study a neural network with axonal delays and postsynaptic summation [6.28]. In all these examples, the limit of strong dilution turned out to be exactly solvable and thus provides a means for analytic discussion of a model which otherwise can only be solved numerically.

Symmetrically diluted neural networks have also been considered. The model is defined by (6.1–4) with the restriction $c_{ij} = c_{ji}$ [6.30]. Some of the results of the asymmetric model carry over to the symmetric version; for example the storage capacity $\alpha_c = 2/\pi$ and the overlap with a single pattern at $T = 0$ are exactly the same in both models.

Another important question is what happens if the dilution is not strong. Symmetrically diluted networks have been studied in the *dense* limit [6.11], when each neuron receives input from $O(N)$ other neurons. The retrieval properties of the network are only mildly affected by weak dilution. Dense asymmetric networks are difficult to study analytically. Amit and Treves have calculated the average number of metastable states as a function of the overlap with one pattern. They find that for moderate dilution there is still an exponentially large number of metastable states which are only weakly correlated with the memories [6.29]. Whether or not this model shows a spin-glass phase is still an open question.

If the average activity of the stored patterns is very low, i.e., $\langle \eta_i^\mu \rangle_\xi = (1 + \langle \xi_i^\mu \rangle_\xi)/2 = a \ll 1$ and ξ_i^μ is replaced by $\eta_i^\mu - a$ in Hebb's learning rule (6.1), the asymmetrically diluted network behaves in a very similar way to a *nondiluted, symmetric* network. In [6.31] it is shown that the self-consistency equations for both networks coincide and that the storage capacity can be considerably enhanced by adding a threshold field h_0. The transition into the retrieval phase at α_c becomes discontinuous and $m(\alpha_c)$ is close to unity.

Other models on sparsely connected random structures have also been studied: ferromagnets [6.32], spin glasses [6.33, 34] and cellular automata [6.17, 21]. In ferromagnets and spin glasses one is mainly interested in the effect of connectivity fluctuations on the ground-state properties of the system. Most studies have focused on symmetric interactions. Automata, on the other hand, are generally not symmetric. In particular, the Kauffman model is on a formal level closely related to asymmetrically diluted neural networks. In fact, it was for this model that Derrida et al. first pointed out the exact solvability [6.17].

Appendix A

In this appendix we show how (6.17) simplifies in the limit of large K and p. In this limit one can introduce continuous variables x and y, such that $n = Kx$ and $s = K(p-1)y$. Using Stirling's formula

$$x! = \sqrt{2\pi x}\, e^{-x} x^x \tag{6.67}$$

we find

$$f_K(m) \simeq \frac{K\sqrt{p-1}}{2\pi} \int_0^1 dx \int_0^1 dy\, e^{-Kg(x,y)} \left[\frac{\tanh \beta(p - 2x - 2(p-1)y)}{(x(1-x)y(1-y))^{1/2}} \right], \tag{6.68}$$

with

$$\begin{aligned} g(x,y) = {} & x \ln x + (1-x)\ln(1-x) + p \ln 2 - (1-x)\ln(1+m) \\ & - x\ln(1-m) + (p-1)y \ln y + (p-1)(1-y)\ln(1-y)\ . \end{aligned} \tag{6.69}$$

In the limit $K \to \infty$, the integral (6.69) is dominated by the saddle point $(\bar{x}, \bar{y})$, determined by

$$\left(\frac{\partial g}{\partial x}\right)_{\bar{x}} = 0 = \ln \bar{x} - \ln(1-\bar{x}) + \ln(1+m) - \ln(1-m)\ , \tag{6.70}$$

$$\left(\frac{\partial g}{\partial y}\right)_{\bar{y}} = 0 = (p-1)\ln \bar{y} - (p-1)\ln(1-\bar{y})\ . \tag{6.71}$$

We expand in fluctuation around the saddle point $\varepsilon = x - \bar{x} = x - (1-m)/2$ and $\delta = y - \bar{y} = y - 1/2$ up to quadratic order and find

$$g(x,y) \simeq \frac{2}{1-m^2}\,\varepsilon^2 + 2(p-1)\delta^2 \tag{6.72}$$

and hence

$$\begin{aligned} f_K(m) \simeq {} & \frac{2K}{\pi}\sqrt{\frac{p-1}{1-m^2}} \int d\varepsilon \int d\delta \exp\left[-K\left(\frac{2}{1-m^2}\varepsilon^2 + 2(p-1)\delta^2\right)\right] \\ & \times \tanh \beta(m - 2\varepsilon - 2(p-1)\delta)\ . \end{aligned} \tag{6.73}$$

With the substitution $x = 2\varepsilon\sqrt{K/(1-m^2)}$, $y = 2\delta(p-1)$ this can be transformed into

$$\begin{aligned} f_K(m) \simeq {} & \int_{-\infty}^{\infty} \frac{dx}{\sqrt{2\pi}} \int_{-\infty}^{\infty} \frac{dy}{\sqrt{2\pi\alpha}} \exp\left(-\frac{x^2}{2} - \frac{y^2}{2\alpha}\right) \\ & \times \tanh \beta \left(m - x\frac{\sqrt{1-m^2}}{K} - y\right) \\ = {} & \int_{-\infty}^{\infty} \frac{dy}{\sqrt{2\pi\alpha}} \exp\left(-\frac{y^2}{2\alpha}\right) \tanh \beta(m-y)\ . \end{aligned} \tag{6.74}$$

Appendix B

The expansion of the exponential in (6.36) up to $O(K^{-2})$ leads to

$$L = L^{(1)} + L^{(2)} = -\frac{K}{N}\sum_{i\neq j}\left[\frac{1}{K}\sum_{\nu}\xi_i^\nu\xi_j^\nu\sum_n \tilde{h}_i(t_n)\sigma_j(t_n)\right.$$

$$\left.+\frac{1}{2K^2}\sum_{\nu\mu}\xi_i^\nu\xi_j^\nu\xi_i^\mu\xi_j^\mu\sum_{n,m}\tilde{h}_i(t_n)\sigma_j(t_n)\tilde{h}_i(t_m)\sigma_j(t_m)\right] . \tag{6.75}$$

We split the second-order term $L^{(2)}$ into its a average (which corresponds to the terms with $\nu = \mu$) and a fluctuation term $\Delta L^{(2)}$, which is of the form

$$2\Delta L^{(2)} = -\frac{1}{\sqrt{K}}\sum_{n,m}\sum_{\nu\neq\mu}^{\alpha K} a_{\nu\mu}(t_n,t_m)\sum_i \xi_i^\nu\xi_i^\mu\tilde{h}_i(t_n)\tilde{h}_i(t_m)$$

$$+\frac{\alpha K(\alpha K-1)}{NK}\sum_i\left(\sum_n \tilde{h}_i(t_n)\sigma_i(t_n)\right)^2 , \tag{6.76}$$

where

$$a_{\nu\mu}(t_n,t_m) = \frac{1}{N\sqrt{K}}\sum_i \xi_i^\nu\xi_i^\mu\sigma_i(t_n)\sigma_i(t_m) . \tag{6.77}$$

Note that $a_{\mu\nu}$ is $O(1/\sqrt{NK})$ for all but a finite number of patterns for which it is $O(1/\sqrt{K})$. We can make use of this fact if we introduce the $a_{\mu\nu}(t_n,t_m)$ as auxiliary variables with the help of the identity

$$1 = \int \mathcal{D}(\tilde{a},a)\exp N\sum_{\nu\neq\mu}\sum_{n,m}\tilde{a}_{\nu\mu}(t_n,t_m)$$

$$\times\left[a_{\nu\mu}(t_n,t_m) - \frac{1}{N\sqrt{K}}\sum_i \xi_i^\nu\xi_i^\mu\sigma_i(t_n)\sigma_i(t_m)\right] , \tag{6.78}$$

where $\mathcal{D}(\tilde{a},a) = \prod_n\prod_m\prod_{\nu\neq\mu}(\mathrm{d}\tilde{a}_{\mu\nu}(t_n,t_m)\mathrm{d}a_{\mu\nu}(t_n,t_m)/2\pi)$ and the integrations over the $\tilde{a}$ variables extend over the imaginary axis. For $N\to\infty$ we perform the $\tilde{a}$, a integrations by the saddle-point method. The saddle-point value of the a variables is given by the expectation value of the right-hand side of (6.77). The contribution to $\Delta L^{(2)}$ arising from the finitely many patterns with $a_{\mu\nu}\sim O(1/\sqrt{K})$ is not larger than $O(N/K)$. For most patterns $a_{\mu\nu}\sim O(1/\sqrt{NK})$ so that their contribution to $\Delta L^{(2)}$ is $O(K)$. Hence both contributions can be neglected, compared to the $O(N)$ terms in the average of $L^{(2)}$. It can be seen directly from (6.76) that the last term on the right-hand side is $\sigma(N)$ and thus it can also be neglected.

Appendix C

We rewrite $\langle \exp(-L) \rangle_\xi$ by introducing auxiliary fields

$$m^\nu(t_n) = N^{-1} \sum_i \xi_i^\mu \sigma_i(t_n) \tag{6.79}$$

and

$$C(t_n, t_m) = N^{-1} \sum_i \sigma_i(t_n) \sigma_i(t_m) \tag{6.80}$$

with the help of the identity

$$\begin{aligned} 1 = \int \mathcal{D}(m, C) & \left[\prod_{\nu=1}^{p} \prod_n \delta \left(m^\nu(t_n) - N^{-1} \sum_i \xi_i^\mu \sigma_i(t_n) \right) \right] \\ \times & \left[\prod_{n,m} \delta \left(C(t_n, t_m) - N^{-1} \sum_i \sigma_i(t_n) \sigma_i(t_m) \right) \right] . \end{aligned} \tag{6.81}$$

From (6.81) L can be written in the form

$$\begin{aligned} L = & -\sum_n \sum_\nu m^\nu(t_n) \sum_i \xi_i^\nu \tilde{h}_i(t_n) \\ & - \frac{\alpha}{2} \sum_{n,m} \sum_i \tilde{h}_i(t_n) C(t_n, t_m) \tilde{h}_i(t_m) , \end{aligned} \tag{6.82}$$

so that different neurons appear uncoupled. The integrations over the auxiliary variables $m^\nu(t_n)$ and $C(t_n, t_m)$ are performed by saddle-point techniques. To this end we represent the δ-functions in (6.81) by Fourier integrals analogous to (6.78). The saddle-point equations for $m^\nu(t_n)$ and $C(t_n, t_m)$ are given by (6.45) and (6.46), whereas the variables $\tilde{m}^\nu$ and $\tilde{C}(t_n, t_m)$, which appear in the Fourier integrals, have to vanish at the saddle point, owing to causality. The argument runs in complete analogy to the case of a dynamics of continuous variables [6.35], to which we refer the reader for further details.

References

6.1 W.A. Little: Math. Biosci. **19**, 101 (1974); W.A. Little, G.L. Shaw: Math. Biosci. **39**, 281 (1978)

6.2 J.J. Hopfield: Proc. Natl. Acad. Sci. USA **79**, 2554 (1982); **81**, 3088 (1984)

6.3 See for example the review by D. Amit in *Heidelberg Colloquium on Glassy Dynamics*, edited by J.L. van Hemmen and I. Morgenstern (Springer, Berlin, Heidelberg 1987)

6.4 W.S. Mc Cullogh and W.A. Pitts: Bull. Math. Biophys. **5**, 115 (1943)

6.5 See for example N.G. van Kampen: *Stochastic Processes in Physics and Chemistry* (North-Holland, Amsterdam 1981); H. Risken: *The Fokker Planck Equation* (Springer, Berlin, Heidelberg 1984)

6.6 R.J. Glauber: J. Math. Phys. (N.Y.) **4**, 294 (1956)

6.7 D.J. Amit, H. Gutfreund, and H. Sompolinsky: Ann. Phys. (N.Y.) **173**, 30 (1987)
6.8 G. Parisi: J. Phys. A **19**, L675 (1986)
6.9 I. Kanter and H. Sompolinsky: Phys. Rev. Lett. **57**, 2861 (1986)
6.10 L. Personnaz, I. Guyon, G. Dreyfus, and G. Toulouse: J. Stat. Phys. **43**, 411 (1986); W. Kinzel in *Heidelberg Colloquium on on Glassy Dynamics*, edited by J.L. van Hemmen and I. Morgenstern (Springer, Berlin, Heidelberg 1987), p. 529; J.A. Hertz, G. Grinstein, and S.A. Solla: **ibid**, p. 538; R. Bausch, H.K. Janssen, R. Kree, and A. Zippelius: J. Phys. C **19**, L779 (1986); M.V. Feigelman, L.B. Ioffe: Int. J. Mod. Phys. B **1**, 51 (1987)
6.11 H. Sompolinsky in *Heidelberg Colloquium on Glassy Dynamics*, edited by J.L. van Hemmen and I. Morgenstern (Springer, Berlin, Heidelberg 1987), p. 485
6.12 B. Derrida, E. Gardner, and A. Zippelius: Europhys. Lett. **4**, 167 (1987)
6.13 R. Kree and A. Zippelius: Phys. Rev. A **36**, 4421 (1987)
6.14 See for example *Principles of Neural Sciences*, 2nd Edition, edited by E.R. Kandel and J.H. Schwartz (Elsevier, New York 1985)
6.15 D.J. Amit, H. Gutfreund, and H. Sompolinsky: Phys. Rev. A **32**, 1007 (1985)
6.16 H. Sompolinsky, A. Crisanti, and H.J. Sommers: Phys. Rev. Lett. **61**, 259 (1988)
6.17 B. Derrida and Y. Pomeau: Europhys. Lett. **1**, 45 (1986); B. Derrida and G. Weisbuch: J. Phys. (Paris) **47**, 1297 (1986)
6.18 H. Sompolinsky and A. Zippelius: Phys. Rev. B **25**, 6860 (1982)
6.19 H.J. Sommers: Phys. Rev. Lett. **58**, 1268 (1987)
6.20 R. Graham: *Statistical Theory of Instabilities in Stationary Nonequilibrium Systems with Applications to Lasers and Nonlinear Optics*, Springer Tracts in Modern Physics, Vol. 66 (Springer, Berlin, Heidelberg 1973)
6.21 H. Hilhorst and M. Nijmeijer: J. Phys. (Paris) **48**, 185 (1987)
6.22 I. Kanter: Phys. Rev. Lett. **60**, 1891 (1988)
6.23 P. Peretto: Biol. Cyb. **50**, 51 (1984)
6.24 B. Derrida: J. Phys. A **22**, 2069 (189)
6.25 B. Derrida and J.P. Nadal: J. Stat. Phys. **49**, 993 (1987)
6.26 H. Gutfreund and M. Mézard: Phys. Rev. Lett. **61**, 235 (1988)
6.27 E. Gardner, S. Mertens, and A. Zippelius: J. Phys. A **22**, 2009 (1989)
6.28 M. Kerszberg and A. Zippelius: Phys. Scr. **T 33**, 54 (1990)
6.29 A. Treves and D.J. Amit: J. Phys. A **21**, 3155 (1988)
6.30 I. Kanter in *Computational Systems – Natural and Artificial*, edited by H. Haken (Springer, Berlin, Heidelberg 1987)
6.31 M.V. Tsodyks: Europhys. Lett. **7**, 203 (1988)
6.32 L. Viana and A.J. Bray: J. Phys. C **18**, 3037 (1985)
6.33 I. Kanter and H. Sompolinsky: Phys. Rev. Lett. **58**, 164 (1987)
6.34 M. Mézard and G. Parisi: Europhys. Lett. **3**, 1067 (1987)
6.35 R. Bausch, H.K. Janssen, and H. Wagner: Z. Phys. B **24**, 113 (1984); C. de Dominicis and L. Peliti: Phys. Rev. B **18**, 353 (1978)

7. Temporal Association

Reimer Kühn and J. Leo van Hemmen

With 22 Figures

Synopsis. Recently proposed models for temporal association in networks of formal neurons are reviewed. Underlying mechanisms are explained, exact results are presented whenever available, and estimates of persistence times are derived. Applications dealing with sequence recognition, with counting, and with the generation of complex sequences are briefly discussed. Finally, a learning mechanism is presented which stores both static patterns and pattern sequences on the basis of one and the same principle, viz., Hebbian learning with a broad distribution of delays. The paper contains a number of previously unpublished results and has been conceived to provide as much detail as required to make it self-contained.

7.1 Introduction

Temporal association refers to the storage and retrieval ("association") of patterns which change *in time*. A typical example is the musical theme B-A-C-H [7.1], each note having a specific duration. Once the network has learnt this theme with its correct timing, it should be able to reproduce it and also recall it when supplied with initial data that might be noisy, incomplete, or distorted versions of the theme. So the recall should be performed in an associative fashion when the network is triggered in a suitable way.

The present paper is devoted to this type of problem. We review and analyze several mechanisms which allow the storage and retrieval of sequences of given (nominated) network states. In so doing we shall restrict most of our attention to formal neurons [7.2], simple threshold automata which can acquire only two states: firing or nonfiring. They are represented by two-valued entities S_i, with $S_i = +1$ and $S_i = -1$ (or $S_i = 1$ and $S_i = 0$) respectively denoting the firing and the nonfiring state of neuron i, and $i = 1, ..., N$ enumerating the neurons of the net. The neurons are connected by a set of synapses J_{ij}, such that $J_{ij}S_j$ gives the contribution of neuron j to the postsynaptic potential of neuron i. The dynamics of the network is then given by the threshold condition

$$S_i(t + \Delta t) = \text{sgn}\left[\sum_j J_{ij}S_j(t)\right] , \qquad (7.1)$$

or a probabilistic version thereof (see below). Updating may be asynchronous or in parallel. Note that in (7.1) we have specialized to the $S_i = \pm 1$ representation.

Networks of formal neurons have received a considerable amount of interest since the discovery of Hopfield [7.3] that the threshold dynamics (7.1) is governed by a Lyapunov (or energy) function if the J_{ij} are symmetric, i.e., $J_{ij} = J_{ji}$ with $J_{ii} = 0$, and if updating in (7.1) is asynchronous. Remarkably, the Lyapunov function governing (7.1) takes the form of an Ising Hamiltonian

$$H_N = -\tfrac{1}{2} \sum_{i,j=1}^{N} J_{ij} S_i S_j \; . \tag{7.2}$$

Its discovery has had a profound effect on further research in the field.

The notion of an energy function gives rise to a clear and extremely productive metaphor of the processes involved in the storage and associative recall of information. Data are stored in the neural net in that by a suitable choice of the J_{ij} several specific (firing) patterns of the S_i are made local minima of H_N. If this can be achieved, the neural net will function as an associative memory. A network state which "somehow resembles" one of the stored prototypes corresponds to a location in the energy landscape which is close enough to the minimum representing that prototype to lie in its basin of attraction. By spontaneously moving downhill, the network reconstructs the prototype.

Certainly no less important is the fact that the physics of neural networks with a Hamiltonian *à la* (7.2) has been made amenable to analysis by tools developed in statistical mechanics. Questions of storage capacity, retrieval quality, size of basins of attraction, and many more have by now been answered for a variety of models. The reader can find an account of this impressive success throughout this book.

On the other hand, *asymmetric* synapses, with $J_{ij} \neq J_{ji}$ for $j \neq i$, are a necessary prerequisite for temporal associations such as would be involved in reciting a poem or a tune, in counting, or in the control of rhythmic motion. To wit, following the dynamics (7.1), a network with symmetric synapses would eventually settle in one of the energy valleys associated with the Hamiltonian (7.2) and moderate amounts of external noise (allowing, with small probability, for uphill motion in the energy landscape created by H_N) would not help it to escape. Perhaps not surprisingly, synaptic interconnections in biological neural networks are, indeed, in general asymmetric.

The first attempt at sequence generation in the context of the Hopfield model [7.3] was undertaken by Hopfield himself. In the Hopfield model, a set of q unbiased binary random patterns $\{\xi_i^\mu; 1 \le i \le N\}$, $1 \le \mu \le q$, having $\xi_i^\mu = \pm 1$ with equal probability, is stored by taking *symmetric* synapses of the form [7.3]

$$J_{ij}^{(1)} = N^{-1} \sum_{\mu=1}^{q} \xi_i^\mu \xi_j^\mu \; . \tag{7.3}$$

In order to recall the patterns in a *sequence*, Hopfield investigated the possibility of adding a set of *asymmetric* forward projections

$$J_{ij}^{(2)} = \varepsilon N^{-1} \sum_{\mu=1}^{q-1} \xi_i^{\mu+1} \xi_j^\mu \tag{7.4}$$

to the symmetric synapses (7.3)[1]. They induce a natural order among the patterns $\{\xi_i^\mu\}$ and are expected to generate transitions between them. To understand why this should be so, it will be instructive first to explain the associative capabilities of the Hopfield model proper, having *only* the symmetric synapses (7.3), and to do so without recourse to the landscape metaphor; the introduction of asymmetric synapses will invalidate that picture anyhow.

To see why the Hopfield model with synapses given by (7.3) has the patterns $\{\xi_i^\mu\}$ as *attractors* of the dynamics (7.1), note that the dynamical law (7.1) embodies a two-step process, the evaluation of the local field or postsynaptic potential (PSP)

$$h_i(t) = \sum_{j(\neq i)} J_{ij}^{(1)} S_j(t) \,, \tag{7.5.a}$$

which is a *linear* operation, and a *non*linear decision process,

$$S_i(t + \Delta t) = \mathrm{sgn}[h_i(t)] \,. \tag{7.5.b}$$

Assuming that the number q of stored patterns is small ($q/N \to 0$), we find that the synapses (7.3) give rise to a local field of the form

$$h_i(t) = \sum_{\mu=1}^{q} \xi_i^\mu m_\mu(t) \,, \tag{7.6}$$

where

$$m_\mu(t) = N^{-1} \sum_{i=1}^{N} \xi_i^\mu S_i(t) \tag{7.7}$$

is the *overlap* of the network state $\{S_i(t)\}$ with the pattern $\{\xi_i^\mu\}$ and measures the proximity of $\{S_i(t)\}$ and $\{\xi_i^\mu\}$. If $S_i(t) = \xi_i^\nu$, then $m_\nu(t) = 1$ while $m_\mu(t) = O(N^{-1/2})$ for $\mu \neq \nu$, since we are dealing with uncorrelated random patterns. If the actual state of the system is close to but not coincident with $\{\xi_i^\nu\}$, then $0 < m_\nu(t) < 1$ and $m_\nu(t) \gg |m_\mu(t)|$ for $\mu \neq \nu$. According to (7.6), the sign of $h_i(t)$ will then "most likely" be that of ξ_i^ν, which in turn through the decision process (7.5b) would "most likely" lead to an *increase* of m_ν, whenever neurons with $S_i(t) = -\xi_i^\nu$ are updated. Repeating this argument, we find $m_\nu(t)$ increasing towards unity as time proceeds, while the $m_\mu(t)$ with $\mu \neq \nu$ tend to values of order $N^{-1/2}$. Away from the limit $q/N \to 0$, mutual interference between many pattern states will modify the above qualitative picture, though not much as long as q/N is sufficiently small.

[1] For the generation of cyclic sequences, the sum in (7.4) is extended to q instead of $q - 1$, and pattern labels are interpreted modulo q.

Returning to the problem of sequence generation, we find that the synapses $J_{ij} = J_{ij}^{(1)} + J_{ij}^{(2)}$ will give rise to a local field of the form

$$h_i(t) = \sum_{\mu=1}^{q} (\xi_i^\mu + \varepsilon \xi_i^{\mu+1}) m_\mu(t) . \tag{7.8}$$

Assuming that the system is in state ν, we get [ignoring $O(N^{-1/2})$ fluctuations]

$$h_i(t) = \xi_i^\nu + \varepsilon \xi_i^{\nu+1} , \tag{7.9}$$

so that $h_i(t)$ consists of two contributions, one stabilizing the current state ν and a second one, generated by the asymmetric synapses (7.4), which favors the next pattern of the sequence, i.e., $\nu + 1$. If $\varepsilon < 1$, the term stabilizing the current state will determine the sign of $h_i(t)$ so that a transition will not occur. If, on the other hand, $\varepsilon > 1$, the second term in (7.9) will through (7.5b) initiate a transition to $\nu + 1$. As soon as $m_{\nu+1}$ becomes sizeable, though, equally sizeable contributions to the local field develop that try to push the system into $\nu+2$. The next transition is initiated *before* the previous one is actually completed so that, before long, the system is mixed up completely. That is, if the number of patterns q is odd, the system relaxes into the q-symmetric state $S_i = \mathrm{sgn}(\sum_{\mu=1}^{q} \xi_i^\mu)$, which is invariant under the dynamics (7.1). On the other hand, if q is even, a fast oscillation around a q-symmetric solution occurs (see Fig. 7.1). But this is not what we want, since the system goes through a sequence of *mixture*-states rather than a sequence of pure patterns. Therefore, the synapses (7.3) and (7.4) as such turn out to be unsuitable for the generation of stable pattern sequences.

Note that this no-go statement applies to *asynchronous* dynamics only. In the case of *parallel* dynamics, each transition will be completed in a *single* synchronous update, and temporal sequences visiting each pattern for a single time step can be reliably generated[2]. Note also that in this case the symmetric synapses could be dispensed with, without destroying the stability of the sequence.

The story of sequence generation in asynchronous neural networks begins as a story of remedies for the above-mentioned mixing phenomenon, remedies

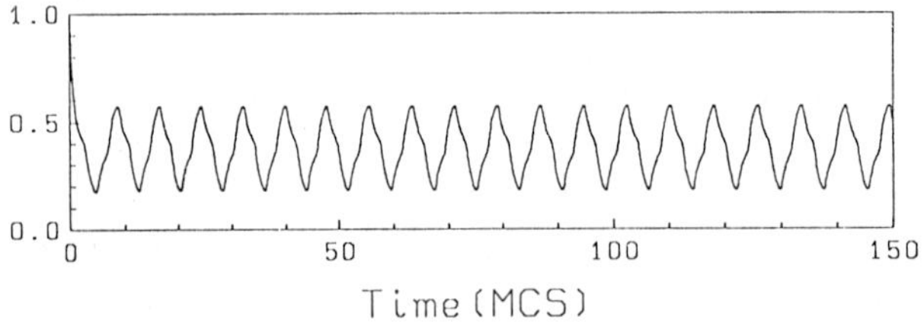

Fig. 7.1. Fast oscillation described in the text. Analytic solution of the dynamics for the thermodynamic limit. Only the overlap with the first of four patterns is shown. Dynamics is sequential at inverse temperature $\beta = 10$. Time is measured in units of Monte Carlo steps per spin (MCS)

[2] Basically the same mechanism would work for sequences of arbitrary patterns, if the couplings (7.3) and (7.4) are replaced by appropriate modifications; see Sect. 7.5.2.

which, by one way or another, allow the system to become quasi-stationary in one pattern before it performs a transition to the next. In what follows, we shall review four basically different approaches to achieve this goal. Two of them, the models of Peretto and Niez [7.4] and of Dehaene et al. [7.5], rely on the introduction of *short-term modifiable* synapses. They will be discussed in Sect. 7.2. A different approach, based on static synapses, has been proposed by Buhmann and Schulten [7.6, 7]. They use a 0-1 representation for the neurons, symmetric connections to stabilize single patterns, a set of forward projections to generate transitions, and carefully tuned pairwise inhibitions to prevent the above-mentioned mixing of states; the performance of this setup will be analyzed in Sect. 7.3. A still different route to sequence generation was proposed by Kleinfeld [7.8] and by Sompolinsky and Kanter [7.9]. They used the synapses (7.3) and (7.4), as originally proposed by Hopfield, but endowed the asymmetric synapses (7.4) with a *transmission delay*, its effect being to induce transitions only when the system has been stable in a given pattern for some time determined by that delay. This idea will be described in Sect. 7.4. In all cases we shall try to explain the basic ideas involved, review whatever exact results are available and supply – where possible – a (self-consistent) estimate of the persistence time Δ_0, i.e., the time the system spends in each pattern of the sequence.

In order to prevent the article from growing without bounds, we had of course to make our choices as to its scope, which unavoidably carry some degree of arbitrariness with them. Two items not included in the present review will serve to indicate its boundaries (and limitations). One is an earlier attempt at sequence generation in neural networks, proposed by Willwacher [7.10]. To produce temporal associations it relies on mechanisms such as fatigue of its neurons and is therefore regarded as being outside the realm of networks of formal neurons to be discussed here. The other is a mechanism for analyzing temporal aspects of external stimuli within a *feed-forward* architecture, proposed by Hopfield and Tank [7.11]. Elementary parts of a sequence, e.g., letters in a word, are recognized by feature detectors. With time delays proportional to the temporal distance between the individual symbol and the end of the prototype sequence they belong to, they are fed into a system of indicator units, where the signals coming from the feature detectors thus arrive "concentrated in time" and can be used for classification. This setup, too, useful as it may turn out to be in technical applications, will readily be recognized as somewhat remote from the problem of sequence generation we have decided to cover in the present article.

Of the four approaches that *are* included in the sequel, the one relying on time delays has, in retrospect, proved to be the most versatile, and we shall accordingly discuss some of its ramifications, such as architectures for the recognition of stored sequences [7.8, 12], for counting [7.13], and for the generation of complex sequences in which a succession of states is not always unambiguously defined [7.14, 15]. This will be done in Sect. 7.5.

Some of the sequence-generating models have been criticized [7.5] for requiring carefully tuned synaptic strengths which have to be put into the models by hand, and thus for not being able to acquire knowledge through a true learning mechanism. In Sect. 7.6, we shall therefore discuss a new view on Hebbian learning [7.16] recently proposed by Herz et al. [7.17] which reconciles the concepts of "learning of static patterns" and "learning of pattern sequences" by incorporating delays as used in the dynamics into the venerable Hebbian learning rule itself. Patterns and pattern sequences can thereby be learnt by *one and the same* principle and without a necessity for prewiring the network.

We shall close this article with an outlook in Sect. 7.7, trying to assess achievements but also pointing out some of the major difficulties and questions that have remained open so far.

7.2 Fast Synaptic Plasticity

In this section, we shall review two approaches to sequence generation which rely on fast synaptic modifications as a mechanism for stabilizing temporal sequences. One has been proposed by Peretto and Niez [7.4], the other by Dehaene et al. [7.5]. The latter is, as we shall see, in some sense orthogonal to Hopfield-type models but is nevertheless included here. It is at the same time also more ambitious than most of the other approaches to be discussed in the present article, in that it not only addresses the question of sequence generation but also that of sequence learning – a problem to which we shall return only later on in Sect. 7.6.

7.2.1 Synaptic Plasticity in Hopfield-Type Networks

As shown in the introduction, the synapses (7.3) and (7.4) as such are unsuitable for generating stable sequences in asynchronous neural networks. If the asymmetry parameter ε is too small, no transition between patterns will occur and the system approaches a stationary state, whereas for larger values of ε transitions *are* induced, but they follow each other so rapidly that the system soon gets mixed up.

As a way out of this dilemma, Peretto and Niez [7.4] propose using the synapses (7.3) and (7.4) with ε less than unity, which by themselves do not generate transitions between the patterns, and to supplement these by a set of short-term modifiable synapses $J^{s}_{ij}(t)$ whose main effect is progressively to *compensate* – and thus weaken – that contribution to the symmetric synapses which stabilizes the pattern the system *is currently in*.

For the short-term modifiable synapses, Peretto and Niez propose a dynamics of the form

$$J^{s}_{ij}(t+\Delta t) = \left(1 - \frac{\Delta t}{T_s}\right) J^{s}_{ij}(t) - \frac{\Delta t}{T_j} S_i(t) S_j(t) , \qquad (7.10)$$

where T_s is a relaxation time controlling the rate of decay of "old" contributions to J^{s}_{ij}, while T_j determines to what extent the current network state will be destabilized. It is proposed that the time constant T_j depends on j in the following manner:

$$T_j^{-1} = \frac{\lambda}{N}\sum_{\mu=1}^{q}(\xi_j^\mu \xi_j^\mu - \varepsilon \xi_j^{\mu+1}\xi_j^\mu) , \qquad (7.11)$$

where λ is some parameter which is at our disposal, and where ε is the asymmetry parameter characterizing the synapses (7.4).

To understand the transition *mechanism*, one solves (7.10) for a given history $\{S_i(t - k\Delta t);\ k \in \mathbb{N}\}$ of the network to obtain [7.18]

$$J_{ij}^{\mathrm{s}}(t) = -\frac{\Delta t}{T_j}\sum_{k=1}^{\infty} R^{k-1} S_i(t - k\Delta t) S_j(t - k\Delta t) , \qquad (7.12)$$

where we have put

$$R = 1 - \frac{\Delta t}{T_{\mathrm{s}}} .$$

Assuming that the system has been stationary in some pattern, say $\{\xi_i^\nu\}$ for the n most recent time steps, $t - \Delta t, t - 2\Delta t, \dots, t - n\Delta t$, we get

$$J_{ij}^{\mathrm{s}}(t) = -\frac{T_{\mathrm{s}}}{T_j}(1 - R^n)\xi_i^\nu \xi_j^\nu - \frac{\Delta t}{T_j}\sum_{k=n+1}^{\infty} R^{k-1} S_i(t - k\Delta t) S_j(t - k\Delta t) . \ (7.13)$$

The first term in (7.13) has a negative weight and *weakens* the contribution to the static synapses that stabilizes pattern ν, i.e., $N^{-1}\xi_i^\nu \xi_j^\nu$. Furthermore, since $R < 1$, this effect becomes stronger with increasing n while the influence of the second contribution to (7.13) is at the same time decreasing. Thus, if parameters are properly tuned, the forward projection $\varepsilon N^{-1}\xi_i^{\nu+1}\xi_j^\nu$ in $J_{ij}^{(2)}$ will eventually be able to induce a transition to $\nu + 1$.

The ideas just outlined can be used to obtain an estimate of the *persistence time* of the sequence generator, i.e., the time the system spends in each pattern of the sequence. The argument is as follows [7.9, 18]. One assumes that the system has gone through a cycle of length q, with persistence time $\Delta_0 = k_0\Delta t$, and that at time t the system has been in pattern q for precisely k_0 elementary time steps so that in the next instant it should undergo a transition to $q + 1 \equiv 1$. The persistence time $\Delta_0 = k_0\Delta t$ is determined from the condition that the above considerations should be *self-consistent*.

Neglecting transition times, the values of the $J_{ij}^{\mathrm{s}}(t)$ corresponding to the above scenario may be computed from (7.12) to give [7.18]

$$J_{ij}^{\mathrm{s}}(t) = (RT_j)^{-1}\left(\Delta t\, \xi_i^q \xi_j^q - T_{\mathrm{s}}\frac{1 - R^{k_0}}{1 - R^{qk_0}}\sum_{\mu=0}^{q-1} R^{\mu k_0}\xi_i^{q-\mu}\xi_j^{q-\mu}\right) . \qquad (7.14)$$

Inserting (7.11) into (7.14) and using the (approximate) orthogonality of the random patterns, we find that (7.14) gives rise to the following contribution to the local field at time t:

$$
\begin{aligned}
h_i^{\mathrm{s}}(t) &= \sum_j J_{ij}^{\mathrm{s}}(t) S_j(t) = \sum_j J_{ij}^{\mathrm{s}}(t) \xi_j^q \\
&= \frac{\lambda q}{R} \Bigg[\left(\Delta t - T_{\mathrm{s}} \frac{1 - R^{k_0}}{1 - R^{q k_0}} \right) \xi_i^q \\
&\quad + \frac{\varepsilon T_{\mathrm{s}}}{q} \, \frac{1 - R^{k_0}}{1 - R^{q k_0}} \left(R^{k_0} \xi_i^{q-1} + R^{k_0 (q-1)} \xi_i^1 \right) \Bigg]
\end{aligned}
\tag{7.15a}
$$

while the static synapses (7.3) and (7.4) contribute

$$
h_i^{1,2}(t) = \sum_j (J_{ij}^{(1)} + J_{ij}^{(2)}) \xi_j^q = \xi_i^q + \varepsilon \xi_i^1 \, . \tag{7.15b}
$$

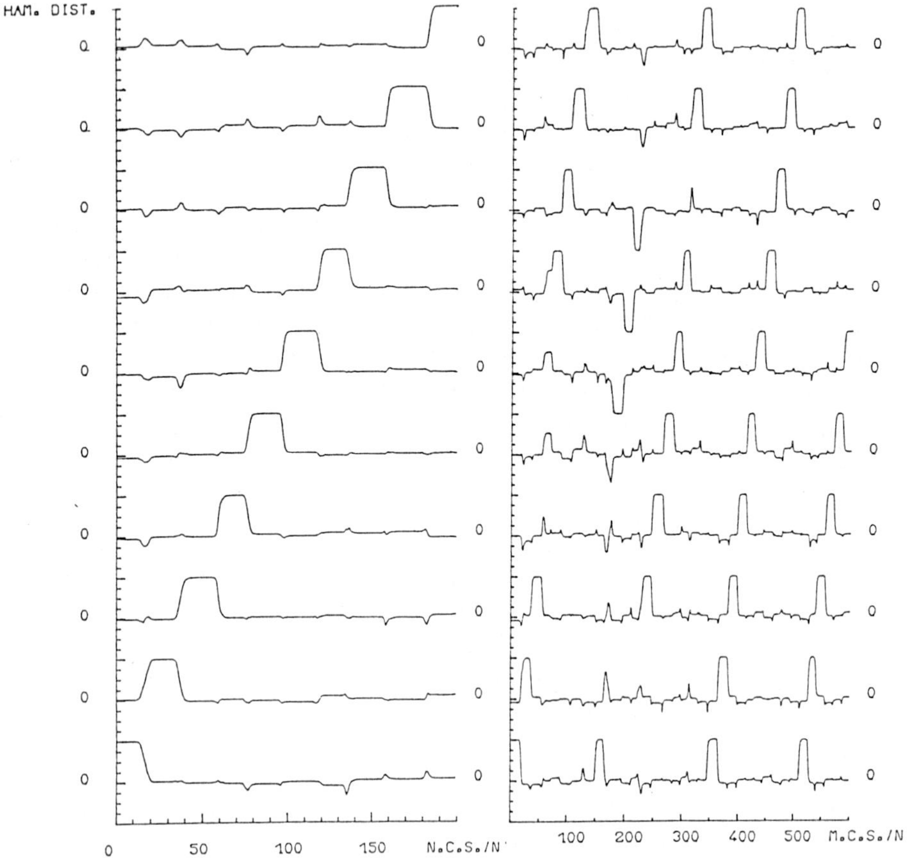

Fig. 7.2. Sequence generator after Peretto and Niez. Parameters of the simulation are $N = 400$, $q = 10$, $\varepsilon = 0.3$, and $\lambda^{-1} = 360$ MCS. Each track represents the time evolution of an overlap with one of the stored patterns. The measured persistence time $\Delta_0 \simeq 22$ MCS agrees reasonably well with the estimate $\Delta_0 \simeq 25$ MCS obtained from (7.17). The second figure has a different set of stored patterns. Fluctuations appear to be larger, and the system hesitates before it settles in the limit cycle. For the second figure one measures $\Delta_0 \simeq 16$ MCS, i.e., the agreement with the estimate is worse. Taken from [7.4]

Specializing, for simplicity, to the limit $q \gg 1$ and assuming $T_s \gg \Delta t$, we get a total local field of the form

$$h_i(t) = h_i^{1,2}(t) + h_i^s(t) \simeq \left(1 - \frac{\lambda q T_s}{R} \frac{1 - R^{k_0}}{1 - R^{q k_0}}\right) \xi_i^q + \varepsilon \xi_i^1 . \tag{7.16}$$

The expected transition into the state $\{\xi_i^1\}$ will indeed be initiated if the two contributions to $h_i(t)$ which stabilize patterns 1 and q, respectively, are of equal size, and the transition begins on those sites for which $\xi_i^1 = -\xi_i^q$. This gives a condition for k_0 which can be solved asymptotically for large q to yield the persistence time [7.18]

$$\Delta_0 = k_0 \Delta t \simeq -T_s \log\left(1 - \frac{1-\varepsilon}{\lambda q T_s}\right) \simeq \frac{1-\varepsilon}{\lambda q} . \tag{7.17}$$

Expression (7.17) is well corroborated by numerical simulations, as long as the number of patterns q is large compared to 1 ($q \simeq 10$ suffices) but still small compared to system size [7.18]. Figure 7.2 gives a typical simulation result, showing that the method of Peretto and Niez [7.4] is able to produce stable sequences and that the persistence time agrees reasonably well with what is expected from (7.17).[3]

As to the question of the *maximum* number of patterns in a cycle that the system can safely store, we are not aware of any systematic studies of this point.

Let us close this section with a brief remark on the modeling. The introduction of the T_j in (7.10) and (7.11) appears to be – and is – an unnecessary complication in the idea of Peretto and Niez. The performance of the model would not be impaired if the T_j^{-1} in (7.10–14) were replaced by the j-independent constant $\tilde{\lambda}/N$, and (7.17) would even remain unaltered (in the large-q limit) if $\tilde{\lambda}$ were of the form $\tilde{\lambda} = \lambda q$. The reader is invited to prove this for him- or herself.

7.2.2 Sequence Generation by Selection

In the majority of neural-network models so far considered, there is a sharp distinction between a learning phase and a so-called reproduction or retrieval phase. During learning, a set of patterns or nominated states is imposed upon the network and is used to establish the values of the synaptic couplings J_{ij} in such a manner that the nominated states become attractors of the neurodynamics during retrieval, when the J_{ij}s are held fixed.

The model of Peretto and Niez [7.4], although equipped with a set of synapses that undergo dynamic modifications during retrieval, does nevertheless *not* provide an exception to this two-phase structure. The point is that, after learning, an initial state at, say, $t = 0$ is completely specified by giving the $S_i(0)$ *and* the $J_{ij}^s(0)$, and *both* sets of variable then evolve in a manner that is completely

3 We thank P. Peretto for having supplied us with original figures and some missing information concerning the parameters of the simulation.

determined by the static synapses $J_{ij}^{(1)}$ and $J_{ij}^{(2)}$ which were fixed during the learning phase[4].

In the sequence-generating model of Dehaene et al. [7.5], no such strict distinction between a learning and a retrieval phase exists: the mechanisms by which a sequence of external stimuli is learnt continue to operate during the retrieval phase when the external stimuli are no longer imposed on the net. Autonomous reproduction of the learnt sequence is then expected to have a training effect leading to further optimization of recall.

The network of Dehaene et al. differs from Hopfield-type networks in that its fundamental entities are *neural clusters*, consisting of groups of neurons with mutually excitatory interactions. The fraction of active neurons in a cluster denotes its activity level which can thus be represented by a continuous variable S_i taking values between 0 and 1. Here i enumerates the clusters. Some of the clusters are assumed to be pattern specific and thus resemble grandmother cells: if active, they represent an item which might, for instance, be a symbol, a syllable, or a note.

Clusters are linked by so-called bundles of intercluster synapses (synaptic bundles for short) which serve to transmit activities between them. Synaptic bundles come in two variants: static and modulated. The static connectivity is such that each cluster excites itself with efficacy $V_{ii} = V_\alpha > 0$ – this mimics the effect of the mutual excitatory interactions between individual neurons within a cluster – and inhibits others with efficacy $V_{ij} = V_\beta < 0$, $i \neq j$. Mutually inhibitory interactions may, but need not, exist between every pair of clusters (i, j). The efficacies of the modulated bundles will be denoted by $W_{ij}(t)$.

The dynamics of the cluster activities S_i is given by

$$S_i(t + \Delta t) = \{1 + \exp[-h_i(t)]\}^{-1} , \tag{7.18}$$

where

$$h_i(t) = \sum_j V_{ij} S_j(t) + \sum_{j \in m(i)} W_{ij}(t) S_j(t) + N_i \tag{7.19}$$

is the local field at the ith cluster. In (7.19), the second sum only extends over the subset $m(i)$ of clusters which have a modulated connection with i, and N_i is a source of random noise, taking values in a band $[-n, n]$.

Dynamic modifications of modulated bundles $W_{ij}(t)$ are accomplished in so-called *synaptic triads* consisting of an anterior cluster j, a posterior cluster i, and a third cluster m of the net acting as modulating cluster (Fig. 7.3). The modulated bundle forms an excitatory connection from anterior to posterior cluster of the triad. Its efficacy is dynamically modified by the activity of the modulating cluster. The efficacy increases towards a maximum value $W_{ij}^m(t)$ if the activity of the modulating cluster exceeds the threshold 0.5, and decreases to zero otherwise.

---

[4] Of course there are also the T_j which enter the dynamic evolution of the $J_{ij}^{\mathrm{s}}(t)$; cf. (7.10). But they, too, are determined during learning alongside the static synapses and are, likewise, held fixed.

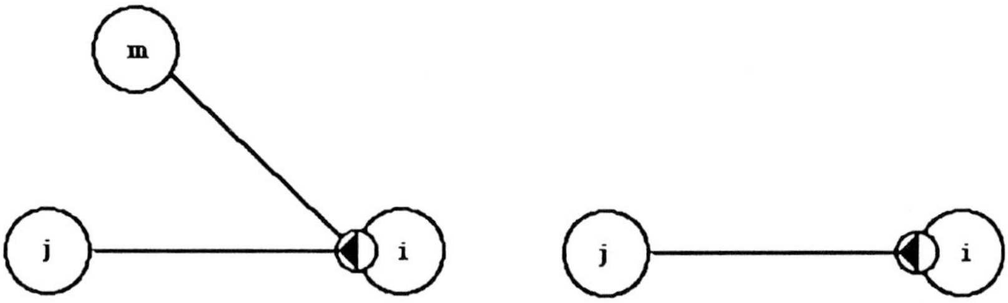

Fig. 7.3. Synaptic triad consisting of anterior cluster j, posterior cluster i and modulating cluster m. If anterior and modulating cluster are the same, one obtains a synaptic "diad"

More precisely,

$$W_{ij}(t+\Delta t) = \begin{cases} \alpha_p W_{ij}(t) + (1-\alpha_p) W_{ij}^m(t) & \text{, if } S_m(t) > 0.5 \text{ ,} \\ \alpha_d W_{ij}(t) & \text{, if } S_m(t) \leq 0.5 \text{ .} \end{cases} \tag{7.20}$$

Here $\alpha_p = \exp(-\Delta t/T_p)$ and $\alpha_d = \exp(-\Delta t/T_d)$ determine the rates of potentiation and decay, respectively. They are parameters of the network architecture.

Finally, the maximum efficacy $W_{ij}^m(t)$ of a modulated bundle in a synaptic triad is itself subject to change through a local Hebbian learning rule. This change proceeds as follows. If the modulated bundle has recently made a significant contribution to the postsynaptic potential of the posterior cluster i – significance being measured relative to the currently possible maximum – then the maximum efficacy of the modulated bundle is modified. It increases towards an absolute maximum W' if the posterior cluster was activated above a threshold, and decreases towards zero otherwise. Formally, if

$$W_{ij}(t-2\Delta t) S_j(t-2\Delta t) > 0.5 W_{ij}^m(t) \text{ ,} \tag{7.21a}$$

then

$$W_{ij}^m(t+\Delta t) = \begin{cases} \beta_1 W_{ij}^m(t) + (1-\beta_1) W' & \text{, if } S_i(t) > 0.5 \text{ ,} \\ \beta_2 W_{ij}^m(t) & \text{, if } S_i(t) \leq 0.5 \text{ .} \end{cases} \tag{7.21b}$$

This rule is expected to select synaptic triads *that stabilize ongoing activity* and eliminate those that perturb it. Again, β_1 and β_2 are fixed rate constants – parameters of the architecture – taking values between zero and one.

The dynamics is *synchronous* in all variables.

It is the synaptic triad which constitutes perhaps the most important element of the network architecture. Understanding it provides the key to understanding the potentialities of the model. A salient feature of the synaptic triads is their ability to *propagate activity* through the net. An active modulating cluster, for instance, gives rise to a potentiation of its modulated bundle. If the current maximum efficacy of this bundle is sufficiently large, the modulated bundle will eventually overcome the inhibitory static connection between anterior and posterior cluster. If the anterior cluster is active, it will be able to excite the posterior cluster. Activity of the latter will then, through lateral inhibition, suppress the activity of the former, and the whole process amounts to a propagation of activity from anterior to posterior cluster.

A modulated bundle whose anterior and modulator cluster are identical operates as a delay line. A ring of such "diadic" connections will propagate activity from one cluster to its successor in a cyclic fashion around the ring, and thus provide a temporal sequence of activity.

More complicated architectures consisting of triads and delay lines can be used to "buffer" previous states of given clusters so that they can be used for reprocessing later on. Such mechanisms allowing for longer memory spans are necessary for the production of complex sequences in which the current state does not uniquely specify its successor – as, for instance, in 1-2-1-3-1-2-... . (We shall return to complex sequences in Sect. 7.5 below.)

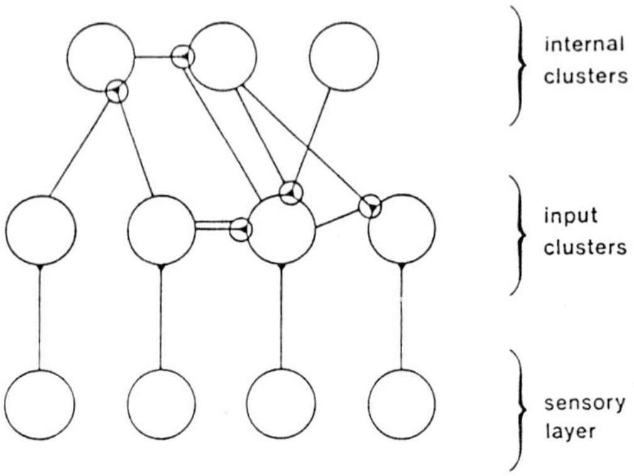

Fig. 7.4. Assembly of neural clusters after Dehaene et al. [7.5]. A structural division into a sensory layer, a layer of pattern-specific input/output clusters, and a layer of internal clusters is assumed. In the two upper layers, only the modulated bundles are shown. Synaptic triads and synaptic diads can be discerned clearly

In general, a set of (randomly chosen) couplings – static as well as dynamic – is assumed to exist before the network is first exposed to external stimuli. Figure 7.4 provides an illustration. These couplings, along with the rules for their dynamical modification, allow for a variety of attractors of the dynamics, called *prerepresentations*. Imposition of external stimuli *selects* among prerepresentations through the learning rule (7.21): Modulated bundles which systematically perturb the externally imposed signal are progressively weakened, while those that are in resonance with the stimulus are enhanced. This will eventually allow autonomous reproduction of the stimulus, *provided* it is compatible with the topology of the initial static and dynamic connectivity and with the time constants governing decay and potentiation of the modulated bundles as well as their maximum efficacies.

Selective learning of simple as well as complex sequences in architectures, compatible (in the above mentioned sense) with the learning task, was demonstrated by Dehaene et al. [7.5] through numerical simulations (Fig. 7.5).

Because of the complicated dynamics (7.18–21), analytical results are in general hard to obtain. Some progress, however, is possible – if only at the cost of further simplifying assumptions.

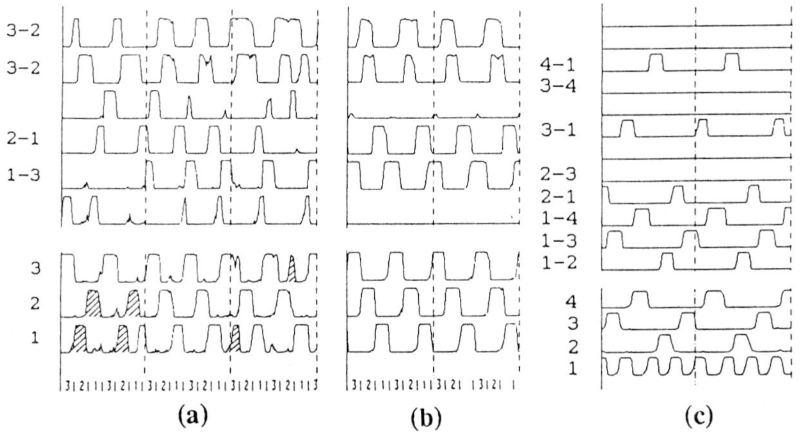

Fig. 7.5a–c. Learning (a) and retrieval (b) of a linear sequence. Each trace represents the time evolution of one cluster activity (vertical lines are drawn every 100 update cycles). *Bottom traces*: input clusters. *Top traces*: internal clusters. Numbers at the bottom represent the sequence imposed on the perceptual layer. During learning, periods of resonance (*plain*) and dissonance (*striped*) are clearly visible in the input/output units. After learning, dissonant responses have been eliminated and the system completes sequences where items are missing.(c) Retrieval of the complex sequence -1-2-1-3-1-4-... . Taken from [7.5]

Sulzer [7.18] has considered a ring of q delay lines, with static connectivity taken – as usual – to be self-excitatory and mutually inhibitory. For suitable initial conditions of the modulated bundles and cluster activities – e.g., $W_{i,i+1}(t=0) = W_0$, $W^m_{i,i+1}(t=0) = W^m_0$, and $S_i(t=0) = \delta_{i,i_0}$, for $1 \le i \le q$ and $W^m_0 > |V_\beta|$ – activity will spontaneously propagate around the ring. An approximation to the asymptotic propagation velocity or, equivalently, the persistence time can be obtained in the limit where the decay constant β_2 for the maximum efficacies $W^m_{ij}(t)$ is very close to unity [7.18]. For $1 - \beta_2 \ll 1$, one readily verifies that, owing to the cyclic propagation of activity, the learning rule (7.21) will soon drive the $W^m_{i,i+1}(t)$ towards an almost constant value very close to the absolute maximum W'. This process manifests itself in an increase of the propagation velocity towards its asymptotic value, which is solely determined by the remaining network parameters α_p, α_d, V_α, V_β, and W' – in colloquial terms: the network gets used to producing the temporal sequence (see Fig. 7.6).

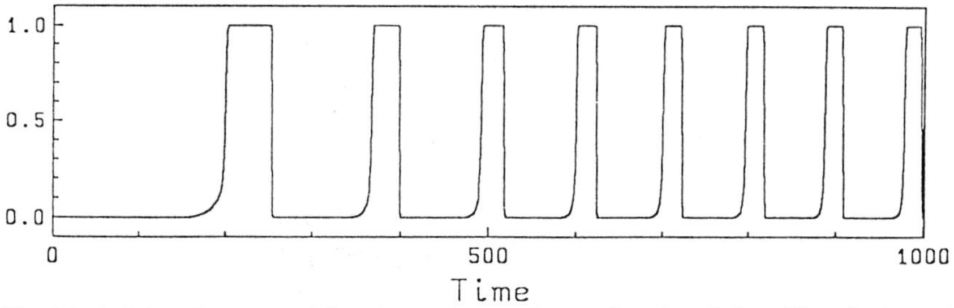

Fig. 7.6. Activity of one out of five clusters in a cycle as a function of time. Note the increasing frequency owing to the increase of the signal propagation velocity. Taken from [7.18]

In the stationary regime, the simplifying assumption $1-\beta_2 \ll 1$ allows us to put $W^m_{i,i+1}(t) \simeq W'$ for all i and thereby to ignore the dynamics of the $W^m_{ij}(t)$ as prescribed by the learning rule (7.21). If the system produces a stable temporal sequence with persistence time Δ_0 for the individual cluster, each modulated bundle must satisfy a periodicity condition of the form

$$W_{i,i+1}(t+q\Delta_0) = W_{i,i+1}(t) \ , \tag{7.22}$$

and an analogous relation must hold for each of the cluster activities S_i.

To proceed, we fix i and approximate the cluster activity $S_i(t)$ by the function

$$S_i(t) \simeq \begin{cases} 0 & \text{, if} \quad nq\Delta_0 < t < [(n+1)q-1]\Delta_0 \ , \\ 1 & \text{, if} \quad [(n+1)q-1]\Delta_0 < t < (n+1)q\Delta_0 \ , \end{cases} \tag{7.23}$$

where $n \in \mathbb{N}$. Equation (7.23) allows us to solve the time evolution (7.20) for the modulated bundle $W_{i,i+1}(t)$ for which S_i acts as the modulating cluster. Let us denote by W_θ the threshold value of $W_{i,i+1}(t)$ at $t = nq\Delta_0$, when $S_i(t)$ decreases below the threshold 0.5. Then, using (7.20) with $W^m_{i,i+1}(t) \equiv W'$ and (7.23), we get

$$W_{i,i+1}((n+1)q\Delta_0) = W_\theta \alpha_{\rm p}^{k_0} \alpha_{\rm d}^{(q-1)k_0} + (1-\alpha_{\rm p}^{k_0})W' = W_\theta \ , \tag{7.24}$$

where k_0 measures the persistence time in units of the elementaty time step; $\Delta_0 = k_0\Delta t$. The second half of the equation follows from the periodicity condition (7.22). If W_θ – the threshold value of $W_{i,i+1}(t)$ at the time when $S_i(t)$ decreases below the value 0.5 – were known, we could solve (7.24) for the unknown k_0. Unfortunately, we do not know W_θ. A way out [7.18] consists in replacing the response function $[1+\exp(-h_i)]^{-1}$ in (7.18) by a piecewise linear one approximating it, which allows the dynamics of the $S_i(t)$ to be followed analytically so that there is no need assuming (7.23). The full analysis is rather involved, so we shall choose a simpler way out. We *measure* k_0 in a simulation for one set of the parameters $\alpha_{\rm p}$, $\alpha_{\rm d}$, V_α, V_β, and W'. This determines W_θ from (7.24). There is no reason to expect a significant dependence of W_θ on $\alpha_{\rm p}$ and $\alpha_{\rm d}$ within reasonable ranges of these parameters, but we may expect W_θ to be proportional to V_β. To see this, we note that the decrease of $S_i(t)$ is caused by the increasing activity of cluster $i+1$ through lateral inhibition. For S_{i+1} to increase, the efficacy of the modulated bundle must have compensated the inhibitory forward connection $V_{i,i+1} = V_\beta$, hence the proportionality $W_\theta \propto V_\beta$. The dependence upon V_α is expected to be less important. Thus a single measurement of the persistence time allows us to predict its variations with $\alpha_{\rm p}$, $\alpha_{\rm d}$, V_α, and V_β. Figure 7.7 checks our heuristics against Sulzer's data [7.18]. The agreement is reasonable.

We now summarize the basic features of the model. Synaptic triads allowing for heterosynaptic regulation of modulated synaptic bundles constitute the key element of the architecture. They have the ability to propagate activity through the net. Learning essentially modifies maximum efficacies of modulated bundles,

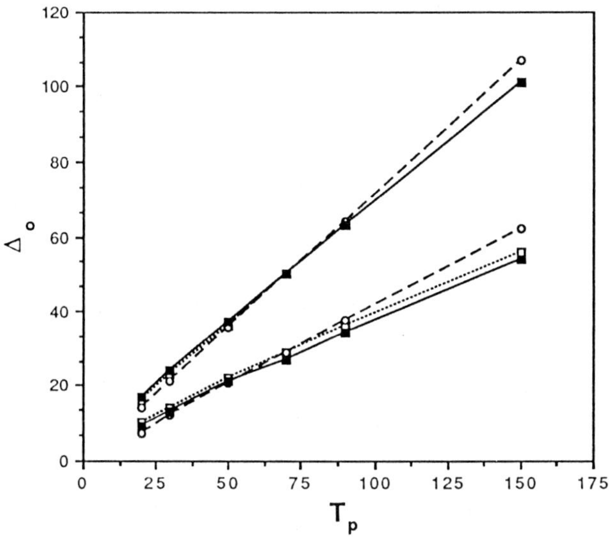

Fig. 7.7. Persistence times for a ring of delay lines. Comparison of simulation results (*full squares*) with our heuristic estimates (*open circles*), based on a single measurement at $T_p = 70$ and $V_\alpha = -V_\beta = 9$, and the refined estimates of [7.18] (*open squares*). The upper set of curves is for $V_\alpha = -V_\beta = 9$, the lower for $V_\alpha = -V_\beta = 6$

not time constants. An initial static and dynamic connectivity is assumed to exist before the network is exposed to external stimuli. This connectivity allows for a variety of attractors of the dynamics called prerepresentations. Learning induced by exposure to external stimuli merely *selects* among them. There is no strict separation between learning and retrieval phases in that the learning mechanisms continue to operate after the teaching stimuli are withdrawn, and autonomous reproduction can have a further training effect. As a result, however, *true learning* in the sense of acquiring genuinely new capabilities *never* occurs. *All* potentialities are prewired – i.e., implicitly present – and are merely made explicit by appropriate external stimuli. On a formal level, therefore, the *combination* of "learning" *and* autonomous reproduction in the present model is closer to what we are used to regard as retrieval phase proper in other models.

Because of its complicated dynamics, the model is not easy to analyze theoretically. We have, for instance, made no comments about the role and nature of the external noise in (7.11), and [7.5] gives only a few hints. With simplifying assumptions about time constants, a ring of delay lines can be analyzed in detail [7.18], and estimates for the signal propagation velocity can be obtained.

Aspects of selective learning can be put in correspondence with observations made, for instance, in the song-acquisition behavior of birds [7.5]. Dehaene et al. also report neurophysiological evidence in favor of mechanisms for heterosynaptic regulation and pattern-specific neurons underlying their modeling. Whether such evidence is compelling, we cannot judge.

7.3 Noise-Driven Sequences of Biased Patterns

We have seen in the previous section that dynamically modifiable synapses can be invoked as a mechanism for producing (and "learning") temporal sequences in associative neural networks. Fast synaptic plasticity, however, does not provide the only possible mechanism for temporal association, and the remainder of the present article is accordingly devoted to alternatives.

While the approaches to be discussed in subsequent chapters utilize asymmetric synapses which transmit activity with time delays to produce stable temporal sequences, the model of Buhmann and Schulten [7.6, 7], to which we now turn, operates with synapses that have *no temporal features at all.*

The model is designed to store and reproduce a sequence of so-called low-activity patterns, i.e., in each pattern of the sequence only a small fraction of the neurons is assumed to be active while the large majority of them remains quiescent. Buhmann and Schulten use a 0-1 representation to describe the non-firing and firing states of the neurons.

Key ingredients of the model are (i) asymmetric interactions and (ii) the presence of (thermal) noise. To incorporate noise, the neurons are updated asynchronously according to the *probabilistic* rule

$$\begin{aligned}&\mathrm{Prob}\{S_i(t+\Delta t)=S\}\\&\quad = S f(h_i(t)) + (1-S)[1-f(h_i(t))]\ , \quad S\in\{0,1\}\ ,\end{aligned} \tag{7.25}$$

where $f(h_i)=\{1+\exp[-(h_i-U)/T]\}^{-1}$ is the probability that neuron i fires ($S_i=1$) at time $t+\Delta t$, if it experiences the local field $h_i(t)=\sum_j J_{ij}S_j(t)$. The parameter U is a threshold potential, while T denotes the temperature, which quantifies the level of noise in the system. Equation (7.25) is a probabilistic version of (7.1), a Glauber dynamics, adapted to a 0-1 representation.

The patterns to be stored will be denoted by[5] $\{\zeta_i^\mu; 1\le i\le N\}$, $1\le\mu\le q$, with $\zeta_i^\mu=1$ ($\zeta_i^\mu=0$) if neuron i is active (quiescent) in pattern μ. Storage of these patterns is achieved by the following sets of synapses. First, there are symmetric, purely excitatory connections between all neurons which fire simultaneously in at least one of the q patterns. These are given by

$$J_{ij}^0=\sum_{\mu=1}^{q}\varepsilon^\mu\zeta_i^\mu\zeta_j^\mu\ , \tag{7.26}$$

where $\varepsilon^\mu=1/n^\mu$, with $n^\mu=\sum_i\zeta_i^\mu$, is a normalization constant. With this choice for the interactions, *all* patterns $\{\zeta_i^\mu\}$ which have a sufficiently large overlap with the initial state of the network will eventually be excited, i.e., the neurons supporting them will fire. In order to prevent such a simultaneous excitation of several patterns, an inhibitory contribution to the couplings is introduced,

[5] We chose a different letter – ζ instead of ξ – in order to indicate the different representation (0-1 instead of ±1).

$$J_{ij}^{\mathrm{I}} = -\gamma \frac{q}{N} \sum_{\mu=1}^{q} \sum_{\substack{\nu=1 \\ |\mu-\nu|>1}}^{q} \zeta_i^\mu \zeta_j^\nu \,, \tag{7.27}$$

where γ denotes the average strength of the inhibition. The particular form of J_{ij}^{I} provides competition between all neurons which fire in different patterns, *except* when these patterns follow each other in the sequence 1-2-...-q. In order to induce transitions between successive patterns in the sequence, an additional set of excitatory forward projections and inhibitory backward interactions

$$J_{ij}^{\mathrm{P}} = \sum_{\mu=1}^{q} \zeta_i^\mu (\varepsilon^{\mu-1} \alpha^{\mu-1} \zeta_j^{\mu-1} - \varepsilon^{\mu+1} \beta^{\mu+1} \zeta_j^{\mu+1}) \tag{7.28}$$

is introduced. Their effect is to excite pattern $\mu + 1$ when the system is in state μ and to inhibit μ when $\mu + 1$ is sufficiently present. The total synaptic strength J_{ij} is then taken to be

$$J_{ij} = \begin{cases} J_{ij}^0 & \text{, if} \quad J_{ij}^0 \neq 0 \,, \\ J_{ij}^{\mathrm{P}} & \text{, if} \quad J_{ij}^0 = 0 \quad \text{and} \quad \sum_{\mu=1}^{q} \zeta_i^\mu (\zeta_j^{\mu-1} + \zeta_j^{\mu+1}) \neq 0 \,, \\ J_{ij}^{\mathrm{I}} & \text{, if} \quad \sum_{\mu=1}^{q} \sum_{\nu=1; |\mu-\nu| \leq 1}^{q} \zeta_i^\mu \zeta_j^\nu = 0 \,. \end{cases} \tag{7.29}$$

That is, all neurons which fire simultaneously in at least one pattern have mutually excitatory synapses; all neurons which do not fire simultaneously in any pattern, but do so in successive patterns, receive excitatory forward and inhibitory backward projections; all neurons which fire only in different nonconsecutive patterns are connected by inhibitory synapses. Finally, neurons which do not fire in any pattern receive no connection at all and can be ignored in what follows.

Buhmann and Schulten verify consecutive recall of a sequence of biased random patterns through Monte Carlo simulations. Figure 7.8 shows the evolution of the magnetizations[6] $m(\mu; t) = \varepsilon^\mu \sum_i \zeta_i^\mu S_i(t)$, which measure the overlap of the current network state $\{S_i(t)\}$ with the pattern $\{\zeta_i^\mu\}$. Note the fluctuations in the persistence times for different patterns. These are due to the random overlaps between consecutive patterns, which in turn are due to finite-size effects.

An analytical description is possible in the case of *disjoint* patterns, where each neuron fires in at most one pattern, so that $\varepsilon^\mu \sum_i \zeta_i^\mu \zeta_i^\nu = \delta_{\mu,\nu}$. In this case the couplings (7.29) can be expressed in a simple closed form,

$$J_{ij} = \sum_{\mu=1}^{q} \zeta_i^\mu \left(\varepsilon^\mu \zeta_j^\mu + \varepsilon^{\mu-1} \alpha^{\mu-1} \zeta_j^{\mu-1} - \varepsilon^{\mu+1} \beta^{\mu+1} \zeta_j^{\mu+1} - \gamma \frac{q}{N} \sum_{\mu=1}^{q} \sum_{\substack{\nu=1 \\ |\mu-\nu|>1}}^{q} \zeta_j^\nu \right). \tag{7.30}$$

[6] Here our notation differs from that of [7.6] and [7.7] This is to make closer contact with a theoretical concept to be introduced in Sect. 7.4, the idea of (sublattices and) sublattice magnetizations, of which the $m(\mu; t)$ only constitute a special case.

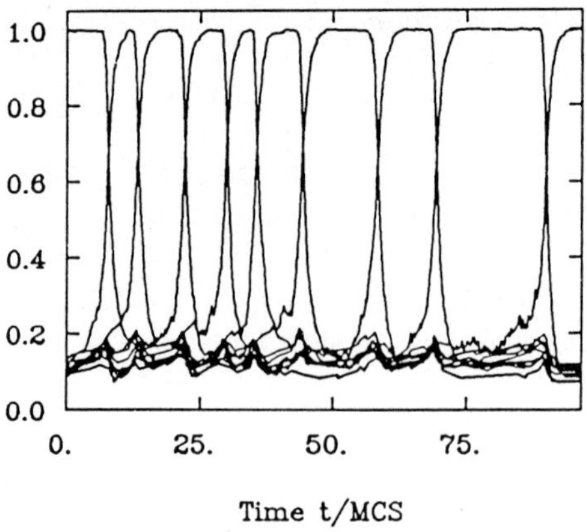

Fig. 7.8. Sequential recall of 10 random biased patterns, each with activity 0.1. The curves show the magnetizations $m(\mu; t)$ as functions of time. The network consists of $N = 3000$ neurons. The remaining network parameters are $\alpha^\mu = 0.15$, $\beta^\mu = 2.0$, $\gamma = 2.0$, $T = 0.075$, and $U = 0.35$. Taken from [7.7]

Besides allowing for a simple closed expression for the J_{ij}, the case of disjoint patterns has two further simplifying features. One is that the system is a *disjoint* decomposition of the subsets (or sublattices) $I(\mu)$ of neurons supporting the different patterns,

$$I(\mu) = \{i; \zeta_i^\mu = 1\}\ , \quad 1 \leq \mu \leq q\ . \tag{7.31}$$

The other simplification is that the local field $h_i(t)$ can be expressed in terms of the corresponding sublattice magnetizations

$$m(\mu; t) = \frac{1}{n^\mu} \sum_{i=1}^{N} \zeta_i^\mu S_i(t) = \frac{1}{n^\mu} \sum_{i \in I(\mu)} S_i(t)\ . \tag{7.32}$$

One finds

$$h_i = \sum_{\mu=1}^{q} \zeta_i^\mu \Bigg[m(\mu; t) + \alpha^{\mu-1} m(\mu-1; t) - \beta^{\mu+1} m(\mu+1; t) - \gamma \frac{q}{N} \sum_{\substack{\nu=1 \\ |\mu-\nu|>1}}^{q} n^\nu m(\nu; t) \Bigg]\ . \tag{7.33}$$

Since the patterns are disjoint, there is for each i only a *single* ζ_i^μ which differs from zero, namely, the one which carries the index of the pattern supported by i. As a consequence, $h_i(t)$ – and thus the spike probability $f(h_i(t))$ – depends on i *only* through the index μ of the sublattice $I(\mu)$ to which i belongs,

$$h_i(t) = h(\mu; t) = m(\mu; t) + \alpha^{\mu-1} m(\mu-1; t) - \beta^{\mu+1} m(\mu+1; t) - \gamma \frac{q}{N} \sum_{\substack{\nu=1 \\ |\mu-\nu|>1}}^{q} n^\nu m(\nu; t)\ , \tag{7.34}$$

whatever $i \in I(\mu)$. Note that the $h_i(t)$ – thus the $f(h_i(t))$ – only depend on the sublattice magnetizations $m(\mu; t)$. The dynamics of the network can therefore be described completely in terms of the sublattice magnetizations alone.

Since the dynamics (7.21) is of a stochastic (Markovian) nature, the appropriate level of description is in terms of a master equation for the probability $\varrho(\boldsymbol{m}; t)$ of finding the system at time t in a state with magnetizations $\boldsymbol{m} = \{m(\mu)\}_{\mu=1}^{q}$. The $m(\mu)$ can assume the values $0, \varepsilon^\mu, 2\varepsilon^\mu, \ldots, 1$. An asynchronous update affects only a single $m(\mu)$ and can change it by at most $\pm\varepsilon^\mu = \pm 1/n^\mu$. Keeping track of the various elementary processes which can change the $m(\mu)$ one finds

$$\begin{aligned}
&\varrho(\boldsymbol{m}; t+\Delta t) - \varrho(\boldsymbol{m}; t) \\
&\quad = \sum_{\mu=1}^{q} \frac{n^\mu}{N} \big[\varrho(\boldsymbol{m}+\boldsymbol{\varepsilon}^\mu; t)(m(\mu)+\varepsilon^\mu)(1 - f^\mu(\boldsymbol{m}+\boldsymbol{\varepsilon}^\mu)) \\
&\quad + \varrho(\boldsymbol{m}-\boldsymbol{\varepsilon}^\mu; t)(1-(m(\mu)-\varepsilon^\mu)) f^\mu(\boldsymbol{m}-\boldsymbol{\varepsilon}^\mu) \\
&\quad - \varrho(\boldsymbol{m}; t) m(\mu)(1 - f^\mu(\boldsymbol{m})) - \varrho(\boldsymbol{m}; t)(1 - m(\mu)) f^\mu(\boldsymbol{m})\big] . \qquad (7.35)
\end{aligned}$$

Here we have introduced the vector $\boldsymbol{\varepsilon}^\mu = \{\varepsilon^\mu \delta_{\mu,\nu}\}_{\nu=1}^{q}$ and $f^\mu(\boldsymbol{m}) = f(h(\mu))$. The terms on the right-hand side of (7.35) correspond to a pair of *ingoing* and a pair of *outgoing* terms, the two terms of each pair corresponding to processes which *de*crease and *in*crease the magnetization, respectively.

The master equation (7.35) can be rewritten in terms of the differential equation [7.6, 7]

$$\begin{aligned}
&\left[\exp\left(\Delta t \frac{\partial}{\partial t}\right) - 1\right] \varrho(\boldsymbol{m}; t) = \sum_{\mu=1}^{q} \frac{n^\mu}{N} \left\{ -\sinh\left(\varepsilon^\mu \frac{\partial}{\partial m(\mu)}\right)(F^\mu \varrho) \right. \\
&\quad \left. + \left[\cosh\left(\varepsilon^\mu \frac{\partial}{\partial m(\mu)}\right) - 1\right](D^\mu \varrho) \right\} \qquad (7.36)
\end{aligned}$$

with

$$F^\mu = -\big(m(\mu) - f^\mu(\boldsymbol{m})\big)$$

and

$$D^\mu = m(\mu) + f^\mu(\boldsymbol{m}) - 2m(\mu) f^\mu(\boldsymbol{m}) .$$

For asynchronous dynamics, the elementary time step Δt must scale with system size N according to $\Delta t = (\Gamma N)^{-1}$ to ensure that each neuron is updated at a finite N-independent rate. If, in the limit of large N ($N \to \infty$) and large patterns ($n^\mu \to \infty$ or $\varepsilon^\mu \to 0$), terms of order $(\varepsilon^\mu)^2$ and terms of order N^{-2} and smaller are neglected, one obtains the second-order differential equation[7]

[7] This equation differs from Eq. (9) of [7.7] by the second-order term on the left-hand side. In general, this term must be kept since its coefficient may be of the same order of magnitude as ε^μ. To see this, note that $\varepsilon^\mu = q/N$ for nonoverlapping patterns of equal size, so that the assertion follows if q remains finite, as $N \to \infty$. Only if the limit $q \to \infty$, $N \to \infty$, with $q/N \to 0$ is considered will there be a range of N where one may drop the second-order time derivative on the left-hand side of (7.37) while keeping the diffusion term on its right-hand side.

$$\frac{\partial \varrho}{\partial t} + (\Gamma N)^{-1}\frac{\partial^2 \varrho}{\partial t^2} = \Gamma \sum_{\mu=1}^{q} \left\{ -\frac{\partial}{\partial m(\mu)}(F^\mu \varrho) + \frac{\varepsilon^\mu}{2}\frac{\partial^2}{\partial m(\mu)^2}(D^\mu \varrho) \right\} . \tag{7.37}$$

In the limit of infinitely large patterns ($\varepsilon^\mu = 1/n^\mu = 0$), the second-order terms in (7.37) vanish so that the sublattice magnetizations $m(\mu;t) \equiv \langle m(\mu) \rangle_{\varrho(m;t)}$ obey the deterministic kinetic equation [7.6, 7, 19]

$$\Gamma^{-1}\frac{\mathrm{d}m(\mu;t)}{\mathrm{d}t} = -\left[m(\mu;t) - f^\mu(\boldsymbol{m}(t))\right] . \tag{7.38}$$

Equation (7.38) is exact in the thermodynamic limit and for nonoverlapping patterns (with $n^\mu \to \infty$). We note in passing that in the case of parallel dynamics one would obtain the discrete evolution equations

$$m(\mu;t+\Delta t) = f^\mu(\boldsymbol{m}(t)) \tag{7.39}$$

instead of (7.38) which for nonoverlapping patterns (with $n^\mu \to \infty$) is likewise exact. (A sketch of the argument, which is based on the strong law of large numbers, can be found in Sect. 5 of [7.20]. There it is formulated for a ± 1 representation of the neurons and for a more general notion of a sublattice, but the argument applies equally well to the case at hand.)

An intuitive understanding of the transition mechanism and a feeling for the required settings of the various parameters, $\alpha^\mu, \beta^\mu, \gamma, U$, and T of the model is in fact easier to obtain in the case of parallel dynamics as described by (7.39). Let us assume that at time t the system is in pattern μ and that it is about to embark upon a transition to $\mu+1$, so that we have $m(\mu;t) \simeq 1$, while $m(\mu+1;t) \ll 1$. Moreover, we assume $m(\nu;t) \simeq 0$ for $\nu \notin \{\mu, \mu+1\}$. With these initial conditions (7.34) gives

$$\begin{aligned}
h(\mu-1;t) &\simeq -\beta^\mu m(\mu;t) - \gamma m(\mu+1;t) \\
h(\mu;t) &\simeq m(\mu;t) - \beta^{\mu+1} m(\mu+1;t) \\
h(\mu+1;t) &\simeq \alpha^\mu m(\mu;t) + m(\mu+1;t) \\
h(\mu+2;t) &\simeq -\gamma m(\mu;t) \\
h(\nu;t) &\simeq -\gamma m(\mu;t) - \gamma m(\mu+1;t) , \quad \nu \notin \{\mu-1, \ldots, \mu+2\} .
\end{aligned} \tag{7.40}$$

Here we have specialized to nonoverlapping patterns of equal size, for which $qn^\mu/N = 1$. According to (7.39), we have $m(\nu;t+\Delta t) = f^\nu(\boldsymbol{m}(t)) = f(h(\nu;t))$, with the $h(\nu;t)$ given above, and $f(h(\nu;t)) = \{1 + \exp[-(h(\nu;t) - U)/T]\}^{-1}$. To achieve a controlled transition we want f^μ to stay close to unity for several time steps, during which $f^{\mu+1}$ is allowed to increase, albeit slowly; the f^ν with ν not involved in the transition should remain close to zero. This requires that the forward excitations α^μ should be small relative to unity so that there is a range for U for which initially (i.e., while $m(\mu;t) \simeq 1$ and $m(\mu+1;t) \ll 1$) one has $h(\mu;t) - U \simeq 1 - U > 0$ while $h(\mu+1;t) - U \simeq \alpha^\mu - U < 0$. For $\alpha^\mu \approx 0.1$, taking $U \approx 0.5$ would do. By choosing the temperature small as compared to unity, say $T \approx 0.1$, one ensures that initially $m(\mu;t)$ and $m(\mu+1;t)$ are indeed, as assumed, close to 1 and 0, respectively. On the other hand, given α^μ, T may not be too small (noise must be sufficiently strong) so that pattern $\mu + 1$ *can* be excited. To see this, assume parameter settings for α^μ and U as indicated and take $T \ll 1$. This would imply that initially $(h(\mu + 1; t) - U)/T \ll 0$ so that $m(\mu + 1; t)$ would *never* get off the ground – the system would simply stay in pattern μ. Finally, to ensure suppression of the patterns not involved in the transition $\mu \to \mu + 1$, it suffices to take $\gamma \approx 1$, and $\beta^\mu \approx 1$ will be appropriate to inhibit μ, once $m(\mu + 1; t)$ has reached the threshold $1 - U$. Figure 7.9 shows a simulation result of [7.7], with parameter settings that agree reasonably well with what is required according to our qualitative reasoning. We now return to asynchronous dynamics.

According to the foregoing discussion, the transition from pattern μ to $\mu + 1$ involves $m(\mu; t)$ and $m(\mu + 1; t)$ only. This allows us to investigate details of the transition by projecting the dynamics (7.38) into the subspace spanned by $m(\mu; t)$ and $m(\mu+1; t)$. Buhmann and Schulten exploit this possibility to obtain an analytical expression for the time needed for the transition from $(m(\mu; t), m(\mu + 1; t)) = (1, 0)$ to (0,1) as follows.

For the first part of the transition, i.e., while $m(\mu; t) \simeq 1$, the right-hand side of the differential equation (7.38) – more specifically, its $\mu + 1^{\text{st}}$ component – is replaced by an expansion of $F^{\mu+1}|_{m(\mu)\equiv 1} = -\big(m(\mu + 1) - f^{\mu+1}(\boldsymbol{m})\big)|_{m(\mu)\equiv 1}$ around its minimum

$$\phi^{\mu+1} \equiv F^{\mu+1}_{\text{min}}|_{m(\mu)\equiv 1} = -m_{\text{min}}(\mu + 1) + \frac{2T}{1 + \sqrt{1 - 4T}} \tag{7.41}$$

at

$$m_{\text{min}}(\mu + 1) = U - \alpha^\mu - 2T \operatorname{arcosh}(1/2\sqrt{T}) \; . \tag{7.41}$$

The approximate differential equation is then integrated up to a time t^* where the assumption $m(\mu; t) = 1$, equivalent to $f^\mu(\boldsymbol{m}) \approx 1$, fails. This gives

$$\begin{aligned} t^* = \frac{\Gamma^{-1}}{\sqrt{\phi^{\mu+1}\omega}} \Bigg[&\arctan\left(\sqrt{\frac{\omega}{\phi^{\mu+1}}}\, m_{\text{min}}(\mu + 1)\right) \\ &+ \arctan\left(\sqrt{\frac{\omega}{\phi^{\mu+1}}}\left(\frac{1 - U - B_0 T}{\beta^{\mu+1}} - m_{\text{min}}(\mu + 1)\right)\right)\Bigg] \end{aligned} \tag{7.43}$$

if we define failure of the condition $f^\mu(\boldsymbol{m}) \approx 1$ by $f^\mu(\boldsymbol{m}) = [1 + \exp(-B_0)]^{-1}$ or, equivalently, by $(h(\mu; t^*) - U)/T|_{m(\mu)\equiv 1} = (1 - \beta^{\mu+1} m(\mu+1; t^*) - U)/T = B_0$ for some threshold B_0. Buhmann and Schulten choose $B_0 = 2$, or equivalently $f^\mu(\boldsymbol{m}) \simeq 0.88$. In (7.43), $\omega = (F^{\mu+1})''/2|_{m(\mu)=1, m_{\text{min}}(\mu+1)} = \sqrt{1 - 4T}/2T$. For times $t > t^*$ the time evolution is then approximated by a free relaxation of $m(\mu; t)$ and $m(\mu + 1; t)$ to their asymptotic values $(f^\mu, f^{\mu+1}) \simeq (0, 1)$ so that $m(\mu; t) = \exp[-\Gamma(t - t^*)]$ and $m(\mu+1; t) = 1 - (1 - m(\mu+1; t^*)) \exp[-\Gamma(t - t^*)]$ where – by definition[8] – $m(\mu + 1; t^*) = (1 - U - B_0 T)/\beta^{\mu+1}$. The transition time

[8] Reference [7.7] contains a minor error here in that the free relaxation of $m(\mu + 1; t)$ is assumed to start at zero instead of at $m(\mu + 1; t^*)$, as would be more adequate.

$t_{\mu\to\mu+1}$ is defined as the time span between the moments when $m(\mu+1;t)$ starts to grow and when, subsequently, $m(\mu+2;t)$ starts to grow as the minimal force

$$F_{\min}^{\mu+2}|_{m(\mu;t),m(\mu+1;t)} = -m_{\min}(\mu+2) + \frac{2T}{1+\sqrt{1-4T}} \tag{7.44}$$

with

$$m_{\min}(\mu+2) = U + \gamma m(\mu;t) - \alpha^{\mu+1} m(\mu+1;t) - 2T \operatorname{arcosh}(1/2\sqrt{T}) \tag{7.45}$$

becomes positive. Inserting the time evolutions of $m(\mu;t)$ and $m(\mu+1;t)$ for $t > t^*$, one obtains[9]

$$\Delta_0 = t_{\mu\to\mu+1} = t^* + \Gamma^{-1} \ln \left[\frac{\gamma + \alpha^{\mu+1}(1 - m(\mu+1;t^*))}{\phi^{\mu+1}} \right] . \tag{7.46}$$

Figure 7.10 shows the transition (or persistence) times as a function of temperature for various α^μ. Given α^μ, the persistence times increase as T decreases, and they diverge at a critical temperature T^* defined by the condition $\phi^{\mu+1} = 0$ (which is in accord with our qualitative analysis given above). For $T < T^*$, no transitions will occur, and the system will remain stable in a given pattern state. It turns out, though, that the predictions of (7.46) are numerically not very precise. For example, the simulation presented in Fig. 7.9 gives $\Delta_0 \simeq 13$ MCS, while (7.46) would predict $\Delta_0 \simeq 4$ MCS. Discrepancies are presumably due to the fact that phases of the transition where both $m(\mu)$ and $m(\mu+1)$ differ appreciably from their asymptotic values before or after the transition are not captured very well by the above analytic description[10]. Orders of magnitude for Δ_0 and the divergence at low noise levels, however, do come out correctly. Figure 7.10 suggests that one possible mode of operation of the present setup might be to induce transitions at will by *externally* regulating the temperature.

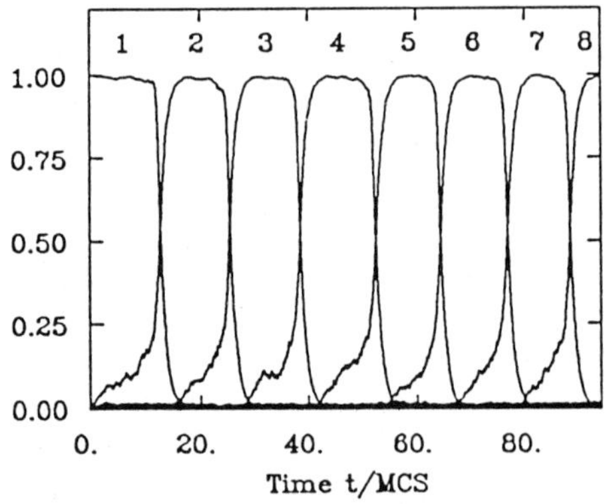

Fig. 7.9. Simulation of a sequence of 8 non overlapping patterns with $n^\mu = 1000$, $\alpha^\mu = 0.1$, $\beta^\mu = 1.0$, $\gamma = 0.1$, $T = 0.1$, and $U = 0.35$. Taken from [7.7]

---

[9] The following equation differs in two repects from the corresponding equation (14) of [7.7]. (i) We have specialized to nonoverlapping patterns of equal size, for which $qn^\mu/N = 1$. (ii) More adequate initial conditions for the relaxation process in the second part of the transition are used; see Footnote 8.

[10] We thank J. Buhmann for helpful correspondence on this point.

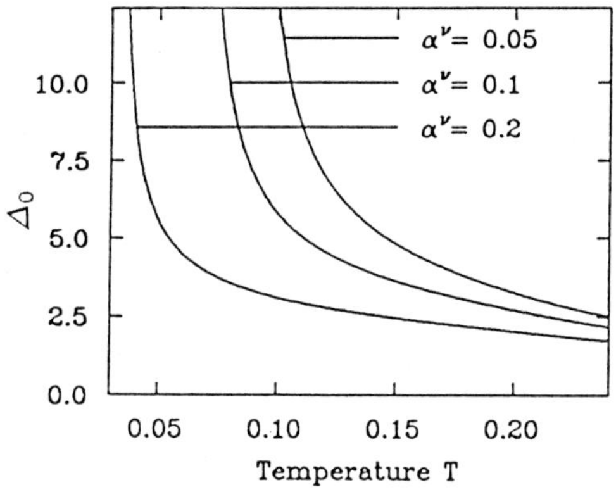

Fig. 7.10. Persistence times as a function of temperature for a network with α^μ = 0.1, β^μ = 1.0, γ = 0.2, and U = 0.35. Taken from [7.7]

The theory for $\varepsilon^\mu = 1/n^\mu = 0$ may be regarded as an approximation for finite systems [7.7]. If ε^μ is small but nonzero, one may return to the Fokker–Planck-type equation (7.37) whose solution (for $q \gg 1$; see Footnote 6) yields a diffusive $\sqrt{t}$ broadening of the probability distribution for the $m(\mu)$. For further details, we refer to [7.6] and [7.7].

7.4 Stabilizing Sequences by Delays

Let us return to the starting point of our journey: unbiased binary random patterns with their ± 1 representation and the synapses (7.3) and (7.4). It was argued in the introduction that, as such, the synapses (7.3) and (7.4) are unsuitable for the generation of stable sequences in asynchronous neural networks. The asymmetric synapses (7.4) always project the *current* network state onto the one that follows in the sequence. So, depending on their strength, they will either not be able to initiate transitions at all, or they will induce them so rapidly that the sequence soon gets smeared out. Kleinfeld [7.8] and Sompolinsky and Kanter [7.9] were the first to realize that this phenomenon could be avoided if the asymmetric synapses would process incoming information with a *time delay* so as to allow the system to settle in a pattern before a transition to the next is induced. In Sect. 7.4.1, the transition mechanism for such a setup is described in detail. Estimates of the persistence time will be given for various delay mechanisms and for zero as well as for nonzero temperatures. Exact analytic descriptions of sequence generators operating with transmission delays have been obtained for two limiting cases: the case of finitely many patterns and the so-called highly dilute limit. They will be presented in Sects. 7.4.2 and 7.4.3, respectively. Architectures for various recognition tasks, which require – among other things – the introduction of an afferent system, and architectures for the generation of complex sequences will be described later on in Sect. 7.5.

7.4.1 Transition Mechanism and Persistence Times

Let us consider the system described in the introduction where a set of q unbiased binary random patterns $\{\xi_i^\mu\}$ is stored in a network of N formal neurons equipped with the synapses (7.3),

$$J_{ij}^{(1)} = N^{-1} \sum_{\mu=1}^{q} \xi_i^\mu \xi_j^\mu \quad .$$

A second set of synapses, the asymmetric forward projections (7.4),

$$J_{ij}^{(2)} = \varepsilon N^{-1} \sum_{\mu=1}^{q} \xi_i^{\mu+1} \xi_j^\mu$$

is introduced so as to recall the patterns in a sequence. (Unless stated otherwise, we shall assume $q + 1 \equiv 1$, i.e., that the sequence is closed to form a cycle.) It was suggested by Kleinfeld [7.8] and by Sompolinsky and Kanter [7.9] that, in order to obtain a working sequence generator, the second set of synapses must be endowed with a *transmission delay* which allows the system to become quasi-stationary in one pattern before the forward projections induce a transition to the next.

Let us assume for simplicity that the number q of stored patterns is small compared to the system size ($q/N \to 0$), so that, given a single time delay τ, the two sets of synapses $\{J_{ij}^{(1)}\}$ and $\{J_{ij}^{(2)}\}$ give rise to a total local field of the form

$$\begin{aligned} h_i(t) &= h_i^{(1)}(t) + h_i^{(2)}(t) = \sum_{j(\neq i)} J_{ij}^{(1)} S_j(t) + \sum_{j(\neq i)} J_{ij}^{(2)} S_j(t-\tau) \\ &= \sum_{\mu=1}^{q} [\xi_i^\mu m_\mu(t) + \varepsilon \xi_i^{\mu+1} m_\mu(t-\tau)] \ . \end{aligned} \tag{7.47}$$

To discuss the transition mechanism in the case of asynchronous dynamics [7.8,9], let us suppose that the system has been in pattern $\nu - 1$ for $-\tau \leq t < 0$ and that at $t = 0$ the system state has changed from $\nu - 1$ to ν. For $0 \leq t < \tau$, therefore, we have

$$m_{\nu-1}(t-\tau) = 1 \tag{7.48}$$

and, ignoring $O(N^{-1/2})$ fluctuations, $m_\mu(t-\tau) = 0$ for $\mu \neq \nu - 1$. Thus, immediately after the transition at $t = 0$, the local field at site i is given by

$$h_i(t) = \xi_i^\nu m_\nu(t) + \varepsilon \xi_i^\nu m_{\nu-1}(t-\tau) = (1+\varepsilon)\xi_i^\nu \ , \tag{7.49}$$

i.e., both the instantaneous *and* the retarded signals, $h_i^{(1)}(t)$ and $h_i^{(2)}(t)$, support pattern ν. Owing to the threshold dynamics (7.1), $S_i(t + \Delta t) = \mathrm{sgn}[h_i(t)]$, this remains unchanged until at time $t = \tau$ the retarded overlap $m_{\nu-1}(t-\tau)$ changes to zero while $m_\nu(t-\tau)$ assumes the value 1, so that

$$h_i(t=\tau) = \xi_i^\nu + \varepsilon\xi_i^{\nu+1} . \tag{7.50}$$

If $\varepsilon > 1$, the sign of $h_i(t = \tau)$ will be determined by $\xi_i^{\nu+1}$, so that, owing to the threshold dynamics (7.1), the system will embark upon a transition to $\nu+1$. During the transition, $m_\nu(t)$ will gradually decrease to zero while $m_{\nu+1}(t)$ increases to unity[11]. Immediately after the transition, both the instantaneous and the retarded field support pattern $\nu+1$, until at $t \simeq 2\tau$ the retarded local field begins to support pattern $\nu+2$ and accordingly initiates a transition; and so on. If, on the other hand ε were less than 1, the sign of the local field at $t=\tau$ would continue to be determined by ξ_i^ν so that no transition would occur.

The above discussion already gives the persistence time Δ_0 for a sequence generator operating with a *single* time delay τ. Neglecting transition times, one obtains $\Delta_0 \simeq \tau$ [7.8, 9]. Moreover, the argument tells us that there is a minimum value $\varepsilon_{\rm min}$ for the relative strength ε of the forward projections below which no transitions would be induced. As far as fluctuations (static – owing to random overlaps between the patterns – or thermal; see below) can be neglected, the argument gives $\varepsilon_{\rm min} = 1$ [7.8, 9]. Figure 7.11 shows a simulation result, demonstrating that the synapses (7.4), if endowed with a transmission delay, are able to produce stable pattern sequences.

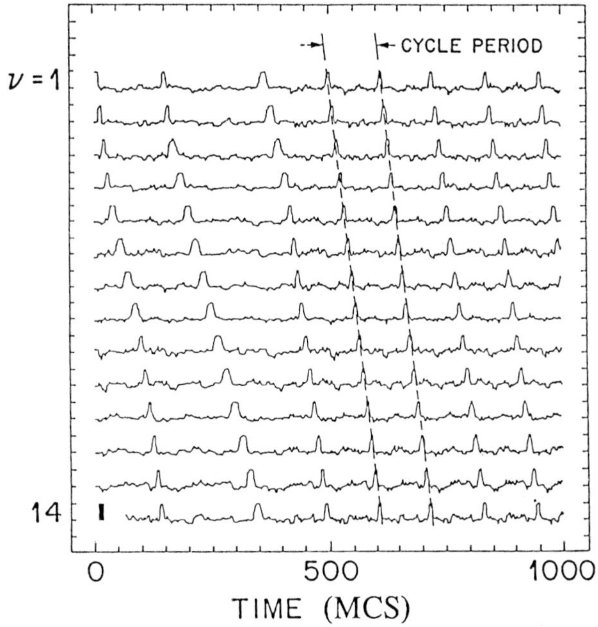

Fig. 7.11. Simulation of a network of $N = 100$ neurons, loaded with 14 patterns to be recalled in a sequence. Each track represents the time evolution of an overlap with one of the embedded patterns. The time delay is $\tau = 6$ MCS. The difference between estimated and actual persistence times is mainly due to the fact that transition times (approximately 2 MCS) are *not* small compared to τ, but also to fluctuations. If transition times are taken into account, the agreement between estimate $\Delta_0 \simeq 8$ MCS and measurement $\Delta_0 \simeq 8.2$ MCS is reasonable. Taken from [7.8]

[11] Closer inspection reveals that overlaps with patterns other than ν and $\nu+1$ also assume small nonzero values during the transition from ν to $\nu+1$. Their influence, via retarded local fields, on transitions to come remains, however, small and can in most cases be neglected.

Delay mechanisms other than the single time delay used in (7.47) may be contemplated, and many of them, too, give rise to stable sequence generators [7.8, 9, 21]. A general delay mechanism can be imagined as a linear mapping w which transforms a family of signals, the "past" $\{S_j(t'), t' \leq t\}$, into

$$\bar{S}_j(t) = \int_0^\infty dt' w(t') S_j(t - t') . \tag{7.51}$$

Here $w(t')$, $t' \geq 0$, is a memory kernel, which is normalized to one,

$$\int_0^\infty dt' w(t') = 1 . \tag{7.52}$$

The single time delay considered above corresponds to a δ-function kernel

$$w(t') = \delta(t' - \tau) . \tag{7.53a}$$

Other popular choices [7.8, 9, 21] are the step-function delay

$$w(t') = \tau^{-1} \theta(\tau - t') \tag{7.53b}$$

and the exponential delay

$$w(t') = \tau^{-1} \exp(-t'/\tau) . \tag{7.53c}$$

The delay functions (7.53) are characterized by a *single* time constant. Plainly, this need not generally be the case.

The persistence time for sequence generators based on the synapses (7.3) and (7.4) with a *general* delay kernel w has been computed by Sompolinsky and Kanter [7.9]. They assume that transitions between successive states ν of the sequence have occurred regularly in time with spacing Δ_0 and consider the local field at i when the system has been in μ for a time span δ. Neglecting transition times and assuming that the system was always perfectly aligned with the patterns it visited, they get

$$\begin{aligned} h_i(t) &= \sum_j J_{ij}^{(1)} S_j(t) + \sum_j J_{ij}^{(2)} \bar{S}_j(t) \\ &= \xi_i^\mu + \varepsilon[\xi_i^{\mu+1} W(0, \delta) + \xi_i^\mu W(\delta, \delta + \Delta_0) \\ &\quad + \xi_i^{\mu-1} W(\delta + \Delta_0, \delta + 2\Delta_0) + \ldots] , \end{aligned} \tag{7.54}$$

where

$$W(a, b) = \int_a^b dt' w(t') . \tag{7.55}$$

The key idea is now to note that the transition from μ to $\mu + 1$ starts on the special sublattice

$$I_\mu = \{i;\ \xi_i^\nu = -\xi_i^\mu \quad \text{for all} \quad \nu \neq \mu\} \tag{7.56}$$

on which the spins have been aligned with $\xi_i^{\mu+1}$ all the time *except* when the system was in state μ. On this sublattice the local field is biased most strongly in favor of $\xi_i^{\mu+1}$, and the transition is imagined to start when $h_i\xi_i^\mu$ becomes negative on I_μ. The corresponding δ in (7.54) must be identified with Δ_0. If the sequence is sufficiently long, so that we have $W(0,\delta)+W(\delta,\delta+\Delta_0)+W(\delta+\Delta_0,\delta+2\Delta_0)+\ldots \simeq 1$ in (7.54), and if transients are neglected, this gives the following equation [7.9]:

$$\begin{aligned} 0 = h_i\xi_i^\mu &= 1 + \varepsilon W(\Delta_0, 2\Delta_0) - \varepsilon[W(0,\Delta_0) + W(2\Delta_0, 3\Delta_0) + \ldots] \\ &\simeq 1 - \varepsilon + 2\varepsilon W(\Delta_0, 2\Delta_0) \end{aligned} \tag{7.57}$$

connecting Δ_0 and ε for a given delay mechanism w. For the delays (7.53), Δ_0 can be determined in terms of ε and τ. The results are [7.9]

$$\Delta_0 = \tau\ , \quad \varepsilon > 1\ , \tag{7.58a}$$

for δ-function delay, and

$$\Delta_0 = \tau(1 + 1/\varepsilon)/2\ , \quad \varepsilon > 1\ , \tag{7.58b}$$

for step-function delay, implying a short period $\Delta_0 = \tau/2$ at large values of ε and a long period $\Delta_0 = \tau$ as $\varepsilon \to \varepsilon_{\min} = 1$. For exponential delay, (7.57) yields

$$\Delta_0 = \tau\{\ln 2 - \ln[1 - (2/\varepsilon - 1)^{1/2}]\}\ , \quad 1 < \varepsilon < 2\ , \tag{7.58c}$$

so that $\Delta_0 \to \infty$ as $\varepsilon \to \varepsilon_{\min} = 1$, while $\Delta_0 \to \tau \ln 2$ as $\varepsilon \to 2$. Note that (7.57) imposes an an upper bound on ε in the case of exponential delay, but not in the case of δ- or step-function delay.

The above argument is valid for the $T = 0$ threshold dynamics (7.1) only. It may be generalized to nonzero temperatures [7.15] by noting that fluctuations will initiate the transition even *before* $h_i\xi_i^\mu$ changes sign on I_μ. The reason is that for $T \neq 0$ ($\beta < \infty$) the dynamics (7.1) is replaced by a probabilistic rule, such as

$$\mathrm{Prob}\{S_i(t + \Delta t) = \pm\xi_i^\mu\} = \tfrac{1}{2}[1 \pm \tanh(\beta h_i\xi_i^\mu)]\ , \tag{7.59}$$

so that a substantial fraction of the spins on I_μ is already flipped from ξ_i^μ to $\xi_i^{\mu+1} = -\xi_i^\mu$ and the transition starts, *once* $\beta h_i\xi_i^\mu$ decreases below a certain threshold B_0 on I_μ. Assuming as before that transition times are negligible and that $m_\nu(t) \simeq 1$ while the system is in state ν, which requires that the temperature be *low*, we conclude that, except for the difference in Δ_0 – which is to be determined – the right-hand side of (7.57) remains unaltered. The left-hand side, however, must be replaced by $\beta^{-1}B_0$ so that we obtain

$$\beta^{-1}B_0 \simeq 1 - \varepsilon + 2\varepsilon W(\Delta_0, 2\Delta_0)\ . \tag{7.60}$$

An immediate consequence of (7.60) is that the ***minimum*** value of ε for which stable sequences occur has a β-dependence of the form $\varepsilon_{\min}(\beta) = 1 - B_0/\beta$. This is not exact [7.21] but it is a reasonable approximation for low temperatures. Applying (7.60) to δ-function delay, we find $\Delta_0 = \tau$ for $\varepsilon > \varepsilon_{\min}(\beta)$. For step-

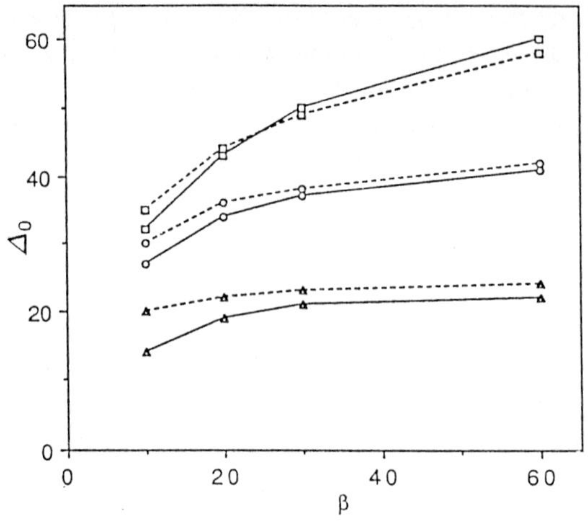

Fig. 7.12. Variation of persistence times with temperature and asymmetry parameter for exponential delay. *Solid lines* are measured values, *dashed lines* are predictions according to the estimates above. From top to bottom, we have $\varepsilon = 1$, $\varepsilon = 1.1$, and $\varepsilon = 1.5$. Taken from [7.15]

function delay, we get $\Delta_0 = \tau[1+\varepsilon_{\min}(\beta)/\varepsilon]/2$, provided that $\varepsilon > \varepsilon_{\min}(\beta)$, while for exponential delay $\Delta_0 = \tau\{\ln 2 - \ln[1-(2\varepsilon_{\min}(\beta)/\varepsilon - 1)^{1/2}]\}$, if $\varepsilon_{\min}(\beta) < \varepsilon < 2\varepsilon_{\min}(\beta)$. Thus, for step-function and exponential delay, (7.60) predicts Δ_0 to be a continuous function of ε and β, which in the case of exponential delay diverges whenever ε approaches $\varepsilon_{\min}(\beta)$ from above. Let us not fail to note that the way $\varepsilon_{\min}$ depends on the inverse temperature β admits a mode of operation where, as in the scheme of Buhmann and Schulten, transitions can be triggered at will by externally regulating the temperature.

The value of B_0 is related to a critical probability $\mathrm{Prob}\{S_i(t+\Delta t) = -\xi_i^\mu\} = p^*$ for i in I_μ which is the minimal fraction of spins that have to be flipped from ξ_i^μ to $-\xi_i^\mu$ on I_μ to induce the transition. Taking $p^* = 0.015$ or, equivalently, $B_0 = 2.09$ reproduces the data rather well at low temperatures; see Fig. 7.12.

Our discussion has been phrased in terms of sequential dynamics. In the case of parallel dynamics, the states of all neurons are updated in synchrony at every time step; time then is discrete and the convolution integrals (7.51) must be replaced by infinite sums, with memory kernels adapted accordingly. In spite of all these changes, the final formulae for Δ_0 remain unaltered.

Attractor neural networks with symmetric synapses are known to be fault tolerant with respect to input data errors as well as internal device failures. It was demonstrated by Kleinfeld [7.8] that the performance of sequence generators as described in the present section enjoys the same sort of robustness. For example, in a network of $N = 100$ neurons, the sequential recall of 14 random patterns remained essentially unchanged if 50% of the connections were removed by cutting a randomly chosen member of each of the pairs $(J_{ij}^{(1)}, J_{ji}^{(1)})$ and $(J_{ij}^{(2)}, J_{ji}^{(2)})$. A similar kind of robustness is observed in the case of random changes of the synaptic strengths or other forms of synaptic dilution. For further details, the reader may consult [7.8].

As they stand, the sequence generators described above would only work for sequences of unbiased binary random patterns. Extensions that work for *any* sequence of *linearly independent* patterns are, however, easily devised, using ideas of error-free recall in networks with symmetric synapses [7.22, 23]. To see this, we consider a sequence of linearly independent patterns and denote by C the correlation matrix of the individual patterns in the sequence, with elements

$$C_{\mu\nu} = N^{-1} \sum_i \xi_i^\mu \xi_i^\nu . \tag{7.61}$$

For linearly independent patterns, its inverse C^{-1} exists so that the synapses

$$\begin{aligned} J_{ij}^{(1)} &= N^{-1} \sum_{\mu,\nu} \xi_i^\mu (C^{-1})_{\mu\nu} \xi_j^\nu , \\ J_{ij}^{(2)} &= \varepsilon N^{-1} \sum_{\mu,\nu} \xi_i^{\mu+1} (C^{-1})_{\mu\nu} \xi_j^\nu , \end{aligned} \tag{7.62}$$

of which the latter are endowed with the same sort of slow dynamic response as above, will drive the system through the sequence, irrespective of the correlations among the patterns or their mean level of activity. To wit, define

$$a_\mu(t) = \sum_\nu (C^{-1})_{\mu,\nu} m_\nu(t) \tag{7.63}$$

and similarly $a_\mu(t-\tau)$, and compute the local field at i,

$$\begin{aligned} h_i(t) &= h_i^{(1)}(t) + h_i^{(2)}(t) = \sum_{j(\neq i)} J_{ij}^{(1)} S_j(t) + \sum_{j(\neq i)} J_{ij}^{(2)} S_j(t-\tau) \\ &= \sum_{\mu=1}^{q} [\xi_i^\mu a_\mu(t) + \varepsilon \xi_i^{\mu+1} a_\mu(t-\tau)] . \end{aligned} \tag{7.64}$$

To elucidate the formal equivalence with (7.47) we have stayed with the essentially finite-q case and with δ-function delay. The $a_\mu(t)$ are constructed such that, as long as the patterns are linearly independent, we have $a_\nu(t) = \delta_{\mu,\nu}$ if the system is in state μ. Consequently, the above discussion of the dynamic features of the network and of the transition mechanism may be repeated word for word, with the $a_\mu(t)$ taking the role of the $m_\mu(t)$.

The *non*local storage prescription (7.62) may be replaced by simpler *local* rules if correlations between the patterns have their origin in a common nonzero level of activity of the stored patterns. Consider, for example, a set of q patterns $\{\xi_i^\mu\}$, for which $\xi_i^\mu = +1$ or -1 with probabilities p and $1-p$, respectively, and assume $p \neq 1/2$ so that the patterns have a nonzero activity $a := \langle \xi_i^\mu \rangle = 2p - 1$. Storage prescriptions for biased patterns [7.24–27] (see also Sect. 1.5 of [7.20]), may be generalized to store sequences of such patterns in a neural network. To avoid introducing constrained dynamics as in [7.24], one may take the couplings proposed by Krogh and Hertz [7.26] and by Bös [7.27]

$$J_{ij}^{(1)} = N^{-1} \left\{ 1 + \sum_\mu \frac{(\xi_i^\mu - a)(\xi_j^\mu - a)}{(1-a^2)} \right\}$$

to embed the individual patterns, and

$$J_{ij}^{(2)} = \varepsilon N^{-1} \left\{ 1 + \sum_{\mu} \frac{(\xi_i^{\mu+1} - a)(\xi_j^{\mu} - a)}{(1 - a^2)} \right\}$$

or, alternatively [7.28],

$$J_{ij}^{(2)} = \varepsilon N^{-1} \sum_{\mu} \frac{\xi_i^{\mu+1}(\xi_j^{\mu} - a)}{(1 - a^2)}$$

to recall them in a sequence. As before, it is understood that the asymmetric synapses are endowed with a delay. We leave it as an exercise for the reader to verify that such low-activity sequence generators will function as desired.

Are the above rather formal considerations relevant to the description of the dynamics in biological neural nets? As far as signal transmission delays are concerned, they *are* omnipresent in the brain [7.29, 30], and their influence on the dynamics of neural assemblies can certainly not be dismissed on a priori grounds. In Sect. 7.6 below, we shall see that they are also of profound importance for *learning* [7.17]. With respect to delay *mechanisms*, it can be said that for axonal signal propagation, a description in terms of δ-function delays seems appropriate [7.8, 29], and that, inasmuch as the synapse itself acts as a low-pass filter [7.29], the associated delay is exponential. One should, however, keep in mind [7.30] that – unlike in the simple models considered above – the time delay (as well as the delay mechanism) for the signal transmission between any pair (i, j) of neurons may, and in general will, depend on i and j. With respect to formal neurons, mean firing rates, stochastic dynamics, and all that, this is an even more fundamental debate, on which here we have little to say.

7.4.2 Analytic Description of the Dynamics

In this section, we shall derive a set of coupled nonlinear differential equations that describe the time evolution of the sequence generators introduced above in the thermodynamic limit, $N \to \infty$, provided the number q of patterns remains finite or essentially finite ($q \ll \log N$). We use a stochastic Glauber dynamics (at inverse temperature β) from the start, i.e., we assume that the neurons are updated asynchronously according to the rule

$$\text{Prob}\{S_i(t + \Delta t) = \pm 1\} = \tfrac{1}{2}\{1 \pm \tanh[\beta h_i(t)]\} \,, \qquad (7.65)$$

if they are excited by the local field $h_i(t)$. The ensuing arguments closely follow Riedel et al. [7.21].

Although we are mainly interested in sequence generators based on the synapses (7.3) and (7.4), it will be instructive to derive the evolution equations in a more general setting. We only require that the symmetric and the nonsymmetric synapses $J_{ij}^{(1)}$ and $J_{ij}^{(2)}$ be determined by the information locally available to the neurons i and j which they connect,

$$J_{ij}^{(1)} = N^{-1}Q^{(1)}(\boldsymbol{\xi}_i;\boldsymbol{\xi}_j) = J_{ji}^{(1)} , \tag{7.66}$$

$$J_{ij}^{(2)} = \varepsilon N^{-1}Q^{(2)}(\boldsymbol{\xi}_i;\boldsymbol{\xi}_j) \neq J_{ji}^{(2)} , \tag{7.67}$$

with

$$\boldsymbol{\xi}_i = (\xi_i^{\mu})_{\mu=1}^{q} , \tag{7.68}$$

for some pair of synaptic kernels $Q^{(1)}$ and $Q^{(2)}$ on $\mathbf{R}^q \times \mathbf{R}^q$. The synapses considered in the previous section have $Q^{(1)}(\boldsymbol{\xi}_i;\boldsymbol{\xi}_j) = \phi^{(1)}(\sum_\mu \xi_i^\mu \xi_j^\mu)$ and $Q^{(2)}(\boldsymbol{\xi}_i;\boldsymbol{\xi}_j) = \phi^{(2)}(\sum_\mu \xi_i^{\mu+1} \xi_j^\mu)$ with $\phi^{(1)}(x) = \phi^{(2)}(x) = x$; they will therefore be called *linear*.

The key idea of our analysis [7.21] is based on the notion of *sublattices* [7.31–33]. Given q binary random patterns $\{\xi_i^\mu\}$, there are only 2^q positions available for the random vectors $\boldsymbol{\xi}_i$ in (7.66) and (7.67), namely the 2^q corners $\boldsymbol{x}$ of the hypercube $[-1,1]^q$. Introducing the corresponding sublattices

$$I(\boldsymbol{x}) = \{i; \boldsymbol{\xi}_i = \boldsymbol{x}\} , \quad \boldsymbol{x} \in \mathcal{C}^q = \{-1,1\}^q , \tag{7.69}$$

and sublattice magnetizations

$$m(\boldsymbol{x};t) = \frac{1}{|I(\boldsymbol{x})|} \sum_{i \in |I(\boldsymbol{x})|} S_i(t) , \tag{7.70}$$

one easily verifies that the local field at i is given by

$$h_i(t) = \sum_{\boldsymbol{y} \in \mathcal{C}^q} \left[Q^{(1)}(\boldsymbol{\xi}_i;\boldsymbol{y}) p_N(\boldsymbol{y}) m(\boldsymbol{y};t) + \varepsilon Q^{(2)}(\boldsymbol{\xi}_i;\boldsymbol{y}) p_N(\boldsymbol{y}) \bar{m}(\boldsymbol{y};t)\right] . \tag{7.71}$$

Here $|I(\boldsymbol{y})|$ denotes the size of $I(\boldsymbol{y})$, $p_N(\boldsymbol{y}) = |I(\boldsymbol{y})|/N$, and $\bar{m}(\boldsymbol{y};t)$ is a convolution of $m(\boldsymbol{y};t)$ with a memory kernel w as in (7.51). Moreover, we have assumed that the $J_{ii}^{(\alpha)}$, $\alpha = 1,2$, tend to zero in the thermodynamic limit, i.e. that $Q^{(\alpha)}(\boldsymbol{y};\boldsymbol{y})/N \to 0$ as $N \to \infty$ – whatever $\boldsymbol{y}$. Since $\boldsymbol{\xi}_i = \boldsymbol{x}$ for *all* i in $I(\boldsymbol{x})$ the local field depends on i only through the label $\boldsymbol{x}$ of the sublattice to which i belongs. To derive a master equation for the probability $\varrho(\boldsymbol{m};t)$ of finding the system at time t in a state with sublattice magnetizations $\boldsymbol{m} = \left(m(\boldsymbol{x}); \boldsymbol{x} \in \mathcal{C}^q\right)$, we note that the $m(\boldsymbol{x})$ can only assume the values $-1, -1+\Delta(\boldsymbol{x}), -1+2\Delta(\boldsymbol{x}), \dots, 1$, with $\Delta(\boldsymbol{x}) = 2/|I(\boldsymbol{x})|$, and that an asynchronous update affects only a single sublattice, changing the corresponding sublattice magnetization by at most $\pm\Delta(\boldsymbol{x})$. Using (7.65) and taking into account the various elementary processes that can change the $m(\boldsymbol{x})$, one obtains

$$\begin{aligned} \varrho(\boldsymbol{m};t+\Delta t) - \varrho(\boldsymbol{m};t) \; = \; & \sum_{\boldsymbol{x} \in \mathcal{C}^q} p_N(\boldsymbol{x}) \left\{ \frac{1 + m(\boldsymbol{x}) + \Delta(\boldsymbol{x})}{4} \right. \\ & \times \left(1 - \tanh[\beta h(\boldsymbol{x}, \boldsymbol{m} + \boldsymbol{\Delta}(\boldsymbol{x}))]\right) \varrho(\boldsymbol{m} + \boldsymbol{\Delta}(\boldsymbol{x});t) \\ & + \frac{1 - m(\boldsymbol{x}) + \Delta(\boldsymbol{x})}{4} \left(1 + \tanh[\beta h(\boldsymbol{x}, \boldsymbol{m} - \boldsymbol{\Delta}(\boldsymbol{x}))]\right) \varrho(\boldsymbol{m} - \boldsymbol{\Delta}(\boldsymbol{x});t) \end{aligned}$$

$$-\frac{1+m(\boldsymbol{x})}{4}\left(1-\tanh[\beta h(\boldsymbol{x},\boldsymbol{m})]\right)\varrho(\boldsymbol{m};t)$$
$$\left.-\frac{1-m(\boldsymbol{x})}{4}\left(1+\tanh[\beta h(\boldsymbol{x},\boldsymbol{m})]\right)\varrho(\boldsymbol{m};t)\right\} . \qquad (7.72)$$

Here we have introduced the vector $\boldsymbol{\Delta}(\boldsymbol{x}) = \left(\Delta(\boldsymbol{x})\delta_{\boldsymbol{x},\boldsymbol{y}};\, \boldsymbol{y} \in C^q\right)$ and explicitly displayed the dependence of the local field h on $\boldsymbol{x}$ and $\boldsymbol{m}$.[12] The terms on the right-hand side of (7.72) describe ingoing and outgoing channels, each with processes that *de*crease and processes that *in*crease the magnetization. Equation (7.72) can be written as a differential equation

$$\left[\exp\left(\Delta t\frac{\partial}{\partial t}\right)-1\right]\varrho(\boldsymbol{m};t)$$
$$=\sum_{\boldsymbol{x}\in C^q} p_N(\boldsymbol{x})\left\{-\frac{1}{2}\sinh\left(\Delta(\boldsymbol{x})\frac{\partial}{\partial m(\boldsymbol{x})}\right)\left(F(\boldsymbol{x},\boldsymbol{m})\varrho(\boldsymbol{m};t)\right)\right.$$
$$\left.+\frac{1}{2}\left[\cosh\left(\Delta(\boldsymbol{x})\frac{\partial}{\partial m(\boldsymbol{x})}\right)-1\right]\left(D(\boldsymbol{x},\boldsymbol{m})\varrho(\boldsymbol{m};t)\right)\right\} , \qquad (7.73)$$

where

$$F(\boldsymbol{x},\boldsymbol{m}) = -\left\{m(\boldsymbol{x})-\tanh[\beta h(\boldsymbol{x},\boldsymbol{m})]\right\}$$

and

$$D(\boldsymbol{x},\boldsymbol{m}) = 1-m(\boldsymbol{x})\tanh[\beta h(\boldsymbol{x},\boldsymbol{m})] .$$

For asynchronous dynamics the elementary time step Δt must be scaled with system size N according to $\Delta t = (\Gamma N)^{-1}$ to ensure that each neuron is updated at a finite N-independent rate. Neglecting, as $N \to \infty$, terms of order N^{-2} and smaller in (7.73), we obtain the second-order differential equation

$$\frac{\partial\varrho}{\partial t}+(\Gamma N)^{-1}\frac{\partial^2\varrho}{\partial t^2} = \Gamma\sum_{\boldsymbol{x}\in C^q}\left\{-\frac{\partial}{\partial m(\boldsymbol{x})}(F\varrho)+\frac{\Delta(\boldsymbol{x})}{2}\frac{\partial^2}{\partial m(\boldsymbol{x})^2}(D\varrho)\right\} . \qquad (7.74)$$

In the thermodynamic limit ($N \to \infty$), the $p_N(\boldsymbol{x})$ converge to $p(\boldsymbol{x})$ with probability one, so that the second-order terms on *both* sides of (7.74) vanish – $\Delta(\boldsymbol{x}) = 2/p_N(\boldsymbol{x})N \sim 2/p(\boldsymbol{x})N$ – and one obtains [7.21] the following set of 2^q coupled deterministic kinetic equations for the sublattice magnetizations $m(\boldsymbol{x};t) = \langle m(\boldsymbol{x})\rangle_{\varrho(m;t)}$

$$\frac{\mathrm{d}m(\boldsymbol{x};t)}{\mathrm{d}t} = -\Gamma\left\{m(\boldsymbol{x};t)-\tanh[\beta h(\boldsymbol{x},\boldsymbol{m},\bar{\boldsymbol{m}})]\right\} . \qquad (7.75)$$

In (7.75), we have restored the dependence of the local fields on the retarded magnetizations $\bar{\boldsymbol{m}}$. Note that because of the delays, (7.75) are not ordinary, but so-called *functional*, differential equations. Analytic solutions are not known in

[12] In order to simplify the notation, the dependence of h on the retarded magnetizations $\bar{\boldsymbol{m}}$ has, however, been suppressed.

general but (7.75) can be solved numerically. The outcomes are presented most conveniently in terms of the overlaps $m_\mu(t)$, which can be expressed in terms of the sublattice magnetizations as

$$m_\mu(t) = \sum_{\boldsymbol{x} \in \mathcal{C}^q} p(\boldsymbol{x}) x^\mu m(\boldsymbol{x}; t) \equiv \langle x^\mu m(\boldsymbol{x}; t) \rangle \ . \tag{7.76}$$

In the case of the linear models discussed in the previous section – more generally if the synaptic kernels are bilinear in the ξs – one can reduce (7.75) to a set of q equations for the overlaps themselves,

$$\frac{\mathrm{d}m_\mu(t)}{\mathrm{d}t} = -\Gamma \left\{ m_\mu(t) - \left\langle x^\mu \tanh \left[\beta \sum_\nu \left(x^\nu m_\nu(t) + \varepsilon x^{\nu+1} \bar{m}_\nu(t) \right) \right] \right\rangle \right\} \ . \tag{7.77}$$

For network models without delayed interactions ($\varepsilon = 0$), analogous equations were later also derived by Coolen et al. [7.34] and by Shiino et al. [7.35]. A heuristic generalization of (7.77), covering the case of extensively many patterns outside the cycle, was proposed by Riedel et al. [7.21] (see also Sects. 7.5.4 of [7.20]. For moderate amounts of synaptic noise, it accounts for the modifications of pulse forms and persistence times fairly well, but it fails to predict reliable phase boundaries.

In the case of parallel dynamics, one obtains the iterative prescriptions

$$m(\boldsymbol{x}; t + \Delta t) = \tanh[\beta h(\boldsymbol{x}, \boldsymbol{m}, \bar{\boldsymbol{m}})] \tag{7.78}$$

and

$$m_\mu(t + \Delta t) = \left\langle x^\mu \tanh \left[\beta \sum_\nu \left(x^\nu m_\nu(t) + \varepsilon x^{\nu+1} \bar{m}_\nu(t) \right) \right] \right\rangle \tag{7.79}$$

instead of the differential equations (7.75) and (7.77). The derivation of (7.78) and (7.79) depends on the strong law of large numbers and can be found in Sect. 7.5.4 of [7.20]. Note that the present analytic theory would also cover the case of overlapping (rather than just disjoint) low-activity patterns as described in Sect. 7.3. The point is that the couplings (7.29) are of the form $J_{ij} = N^{-1} Q(\zeta_i; \zeta_j)$ with $\zeta_i = (\zeta_i^\mu)_{\mu=1}^q \in \{0,1\}^q$. The remainder of the above derivation carries over and yields

$$\frac{\mathrm{d}m(\boldsymbol{z}; t)}{\mathrm{d}t} = -\Gamma \{ m(\boldsymbol{z}; t) - f(h(\boldsymbol{z}; t)) \} \ , \quad \boldsymbol{z} \in \{0,1\}^q$$

with

$$h(\boldsymbol{z}; t) = \sum_{\boldsymbol{z}'} Q(\boldsymbol{z}; \boldsymbol{z}') p(\boldsymbol{z}') m(\boldsymbol{z}'; t) \ .$$

Riedel et. al. [7.21] have studied (7.75) and (7.77) for the three delay functions w introduced in the previous section and for various choices of the synaptic

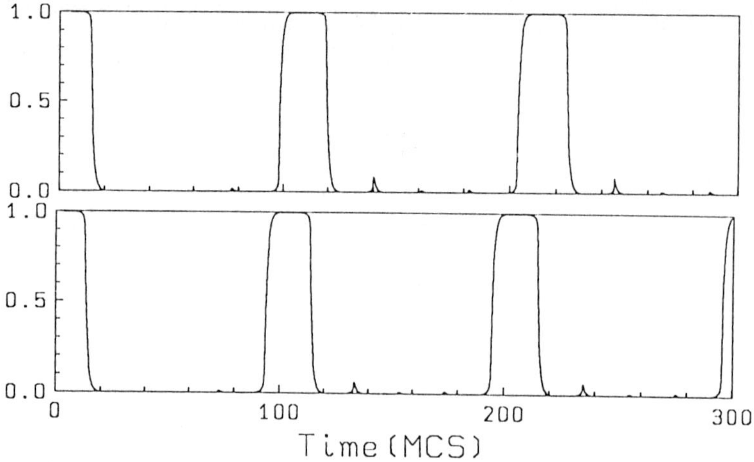

Fig. 7.13. Temporal sequences in a system with linear synapses (*top*), and in a system with clipped synapses (*bottom*). The number of patterns in the sequence is 5, the inverse temperature is $\beta = 10$, and the asymmetry parameter $\varepsilon = 1.1$. The delay is exponential. Only the overlap with the first pattern of the sequence is shown. The persistence time for the system with clipped synapses is slightly lower than that for the linear model

kernels $Q^{(1)}$ and $Q^{(2)}$, such as the *linear* synapses (7.3) and (7.4) and their *clipped* variants $Q^{(1)}(\boldsymbol{\xi}_i; \boldsymbol{\xi}_j) = \sqrt{q}\,\mathrm{sgn}(\sum_\mu \xi_i^\mu \xi_j^\mu)$ and $Q^{(2)}(\boldsymbol{\xi}_i; \boldsymbol{\xi}_j) = \sqrt{q}\,\mathrm{sgn}(\sum_\mu \xi_i^{\mu+1} \xi_j^\mu)$. The latter may be important for hardware realizations of sequence generators and were found to perform essentially like their linear counterparts; see Fig. 7.13.

The various types of solutions of (7.75) or (7.76) can be subjected to bifurcation and stability analyses [7.36] so as to obtain, for instance, the parameter ranges for which time-periodic solutions exist (Fig. 7.14). In this vein, the upper limit $\varepsilon_{\max}(\beta) = 2\varepsilon_{\min}(\beta)$ for the asymmetry parameter ε in the case of exponential delay can be shown to be associated with a secondary bifurcation from periodic to *quasi*-periodic solutions of (7.77) [7.21]. In the case of δ-function delay, stable sequences can be demonstrated to exist for arbitrarily high temperatures, provided the asymmetry parameter ε is sufficiently large; this is equivalent to showing that the symmetric synapses can be dispensed with but that, nevertheless, the sequence is stored as a stable limit cycle. Incidentally, this implies that one of the standard interpretations of the operating mechanism of these sequence generators, according to which the sole effect of $J_{ij}^{(2)}$ is to induce transitions between preexisting valleys of a free-energy surface created by the symmetric synapses $J_{ij}^{(1)}$, cannot be completely adequate.

In the derivation of (7.75), we have assumed that the $J_{ii}^{(\alpha)}$, $\alpha = 1,2$, of (7.66) and (7.67) could be neglected in the bulk limit. However, macroscopic evolution equations can still be derived, if the $J_{ii}^{(1)}$ have a nonzero limit as $N \to \infty$. The derivation requires an extension of the sublattice idea that includes *dynamic* features, viz.

$$I(\boldsymbol{x}, \sigma; t) = \{i;\ \boldsymbol{\xi}_i = \boldsymbol{x},\ S_i(t) = \sigma\}\ , \quad \boldsymbol{x} \in C^q\ , \quad \sigma = \pm 1\ . \tag{7.80}$$

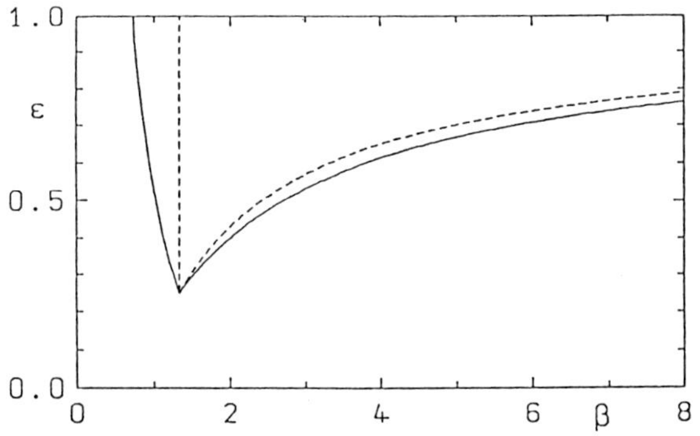

Fig. 7.14. Dynamical phase diagram for a two-cycle with δ-function delay (*solid curve*) and exponenetial delay (*dashed curve*). Within the V-shaped region, a pure cycle exists. Taken from [7.21]

Expressed in terms of the sublattice magnetizations $m(\boldsymbol{x};t)$ and the $p_N(\boldsymbol{x})$, the probability for some i to belong to $I(\boldsymbol{x},\sigma;t)$ is $p_N(\boldsymbol{x})(1+\sigma m(\boldsymbol{x};t))/2$. One can now set up a master equation that keeps track of the elementary processes which change magnetizations on the dynamically evolving sublattices $I(\boldsymbol{x},\sigma;t)$. Following the reasoning that led us to (7.75), we obtain again a differential equation describing the evolution of the sublattice magnetizations $m(\boldsymbol{x};t)$ in the thermodynamic limit,

$$\frac{\mathrm{d}m(\boldsymbol{x};t)}{\mathrm{d}t} = -\Gamma\left\{m(\boldsymbol{x};t) - \frac{1}{2}\sum_{\sigma=\pm 1}(1+\sigma m(\boldsymbol{x};t)) \times \tanh[\beta h(\boldsymbol{x},\boldsymbol{m},\bar{\boldsymbol{m}}) - \sigma a(\boldsymbol{x})]\right\}. \tag{7.81}$$

Here $h(\boldsymbol{x},\boldsymbol{m},\bar{\boldsymbol{m}})$ is the local field including (as above) contributions from the self-interactions. Since these are now assumed to have a nonzero limit, $J_{ii}^{(1)} \to a(\boldsymbol{x})$, $i \in I(\boldsymbol{x})$, this limit must be subtracted in (7.81). Note that a reduction of (7.81) to a set of q equations for the overlaps is impossible even for the linear models.

Before turning to the other analytically tractable limit – the case of very low random connectivity – let us mention an observation that turns out to be useful in the numerical solution and in analytic investigations of the differential equations (7.75) or (7.77). In the case of exponential delay, the *functional* differential equations (7.75) and (7.77) can be transformed into sets of *ordinary* differential eqations of dimension 2^{q+1} and $2q$, respectively. The reason is that for this type of delay the retarded sublattice magnetizations or overlaps obey simple first-order differential equations themselves, namely

$$\frac{\mathrm{d}\bar{m}(\boldsymbol{x};t)}{\mathrm{d}t} = -\tau^{-1}[\bar{m}(\boldsymbol{x};t) - m(\boldsymbol{x};t)], \tag{7.82}$$

and similarly for the overlaps. At the cost of simply doubling the dimension of the system of equations, one can therefore treat the ms and $\bar{m}$s as separate variables; (7.82) introduce no further unknowns into the game.

7.4.3 Extreme Dilution of Synapses

In the previous section we have solved the dynamics of *fully* connected networks in the low loading limit. The other analytically tractable limit is that of extreme dilution of synaptic efficacies [7.12, 37] (for further information, see also Chap. 6). To be specific, one assumes that in analogy with (7.3) and (7.4) the couplings are given by

$$J_{ij}^{(1)} = \frac{C_{ij}}{C} \sum_{\mu=1}^{q} \xi_i^\mu \xi_j^\mu \tag{7.83}$$

and

$$J_{ij}^{(2)} = \varepsilon \frac{C_{ij}}{C} \sum_{\mu=1}^{q} \xi_i^{\mu+1} \xi_j^\mu . \tag{7.84}$$

As before, the $J_{ij}^{(2)}$ are endowed with a delay τ. The C_{ij} denote the absence or presence of the bond $j \to i$. They are randomly chosen to take the values 0 or 1 according to the distribution

$$\varrho(C_{ij}) = \left(1 - \frac{C}{N}\right) \delta(C_{ij}) + \frac{C}{N} \delta(C_{ij} - 1) . \tag{7.85}$$

The neurons are updated according to the probabilistic rule (7.65), either sequentially or in parallel. The state of neuron i at time $t + \Delta t$ is determined by the value of its local field $h_i(t)$ and thus depends on the current and delayed states of the group $\{j\}$ of neurons which are actually connected to i, i.e., for which $C_{ij} \neq 0$. The local field may therefore be written

$$h_i(t) = \frac{1}{C} \sum_{\mu=1}^{q} \sum_{r=1}^{K} \xi_i^\mu \xi_{j_r}^\mu S_{j_r}(t) + \frac{\varepsilon}{C} \sum_{\mu=1}^{q} \sum_{r=1}^{K} \xi_i^{\mu+1} \xi_{j_r}^\mu S_{j_r}(t - \tau) , \tag{7.86}$$

with $j_1, j_2, \ldots, j_K$ denoting the sites connected to i. The states of the neurons that determine $h_i(t)$ in turn depend on previous and delayed states of other groups of neurons to which *they* are connected; and so on. In this manner, the calculation of $S_i(t + \Delta t)$ involves a tree-like set of ancestors which connect $S_i(t + \Delta t)$ to the initial conditions at $t = 0$ [7.37]. The typical number of sites in this set grows as C^t, and as long as $C^t \ll N^{2/3}$ all sites in the set are different[13]. More

[13] To see this, note that the probability of selecting randomly one out of N sites is $1/N$. The probability of selecting it k times in C^t trials is described by the corresponding binomial distribution. Demanding that the probability

$$P_{>1} = C^t \sum_{k>1} \binom{C^t}{k} N^{-k} (1 - N^{-1})^{C^t - k}$$

for selecting *any* of the $O(C^t)$ sites in the set of ancestors *more than once* satisfies $P_{>1} \ll 1$, one obtains the desired estimate. The derivation of the evolution equations below will require, in addition, that the sets of ancestors belonging to *different* sites be independent. This calls (at least) for the stronger constraint that $C^t \ll N^{1/2}$.

precisely, the probability for the occurrence of loops is vanishingly small so that the set of ancestors is really a tree. As a consequence, if the initial conditions are random and uncorrelated, the $S_{j_1}(t), S_{j_2}(t), \ldots, S_{j_K}(t)$ will also be uncorrelated for almost all sites i. If – as a first step – one chooses to consider the dynamics for $t < \tau$ only [7.12], and if one imposes random initial conditions for $-\tau \le t < 0$, the same can be said about the delayed states $S_{j_1}(t-\tau), S_{j_2}(t-\tau), \ldots, S_{j_K}(t-\tau)$.

In the thermodynamic limit, the set of ancestors will retain its tree structure *for all finite times* [7.37, 38] provided that

$$C \ll \ln N ,$$

as $N \to \infty$. For $t > \tau$, however, dynamical correlations between the neuron states which determine $h_i(t)$ will evolve, because $S_{j_r}(t)$ and $S_{j_r}(t-\tau)$ can no longer be said to be independent.

Under the condition of statistical independency of the dynamical variables, which in the case at hand requires that t be less than τ, it is easy to obtain the probability distribution of the local fields $h_i(t)$ and thereby derive an exact evolution equation for the overlaps with the stored patterns [7.12, 37]. This is what we now do. In the case of Glauber dynamics and for parallel updating one obtains, after averaging over (7.65),

$$m_\nu(t+\Delta t) = \frac{1}{N}\sum_{i=1}^{N} \tanh[\beta \xi_i^\nu h_i(t)] = \langle \tanh[\beta \xi_i^\nu h_i(t)] \rangle , \tag{7.87}$$

where the angular brackets denote an average over the local fields and over the ξ_i^ν. If the initial state for $-\tau \le t < 0$ has a macroscopic overlap Q^μ with pattern μ only and if for $0 \le t < \tau$ the system state has nonzero overlaps $m_\mu(t)$ and $m_{\mu+1}(t)$ with patterns μ and $\mu+1$ only (the consistency of these assumptions should be checked), then [7.12, 37]

$$\begin{aligned}
&\mathrm{Prob}\{\xi_{j_r}^\nu S_{j_r}(t) = \pm 1\} = (1 \pm m_\nu(t)/2 , \quad \nu \in \{\mu, \mu+1\} , \\
&\mathrm{Prob}\{\xi_{j_r}^\nu S_{j_r}(t) = \pm 1\} = \tfrac{1}{2} , \quad \nu \notin \{\mu, \mu+1\} , \\
&\mathrm{Prob}\{\xi_{j_r}^\nu S_{j_r}(t-\tau) = \pm 1\} = \tfrac{1}{2}(1 \pm Q^\mu \delta_{\nu,\mu}) .
\end{aligned} \tag{7.88}$$

In the limit $C/N \to 0$, the probability that a given site i is connected to K other sites is expressed by the Poisson distribution $C^K \mathrm{e}^{-C}/K!$. For $t < \tau$, the $\xi_{j_r}^\nu S_{j_r}(t)$ and the $\xi_{j_r}^\nu S_{j_r}(t-\tau)$, with $1 \le r \le K$ and $1 \le \nu \le q$, are *independent* by assumption. The probability $P(K, n_\mu, n_{\mu+1}, \bar{n}_\mu, s, \bar{s})$ that n_μ of the K random variables $\{\xi_{j_r}^\mu S_{j_r}(t); r = 1, \ldots, K\}$ are negative and, similarly, that $n_{\mu+1}$ of the $\{\xi_{j_r}^{\mu+1} S_{j_r}(t); r = 1, \ldots, K\}$, $\bar{n}_\mu$ of the $\{\xi_{j_r}^\mu S_{j_r}(t-\tau); r = 1, \ldots, K\}$, and s ($\bar{s}$) of the $\{\xi^\mu \xi_{j_r}^\nu S_{j_r}(t); r = 1, \ldots, K; \nu \notin \{\mu, \mu+1\}\}$ ($\{\xi^\mu \xi_{j_r}^\nu S_{j_r}(t-\tau); r = 1, \ldots, K; \nu \ne \mu\}$) are negative, so that

$$\begin{aligned}
\xi_i^\mu h_i(t) = C^{-1}[&K - 2n_\mu + \xi_i^\mu \xi_i^{\mu+1}(K - 2n_{\mu+1}) + \varepsilon \xi_i^\mu \xi_i^{\mu+1}(K - 2\bar{n}_\mu) \\
&+ K(q-2) - 2s + \varepsilon(K(q-1) - 2\bar{s})] ,
\end{aligned} \tag{7.89}$$

is therefore described by a *product* of binomial distributions with parameters given in (7.88), multiplied by the Poisson probability that the connectivity of i is indeed K. That is,

$$P(K, n_\mu, n_{\mu+1}, \bar{n}_\mu, s, \bar{s}) = \frac{C^K}{K!} \mathrm{e}^{-C} p^K_{m_\mu(t)}(n_\mu)\, p^K_{m_{\mu+1}(t)}(n_{\mu+1})\, p^K_{Q^\mu}(\bar{n}_\mu) \times p_0^{K(q-2)}(s)\, p_0^{K(q-1)}(\bar{s}) \tag{7.90}$$

with

$$p^K_m(n) = 2^{-K}\binom{K}{n}(1+m)^{K-n}(1-m)^n \,.$$

A similar expression holds for the probability distribution of $\xi_i^{\mu+1} h_i(t)$. Moreover, if ν does not belong to $\{\mu, \mu+1\}$, the distribution of $\xi_i^\nu h_i(t)$ is symmetric with respect to zero so that the corresponding $m_\nu(t)$ remain zero, as assumed.

For the two nonvanishing overlaps we thus have

$$\begin{aligned} m_\mu(t+\Delta t) = \frac{1}{2} \sum_{\sigma=\pm 1} \sum_{K=0}^{\infty} \sum_{n_\mu, n_{\mu+1}, \bar{n}_\mu, s, \bar{s}} & P(K, n_\mu, n_{\mu+1}, \bar{n}_\mu, s, \bar{s}) \\ & \times \tanh\left\{\beta C^{-1}[K - 2n_\mu + \sigma(K - 2n_{\mu+1}) + \varepsilon\sigma(K - 2\bar{n}_\mu) \right. \\ & \left. + K(q-2) - 2s + \varepsilon(K(q-1) - 2\bar{s})]\right\} \end{aligned} \tag{7.91}$$

and

$$\begin{aligned} m_\mu(t+\Delta t) = \frac{1}{2} \sum_{\sigma=\pm 1} \sum_{K=0}^{\infty} \sum_{n_\mu, n_{\mu+1}, \bar{n}_\mu, s, \bar{s}} & P(K, n_\mu, n_{\mu+1}, \bar{n}_\mu, s, \bar{s}) \\ & \times \tanh\left\{\beta C^{-1}[\sigma(K - 2n_\mu) + K - 2n_{\mu+1} + \varepsilon(K - 2\bar{n}_\mu) \right. \\ & \left. + K(q-2) - 2s + \varepsilon(K(q-1) - 2\bar{s})]\right\} \end{aligned} \tag{7.92}$$

In the large-C limit, the Poisson distribution $C^K \mathrm{e}^{-C}/K!$ gives appreciable weight only to K-values in the immediate vicinity of C. Approximating the K-summations in (7.91) and (7.92) by integrals over the variable $x = K/C$, one transforms the Poisson distribution into a probability measure which converges, as $C \to \infty$, to a Dirac measure at $x = 1$. Taking the limit $C \to \infty$ and $q \to \infty$ at fixed $\alpha = q/C$, one may approximate the summations over the ns and ss in (7.91) and (7.92) by integrals in a similar fashion, with probability measures converging, as $C \to \infty$, either to Dirac or to Gaussian measures (the latter arising from the *symmetric* binomial distributions $p_0^{K(q-2)}$ and $p_0^{K(q-1)}$ related to the extensively many uncondensed states). As a result, one obtains [7.12]

$$\begin{aligned} m^{\pm}_\mu(t+\Delta t) &= \int \frac{\mathrm{d}z}{\sqrt{2\pi}} \mathrm{e}^{-z^2/2} \tanh[\beta(m^{\pm}_\mu(t) \pm \varepsilon Q^\mu + \sqrt{w}z)] \\ &\equiv F_w(m^{\pm}_\mu(t) \pm \varepsilon Q^\mu) \,, \end{aligned} \tag{7.93}$$

where $m^{\pm}_\mu = m_\mu \pm m_{\mu+1}$, $w = \alpha(1+\varepsilon^2)$, and where Q^μ is the value of the only nonzero initial overlap; cf. (7.88). Equations (7.91–93) describe the dynamics only up to $t = \tau$. For later times, the statistical independence of the instantaneous and the retarded contributions to the local fields, on which the derivation of

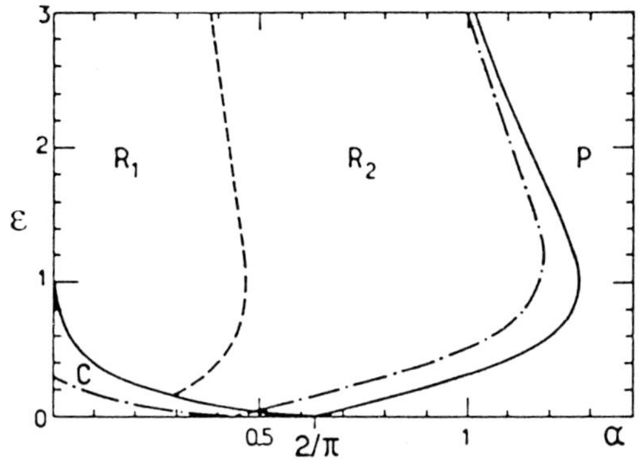

Fig. 7.15. Dynamic phase diagram for the dilute system described in the text. The constant-temperature curves, $T = 0$ (*full lines*) and $T = 0.5$ (*dot-dashed lines*) separate three regions of qualitatively different system behavior. In region C, the system remains in the previous state. This is the region of interest for the counting scheme. In region P, the system converges asymptotically to the paramagnetic state. The entire sequence is correctly retrieved in regions R_1 and R_2, where in the former the overlap with successive patterns exceeds 0.95, at $T = 0$. The C–R boudary has been computed for an initial overlap $Q^1 = 0.9$. Taken from [7.12]

(7.91–93) was based, is no longer guaranteed, as discussed above. However, if τ is sufficiently large, the system will settle into a *quasi*-stationary state well before $t = \tau$, the limit of validity of (7.91–93). Investigating the *quasi*-stationary values of the $m_\mu^\pm$ which are reached from the initial conditions $m_\mu^\pm(t) = Q^\mu$, $-\tau \le t \le 0$, one obtains the phase diagram shown in Fig. 7.15 [7.12].

It has been argued [7.12] that, because of the fact that for large values of the delay τ the system settles into a *quasi*-fixed point $m_\mu \simeq 0$, $m_{\mu+1} = Q^{\mu+1}$ before the asymmetric interactions induce another transition, the second transition could be described by same equations as the first, except that pattern labels are incremented by one. So, superimposed on the short-time dynamics (7.93) which describes the transitions between patterns μ and $\mu + 1$, there is a dynamics on a large timescale set by τ which connects the values of the quasi-stationary overlaps Q^ν of successive patterns in the sequence. From the m_μ^+ branch of (7.91), one finds that this long-time dynamics is described by the mapping

$$Q^{\nu+1} = F_w(Q^{\nu+1} + \varepsilon Q^\nu) \,. \qquad (7.94)$$

The Q^ν converge asymptotically to a nonzero value Q^* if $(1 + \varepsilon)F'_w(0) \ge 1$, i.e., if the noise is not too strong. Here noise refers both to thermal noise governed by the temperature and the static synaptic noise quantified by α. Note that it is possible to store more patterns in a sequence ($\varepsilon > 0$) than separately. As the noise increases, $Q^* \to 0$ continuously [7.12].

A related model operating with a range of equidistant time delays, the delay $n\tau$ belonging to synaptic couplings of the form $J_{ij}(n\tau) = (C_{ij}/C)\sum_\mu \xi_i^{\mu+n}\xi_j^\mu$, has been studied recently by Mato and Parga [7.39]. They find that the storage capacity for cycles depends on the delay distribution, the optimum being attained for delay distributions which are uniform over the range of delays.

Finally, it is perhaps worth pointing out that the conceptual separation between short- and long-time dynamics can be avoided if one assumes that the random connectivities for the instantaneous and the delayed signal paths are *independent*, thereby guaranteeing statistical independence of instantaneous and delayed contributions to each neuron's local field beyond $t = \tau$. For such a system, evolution equations in the manner of (7.93), which remain valid for $t > \tau$ and which can thus describe many consecutive transitions, can be derived fairly easily. The verification of this statement is left to the reader.

7.5 Applications: Sequence Recognition, Counting, and the Generation of Complex Sequences

In previous sections, we have studied several mechanisms for the storage and retrieval of temporal sequences in neural networks. In all cases, retrieval was required to be an autonomous process, i.e., given suitable initial conditions, the network was to reproduce a given sequence unaided by further input from an afferent system. With the provisional exception of the model of Dehaene et al., a second invariant of our considerations was that the sequences to be handled were required to be *simple* or *linear* in the sense that each pattern of the sequence had a uniquely specified successor. Plainly, this is a very severe restriction, and it is not met with, except by the most elementary variants of temporal sequence production encountered in biological systems. One may take a song with a refrain or the present text (as a sequence of letters) or whatever else comes to mind as examples to illustrate this point.

The present section is devoted to relaxing the above-mentioned constraints, albeit – as a first step – only separately[14]. Relaxing the requirement of *autonomous* sequence reproduction, we are led, in Sect. 7.5.1, to the problems of sequence recognition [7.8, 12] and counting [7.12, 13]. Dropping the linearity constraint on the sequences to be handled, we enter, in Sect. 7.5.2, the endless world of complex sequences [7.14, 15].

Our discussion will be in terms of the delay mechanism for sequence stabilization. It should be relatively easy to implement the recognition and the counting tasks in the models of Peretto and Niez [7.4] or Buhmann and Schulten [7.6, 7] as well, whereas it appears far less obvious how to deal with these problems in the setup of Dehaene et al. [7.5]. Complex sequences, on the other hand, can be generated in the model of Dehaene et al. but presumably not so easily in the models of Peretto and Niez or Buhmann and Schulten. The capablility to deal simply and efficiently with both sets of problems more or less singles out the delay mechanism.

[14] The avid reader will find it easy to reassemble the bits of the two puzzles in a fashion that will enable him or her to treat both cases together.

7.5.1 Sequence Recognition and Counting

Having discussed in great detail in the previous section autonomous sequence production and the transition mechanism for sequence generators operating with time delays, we can now afford to be fairly brief in describing the recognition and the counting task. In both cases, one assumes a sequence generator of the form introduced in Sect. 7.4.1, coupled to an afferent system that feeds an external signal $h_i^{\rm ext}(t)$ into the system. Assuming that the number of patterns is small compared to system size, and a δ-function delay for the asymmetric forward projections (7.4), we have a local field at i which is of the form

$$h_i(t) = \sum_{\mu=1}^{q} \left[\xi^\mu m_\mu(t) + \varepsilon \xi_i^{\mu+1} m_\mu(t-\tau)\right] + h_i^{\rm ext}(t) . \tag{7.95}$$

In Sect. 7.4.1 we have seen that there exists an $\varepsilon_{\rm min}(\beta)$ such that the sequence is autonomously reproduced if $\varepsilon > \varepsilon_{\rm min}(\beta)$. For $\varepsilon < \varepsilon_{\rm min}(\beta)$, however, the sequence is stored in the net but cannot be retrieved if the external signal vanishes.

In the recognition problem [7.8, 12], one assumes that the external signal is correlated with the patterns of the sequence to be recognized. That is, if the system has been in pattern μ for a sufficiently long time, the external signal which is to trigger the transition to $\mu+1$ is assumed to be of the form $h_i^{\rm ext}(t) = h_{\rm r}\xi_i^{\mu+1}$. To simplify the discussion further, we assume a parallel dynamics at $T = 0$, for which $\varepsilon_{\rm min} = 1$. If then at time t the system has been in pattern μ for a period exceeding the delay τ, so that the internally produced local field is of the form $h_i^{\rm int}(t) = \xi^\mu + \varepsilon\xi_i^{\mu+1}$, and an external signal correlated with $\xi_i^{\mu+1}$ arrives, we have

$$h_i(t) = \xi^\mu + (\varepsilon + h_{\rm r})\xi_i^{\mu+1} . \tag{7.96}$$

Thus, provided that $\varepsilon + h_{\rm r} > 1$, the system will perform a transition to $\mu + 1$ in the next synchronous update. If $h_{\rm r} < 1$, the transition is accomplished *only* through the combination of internal transition mechanism and external triggering field. Moreover, for the recognition scheme to work, the time spacing between the arrival of external signals must be greater than the delay τ of the asymmetric interactions.

In the counting problem [7.13], the pattern states ξ_i^μ are interpreted as *number* states. Transitions between successive number states should be induced by a sequence of identical external stimuli – *uncorrelated* with the number states themselves. One might think of the external stimuli as, for instance, chimes of a church bell, so that by counting them one would be able to tell the hour. To analyze the counting mechanism, let us assume that at time t the system has been in the number state μ for some time exceeding the delay τ, and a chime η_i, uncorrelated with the number states ξ^μ, arrives, so that the local field at i is

$$h_i(t) = \xi^\mu + \varepsilon\xi_i^{\mu+1} + h_{\rm c}\eta_i . \tag{7.97}$$

As in the recognition scheme, ε is assumed to be less than 1 so that a transition – if any – can only be induced by the external signal. To check whether a transition will indeed occur, we have to investigate the fate of those neurons for which $\xi_i^{\mu+1}$ differs from ξ^μ. Among these, we consider the subset I_μ where the chime η_i favors $\xi_i^{\mu+1}$, i.e.,

$$I_\mu = \{i;\ \eta_i = \xi_i^{\mu+1} = -\xi^\mu\}\ . \tag{7.98}$$

On I_μ the local field is given by

$$h_i(t) = \xi^\mu + \varepsilon\xi_i^{\mu+1} + h_c\eta_i = \xi_i^{\mu+1}(-1 + \varepsilon + h_c)\ . \tag{7.99}$$

Thus, in a synchronous update, the neurons in I_μ will change their state from ξ_i^μ to $\xi_i^{\mu+1} = -\xi_i^\mu$, *provided* that $h_c + \varepsilon > 1$. Since the number states and the chime were assumed to be uncorrelated, this change of state affects about a quarter of all neurons. Computing the overlaps at time $t + \Delta t$, we thus find

$$m_\mu(t + \Delta t) = m_{\mu+1}(t + \Delta t) = \tfrac{1}{2}\ , \quad m_\mu(t + \Delta t - \tau) = 1\ ,$$

while the remaining (instantaneous and retarded) overlaps are zero. Assuming that $h_i^{\rm ext}(t + \Delta t) = 0$ and using (7.95), we get

$$h_i(t + \Delta t) = \tfrac{1}{2}\xi^\mu + \xi_i^{mu+1}\left(\tfrac{1}{2} + \varepsilon\right) \tag{7.100}$$

for *all* i. The next update will therefore fully align the system with the number state $\xi_i^{\mu+1}$, i.e., the chime has incremented the counter by 1 as desired.

To construct a fully operative counting network one needs one additional element of architecture [7.13]: an attractor strongly correlated with the chime which is appended at the beginning of the sequence of number states and which is endowed with a relatively fast forward projection into the first number state. Its role is to recognize the first chime as special and to initialize the counting network. For further details concerning the embedding strength of this additional attractor and the strength of the forward projection into the first number state, the reader may consult [7.13].

Both the recognition and the counting scheme put critical conditions on the strength of the external stimulus. In general, the critical values for the counting and the recognition fields, h_c^* and h_r^*, will depend on the amount of thermal or static noise in the system, the latter being generated, for instance, by an extensive storage ratio α. That is, $h_r^* = h_r^*(\varepsilon, \beta, \alpha)$ and $h_c^* = h_c^*(\varepsilon, \beta, \alpha)$. In the noiseless limit considered above, the critical conditions for the recognition field, $h_r + \varepsilon > 1$, and the counting field, $h_c + \varepsilon > 1$, are identical, so that the system cannot differentiate whether an external signal is correlated with the next state of the sequence or not. However, the fact that the counting transition takes *two* time steps, whereas the recognition transition is completed in a *single* time step, already indicates that the two processes differ. In fact, it has been found [7.12] that under noisy conditions the critical value $h_c^*(\varepsilon, \beta, \alpha)$ for the counting field is generally *larger* than that for the recognition field, $h_r^*(\varepsilon, \beta, \alpha)$

Before turning to complex sequences, let us briefly mention a second variant of the recognition problem proposed by Gutfreund [7.40]. In its simplest form, it may be described in terms of two independent sequences of uncorrelated states $\{\xi^\mu\}$ and $\{\eta_i^\mu\}$, $1 \le \mu \le q$, where in addition to the usual synapses that store these sequences, one assumes (delayed) cross-projections of the form $\varepsilon N^{-1}\sum_{\mu=1}^{q}\xi_i^{\mu+1}\eta_j^\mu$ and $\varepsilon N^{-1}\sum_{\mu=1}^{q}\eta_i^{\mu+1}\xi_j^\mu$. The asymmetry parameter ε is taken to be small enough to prevent spontaneous transitions, either within each sequence or following the cross-projections. The external field, if properly chosen, now serves two purposes, namely to *trigger* transitions *and* to *select* which of the two transitions that are open for choice at any time is actually to be taken [7.12, 40]. Specifically, if the system has been in ξ_i^μ or η_i^μ for a time exceeding the delay τ, an external signal of the form $h_i^{\text{ext}}(t) = h_\tau \xi_i^{\mu+1}$ would trigger a transition into $\xi_i^{\mu+1}$, whereas an external signal of the form $h_i^{\text{ext}}(t) = h_\tau \eta_i^{\mu+1}$ would provoke a transition into $\eta_i^{\mu+1}$. Which of the 2^q possible paths then is actually traced out by the system depends on the *complete* sequence of external signals.

7.5.2 Complex Sequences

Complex-sequence generation in neural networks deals with the storage and retrieval of pattern sequences such as *"Complex-Sequence Generation in Neural Networks"*. In pattern sequences of this type, a state or pattern, here a letter, does not uniquely specify its successor. The letter *"e"*, for instance, is followed by *"x"* in *"complex"*, by *"q"* or *"n"* (or a word boundary) in *"sequence"*, by *"n"* or *"r"* in *"generation"*, and so on, so that there is an ambiguity whenever the system is in state *"e"*. Complex-sequence generation shares an important aspect with the second variant of the recognition task just discussed, namely, the necessity for path selection at points of the sequence where a succession of states is not unambiguously defined. In the recognition problem, paths are selected by the sequence of external signals, whereas in the case of complex-sequence generation, path selection must be based on past experience of the network itself. The dynamics must therefore be endowed with a *sufficiently long* memory span, so that ambiguities in the succession of states can be resolved.

Methods for complex-sequence generation were independently developed by Personnaz et al. and by Guyon et al. [7.14] – with the emphasis on exact (nonlocal) storage prescriptions and parallel dynamics – and by Kühn et al. [7.15] – closer in spirit to the models proposed by Kleinfeld [7.8] and by Sompolinsky and Kanter [7.9].

In the context of exact storage prescriptions based on pseudoinverse solutions, one stores a set of patterns $\{\xi^\mu; i = 1, \ldots, N\}$, $1 \le \mu \le q$, as fixed points of the threshold dynamics (7.1) by imposing on the J_{ij} instead of $\xi^\mu = \text{sgn}(\sum_j J_{ij}\xi_j^\mu)$ the stronger condition that

$$\xi^\mu = \sum_j J_{ij}\xi_j^\mu \tag{7.101}$$

should hold for all i and μ [7.22]. The virtue of this stronger condition is that it has transformed the nonlinear fixed-point conditions into a problem of linear algebra, which has been studied by Kohonen in the context of linear matrix memories [7.41]. Denoting by Ξ the $N \times q$ matrix of column vectors $\boldsymbol{\xi}^\mu = (\xi^\mu)_{i=1}^N$, one can rewrite the set of conditions (7.101) in terms of a matrix equation:

$$J\Xi = \Xi \, . \tag{7.102}$$

The solution can be formulated in terms of the Moore–Penrose pseudoinverse Ξ^I of Ξ as [7.22, 41, 42]

$$J = \Xi\Xi^{\mathrm{I}} + B(I - \Xi\Xi^{\mathrm{I}}) \, , \tag{7.103}$$

where B is an arbitrary $N \times N$ matrix. The matrix $\Xi\Xi^{\mathrm{I}}$ is just the minimal projection onto the space spanned by the $\boldsymbol{\xi}^\mu$, i.e., it satisfies $\Xi\Xi^{\mathrm{I}}\Xi = \Xi$. If the $\boldsymbol{\xi}$s are linearly independent, an explicit representation of $\Xi\Xi^{\mathrm{I}}$ can be given in terms of the correlation matrix (7.61),

$$(\Xi\Xi^{\mathrm{I}})_{ij} = N^{-1} \sum_{\mu,\nu} \xi^\mu (C^{-1})_{\mu\nu} \xi_j^\nu \, . \tag{7.104}$$

If the $\boldsymbol{\xi}$s are *not* linearly independent, one may take *any* maximal linearly independent subset, compute the correlation matrix for this set, and use (7.104) again, extending the sum only over the indices corresponding to the linearly independent subset chosen.

In a network operating with parallel dynamics, the same idea may be used to store a set of transitions $\boldsymbol{\xi}^\mu \to \boldsymbol{\xi}_i^{\mu+1}$, if $\boldsymbol{\xi}^\mu$ always has a uniquely specified succesor $\boldsymbol{\xi}_i^{\mu+1}$. The solution is

$$J = \Xi^+\Xi^{\mathrm{I}} + B(I - \Xi\Xi^{\mathrm{I}}) \, , \tag{7.105}$$

where Ξ^+ is the column matrix of the $\boldsymbol{\xi}_i^{\mu+1}$. The setup is such that transitions occur at every time step and is a straightforward generalization of the setup for unbiased binary random patterns discussed in the introduction. Again, an explicit representation of $\Xi^+\Xi^{\mathrm{I}}$ can be given in terms of the correlation matrix C of any maximal linearly independent subset of the $\boldsymbol{\xi}^\mu$ as

$$(\Xi^+\Xi^{\mathrm{I}})_{ij} = N^{-1} \sum_{\mu,\nu} \xi_i^{\mu+1} (C^{-1})_{\mu\nu} \xi_j^\nu \, . \tag{7.106}$$

However, the sequence must be free of bifurcation points for the above idea to work: $\boldsymbol{\xi}_i^{\mu+1}$ must always be uniquely determined by $\boldsymbol{\xi}^\mu$.

To resolve ambiguities occurring at bifurcation points of the sequence, it is necessary to have a nonzero memory span. The *minimal* memory span necessary for uniquely prescribing the future course of the network dynamics at any time is called the *order* or *complexity* of the sequence [7.5, 14]. In general, the order of a sequence should be measured in terms of the minimal number of states that must be traced back in time in order to specify each transition. In the context of exact storage prescriptions and parallel dynamics where transitions occur at every synchronous update, the order is thus equivalent to the memory span as

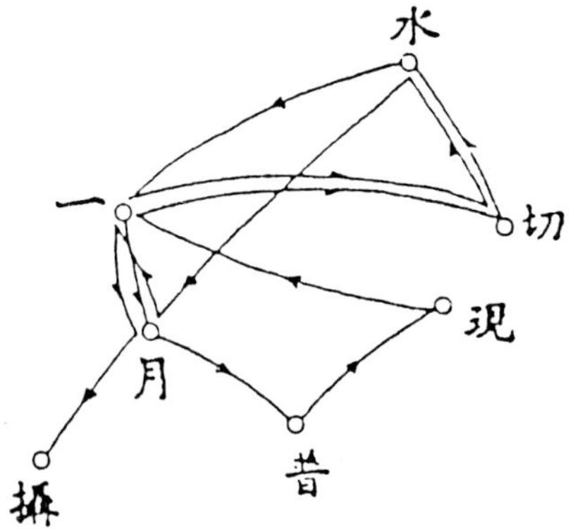

Fig. 7.16. Chinese poem representing a complex sequence of order three. Taken from [7.14]. A free translation of the poem is as follows:

The still water of the lake is bathed in moonlight,
Shining and reflecting the moon's face;
Yet the water and moon's image are visible
Only because of the full moon in the sky

(The authors are indebted to Lian-Ming Bo for the translation.)

measured in time steps. If, on the other hand, persistence times of individual states are controlled by time delays themselves – as is necessary for sequence generation in asynchronous neural nets – one must resort to the general prescription of counting states. Linear or cyclic sequences – more generally, sequences *down* a tree structure, possibly ending in turn in linear or cyclic sequences – have complexity zero. The Chinese poem depicted in Fig. 7.16 is a sequence of complexity three.

A general expression for a local field which has a memory span g, i.e., which depends on the system states $t, t-\Delta t, \ldots, t-g\Delta t$, is of the form [7.14]

$$\begin{aligned} h_i(t) = &\sum_{l=0}^{g} \sum_{j} J_{i,j}^{(l)} S_j(t-l) \\ &+ \sum_{l,l'} \sum_{j,j'} J_{i,jj'}^{(l,l')} S_j(t-l) S_{j'}(t-l') + \ldots \\ &+ \sum_{j_0,\ldots,j_g} J_{i,j_0,\ldots,j_g}^{(0,\ldots,g)} S_{j_0}(t) S_{j_1}(t-1) \cdots S_{j_g}(t-g) \,, \end{aligned} \tag{7.107}$$

where, to simplify the notation, we have put $\Delta t = 1$. This may be written in matrix form:

$$h_i(t) = \sum_a J_{i,a} \gamma_a(t) \,, \tag{7.108}$$

the vector $\gamma(t) = (\gamma_a(t))$ being obtained by a concatenation of the vectors $\boldsymbol{S}(t), \boldsymbol{S}(t-1), \ldots, \boldsymbol{S}(t-g), \boldsymbol{S}(t) \otimes \boldsymbol{S}(t), \ldots, \boldsymbol{S}(t) \otimes \boldsymbol{S}(t-1) \otimes \ldots \otimes \boldsymbol{S}(t-g)$.

To encode a sequence of global order g one proceeds as follows. For each transition $\mu \to \mu_+$, μ_+ denoting one out of *several* possible successors of μ, one constructs the vectors γ^{μ,μ_+} by a concatenation of those vectors $\boldsymbol{\xi}^\nu$ which lie within a memory span of length g of $\boldsymbol{\xi}^\mu$ and which represent the *unique* trajectory of network states that heads for the transition under consideration. One then demands that the matrix J solves

$$J\Gamma = \Xi^+ \,, \tag{7.109}$$

where Γ is the column matrix of the vectors $\gamma^{\mu,\mu+}$ and where Ξ^+ is the column matrix of the corresponding *target* states $\boldsymbol{\xi}^{\mu+}$, which are now uniquely specified by the $\gamma^{\mu,\mu+}$. A general solution to (7.109) can be given in terms of the pseudo-inverse Γ^{I} of Γ as

$$J = \Xi^+\Gamma^{\mathrm{I}} + B(I - \Gamma\Gamma^{\mathrm{I}}) . \tag{7.110}$$

At this point we will not discuss the problem of *learning* the matrix $\Xi^+\Gamma^{\mathrm{I}}$ needed for the general solution (7.110) in any detail. Suffice it to mention here that learning algorithms for static patterns can be generalized in a straightforward manner to encode a set of transitions *à la* (7.110). For details the reader may consult [7.14].

Because of the excessively high dimension of the vectors γ, the solution (7.110) turns out to be of little practical value in general. Moreover, the highest-order terms in (7.110) represent contributions to the local field produced by $(g+2)$-neuron interactions. In a fully connected network such interactions would raise severe wiring problems[15]. The freedom of choice of the matrix B in (7.110) may, however, be exploited to avoid as many of the higher-order terms in (7.107) and (7.110) as possible while still being able to resolve every ambiguity that might occur as the complex sequence is traced out. Guyon et al. [7.14], for instance, have compiled a table which summarizes the various possible solutions for sequences of global order 0 and 1.

Solutions avoiding higher-order interactions may be constructed by inspection if they exist. The strategy is based on a *local* complexity analysis of the sequence to be stored [7.14, 15], and it is perhaps most succinctly explained by describing transitions at bifurcation points in terms of their target and control states [7.15]. For simplicity, we shall restrict the ensuing discussion to the case where the set of patterns is linearly independent.

Let μ be a bifurcation point of local complexity g, and let us denote by R_μ the set of target states of the transitions that have μ as initial state. Once the system is in state μ, extra information is necessary for the decision which of the possible target sates in R_μ is to be selected. This piece of extra information is provided by past experience of the network. A subset G_μ of the set of network states is selected to function as control states as follows. The information which out of them were visited at specified times before the decision where to go from μ is on the agenda determines which of the possible target states in R_μ is to be selected; this information must enter the local field in such a manner that it initiates the selected transition at the appropriate time. So far, this prescription does not differ in principle from that embodied in the general solution (7.110). The point to note, however, is that the set of control states is quite often much smaller than the set of states which completely specify the $|R_\mu|$ different trajectories of length $g^* + 2$ that pass through μ, each ending in one of the target states in R_μ, which is used

[15] The problem may be less severe in nature than for artificial devices. In biological neural networks, the mutual influence of simple synapses on a thin dendrite, for instance, can provide higher-order interactions *without* additional long-distance wiring.

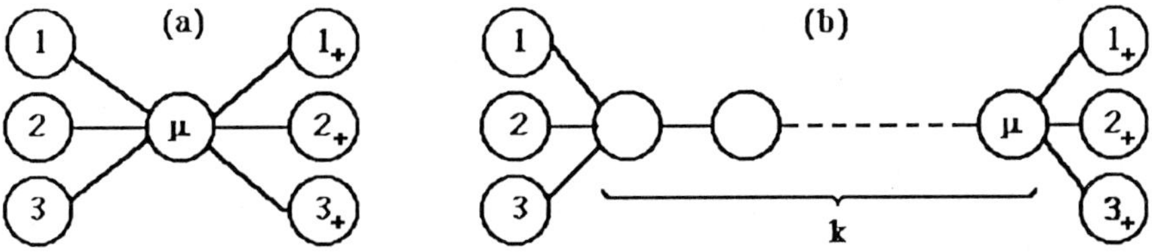

Fig. 7.17. (a) Bifurcation point of order 1. If μ is entered via ν, the next transition is $\mu \to \nu_+$. (b) Bifurcation point of order k. Here the transition from μ is into ν_+, if the common path of k states is entered via ν

to construct the general solution if the sequence is of global order $g^* \geq g$. (This holds in particular, if the local complexity of μ was zero to begin with, so that μ itself can act as control state for the transition $\mu \to \mu_+$.)

A bifurcation point μ of order 1, for instance, is generally of the form depicted in Fig. 7.17a The transitions emanating from μ are encoded in the synapses

$$J_{ij}^{(\mu,1)} = N^{-1} \sum_{\gamma \in G_\mu} \sum_{\nu} \xi_i^{\gamma+} (C^{-1})_{\gamma,\nu} \xi_j^{\nu} , \tag{7.111}$$

which act with a delay of *one* time step and thus give rise to a contribution $h_i^{(\mu,1)}$ of the local field which is of the form

$$h_i^{(\mu,1)} = \sum_{\gamma \in G_\mu} \sum_{\nu} \xi_i^{\gamma+} (C^{-1})_{\gamma,\nu} m_\nu(t-1) = \sum_{\gamma \in G_\mu} \xi_i^{\gamma+} a_\gamma(t-1) ; \tag{7.112}$$

cf. (7.63). Second-order interactions, as would occur in the general solution (7.110) if the sequence were of global order 1, have thus been avoided. The benefits of local coding are even more substantial in a situation such as that shown in Fig. 7.17b. Here the bifucation point μ is of order $g = k$, so that the general solution would involve $k+2$-neuron interactions (at least), creating local fields of order $k+1$ (at least). The solution by inspection encodes the transitions emanating from μ in coupling constants of the same structure as (7.111),

$$J_{ij}^{(\mu,k)} = N^{-1} \sum_{\gamma \in G_\mu} \sum_{\nu} \xi_i^{\gamma+} (C^{-1})_{\gamma,\nu} \xi_j^{\nu} , \tag{7.113}$$

which now transmit information with a time delay of k time steps, so that their contribution to the local field is

$$h_i^{(\mu,k)} = \sum_{\gamma \in G_\mu} \xi_i^{\gamma+} a_\gamma(t-k) . \tag{7.114}$$

Two remarks are in order. First, a solution in terms of pair interactions or, equivalently, local fields linear in the a_γ, may not always exist, as the reader will verify by studying the example shown in Fig. 7.18a. That is, *the sequences that can be stored depend on the architecture of the network*, where architecture refers to the maximally available order of the interactions as well as to the spectrum of available transmission delays [7.5, 14, 15, 17]. Second, in constructing solutions

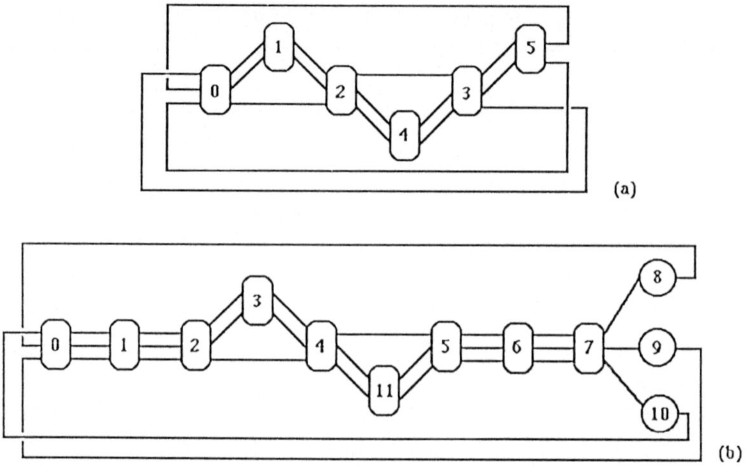

Fig. 7.18. (**a**) A Complex sequence which cannot be encoded by pair interactions. The sequence is traced out in the order 0-1-2-3-5-0-1-2-4-3-5-0-2-, and so on. (**b**) Complex sequence illustrating the antagonistic optimization criteria mentioned in the main text. The order in which the states are visited can be read off from the picture in analogy to (a)

by inspection, one encounters two antagonistic optimization criteria, namely, keeping delays as small as possible versus keeping the order of interactions as low as possible. Fig. 7.18b provides an example where transitions emanating from state 7 are correctly induced by a local field of the form

$$h_i^{(7,9)} = \xi_i^8 a_{10}(t-8) + \xi_i^9 a_8(t-9) + \xi_i^{10} a_9(t-8) ,$$

using the states 8, 9, and 10 as control states and involving delays as long as 9 time steps. Alternatively one may use 2, 3, 4, and 5 as control states and induce the transitions emanating from 7 by a local field which is now *bilinear* in the as, viz.,

$$h_i^{(7,5)} = \xi_i^8 a_5(t-2)a_4(t-3) + \xi_i^9 a_5(t-2)a_3(t-5) + \xi_i^{10} a_4(t-4)a_2(t-5) .$$

If one uses asynchronous instead of parallel dynamics, the persistence times of the individual quasi-attractors are controlled by time delays themselves and the transmission delays used above have to be adapted accordingly. Finally, if delay mechanisms other than the δ-function delay are used, one must introduce additional symmetric synapses *à la* (7.103) to stabilize the individual patterns as quasi-attractors of the dynamics, and higher-order interactions will generally be needed to control transitions at bifurcation points, even in those cases where pair interactions can suffice if the delay is sharp [7.15]. It goes without saying (cf. Sect. 7.4.1) that the formulae for the synapses used in the present section simplify if the sequences consist of unbiased binary random patterns or if correlations between patterns have their sole origin in a common nonzero level of activity of the (otherwise random) patterns.

We have seen that complex-sequence generation requires time delays of sufficient length to memorize as much of the history of the network evolution as is necessary to resolve any ambiguity that might come along as the complex sequence is traced out. Dehaene et al. [7.5], on the other hand, use dynamic prescriptions which embody a memory span of no more than two elementary time steps, $2\Delta t$, but are still able to produce complex sequences which seem to require a much longer memory span than that, persistence times of individual cluster activities already being generally much larger than $2\Delta t$. Is there a contradiction? The answer is of course no, once we note that the sequences produced in the network of Dehaene et al. are *complex only at the level of the output clusters*, and that – taking internal clusters into account – the network as a whole only goes through linear sequences.

As it stands, it appears that our discussion of complex-sequence generation would be relevant to *artificial* devices only. One may, however, introduce minor modifications of architecture that would render some of our considerations even biologically plausible [7.15]. On the other hand, it seems that interesting and important examples of complex-sequence production in nature (the generation of speech, for instance) would require time delays on a scale not attainable in neural systems. To cope with these, one will certainly have to leave the realm of *unstructured* nets that has been underlying all our modeling so far and invest in *structure* for obtaining higher levels of neural processing.

7.6 Hebbian Learning with Delays

Stepping back for an overview, we cannot fail to notice that the models for sequence generation described so far still lack a very important quality, i.e., the possibility of acquiring knowledge through a true learning mechanism. Values for the synaptic efficacies have to be computed separately and are put into the models by hand, the most delicate – if not fragile – version perhaps being that of Buhmann and Schulten, discussed in Sect. 7.3. Certain forms of synaptic plasticity as well as time delays, on the other hand, were also introduced more or less *ad hoc* as mechanisms to stabilize sequential recall, *given* a certain structure of the couplings. Superficially, the model proposed by Dehaene et al. would seem to constitute an exception to such a "no-learning" verdict. Upon closer inspection, however, we were led to conclude that, at a deeper level, the learning phase in this model rather serves as a more elaborate version of setting initial conditions for the retrieval dynamics than as a phase where the network acquires persistent knowledge about the outside world.

A learning mechanism, if it is to be convincing, should enjoy at least the same degree of neurophysiological plausibility as, for instance, the Hopfield–Cooper [7.3, 43] prescription (7.3), which can be said to constitute a particular formalization of Hebb's neurophysiological postulate for learning [7.16]. Moreover, it should store static patterns – if then they are perceived as such – and pattern sequences *on the basis of one and the same principle.*

Such a learning mechanism has been proposed by Herz et al. [7.17,44,45] and, independently, by Coolen and Gielen [7.46] and by Kerszberg and Zippelius [7.47]. The proposal is the result of a scrutiny of nothing but the venerable Hebbian learning rule itself[16]. A careful analysis of this rule reveals that it is not formulated without implicit reference to the underlying network dynamics. For the sake of internal consistency it is then imperative to describe the retrieval properties of the network in terms of the *same* dynamical rules that were assumed to govern the system behavior while synaptic modifications *à la* Hebb were taking place, i.e., during the learning phase. There is an important second ingredient which finally sets the stage for the proposed refinement of the Hebb rule, namely, the observation that signal transmission delays in neural nets not only exist but are nonnegligible compared to absolute or relative refractory periods, *and* that their distribution is broad [7.29, 30]. With respect to this observation, the obvious conclusion is of course that transmission delays, as they occur in the neurodynamics, should be taken into account during the learning phase as well. Actually, this conclusion is not brand-new. Remarks to its end can already be found in Hebb's book [7.16]. It appears quite succinctly in Caianiello's early work on thought processes and thinking machines [7.48] and from there found its way into the work of Grossberg [7.49], Amari [7.50], and Fukushima [7.51]. More recently, it has subcutaneously reappeared in the work of Peretto and Niez [7.4] and Amit [7.13], but its full consequences have apparently never been explored, either qualitatively or quantitatively.

We shall find that transmission delays play a decisive role in encoding the temporal aspects of external stimuli to be learnt – no matter what particular formalization of the learning rule is chosen.

Let us then embark upon a careful analysis of Hebb's neurophysiological postulate for learning in the context of *formal* neural networks and let us try to be as explicit as possible about the underlying assumptions concerning transmission delays and dynamics.

Dynamics first. Neurons are assumed to change their state according to the value of the total input potential (postsynaptic potential or local field). Formally,

$$S_i(t + \Delta t) = g\big(h_i(t)\big) \ , \tag{7.115}$$

where g is some nonlinear input–output relation whose precise form depends on the way one chooses to represent the various levels of neural activity. In most of the present paper, we have opted for the ± 1 representation, for which $g(x) = \mathrm{sgn}(x)$; in Sect. 7.3 we also considered the 0-1 representation, with $g(x) = \theta(x)$, and in Sect. 7.2.2 even a continuous input–output relation. Updating in (7.115) may be asynchronous or in parallel, and the elementary time step Δt must be adapted accordingly; cf. Sects. 7.3 and 7.4. Finally, one may consider replacing the deterministic rule (7.115) by a probabilistic version thereof, so as to take noise effects into account.

[16] A citation of Hebb's own paraphrase of this rule can be found in Sect. 7.5.4 of [7.20].

The total input potential of a given neuron is a sum of the signals it receives from all other neurons having synaptic connections with it. This leads us, as a first step, to *axonal* transmission delays. Taking them into account, we find

$$h_i(t) = \sum_j J_{ij} S_j(t - \tau_{ij}) \ , \tag{7.116}$$

where τ_{ij} is the time needed for a signal emitted at j to travel along j's axon before it arrives at i. Nondispersive – soliton-like – signal propagation described by a δ-function delay is implicitly assumed in (7.116). More general delay mechanisms – including capacitive (post)synaptic delays at i – may be contemplated and can be included in the scheme in a straightforward manner [7.17,44,45]. In (7.116), J_{ij} denotes the synaptic efficacy for the information transport from j to i. It is at the synapses where – according to Hebb – information is stored, in that each synaptic efficacy is changed in a way that depends upon a piece of *correlation* information, namely, on whether the presynaptic neuron *has contributed to firing* the postsynaptic neuron or not.

Recalling that the dynamical rule (7.115) specifies the firing state of neuron i at time $t + \Delta t$ according to the value of its postsynaptic potential h_i at time t, and that the contribution of neuron j to $h_i(t)$ is given by its firing state at $t - \tau_{ij}$, we conclude that the relevant correlation information that determines the change of the synaptic efficacy J_{ij} depends on the states of the pre- and postsynaptic neurons at times $t - \tau_{ij}$ and $t + \Delta t$, respectively[17]:

$$\Delta J_{ij}(\tau_{ij};t) = \phi_{ij}\big(S_i(t+\Delta t)\, ,\, S_j(t-\tau_{ij})\, ,\, J_{ij}(\tau_{ij};t)\big) \ . \tag{7.117}$$

Equation (7.117) provides a general though rather formal description of Hebbian learning in networks with delays. The dependence of the *mnemodynamic* updating function ϕ on i and j can serve to take morphological characteristics of the connection $j \to i$ into account. The inclusion of $J_{ij}(\tau_{ij};t)$ in the list of its arguments appears to be necessary if one wants to model saturation effects so as to get synaptic efficacies of bounded strength, thereby modeling forgetfulness.

In what follows, we consider only the simplest case, also studied by Herz et al. [7.17], where ϕ describes a simple imprinting process without saturation effects, and where at each instant of time J_{ij} changes by an amount proportional to[18] $S_i(t+\Delta t)S_j(t-\tau_{ij})$. Starting with a *tabula rasa* ($J_{ij} = 0$), and assuming a continuous time sequential dynamics ($\Delta t \propto N^{-1} \to 0$), we get, adding partial increments,

$$J_{ij}(\tau_{ij}) = \frac{\varepsilon_{ij}}{N}\frac{1}{T_0}\int_0^{T_0} S_i(t)S_j(t-\tau_{ij})\mathrm{d}t \tag{7.118a}$$

for simple δ-function delay, and the same expression with $S_j(t - \tau_{ij})$ replaced

[17] We owe the $+\Delta t$ bit of this observation to A. Herz.

[18] Strictly, this choice is adapted to (and optimal only for) a network storing unbiased binary random patterns.

by $\bar{S}_j(t)$ for a more general delay mechanism (cf.(7.51)). Here T_0 denotes the duration of the learning session, and $\varepsilon_{ij} = \varepsilon_{ij}(\tau_{ij})$ is a weight which may carry further detailed information about the connection $j \to i$.

In the case of parallel dynamics, time is discrete. Without loss of generality, we may put $\Delta t = 1$ and obtain

$$J_{ij}(\tau_{ij}) = \frac{\varepsilon_{ij}}{N} \frac{1}{T_0} \sum_{t=1}^{T_0} S_i(t+1) S_j(t - \tau_{ij}) \qquad (7.118b)$$

instead of (7.118a). Note the shift by one time step in the argument of S_i, which reflects the important role of dynamics in a proper formulation of the Hebb rule. We shall have occasion to elaborate on this point later on.

Hebbian synapses $J_{ij}(\tau)$ encode correlations of external signals in space (ij) and time (τ). The coding is distributed. In systems with a large connectivity, it is also highly redundant. If then the network dynamics described by (7.115) and (7.116) is at all able to extract the spatio-temporal information encoded in the $J_{ij}(\tau)$, it can be expected to do so in a robust and fault-tolerant fashion, which is largely independent of details of the distribution of the delays τ. To illustrate this point, Herz et al. have considered three models which differ substantially in the way the connections and their associated delays are organized [7.17, 44].

(A) For each pair (i, j) there is a *large* number of axons (possibly, interneurons) whose delays τ have a distribution independent of i and j. Summing over incoming signals, one obtains the postsynaptic potential

$$h_i(t) = \sum_j \sum_\tau J_{ij}(\tau) S_j(t - \tau) \ . \qquad (7.119a)$$

The weights $\varepsilon_{ij} = \varepsilon_{ij}(\tau)$ are assumend to be independent of i and j and are chosen according to a given distribution of the delays τ.

Of course, a more standard assumption would be to have only a single connection for each pair (i, j). This leads to (at least) two further models.

(B) In addition to the axonal delay, which is assumed to depend on j only, there is a (post)synaptic exponential delay at i. Thus, the total delay τ_{ij} is split into τ_i and τ_j so that

$$h_i(t) = \sum_j J_{ij}(\tau_i, \tau_j) \bar{S}_j(t) \ , \qquad (7.119b)$$

where $\bar{S}_j(t)$ incorporates the exponential and δ-function delays with time constants τ_i and τ_j, respectively. The τ_i and τ_j are drawn from given distributions, and $\varepsilon_{ij}(\tau_i, \tau_j) = 1$.

(C) For each pair (i, j) there is a single axon with delay τ_{ij} which is sampled from a given distribution *independent* of i and j. This gives rise to

$$h_i(t) = \sum_j J_{ij}(\tau_{ij}) S_j(t - \tau_{ij}) \ . \qquad (7.119c)$$

Again, $\varepsilon_{ij}(\tau_{ij}) = 1$.

Models A and B are random-site problems; model C is a random-bond problem.

As may perhaps be anticipated, the problem of encoding and recalling a temporal pattern sequence may serve as a paradigm to bring out the most salient features of the Hebbian learning rule (7.118). For the sake of definiteness we consider a cycle of q unbiased binary random patterns $\{\xi_i^\mu; 1 \le i \le N\}$ with $1 \le \mu \le q$, each of duration Δ. That is, during the learning session, the system is exposed to an external stimulus of the form

$$S_i(t) = \xi_i^{\nu(t)} , \quad \nu(t) = \mu \bmod q \quad \text{for} \quad (\mu - 1)\Delta \le t < \mu\Delta , \tag{7.120}$$

which amounts to a *clamped* learning scenario.

To compute the synaptic efficacies $J_{ij}(\tau)$ which are generated according to the Hebb rule (7.118) in response to the external stimulus (7.120), one splits τ into integer and fractional parts of Δ, $\tau = (n_\tau + d_\tau)\Delta$, with n_τ a nonnegative integer and $0 \le d_\tau < 1$. For a learning session that extends over one or several complete sweeps through the cycle, one obtains after a little algebra

$$J_{ij}(\tau) = \frac{\varepsilon(\tau)}{N}\left\{(1 - d_\tau)\sum_{\mu=1}^{q} \xi_i^{(\mu+n_\tau)\bmod q}\xi_j^\mu + d_\tau \sum_{\mu=1}^{q} \xi_i^{(\mu+n_\tau+1)\bmod q}\xi_j^\mu\right\} \tag{7.121a}$$

for sequential dynamics, and similarly

$$J_{ij}(\tau) = \frac{\varepsilon(\tau)}{N}\left\{(1 - d_\tau^+)\sum_{\mu=1}^{q} \xi_i^{(\mu+n_\tau)\bmod q}\xi_j^\mu + d_\tau^+ \sum_{\mu=1}^{q} \xi_i^{(\mu+n_\tau+1)\bmod q}\xi_j^\mu\right\} \tag{7.121b}$$

for parallel dynamics. Here $d_\tau^+ = d_\tau + \Delta^{-1}$. These expressions may be used for models A and C. The relevant expression for model B, which combines the effects of exponential and δ-function delays, is voluminous, and may be found in [7.17]. In all cases, however, the J_{ij} are bilinear in the ξs and may therefore be written

$$J_{ij}(\tau) = N^{-1} \sum_{\mu,\nu} \xi_i^\mu Q_{\mu,\nu}^{(\tau)} \xi_j^\nu \tag{7.122}$$

for some matrix $Q^{(\tau)}$, where in the case of model B, τ stands for a pair (τ', τ); cf. [7.17].

Before proceeding to an analysis of the retrieval properties, let us point out the following. First, apart from the a priori weights $\varepsilon(\tau)$, (7.121) clearly exhibit a resonance phenomenon: the $J_{ij}(\tau)$ with delays which are *integer* multiples of Δ and thus match the timing of the external stimulus are the ones that receive maximum strength[19] (see Fig. 7.19). Note that they are also the ones that would support a stable cycle of exactly the same period (neglecting transition times). Thus, through a subtle interplay between external stimulus and internal architecture (distribution of the τs), the Hebb rule (7.118), though at the heart of it instructive in character, has in fact also pronounced selective aspects. Second, in

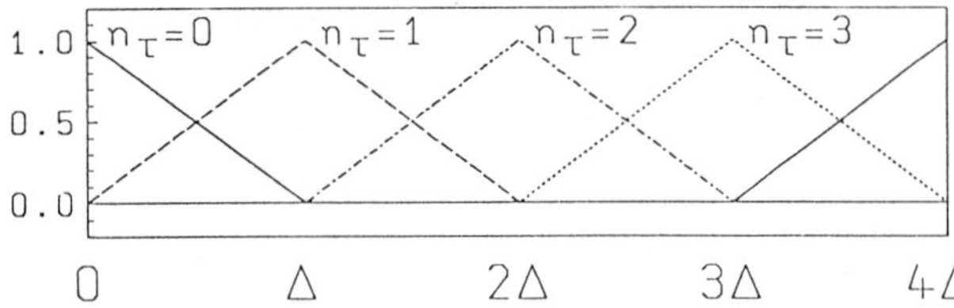

Fig. 7.19. Contributions to the synaptic efficacies $J_{ij}(\tau)$ as a function of the delay τ. The *solid line* gives the weight of contributions of the form $\xi_i^\mu \xi_j^\mu$, corresponding to $n_\tau = 0$. The *dashed line* shows the weight of contributions of the form $\xi_i^{\mu+1} \xi_j^\mu$, for which $n_\tau = 1$; and so on. The weights are computed for asynchronous neural nets. Taken from [7.17]

a neural net with some given distribution of transmission delays τ, the external stimulus encoded in the $J_{ij}(\tau)$ will enjoy a rather multifaceted representation. For instance, the $J_{ij}(\tau)$ with $\tau/\Delta \ll 1$ will be almost symmetric, and they will therefore encode the individual patterns of the sequence *as unrelated static objects*. On the other hand, synapses with transition delays of the order of Δ (or more) will be able to detect the transitions between the patterns in the sequence. The corresponding synaptic efficacies will be asymmetric and establish various temporal relations between the patterns of the cycle, *thereby representing the complete cycle as a dynamic object.*

Since a full characterization of the models introduced above requires the specification of a large – in the case of continuous-delay distributions even an infinite – set of parameters, an exhaustive description of their dynamic features is out of the question, and the presentation of a number of generic results must suffice. These can be obtained by way of simulations or – in the low loading limit – analytically [7.17, 44, 47]. For models of type A and B, the sublattice techniques introduced in Sect. 7.4.2 are appropriate (model B requiring only a mild extension of the sublattice concept [7.17]); models of type C were solved by Kerszberg and Zippelius [7.47] using dynamic mean-field techniques.

It is found that the behavior of the systems described by (7.120–122) hardly depends on the way the connections and their associated delays are organized, i.e., on whether the synapses (7.122) are put into models of type A, B, or C; see Fig. 7.20. This holds at least for low and moderate levels of loading. The dependence on the range and distribution of delays is dichotomic. Either the learnt sequence is reproduced fairly accurately, and a dependence on the τ-distribution is hard to discern, or the cycle does not run at all because, as a rule of thumb, the duration Δ of each pattern exceeds $\tau_{\max}$, so that the system has no means of measuring the patterns' lifetime; see Fig. 7.21. Note also that the distribution of delays may be such that symmetric synapses are completely absent, but nevertheless the learnt sequence is stored as a stable limit cycle. In the case of model B, there is an upper bound for the (post)synaptic exponential delays beyond which the system will not be able to learn and reproduce an

[19] The situation is slightly different in the case of parallel dynamics.

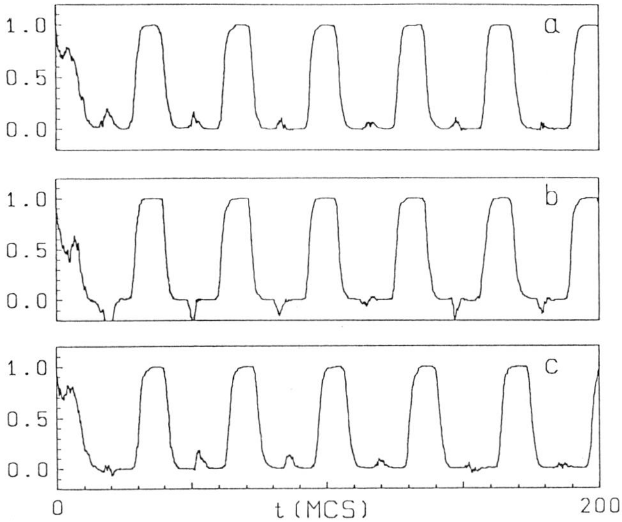

Fig. 7.20a–c. Model dependence. The overlap with the first pattten of a 3-cycle is shown as a function of time. (a) model A. (b) model B. (c) model C. In all three cases the number of neurons is $N = 256$. Dynamics is sequential at $\beta = 10$. The initial conditions are $m_1(t) = 1$ for $-1\,\text{MCS} \leq t \leq 0$ and $m_\mu(t) = 0$ otherwise. The distribution of axonal δ-function delays is discrete with support at $\tau = 0, 1, \ldots, 30$ MCS. There are no (post)synaptic exponential delays. The weights $\varepsilon(\tau)$ are uniform and normalized to 1. During the learning session each pattern lasted $\Delta = 10\,\text{MCS}$. Taken from [7.17]

external stimulus of the form (7.120). If the delay is too large – relative to the presentation times of the individual patterns – the signals to be learnt are washed out too much and their temporal correlations are no longer faithfully represented in the synaptic code, since their variations on short timescales are filtered away; see Fig. 7.22. Sequences with a *range* of persistence times for the individual patterns can also be learnt and are – within limits, of course – faithfully reproduced.

The above results pertain to simple sequences of unbiased binary random patterns. To remove such restrictions on the sequences to be learnt, one has to go beyond the simple *one-shot* learning algorithm described by (7.118), a task that was recently attacked by Bauer and Krey. They have considered supervised variants of learning networks with delays, and they have obtained encouraging first results for complex sequences of correlated patterns [7.52].

Systems endowed with Hebbian synapses *à la* (7.118) function as pattern recognizers for the learnt patterns, where patterns are now, in general, spatio-temporal objects. They may, but need not, be static. Whether they are perceived as such will depend on the system architecture (its distribution of delays)[20]. During retrieval, error correction is performed, and recognition will be successful, if the triggering stimulus is close enough in space *and* time to one of the

[20] We regard this aspect as natural from a psychological point of view. Interestingly, the range of a memory span in a society, as embodied e.g. in its cultural and intellectual traditions, will also influence whether, in a historical perspective, things are perceived as static or dynamic.

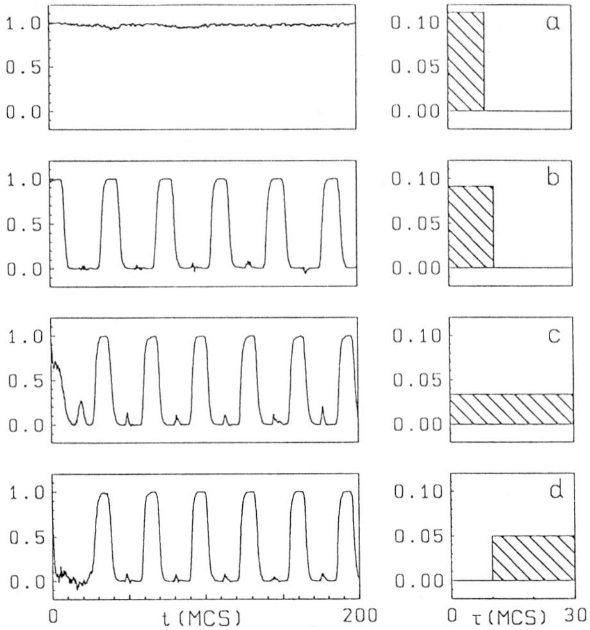

Fig. 7.21. Dependence on the delay distribution for model B. Each track represents the time evolution of the overlap with the first pattern of a 3-cycle. The delay distribution is displayed in a box to the right of the $m_1(t)$ plot to which it belongs. All delay distributions are discrete with a spacing of one MCS. The sequence can be retrieved provided that $\Delta < \tau_{\max}$. Except for this proviso, variations in the τ distribution hardly change the long time behavior, though they do affect the transients. The system size is $N = 512$, dynamics is sequential, and β and Δ are as in Fig. 7.20. Taken from [7.17]

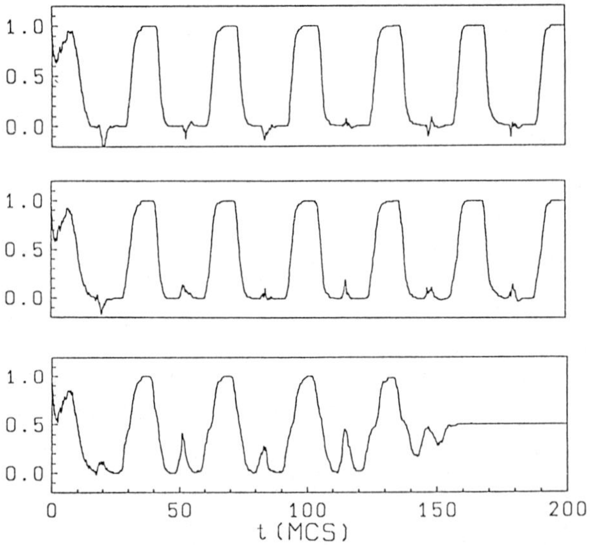

Fig. 7.22. Influence of exponential (post)synaptic delays for model B. From top to bottom we have $\tau' = 0.15\Delta$, 0.20Δ, and 0.25Δ. Except for $\beta = 20$, other network parameters are as in Fig. 7.20

stored prototypes. Domains of attraction are generally found to be much greater than, for instance, in systems operating without delays or with a single delay [7.17,44,45,47,52].

The problem of Hebbian learning in networks with transmission delays was also considered by Coolen and Gielen [7.46]. They concentrated on networks operating with *parallel* dynamics and studied sequences of patterns $\{\xi_i^{\nu(t)}\}$ which change at every synchronous update – so that one may write $\xi_i^{\nu(t)} = \xi_i(t)$ – but which may have a finite correlation time, entailing

$$N^{-1} \sum_i \xi_i(t)\xi_i(t') = \Gamma(t - t') \tag{7.123}$$

for some function Γ, as $N \to \infty$. Patterns of this sort may be generated, for instance, by a *stationary* homogeneous Markov process. Since they change at every synchronous time step, a network recalling such a sequence of patterns does not allow a meaningful distinction between transient and persistent states. The latter, which have been hailed as signifying recognition or cognitive states in neural networks, simply do not exist.

To store such a sequence of patterns, Coolen and Gielen use a modification of Hebb's rule which accounts for transmission delays,[21]

$$J_{ij}(\tau_{ij}) = \frac{\varepsilon_{ij}}{N}\frac{1}{T_0}\sum_{t=1}^{T_0} S_i(t)S_j(t - \tau_{ij}) \ , \tag{7.124}$$

but which is non-Hebbian in that it is incompatible with the underlying parallel dynamics. The reader will recall our discussion of the dynamical aspects of Hebbian learning and is invited to compare (7.118b) and (7.124). Indeed, if one specializes to orthogonal patterns with $\Gamma(t-t') = \delta_{t,t'}$ in (7.123), it is found that under the zero-temperature dynamics

$$S_i(t + 1) = \mathrm{sgn}[h_i(t)] \tag{7.125}$$

a sequence of such patterns, if encoded *à la* (7.124), is *not* reproduced as expected. To get at least consistent results, Coolen and Gielen had to sacrifice dynamics, replacing (7.125) by the *self-consistency condition* $S_i(t) = \mathrm{sgn}[h_i(t)]$. Remarkably, however, correct variants of their main results can be established if the correct learning rule for parallel dynamics, i.e. (7.118b), is chosen so that dynamics can be reinstantiated[22]. The results are the following.

(i) Sequences of mutually orthogonal patterns can be learnt and recalled by (7.118b), independently of the delay distribution. Remember that we have seen in the introduction, that such sequences can already be stored and retrieved by

[21] In what follows, our notation, but also the underlying physical content, differs slightly from [7.46], so as to avoid some minor inconsistencies contained in that paper.

[22] There is also need for modification at a technical level, mainly related to replacing time integrals (some of them ill-defined) by discrete sums as appropriate for parallel dynamics.

the Hopfield prescription (7.4) – without any delays. Indeed, (7.4) is precisely the limit of (7.118b) if all delays vanish. The arguments of [7.46] and our introduction can be supplemented by a standard signal-to-noise-ratio analysis to find that a sequence of q unbiased binary random patterns can be stored and recalled without any error (as $N \to \infty$), provided that $q < q_c = N/4 \ln N$, again independently of the delay distribution.

(ii) For ensembles of patterns described by (7.123), storage and recall correlated at any time with the encoded sequence is possible. Recall does not so much depend on the delay distribution as on the correlation properties of the embedded sequence, as expressed through (7.123). This result can be established under the condition that there are only finitely many different delays.

Except for the necessary modifications indicated above, the ensuing arguments closely follow [7.46].

Let us first consider the case of mutually orthogonal patterns, for which $\Gamma(t-t') = \delta_{t,t'}$ in (7.123). Let us also assume that such a sequence has been encoded in synapses of the form (7.118b) and that the system was initialized with the correct sequence, i.e., that $S_i(t') = \xi_i(t'-t_0)$ for $t' \leq t$ and some t_0. We then have

$$h_i(t) = \frac{1}{T_0} \sum_{t'=1}^{T_0} \xi_i(t'+1) \frac{1}{N} \sum_{j(\neq i)} \xi_j(t'-\tau_{ij}) \xi_j(t-t_0-\tau_{ij}) \, . \tag{7.126}$$

Because of the orthogonality condition for the patterns this gives

$$h_i(t) = \frac{1}{T_0} \{\xi_i(t+1-t_0) + \mathrm{O}(T_0/N)\} \, . \tag{7.127}$$

For $T_0 \ll N$, the $\mathrm{O}(T_0/N)$ contribution originating from the $j = i$ terms in (7.118b) can safely be ignored so that the next parallel update will align the system with $\xi_i(t+1-t_0)$ as desired, given the above initialization. Note that, because of the way the delays entered the learning rule and the retrieval dynamics, they have completely dropped out of the final result. If the patterns had been assumed to be random and unbiased rather than strictly orthogonal, the same initialization would give rise to a local field of the form

$$h_i(t) = \frac{1}{T_0} \xi_i(t+1-t_0)\{1 + \delta_i + \mathrm{O}(T_0/N)\} \, , \tag{7.128}$$

where δ_i now is a random-noise term of mean zero. For large T_0 its distribution is a Gaussian of width $\sqrt{T_0/N}$, dominating the $\mathrm{O}(T_0/N)$ contribution. A standard signal-to-noise-ratio analysis then reveals that the system will recover the next pattern of the sequence without errors (as $N \to \infty$) if $T_0 < N/2 \ln N$, while to recover the complete sequence without errors one needs $T_0 < N/4 \ln N$. Note that T_0 equals the number q of distinct new patterns encountered during the learning session.

Let us next consider the case of a sequence of nonorthogonal patterns with time invariant correlations as expressed through (7.123), and let us relax the condition of perfect initialization. Following [7.46], we shall write

$$S_i(t') = \varepsilon_i(t')\xi_i(t' - t_0)\ , \quad t' \le t\ , \tag{7.129}$$

where the $\varepsilon_i(t')$ are random and ± 1 with probabilities depending on the quality of initialization. We shall assume that during initialization there were always macroscopic positive correlations with the embedded sequence, i.e.,

$$Q(t') = N^{-1}\sum_i \varepsilon_i(t') > 0\ , \quad t' \le t\ . \tag{7.130}$$

One then may compute

$$\begin{aligned}\varepsilon_i(t+1) &= \mathrm{sgn}[h_i(t)\xi_i(t+1-t_0)] \\ &= \mathrm{sgn}\left[\xi_i(t+1-t_0)\frac{1}{T_0}\sum_{t'=1}^{T_0}\xi_i(t'+1)\right. \\ &\quad \left.\times\frac{1}{N}\sum_{j(\neq i)}\xi_j(t'-\tau_{ij})\varepsilon_j(t-\tau_{ij})\xi_j(t-t_0-\tau_{ij})\right]\ .\end{aligned} \tag{7.131}$$

To evaluate (7.131), Coolen and Gielen introduce the sublattices

$$I_i(t,\tau,\sigma) = \{j;\ \tau_{ij} = \tau,\ \varepsilon_j(t-\tau) = \sigma\};\ .$$

If there are only finitely many delays (or if their number is $o(N)$, as $N \to \infty$), the $I_i(t,\tau,\sigma)$ are all of macroscopic size. Moreover, the correlation properties of the sequence expressed through (7.123) will hold separately for each of the sublattices, so that

$$\frac{1}{|I_i(t,\tau,\sigma)|}\sum_{j\in I_i(t,\tau,\sigma)}\xi_j(t'-\tau)\xi_j(t-t_0-\tau) = \Gamma(t'-t+t_0) \tag{7.132}$$

as $N \to \infty$. Thus

$$\begin{aligned}\varepsilon_i(t+1) = \mathrm{sgn}&\left[\xi_i(t+1-t_0)\frac{1}{T_0}\right. \\ &\left.\times\sum_{t'=1}^{T_0}\sum_{\tau,\sigma}\frac{|I_i(t,\tau,\sigma)|}{N}\sigma\xi_i(t'+1)\Gamma(t'-t+t_0)\right]\ .\end{aligned} \tag{7.133}$$

By the same token, again as $N \to \infty$,

$$\frac{1}{N}\sum_j \varepsilon_j(t-\tau_{ij}) = \sum_{\tau,\sigma}\frac{|I_i(t,\tau,\sigma)|}{N}\sigma = \langle Q(t-\tau)\rangle_\tau \tag{7.134}$$

is almost surely independent of i and equal to the average of $Q(t-\tau)$ over the τ-distribution. This, together with the fact that correlations between patterns depend only on the *separation* of their time arguments, which was exploited in (7.132) and (7.133), gives rise to a remarkable factorization in (7.133),

$$\varepsilon_i(t+1) = \mathrm{sgn}[\langle Q(t-\tau)\rangle_\tau]$$

$$\times\ \mathrm{sgn}\left[\frac{1}{T_0}\sum_{s=1-(t-t_0)}^{T_0-(t-t_0)} \xi_i(t+1-t_0)\xi_i(t+1-t_0+s)\Gamma(s)\right]. \quad (7.135)$$

Since we had assumed that during initialization the system was always positively correlated with the embedded sequence, the first factor in (7.135) is unity, and $\varepsilon_i(t+1)$ is *solely* determined by the correlation properties of the embedded sequence.

For random patterns, the $\varepsilon_i(t+1)$ will be random themselves. If they turn out to be predominantly positive, one may conclude that the sequence remains positively correlated with the embedded sequence and, moreover, that $Q(t')$ remains constant and equal to

$$Q(t+1) = 2\mathrm{Prob}\{\varepsilon_i(t+1) > 0\} - 1 \quad (7.136)$$

for all $t' > t$. To estimate $\mathrm{Prob}\{\varepsilon_i(t+1) > 0\}$, Coolen and Gielen argue that, by virtue of the central-limit theorem, the argument of the second sign function in (7.135) – let us call it $X_i(t+1)$ – should be a Gaussian random variable of mean

$$\langle X_i(t+1)\rangle = \frac{1}{T_0}\sum_{s=1-(t-t_0)}^{T_0-(t-t_0)} \Gamma^2(s) \quad (7.137)$$

and variance

$$\langle (X_i(t+1) - \langle X_i(t+1)\rangle)^2\rangle$$
$$= \frac{1}{T_0^2}\sum_{s,s'}[\Gamma(s-s') - \Gamma(s)\Gamma(s')]\Gamma(s)\Gamma(s'), \quad (7.138)$$

and they use this observation to evaluate $\mathrm{Prob}\{\varepsilon_i(t+1) > 0\}$ in terms of error functions. Such a line of reasoning can, however, not be correct in general, since – precisely because of the correlation properties of the patterns expressed through Γ – the $\xi_i(t)$ for different t are not independent[23]. Moreover, the number of terms contributing to $X_i(t+1)$ is not necessarily large, as, for example, in the case finally studied by Coolen and Gielen, where Γ has a finite support.

A criterion for obtaining recall positively correlated with the embedded sequence for all $t' > t$ is the requirement that $Q(t+1) = 2\mathrm{Prob}\{\varepsilon_i(t+1) > 0\} - 1$ be greater than zero. To verify this condition, one has to specify further details about the way the patterns are generated. A sufficient condition would be that the distribution of the $X_i(t+1)$ is even about its mean, which is positive by virtue of (7.137).

If the patterns were such that

$$\sum_{s\neq 0} |\Gamma(s)| < \Gamma(0) = 1\ , \quad (7.139)$$

[23] The third moment of the $X_i(t+1)$-distribution, for instance, cannot be expressed in terms of Γ alone, and it will not generally be zero.

one would obtain $Q(t') = Q(t + 1) = 1$ for all $t' > t$, i.e., perfect alignment of recalled and embedded sequences. Orthogonal patterns, for instance, are among those satisfying (7.139).

Remarkably, the fact that the correlation properties of the embedded patterns were assumed to be invariant with respect to time translations entails that the quality of retrieval – upon positively correlated initialization – does *not* depend on the distribution of the delays τ [7.46].

Sequences of patterns with finite persistence times are unfortunately *not* among those having the correlation properties required for such a general result to hold. In fact, we have demonstrated above that the system architecture, as embodied in the range and distribution of delays, will determine whether a sequence of patterns, each having a finite duration, is encoded as a single spatio-temporal object or rather as a set of unrelated static objects. That is, for such patterns we cannot even expect that the system's performance will be independent of the distribution of the delays. However, following ideas of Zhaoping Li [7.53], A. Herz [7.54] has very recently been able to show that in the case of parallel dynamics, and for delay distributions which respect certain symmetries relative to the period of the embedded cycle, one can construct a Lyapunov functional which decreases as a function of time, until the system has approached a limit cycle of *exactly* the same period as that of the embedded sequence. Under relatively mild additional assumptions (about the level of storage and correlations between patterns), positively correlated retrieval with exact timing can be inferred. While the symmetry assumptions about the delay distribution – though somehow contrary to the spirit of the present modeling – have been necessary to arrive at the desired result, our experience with simulations and with analytical solutions of the evolution equations in the low loading limit suggests that they are unnecessarily restrictive; retrieval with very accurate timing has been found for a large variety of delay distributions violating such symmetry conditions.

7.7 Epilogue

In the foregoing chapters we have studied several mechanisms for the generation of temporal sequences in networks of formal neurons. We have concentrated on the approaches proposed during the recent wave of interest in neural-network models spurred on by the work of Hopfield [7.3], and among these we have selected the ones which have originated in or which have had impact on the physics community engaged in neural-network research. This strategy has certainly implied a sacrifice of scope (in favor of depth) of the present article. It may perhaps be excused by the fact that the approaches covered are quite varied, employing architectural elements which range from short-term synaptic plasticity [7.4, 5] over static synapses without temporal features, temporal association being aided by fast stochastic noise [7.6, 7], to static synapses in conjunction with transmission delays [7.8, 9][24].

In all cases the *mechanism* for the transition between successive patterns of a sequence, which underlies the occurrence of temporal association, is well understood and estimates for the persistence times of the individual patterns could be obtained – at least in those limits where fluctuations could be ignored. Macroscopic evolution equations have been derived for fully connected systems in the low loading limit – for noise-driven sequence generators [7.6, 7] and for sequence generators operating with transmission delays [7.21] – and for systems with extreme synaptic dilution [7.12]. In this latter case, the limit of extensive loading (measured relative to the average connectivity) was also accessible to explicit analysis [7.12, 39]. In the case of fully connected systems, as discussed in Sects. 7.2.1, 7.3, and 7.4.1 the limit of extensive levels of loading is less well understood. Up to now, it has been accessible to approximations and numerical simulations only.

Remarks scattered in the literature [7.9, 12, 13] indicate that, for systems operating with the delay mechanism, the storage capacity for sequences of unbiased binary random patterns is larger than that of the Hopfield model, which stores the patterns separately, the precise value depending on the the relative strength ε of the asymmetric interactions. Systematic studies of this point, which include the analysis of finite-size effects, have to the best of our knowledge not been performed yet. Straatemeier and Coolen [7.56] have done simulations on a model endowed with the forward projections (7.4) and operating with parallel dynamics. Their results indicate that, for this model too, the storage capacity for patterns connected to a cycle might be larger by a factor of about 1.8 than the capacity of the Hopfield model. Again it is unclear how much of this must be attributed to finite-size effects. Remember that the condition for error-free retrieval of a sequence in this model was found to be *identical* with the corresponding condition known from the Hopfield model.

In the case of the model of Dehaene et al. [7.5], the learning session merely serves to select one of the preexisting attractors of the network dynamics. Its storage capacity could therefore be defined as the number of such preexisting attractors, which might be fixed points or limit cycles. This number, however, will more strongly depend on the network architecture than on its size, and it will be hard to estimate it, except perhaps for the simplest cases.

All results discussed in the present article were obtained for networks of formal neurons. To what extent they may be relevant for biological systems is still very much under debate.

This is only partly due to the fact that the simplifications contained in the models are quite radical, which appears to render formal neural networks as rather

---

[24] An approach not discussed in the present paper, which should, however, be mentioned for the sake of completeness, is that of Nebenzahl [7.55]. The model employs the synapses (7.3) and (7.4) originally proposed by Hopfield, and achieves stability of sequential recall by *externally* regulating the relative strength ε of the forward projections (7.4).

remote from biology at first sight. The performance of these models has, after all, in many cases been shown to be rather insensitive to removal of simplifications as well as to other disturbances, be they random or systematic.

It is also partly due to problems on the experimental side. Biological neural networks are in general too complicated to have their performance monitored by way of measurements in a manner that may be called even nearly complete. It is also for this reason that it appears hard to confront models with experimental evidence which may have something to say about their biological plausibility.

The situation is of course different if the biological system is relatively simple. Such is the case, for instance, for the neural circuit governing the escape swimming reflex in the mollusc *Tritonia diomedea*. The circuit is well identified and under fairly complete experimental control [7.57]. From measurements of neural activities, relative synaptic strengths and response times could be inferred. On the basis of this information, Kleinfeld and Sompolinsky have been able to construct a formal neural-network model whose performance comes remarkably close to that of its biological counterpart [7.58].

In retrospect, it appears that among the mechanisms for sequence generation the delay mechanism has proved to be the most versatile[25]. This is a relevant observation if one considers building artificial devices, but it does not, of course, necessarily imply that the delay mechanism is the one and only mechanism for sequence generation in biological neural nets.

We have, however, seen that delays in biological neural systems do exist and, moreover, that they are too large to be ignored. With respect to this observation, two possible attitudes come to mind. One is the *concern* that delays – implying a certain sloppiness in the synchronization of various neural processes – may induce a loss of the associative capabilities a network might have if synchronization were perfect. This is a central issue of a paper by Kerszberg and Zippelius [7.47]. Their finding is that associative capabilities of neural networks are not only resistent to sloppy synchronization but that they are sometimes even slightly improved. (Related problems, though for networks of analog neurons, were also considered by Marcus and Westervelt [7.59].)

The other attitude is contained in the surmise that – given that delays in neural nets abound – nature may have opted to make constructive use of them. Such is obviously the attitude that has motivated some of the modeling described in Sects. 7.4–6. Moreover, the considerations about Hebbian learning with delays in Sect. 7. 6 show that delays would add a new dimension, time, to *any* learning mechanism based on correlation measurements of pre- and postsynaptic events. This extra dimension is then naturally added to the domain of objects representable in the synaptic code generated by such a learning mechanism, thereby considerably boosting its potential power. For learning and generating genuinely complex sequences, for instance, delays are a *sine qua non*, and we have seen

[25] The reader is cautioned that this and the following statements are most probably biased, considering where our own involvement in the field lies.

in Sect. 7.5 that the range of representable objects increases with the range of delays.

On the other hand, the delay mechanism can by no means account for any form of (complex-) sequence generation encountered in nature. For instance, playing *The Art of Fugue* [7.1], with its variations on the B-A-C-H theme mentioned in the introduction, would require delays of the order of tens of seconds – far too long to be attainable in neural systems. For such a problem, other mechanisms would definitely have to come in to help. It seems, however, that none of those discussed in the present article would do the job, except perhaps a sufficiently richly structured network *à la* Dehaene et al. [7.5]. A problem with this model, though, is that everything must be prewired and that nothing can be learnt, and the idea that the ability to command playing *The Art of Fugue* on the organ should be among the stable-limit cycles of any person's neurodynamics *before* the person was first exposed to this fine piece of music does not seem very plausible.

There is nevertheless a positive message here, namely, that *structure* must be invested in obtaining higher levels of cognitive processing in neural networks. Incidentally, the same message could also be conveyed by a closer look at the cerebral cortex. It seems, therefore, that there are still fascinating open fields for further research.

Acknowledgements. It is a pleasure to thank A. Herz, B. Sulzer, and U. Riedel for their help and advice, and for having shared with us their thoughts, insights, and ideas on neural networks in general, and sequence generation in particular. A. Herz and B. Sulzer have also generously provided us with some of their recent unpublished material. Moreover, we gratefully acknowledge valuable discussions with S. Bös, H. Horner, M. Vaas, K. Bauer, U. Krey, and G. Pöppel. This work has been supported by the Deutsche Forschungsgemeinschaft (Bonn).

Note Added

The following contains a comment on papers which have come to our attention since the completion of the manuscript.

Nakamura and Nishimori [7.60] have introduced a slight simplification in the coupling scheme of Buhmann and Schulten for nonoverlapping patterns [7.6, 7], and were able to identify a regime in the space of coupling constants and thresholds where sequential recall is possible even at $T = 0$, i.e., in the noiseless limit.

A coupling scheme allowing sequential retrieval of finitely many unbiased binary random patterns for asynchronous dynamics in the ± 1-representation, unaided by stochastic noise or time delays but restricted to odd numbers of patterns, was presented by Nishimori et al. [7.61]; see also [7.35].

Fukai and Shiino [7.62] have considered an asymmetric neural network with a coupling matrix satisfying Dale's law, and found chaotic attractors for suitable, non-Hebbian connection matrices. The analyses of [7.61] and [7.62] closely follow the general scheme introduced by Riedel et al. [7.21], as described in Sect. 7.4.2. The enforcement of Dale's law in [7.62] duplicates the number of sublattices to be considered; in the linear models, this leads to one additional overlap in the scheme.

Li and Herz [7.63] have discovered a Lyapunov functional for parallel dynamics in neural networks with delayed interactions exhibiting a certain form of extended synaptic symmetry in space *and* time; see also [7.54]. This permits the application of powerful free-energy techniques in the analysis of such systems, allowing for the first time the computation of storage capacities for cycles in fully connected networks whose connections have a broad distribution of transmission delays.

References [7.64] and [7.65] are reviews on Hebbian learning (see Sect. 7.6) and unlearning (see Sect. 1.5.7). The latter contains prescriptions for storage and associative retrieval of pattern sequences which are correlated in space *and* time.

Finally, Watkin and Sherrington [7.66] have recently presented a sequence generator for Boolean networks – an extension into the time domain of previous work on static pattern storage in Boolean networks conducted by Wong and Sherrington [7.67].

References

7.1 J.S. Bach: *The Art of Fugue*, Contrapunctus XVIII, (Leipzig, 1750/51). B-A-C-H is the last of three themes in the *"fuga à 3 soggetti et à 4 voci"*, which might well have been the crowning piece of *The Art of Fugue*. Bach died before its completion

7.2 W.S. Mc Culloch and W.S. Pitts: A Logical Calculus of the Ideas Immanent in Nervous Activity, Bull. Math. Biophys. **5**, 115–133 (1943)

7.3 J.J. Hopfield: Neural Networks and Physical Systems with Emergent Computational Abilities, Proc. Natl. Acad. Sci. USA **79**, 2554–2559 (1982)

7.4 P. Peretto and J.J. Niez: Collective Properties of Neural Networks, in *Disordered Systems and Biological Organization*, edited by E. Bienenstock, F. Fogelman-Soulié, and G. Weisbuch, (Springer, Berlin, Heidelberg, New York 1986) pp. 171–185

7.5 S. Dehaene, J.P. Changeux, and J.P. Nadal: Neural Networks that Learn Temporal Sequences by Selection, Proc. Natl. Acad. Sci. USA **84**, 2727–2731 (1987)

7.6 J. Buhmann and K. Schulten: Noise Driven Temporal Association in Neural Networks, Europhys. Lett. **4**, 1205–1209 (1987)

7.7 J. Buhmann and K. Schulten: Storing Sequences of Biased Patterns in Neural Networks with Stochastic Dynamics, in: NATO ASI Series Vol. F41, *Neural Computers*, edited by R. Eckmiller and Ch. v.d. Malsburg, (Springer, Berlin, Heidelberg, New York 1988) pp. 231–242

7.8 D. Kleinfeld: Sequential State Generation by Model Neural Networks, Proc. Natl. Acad. Sci. USA **83**, 9469–9473 (1986)

7.9 H. Sompolinsky and I. Kanter: Temporal Association in Asymmetric Neural Networks, Phys. Rev. Lett. **57**, 2861–2864 (1986)

7.10 G. Willwacher: Fähigkeiten eines assoziativen Speichersystems im Vergleich zu Gehirnfunktionen, Biol. Cybern. **24**, 191–198 (1976); Storage of a Temporal Pattern Sequence in a Network, Biol. Cybern. **43**, 115–126 (1982)

7.11 D.W. Tank and J.J. Hopfield: Neural Computation by Concentrating Information in Time, Proc. Natl. Acad. Sci. USA **84**, 1896–1900 (1987)

7.12 H. Gutfreund and M. Mézard: Processing of Temporal Sequences in Neural Networks, Phys. Rev. Lett. **61**, 235–238 (1988)

7.13 D.J. Amit: Neural Networks Counting Chimes, Proc. Natl. Acad. Sci. USA **85**, 2141–2145 (1987)

7.14 L. Personnaz, I. Guyon, and G. Dreyfus: Neural Networks for Associative Memory Design, in: *Computational Systems – Natural and Artificial*, edited by H. Haken (Springer, Berlin, Heidelberg, New York 1987) pp. 142–151; I. Guyon, L. Personnaz, J.P. Nadal, and G. Dreyfus: Storage and Retrieval of Complex Sequences in Neural Networks, Phys. Rev. A **38**, 6365–6372 (1988)
7.15 R. Kühn, J.L. van Hemmen, and U. Riedel: Complex Temporal Association in Neural Nets, in: *Neural Networks, From Models to Applications*, Proceedings of the nEuro'88 Conference, edited by G. Dreyfus and L. Personnaz (I.D.S.E.T., Paris, 1989) pp. 289–298; Complex Temporal Association in Neural Networks, J. Phys. A **22**, 3123–3135 (1989)
7.16 D. O. Hebb: *The Organization of Behavior* (Wiley, New York 1949)
7.17 A. Herz, B. Sulzer, R. Kühn, and J.L.van Hemmen: The Hebb Rule: Storing Static and Dynamic Objects in an Associative Neural Network, Europhys. Lett. **7**, 663–669 (1988); Hebbian Learning Reconsidered: Representation of Static and Dynamic Objects in Associative Neural Nets, Biol. Cybern. **60**, 457–467 (1989)
7.18 B. Sulzer: Modelle zur Speicherung von Sequenzen in Neuronalen Netzwerken, diploma thesis, (Heidelberg 1989)
7.19 N.G. van Kampen: *Stochastic Processes in Physics and Chemistry* (North-Holland, Amsterdam 1981)
7.20 J.L. van Hemmen and R. Kühn: Collective Phenomena in Neural Networks, this volume, Chap. 1
7.21 U. Riedel, R. Kühn, and J.L. van Hemmen: Temporal Sequences and Chaos in Neural Networks, Phys Rev A **38**, 1105–1108 (1988); U. Riedel: Dynamik eines Neuronalen Netzwerks – Musterfolgen durch asymmetrische Synapsen, diploma thesis (Heidelberg 1988)
7.22 L. Personnaz, I. Guyon, and G. Dreyfus: Information Storage and Retrieval in Spin-Glass Like Neural Networks, J. Physique Lett. **46**, L359–L365 (1985)
7.23 I. Kanter and H. Sompolinsky: Associative Recall of Memory without Errors, Phys. Rev. A **35**, 380–392 (1987)
7.24 D.J. Amit, H. Gutfreund, and H. Sompolinsky: Information Storage in Neural Networks with Low Levels of Activity, Phys. Rev. A **35**, 2293–2303 (1987)
7.25 M. Feigel'man and L.B. Ioffe: The Augmented Models of Associative Memory: Asymmetric Interaction and Hierarchy of Patterns, Int. J. Mod. Phys. **1**, 51–68 (1987)
7.26 A. Krogh and J. Hertz: Mean-Field Analysis of Hierarchical Neural Nets with Magnetization, J. Phys. A: Math. Gen. **21**, 2211–2224 (1988)
7.27 S. Bös, R. Kühn, and J.L. van Hemmen: Martingale Approach to Neural Networks with Hierarchically Structured Information, Z. Phys. B **71**, 261–271 (1988); S. Bös: Neuronales Netzwerk mit hierarchisch strukturierter Information, diploma thesis, (Heidelberg 1988)
7.28 W.Gerstner, J.L. van Hemmen, and A. Herz: in preparation
7.29 A.C. Scott: *Neurophysics* (Wiley, New York 1977)
7.30 V. Braitenberg: On the Representation of Objects and their Relations in the Brain, in: *Lecture Notes in Biomathematics*, edited by M. Conrad, W. Güttinger, and M. Dal Cin (Springer, Berlin, Heidelberg, New York 1974) pp. 290–298; Two Views on the Cerebral Cortex, in: *Brain Theory*, edited by G. Palm and A. Aertsen (Springer, Berlin, Heidelberg, New York 1986) pp. 81–96
7.31 D. Grensing and R. Kühn: Random-Site Spin Glass Models, J. Phys. A: Math. Gen. **19**, L1153–L1157 (1986)
7.32 J.L. van Hemmen and R. Kühn: Nonlinear Neural Networks, Phys. Rev. Lett. **57**, 913–916 (1986)
7.33 J.L. van Hemmen, D. Grensing, A. Huber, and R. Kühn: Nonlinear Neural Networks I. General Theory, J. Stat. Phys. **50**, 231–257 (1988); Nonlinear Neural Networks II. Information Processing, J. Stat. Phys. **50**, 259–293 (1988)
7.34 A.C.C. Coolen and Th.W. Ruijgrok: Image Evolution in Hopfield Networks, Phys. Rev. A**38**, 4253–4255 (1988)
7.35 M. Shiino, H. Nishimori, and M. Ono: Nonlinear Master Equation Approach to Asymmetrical Neural Networks of the Hopfield-Hemmen Type, J. Phys. Soc. Japan **58**, 763–766 (1989)
7.36 G. Ioos and D. Joseph: Elementary Stability and Bifurcation Theory (Springer, New York, Heidelberg, Berlin 1980)
7.37 B. Derrida, E. Gardner, and A. Zippelius: An Exactly Solvable Asymmetric Neural Network Model, Europhys. Lett. **4**, 167–173 (1987)

7.38 B. Derrida and G. Weisbuch: Evolution of Overlaps between Configurations in Random Boolean Networks, J. de Physique **47**, 1297–1303 (1986)
7.39 G. Mato and N. Parga: Temporal Sequences in Strongly Diluted Neural Networks, in: *Neural Network and Spin Glasses*, edited by W.K. Theumann and R. Köberle (World Scientific, Singapore 1990) pp. 114–126
7.40 H. Gutfreund, cited as Ref. 14 in [7.12], and private communication
7.41 T. Kohonen: Content-Addressable Memories (Springer, Berlin, Heidelberg, New York 1980)
7.42 A. Albert: *Regression and the Moore–Penrose Pseudoinverse* (Academic, New York 1972)
7.43 L.N. Cooper, in: *Nobel Symposia*, Vol. **24**, edited by B. Lundqvist and S. Lundqvist (Academic, New York 1973) pp. 252–264
7.44 A. Herz, B. Sulzer, R. Kühn, and J.L.van Hemmen: Hebbian Learning – A Canonical Way of Representing Static and Dynamic Objects in an Associative Neural Network, in: *Neural Networks, From Models to Applications*, Proceedings of the nEuro'88 Conference, edited by G. Dreyfus and L. Personnaz (I.D.S.E.T., Paris 1989) pp. 307–315
7.45 A. Herz: Representation and Recognition of Spatio-Temporal Objects within a Generalized Hopfield Scheme, in: *Connectionism in Perspective*, edited by R. Pfeiffer, Z. Schreter, F. Fogelman-Soulié, and L. Steels (North-Holland, Amsterdam 1989)
7.46 A.C.C. Coolen and C.C.A.M. Gielen: Delays in Neural Networks, Europhys. Lett. **7**, 281–285 (1988)
7.47 M. Kerszberg and A. Zippelius: Synchronization in Neural Assemblies, Physica Scripta T **33**, 54–64 (1990)
7.48 E. Caianiello: Outline of a Theory of Thought Processes and Thinking Machines, J. Theor. Biol. **1**, 204–235 (1961)
7.49 S. Grossberg: Prediction Theory for Some Nonlinear Functional Differential Equations I. Learning of Lists, J. Math. Anal. Appl. **21**, 643–694 (1968)
7.50 S.I. Amari: Learning Patterns and Pattern Sequences by Self-Organizing Nets of Threshold Elements, IEEE Trans. Comp. **C-21**, 1197–1206 (1972)
7.51 K. Fukushima: A Model of Associative Memory in the Brain, Kybernetik **12**, 58–63 (1973)
7.52 K. Bauer and U. Krey: Learning and Recognition of Temporal Sequences of Correlated Patterns – Numerical Investigations, Z. Phys. B **79**, 461–475 (1990)
7.53 Zhaoping Li: private communication
7.54 A. Herz: Untersuchungen zum Hebbschen Postulat: Dynamik und statistische Physik raumzeitlicher Assoziation, PhD thesis (Heidelberg 1990)
7.55 I. Nebenzahl: Recall of Associated Memories, J. Math. Biol. **25**, 511–519 (1987)
7.56 D. Straatemeier and A.C.C. Coolen: Capacity of a Neural Network to Store a Pattern Cycle, preprint, University of Utrecht (1989)
7.57 P. A. Getting: Mechanism of Pattern Generation Underlying Swimming in Tritonia I. Neuronal Network Formed by Monosynaptic Connections, J. Neurophys. **46**, 65–79 (1981); Mechanism of Pattern Generation Underlying Swimming in Tritonia II. Network reconstruction, **49**, 1017–1035 (1983); Mechanism of Pattern Generation Underlying Swimming in Tritonia III. Intrinsic and Synaptic Mechanisms for Delayed Excitation, **49**, 1036–1050 (1983)
7.58 D. Kleinfeld and H. Sompolinsky: Associative Neural Network Model for the Generation of Temporal Patterns: Theory and Application to Central Pattern Generators, Biophys. J. **54**, 1039–1051 (1988)
7.59 C.M. Marcus and R.M. Westervelt: Stability of Analog Neural Networks with Delay, Phys. Rev. A **39**, 347–359 (1989)
7.60 T. Nakamura and H. Nishimori: Sequential Retrieval of Non-Random Patterns in a Neural Network, J. Phys. A: Math. Gen. **23**, 4627–4641 (1990)
7.61 H. Nishimori, T. Nakamura and M. Shiino: Retrieval of Spatio-Temporal Sequence in Asynchronous Neural Networks, Phys. Rev. A **41**, 3346–3354 (1990)
7.62 T. Fukai and M. Shiino: Asymmetric Neural Networks Incorporating the Dale Hypothesis and Noise-Driven Chaos, Phys. Rev. Lett. **64**, 1465–1468 (1990)
7.63 Z. Li and A.V.M. Herz: Lyapunov Functional for Neural Networks with Delayed Interactions and Statistical Mechanics of Temporal Association, in *Neural Networks*, Proceedings of the XIth Sitges Conference, edited by L. Garrido, Springer Lecture Notes in Physics 368 (Springer, Berlin, Heidelberg, New York 1990) pp. 287–302; A.V.M. Herz, Z. Li, and J.L. van Hemmen: Statistical Mechanics of Temporal Association in Neural Networks with Transmission Delays, Phys. Rev. Lett. **66**, 1370–1373 (1991)

7.64 J. L. van Hemmen: Hebbian Learning and Unlearning, in *Neural Networks and Spin Glasses*, Proceedings of a Workshop held at Porto Allegre (1989), edited by W. K. Theumann and R. Köberle (World Scientific, Singapore 1990) pp. 91–114
7.65 J. L. van Hemmen, W. Gerstner, A. Herz, R. Kühn, and M. Vaas: Encoding and Decoding of Patterns which are Correlated in Space and Time, in *Konnektionismus in Artificial Intelligence und Kognitionsforschung*, edited by G. Dorffner (Springer, Berlin, Heidelberg 1990) pp. 153–162
7.66 T. L. H. Watkin and D. Sherrington: Temporal Sequences in Boolean Networks, Europhys. Lett. **14**, 621–625 (1991)
7.67 K. Y. M. Wong and D. Sherrington: Theory of Associative Memory in Randomly Connected Boolean Neural Networks, J. Phys. A: Math. Gen. **22**, 2233–2263 (1989)

8. Self-organizing Maps and Adaptive Filters

Helge Ritter, Klaus Obermayer, Klaus Schulten and Jeanne Rubner

With 13 Figures

Synopsis. Topographically organized maps and adaptive filters fulfill important roles for information processing in the brain and are also promising to facilitate tasks in digital information processing. In this contribution, we report results on two important network models. A first network model comprises the "self-organizing feature maps" of Kohonen. We discuss their relation to optimal representation of data, present results of a mathematical analysis of their behavior near a stationary state, demonstrate the formation of "striped projections", if higher-dimensional feature spaces are to be mapped onto a two-dimensional cortical surface, and present recent simulation results for the somatosensory map of the skin surface and the retinal map in the visual cortex. The second network model is a hierarchical network for principal component analysis. Such a network, when trained with correlated random patterns, develops cells the receptive fields of which correspond to Gabor filters and resemble the receptive fields of "simple cells" in the visual cortex.

8.1 Introduction

One essential task of neural-network algorithms is optimal storage of data. Different criteria for optimal storage are conceivable, and correspondingly different neural-network algorithms have been derived. Many of them fall into one of two major, and to some extent complementary, categories.

The first category is that of so-called *attractor networks* [8.1]. Such networks are fully connected: information is stored in a distributed way and retrieved by a dynamical relaxation process. The distributed storage mechanism makes these systems very tolerant to partial damage or degradation in their connectivity but also introduces a tendency for "crosstalk" between similar patterns [8.2, 3]. This type of storage does not reduce the information content of patterns stored. In fact, it stores prototype patterns completely, e.g. as pixel images, and allows classification of presented patterns according to the stored prototypes.

The second category is formed by so-called *competitive learning networks*, in which a set of "grandmother cells" is used for storage of the presented patterns [8.4, 5]. Such networks involve an input and output layer of "grandmother cells" and storage is achieved through the development of receptive fields which resemble stored patterns. The receptive fields act as filters: when a pattern similar to one of the patterns in a training set is offered, the output cell the receptive field of which best matches the input becomes activated. Since a single cell provides the network response, such systems lack any tolerance against hardware failure, but they avoid crosstalk between patterns of even very high overlap. Although each "grandmother cell" might appear as a fully localized storage device, part of the information is actually distributed: the "grandmother cell" selected by the input pattern is only determined by competition among many candidate cells and, therefore, depends crucially on information from many different cells in the network.

The usable storage capacity of both types of network is similar and can be brought close to the information inherent in the required weight values (see e.g. [8.6, 7]). Generalization or "associative completion" of partial inputs is also very similar: in the absence of any special preprocessing the stored pattern of maximal overlap with the presented input is usually retrieved.

While attractor networks have been investigated very much in recent years, competitive learning networks have received less attention. There are many non-trivial and interesting properties of competitive networks that deserve more study. This is particularly true for a generalization of these networks where weight adjustments of "grandmother cells" lose their independence and are mutually coupled in some prespecified way. These networks, introduced by Kohonen under the name *"self-organizing feature maps"* [8.8–10], possess properties which make them particularly interesting for both understanding and modeling the biological brain [8.11, 56–58] and for practical applications such as robotics [8.12, 13].

In the following, we will present several mathematical results pertaining to Kohonen networks and review some work concerning the application of Kohonen networks to modeling of neural tissue in the cortex. A more comprehensive account can be found in [8.14].

Another important issue for optimal storage, relevant to both types of model discussed above, is efficient preprocessing of information. It is, of course, most desirable to achieve dense information storage through filters which rapidly discern the important features of input data and restrict storage to these features. In the visual system of biological species such filters operate on the lowest levels of the system in the optical cortex and extract important visual features such as edges and bars (see e.g. [8.15, 16]). An answer to the question how the brain achieves the neural connectivity which establishes optimal filters for preprocessing is extremely desirable for the development of neural-network algorithms for computer vision and other information-processing tasks characterized by huge amounts of data. Only a small part of the architecture of the brain is genetically specified; most of the brain's synaptic connections are achieved through self-organization. Postnatal visual input plays an essential role in the organization of synaptic patterns of the optical cortex of mature animals (see e.g [8.17]). These observations raise the question of how a sensory system, in response to input information, can organize itself so as to form feature detectors which encode mutually independent aspects of the information contained in patterns presented to it. In Sect. 8.6 we will present a two-layered network as a model for such system. It will be demonstrated that simple local rules for synaptic connectivities allow the model to learn in an unsupervised mode. Presented with a set of input patterns the network learns to discern the most important features defined as the principal components of the correlation matrix of a set of training patterns. The network described generalizes a model of Linsker [8.18] which established the possibility of self-organized formation of feature detectors.

8.2 Self-organizing Maps and Optimal Representation of Data

The basic aim of "competitive networks" as well as of "self-organizing maps" is to store some, usually large, set V of patterns, encoded as "vectors" $\boldsymbol{v} \in V$, by finding a smaller set W of "prototypes" $\boldsymbol{w}_{\boldsymbol{r}}$ such that the set $W := \{\boldsymbol{w}_{\boldsymbol{r}_1}, \boldsymbol{w}_{\boldsymbol{r}_2}, \ldots \boldsymbol{w}_{\boldsymbol{r}_N}\}$ of prototypes provides a good approximation of the original set V. Intuitively, this should mean that for each $\boldsymbol{v} \in V$ the distance $||\boldsymbol{v} - \boldsymbol{w}_{\boldsymbol{s}(\boldsymbol{v})}||$ between $\boldsymbol{v}$ and the closest prototype $\boldsymbol{w}_{\boldsymbol{s}(\boldsymbol{v})}$ in the set W shall be small. Here, the "mapping function" $\boldsymbol{s}(.)$ has been introduced to denote for each $\boldsymbol{v} \in V$ the (index of the) closest prototype in W. (Therefore, the function $\boldsymbol{s}(.)$ depends on all prototypes $\boldsymbol{w}_{\boldsymbol{r}}$ and is equivalent to a full specification of their values: given $\boldsymbol{s}(.)$, one can reconstruct each $\boldsymbol{w}_{\boldsymbol{r}}$ as the centroid of the subset of all $\boldsymbol{v} \in V$ for which $\boldsymbol{s}(\boldsymbol{v}) = \boldsymbol{r}$.)

For a more precise formulation of the notion "good approximation of V" we assume that the pattern vectors $\boldsymbol{v}$ are subject to a probability density $P(\boldsymbol{v})$ on V, and then require that the set of prototypes should be determined such that the *expectation value* E of the square error,

$$E[\boldsymbol{w}] = \int ||\boldsymbol{v} - \boldsymbol{w}_{\boldsymbol{s}(\boldsymbol{v})}||^2 P(\boldsymbol{v}) \mathrm{d}^d \boldsymbol{v} \tag{8.1}$$

is minimized. Here $\boldsymbol{w} := (\boldsymbol{w}_{\boldsymbol{r}_1}, \boldsymbol{w}_{\boldsymbol{r}_2}, \ldots \boldsymbol{w}_{\boldsymbol{r}_N})$ represents the vector of prototypes in W.

Minimization of the functional $E[\boldsymbol{w}]$ is the well-known problem of *optimal vector quantization* [8.19, 20] and is related to *data compression* for efficient transmission of the pattern set V: if an average error $E[\boldsymbol{w}]$ can be tolerated, sender and receiver can agree to use the mapping function $\boldsymbol{s}(.)$ to transmit only the (usually much shorter) "labels" $\boldsymbol{s}(\boldsymbol{v})$ of the approximating prototypes $\boldsymbol{w}_{\boldsymbol{s}}$ instead of the complete pattern vectors $\boldsymbol{v}$ themselves.

A straightforward approach to find a local minimum of (8.1) is gradient descent for the functional $E[.]$, i.e. the prototypes are changed according to

$$\dot{\boldsymbol{w}}_{\boldsymbol{r}} = \int_{\boldsymbol{s}(\boldsymbol{v})=\boldsymbol{r}} (\boldsymbol{v} - \boldsymbol{w}_{\boldsymbol{s}(\boldsymbol{v})}) P(\boldsymbol{v}) \mathrm{d}^d \boldsymbol{v} \; . \tag{8.2}$$

Equation (8.2) is equivalent to the discrete "learning rule"

$$\Delta \boldsymbol{w}_{\boldsymbol{r}} = \varepsilon \delta_{\boldsymbol{r},\boldsymbol{s}(\boldsymbol{v})} (\boldsymbol{v} - \boldsymbol{w}_{\boldsymbol{r}}) \; , \tag{8.3}$$

applied for a sequence of random "samples" $\boldsymbol{v} \in V$ that are distributed according to the probability density $P(\boldsymbol{v})$ in the limit of vanishing "step size" ε. Equation (8.3) is a well-known "learning rule" found in many competitive networks: for each "presentation" of an input $\boldsymbol{v}$, $\boldsymbol{s}(\boldsymbol{v})$ "selects" the best-matching prototype vector $\boldsymbol{w}_{\boldsymbol{s}(\boldsymbol{v})}$, for an adjustment towards $\boldsymbol{v}$. For rapid convergence, one usually starts with a larger initial learning step size $\varepsilon_\mathrm{i} < 1$, which is gradually lowered to a final, small value $\varepsilon_\mathrm{f} \geq 0$.

The *self-organizing feature maps* [8.8–10] generalize this scheme for optimal storage (in the sense of minimal average error $E[\boldsymbol{w}]$, (8.1)) by considering the prototypes $\boldsymbol{w}_{\boldsymbol{r}}$ to be associated with points $\boldsymbol{r}$ in some "image domain" A and requiring that *a "structured representation" or "map" of the data V is created on A during the storage process.* The "map" arises through the selection function $\boldsymbol{s}(\boldsymbol{v})$, which maps each pattern vector to a point $\boldsymbol{s} \in A$. The discrete set A is endowed with some *topology*, e.g. by arranging the points as a (often two-dimensional) lattice. The aim of the algorithm of self-organizing feature maps then is, besides approximating V by the prototypes $\boldsymbol{w}_{\boldsymbol{r}}$, also to arrange the $\boldsymbol{w}_{\boldsymbol{r}}$ in such a way that the associated mapping $\boldsymbol{s}(.)$ from V to A maps the topology of the set V, defined by the metric relationships of its vectors $\boldsymbol{v} \in V$, onto the topology of A in a least distorting way. This requires that (metrically) *similar patterns $\boldsymbol{v}$ are mapped onto neighboring points in A.* The desired result is a (low-dimensional) image of V in which *the most important similarity relationships among patterns from V are preserved and transformed into spatial neighborhood relationships in the chosen "image domain"*.

To achieve this, the adjustments of the prototypes $\boldsymbol{w}_{\boldsymbol{r}}$ must be coupled by replacing the Kronecker $\delta_{\boldsymbol{rs}}$ in (8.3) by a "neighborhood function" $h_{\boldsymbol{rs}}$,

$$\Delta \boldsymbol{w}_{\boldsymbol{r}} = \varepsilon h_{\boldsymbol{rs}(\boldsymbol{v})}(\boldsymbol{v} - \boldsymbol{w}_{\boldsymbol{r}}) \,. \tag{8.4}$$

The function $h_{\boldsymbol{rs}}$ is (usually) a unimodal function of the lattice distance $d = ||\boldsymbol{r} - \boldsymbol{s}||$ in A, decaying to zero for $d \to \infty$ and with maximum at $d = 0$. A suitable choice, for example, is a Gaussian $\exp(-d^2/2\sigma^2)$. Therefore, vectors $\boldsymbol{w}_{\boldsymbol{r}}$, $\boldsymbol{w}_{\boldsymbol{s}}$ associated with *neighboring* points $\boldsymbol{r}$, $\boldsymbol{s} \in A$ are coupled more strongly than vectors associated with more distant points and tend to converge to more similar patterns during learning. This mechanism enforces a good "match" between the topologies of V and A on a local scale. Consequently, it is no longer $E[\boldsymbol{w}]$ but instead the functional

$$F[\boldsymbol{w}] = \sum_{\boldsymbol{rr}'} h_{\boldsymbol{rr}'} \int\limits_{\boldsymbol{s}(\boldsymbol{v})=\boldsymbol{r}'} ||\boldsymbol{v} - \boldsymbol{w}_{\boldsymbol{r}}||^2 P(\boldsymbol{v}) \mathrm{d}^d \boldsymbol{v} \tag{8.5}$$

that is (approximately) minimized by the new process. Equation (8.5) for a discrete set V has been stated in [8.21] and there it was shown that its minimization is related to the solution of the "traveling-salesman problem". A more general interpretation was subsequently given by Luttrell [8.22]. He considers again the case when the prototypes in W are used to obtain "labels" $\boldsymbol{s}(\boldsymbol{v})$ to compress the pattern set V for the purpose of transmission, but in addition assumes that this transmission is "noisy", i.e. there is a probability of h_{rs} that label s is confused with label r as a result of the transmission process. Then the reconstruction of a transmitted pattern v will not always be the closest prototype $w_{s(v)}$ but will be w_r with a probability of $h_{rs(v)}$. Hence the expected mean square error on transmitting a pattern v will be given by $\sum_r h_{rs(v)}(v - w_r)^2$ and $F[w]$ can thus be seen to represent *the expected mean square transmission error over the noisy channel*, averaged over the whole pattern set V.

In Sect. 8.4 we will return to this interpretation and discuss the relationship between some aspects of brain function and optimization of $F[w]$.

8.3 Learning Dynamics in the Vicinity of a Stationary State

After clarifying the optimization task underlying the formation of self-organizing maps, the next task is to characterize the convergence properties of the map formation process, based on (8.4). For a more detailed account of the mathematical analysis the reader is referred to [8.23].

First, one needs to address the question how far convergence to a global minimum can be achieved. Even for simple distributions $P(v)$, the functional $F[w]$ can exhibit many different local minima [8.59, 60]. As the adaptation equation (8.4) is based on a gradient descent procedure, one generally cannot hope to find the global optimum but must be content with some more or less optimal local minimum. The function h_{rs} plays an important role in finding a good minimum [8.59, 60]. Inspection of (8.4) shows that in the long-range limit (i.e. $h_{rs} = \text{const}$) the functional $F[w]$ approaches a simple quadratic function with a single minimum. Therefore, by starting an adaptation process with a long-ranged h_{rs} and then reducing the range of h_{rs} slowly to the intended, smaller final value, one gradually deforms the "optimality landscape" from a simple, convex shape to a final, multi-minimum surface. Such a strategy facilitates convergence to a good minimum of $F[.]$. The choice of a good starting configuration is also helpful in this respect. As will be pointed out in Sect. 8.4, the optimization process can be interpreted as a model for the formation of topographically ordered neural projections in the brain. In this case, a good starting configuration is provided by an initial coarse ordering of the neural connectivity.

We focus in the following on the behavior of the optimization process in the vicinity of a good minimum of F. For the mathematical analysis we consider an ensemble of systems, each characterized by a set $w = (w_{r_1}, w_{r_2}, \ldots w_{r_N})$, $r_j \in A$, of prototype vectors. $S(w, t)$ shall denote the distribution of the ensemble, after t adaptation steps, in the state space Ω spanned by the prototype sets w. We make the assumption that the pattern vectors v are statistically independent samples from V, subject to the probability density $P(v)$. In this case (8.4) defines a Markov process in the space Ω with transition probability

$$Q(w, w') = \sum_r \int_{F_r(w')} dv\, \delta\Big(w - w' - \varepsilon h_{r,s'(v)}(v - w')\Big) P(v) \tag{8.6}$$

for the transition from a state $\boldsymbol{w}'$ to a state $\boldsymbol{w}$. $F_{\boldsymbol{r}}(\boldsymbol{w}')$ denotes the set of all $\boldsymbol{v}$ which are closer to $\boldsymbol{w}'_{\boldsymbol{r}}$ than to any other $\boldsymbol{w}'_{\boldsymbol{s}}$, $\boldsymbol{s} \neq \boldsymbol{r}$, and $\boldsymbol{s}'(\boldsymbol{v})$ is defined in analogy to $\boldsymbol{s}(\boldsymbol{v})$, but using the primed reference vectors $\boldsymbol{w}'_{\boldsymbol{r}}$ instead. $F_{\boldsymbol{r}}(\boldsymbol{w}')$ and $\boldsymbol{s}'(.)$ are related: the former is the set of patterns $\boldsymbol{v} \in V$ mapped onto the same point $\boldsymbol{r} \in A$ by the latter.

Assuming a small learning step size ε and a distribution function $S(\boldsymbol{w}, t)$ that is concentrated in the vicinity of the selected local optimum $\bar{\boldsymbol{w}}$ of $F[.]$, one can derive the following Fokker–Planck equation for the time development of S:

$$\frac{1}{\varepsilon}\partial_t S(\boldsymbol{u}, t) = \sum_{rmr'n} \frac{\partial}{\partial \mathrm{u}_{rm}} \boldsymbol{B}_{rmr'n} \mathrm{u}_{r'n} S(\boldsymbol{u}, t) + \frac{\varepsilon}{2} \sum_{rmr'n} D_{rm,r'n} \frac{\partial^2 S(\boldsymbol{u}, t)}{\partial \mathrm{u}_{rm} \partial \mathrm{u}_{r'n}}, \tag{8.7}$$

where we have tacitly shifted the origin of $S(., t)$ to the selected state $\bar{\boldsymbol{w}}$, using now the new argument variable $\boldsymbol{u} = \boldsymbol{w} - \bar{\boldsymbol{w}}$ instead of $\boldsymbol{w}$. $\boldsymbol{B}$ is a matrix given by

$$\boldsymbol{B}_{rmr'n} := \left(\frac{\partial V_{rm}(\boldsymbol{w})}{\partial \mathrm{w}_{r'n}} \right)_{\boldsymbol{w}=\bar{\boldsymbol{w}}}, \tag{8.8}$$

and the quantities V_{rm} and $D_{rmr'n}$ are the expectation values $\langle -\delta \boldsymbol{w}_{rm} \rangle$ and $\langle \delta \boldsymbol{w}_{rm} \delta \boldsymbol{w}_{r'n} \rangle$ under one adaptation step, where $\delta \boldsymbol{w}_{rm}$ is the change of the mth component of prototype $\boldsymbol{w}_{\boldsymbol{r}}$, but scaled to $\varepsilon = 1$. Their explicit expressions are are given in [8.23].

With the help of (8.7) one can answer an important question about the convergence of the process: Which control of the learning step size ε guarantees that each ensemble member converges (with probability 1) to the optimum $\bar{\boldsymbol{w}}$? It turns out that, if $\bar{\boldsymbol{w}}$ is a stable stationary state (a necessary and sufficient condition for this is that $\boldsymbol{B} + \boldsymbol{B}^{\mathrm{T}}$ be a positive definite matrix), the two conditions

$$\lim_{t \to \infty} \int_0^t \varepsilon(t')\mathrm{d}t' = \infty, \tag{8.9}$$

$$\lim_{t \to \infty} \varepsilon(t) = 0 \tag{8.10}$$

provide the desired answer [8.23]. Equation (8.10) is rather obvious, but (8.9) sets a limit on the rate of reduction of the learning step size. If ε decreases faster than allowed according to condition (8.9) there is some nonzero probability that the adaptation process "freezes" before the local optimum is reached.

In reality, of course, (8.9) can never be met exactly; however, for $\varepsilon_{\mathrm{f}} = 0$, (8.7) can be used to show that the remaining deviations from the optimum are exponentially small in the quantity $\int \varepsilon(t)\mathrm{d}t$ [8.23].

A second important question concerns the *statistical fluctuations* that are present for a non-vanishing learning step size and that are due to the randomness of the sequence of input patterns $\boldsymbol{v}$. If ε = const, (8.7) admits as a stationary solution a Gaussian with correlation matrix

$$\langle \boldsymbol{u}_{rm} \boldsymbol{u}_{sn} \rangle = \langle (\boldsymbol{w}_{rm} - \bar{\boldsymbol{w}}_{rm})(\boldsymbol{w}_{sn} - \bar{\boldsymbol{w}}_{sn}) \rangle_S = \varepsilon \left[(\boldsymbol{B} + \boldsymbol{B}^{\mathrm{T}})^{-1} \boldsymbol{D} \right]_{rm,sn} \tag{8.11}$$

where $\langle \ldots \rangle$ denotes averaging over the ensemble. For an explicit evaluation of the matrices $\boldsymbol{B}$ and $\boldsymbol{D}$ one needs to know the configuration $\bar{\boldsymbol{w}}$ of the local optimum chosen for the discussion. Even for simple distributions $P(\boldsymbol{v})$ there can be numerous and complex configurations $\bar{\boldsymbol{w}}$ leading to a local minimum of $F[.]$. To study a simple, but still interesting, case we assume that A is a two-dimensional lattice of $N \times N$ points and that the set V is continuous and of higher dimensionality. We choose V as the three-dimensional volume $0 \leq x, y \leq N$, $-s \leq z \leq s$ and assume a constant probability density $P(\boldsymbol{v}) = [2sN^2]^{-1}$. Clearly, for a topology-preserving mapping between V and A there exists some "conflict" because of the different dimensions of A and V. The parameter s can be considered a measure of this conflict: if s is small (i.e. $2s \ll N$), a "perpendicular" projection obviously provides a very good match between the topologies of V and of A (Fig. 8.1a). However, this is no longer the case if s becomes larger. Then minimization of $F[.]$ is expected to require a more complicated mapping from V to A (Fig. 8.1b).

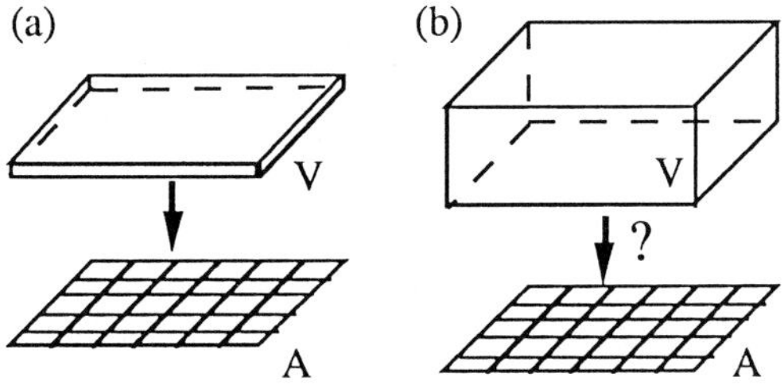

Fig. 8.1. **(a)**: For small height $2s$ of the volume V, a "perpendicular" projection of V onto the lattice A provides a good match between the topologies of V and A. **(b)** For larger values of s the mapping $V \mapsto A$ required to minimize the functional $F[.]$ (8.5) is no longer obvious

We want to find the limiting value of s for which the "perpendicular" projection loses its optimality and to discuss in which way the mapping can be "improved" then.

To avoid any edge effects, we assume periodic boundary conditions for the lattice A and for V along the x- and y-axes. One can then evaluate (8.11) for $\bar{\boldsymbol{w}}_r = \boldsymbol{r}$, $\boldsymbol{r} = m\boldsymbol{e}_x + n\boldsymbol{e}_y$, which, by symmetry, must be a stationary solution for F (as well as any other configuration obtained by translations or rotations of $\bar{\boldsymbol{w}}$). This choice for $\bar{\boldsymbol{w}}$ corresponds to the mapping $\bar{\mathbf{s}}(\boldsymbol{v}) = \mathrm{nint}(v_x)\boldsymbol{e}_x + \mathrm{nint}(v_y)\boldsymbol{e}_y$ ($\mathrm{nint}(x)$ = nearest integer to x), i.e. to a perpendicular projection suppressing the v_z-coordinate. Besides $P(\boldsymbol{v})$ and $\bar{\boldsymbol{w}}$, the remaining important determinant of the behavior of the system is the function h_{rs} that defines the coupling between different lattice points. A simple choice is the Gaussian

$$h_{rr'} = \sum_{\boldsymbol{s}} \delta_{r+s,r'} \exp\left(-\frac{s^2}{2\sigma^2} \right), \tag{8.12}$$

with lateral width σ, for which we will require $1 \ll \sigma \ll N$. Owing to the translational invariance, both $D_{rmr'n}$ and $\boldsymbol{B}_{rmr'n}$ depend only on the difference

$\boldsymbol{r} - \boldsymbol{r}'$ and on m, n. Therefore, we can decouple (8.7) if we represent $S(\boldsymbol{u}, t)$ in terms of the Fourier mode amplitudes

$$\hat{\boldsymbol{u}}(\boldsymbol{k}) = \frac{1}{N} \sum_{\boldsymbol{r}} \mathrm{e}^{\mathrm{i}\boldsymbol{k}\cdot\boldsymbol{r}} \boldsymbol{u}_{\boldsymbol{r}} \tag{8.13}$$

of $\boldsymbol{u}$, where $\boldsymbol{k} = (l/2\pi N, m/2\pi N)$ is a two-dimensional wave vector of the lattice A. Each mode amplitude turns out to be distributed independently, and its fluctuations can be calculated explicitly by separating (8.7) into a set of independent equations for each mode. The exact result is fairly complicated (for details, see [8.23]), but if one neglects "discretization effects" due to the finite lattice spacing and uses axes parallel (||) and perpendicular ($\perp$) to the wave vector instead of the fixed x- and y-directions, one can bring the mean square value of the equilibrium fluctuations of the different Fourier modes into a simpler form:

$$\langle \hat{u}_\perp(\boldsymbol{k})^2 \rangle = \varepsilon\pi\sigma^2 \frac{\exp(-k^2\sigma^2)}{12(1 - \exp(-k^2\sigma^2/2))}\,, \tag{8.14}$$

$$\langle \hat{u}_{||}(\boldsymbol{k})^2 \rangle = \varepsilon\pi\sigma^2 \frac{(12k^2\sigma^4 + 1)\exp(-k^2\sigma^2)}{12 - 12(1 - k^2\sigma^2)\exp(-k^2\sigma^2/2)}\,, \tag{8.15}$$

$$\langle \hat{u}_3(\boldsymbol{k})^2 \rangle = \varepsilon\pi\sigma^2 \frac{s^2 \exp(-k^2\sigma^2)}{3 - s^2k^2 \exp(-k^2\sigma^2/2)}\,. \tag{8.16}$$

Figures 8.2–4 compare the theoretical prediction (curves) and data points from a Monte Carlo simulation of the process on a 32×32-lattice for the square roots $f_{1,2,3} = \langle \hat{u}^2_{\perp,||,3} \rangle^{1/2}$ of the mode fluctuations for $\varepsilon = 0.01$. To make the Monte Carlo simulation computationally more feasible, $h_{\boldsymbol{rs}}$ was not chosen according to (8.12), but instead as unity for $\boldsymbol{r} = \boldsymbol{s}$ and all nearest-neighbor pairs $\boldsymbol{r},\boldsymbol{s}$ in the lattice and zero otherwise. This corresponds roughly to $\sigma = 1$ in (8.12), but the corresponding theoretical predictions are somewhat different from (8.14–16) (they are given in [8.23]); however, all essential features discussed below remain. Each mode can be interpreted as a periodic distortion of the equilibrium mapping. The first set of modes ($\hat{\boldsymbol{u}}_\perp$, Fig. 8.2) represents distortions that are "transverse" to their direction of periodicity, while the second set of modes ($\hat{\boldsymbol{u}}_{||}$, Fig. 8.3) represents distortions that are "longitudinal". For both kinds of mode, the fluctuations increase with increasing wavelength. This is to be expected, since the "restoring force" for modes with very long wavelengths is determined by the boundary conditions, which are assumed periodic, and, therefore, allow an arbitrary translational shift ($k = 0$-mode).

The most interesting set of modes are those perpendicular to the xy-directions ($\hat{\boldsymbol{u}}_3$). These modes describe fluctuations of the prototypes $\boldsymbol{w}_{\boldsymbol{r}}$ along the additional dimension, which is "lost" in the "perpendicular" projection $\bar{\boldsymbol{s}}(.)$ associated with the equilibrium configuration $\bar{\boldsymbol{w}}$. For values $s \ll \sigma$, inspection of (8.16) shows that the amplidude of these modes is of the order of s and, therefore, is small for any $\boldsymbol{k}$. This indicates that, although some information is lost, for $s \ll \sigma$ the mapping defined by $\bar{\boldsymbol{w}}$ cannot be improved by small distortions. However,

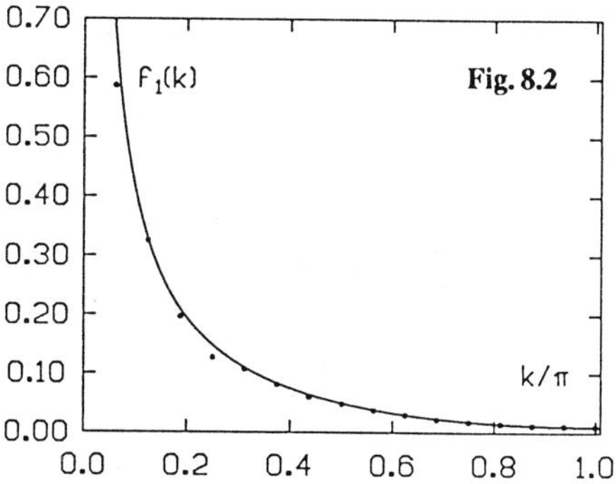

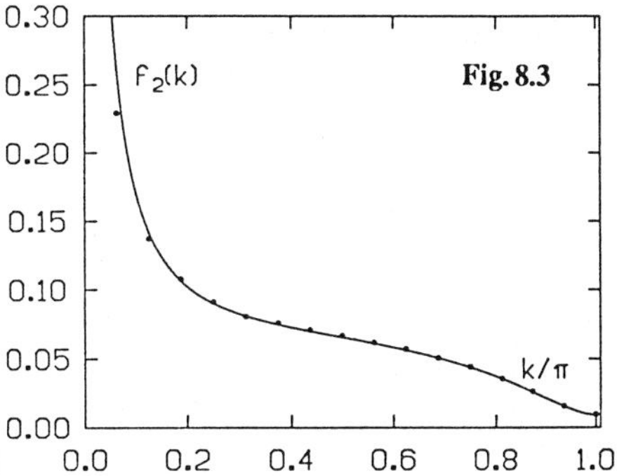

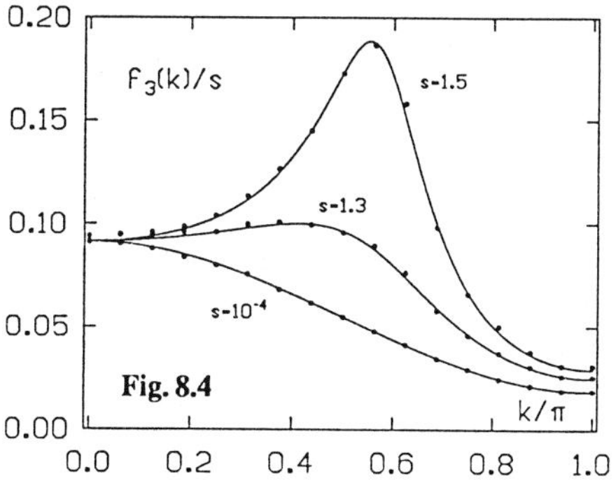

Fig. 8.2. Dependence of fluctuations of "transverse mode" $u_\perp$ on the wave number k. The "neighborhood function" was chosen to be $h_{\boldsymbol{rs}} = 1$ for $\boldsymbol{r} = \boldsymbol{s}$ and all nearest-neighbor pairs $\boldsymbol{r}$, $\boldsymbol{s}$ and zero otherwise. The data points were obtained from a Monte Carlo simulation with 10 000 samples of the Markov process (8.4) for fixed $\varepsilon = 0.01$ and $s = 10^{-4}$. The curve represents the analytical result

Fig. 8.3. Fluctuations of the "longitudinal mode" $u_{||}$ of the same simulation as in Fig. 8.2 above. For small wave numbers the fluctuations are smaller than for u_1

Fig. 8.4. Fluctuations of the "perpendicular mode" u_3 for three different values of the thickness parameter s: for $s = 10^{-4}$, i.e. essentially a two-dimensional input distribution, only small fluctuations arise. For $s = 1.3$ fluctuations begin to exhibit a broad maximum at about $k^* = 0.58\pi$, which becomes very pronounced for $s = 1.5$, i.e. a value closely below the critical value $s^* = \sqrt{12/5}$

as s increases, (8.16) shows that for s close to a "threshold value" of $s^* = \sigma\sqrt{3e/2} \approx 2.02\sigma$ the denominator can become very small for $||\boldsymbol{k}||$-values in the vicinity of $||\boldsymbol{k}|| = k^* = \sqrt{2}/\sigma$, and correspondingly large fluctuations are exhibited by these modes. Finally, at $s = s^*$, all modes with $||\boldsymbol{k}|| = k^*$ become unstable: s has become so large that the mapping $\bar{\mathbf{s}}(.)$ has lost its optimality and can be further optimized if the prototypes $\boldsymbol{w}_r$ assume a wavelike "modulation" along their $\boldsymbol{w}_{r3}$-direction. The characteristic wavelength of this modulation is $\lambda^* = \sigma\pi\sqrt{2} \approx 4.44\sigma$ [8.21]. For $s > s^*$, a whole "window" of unstable modes appears. This is also discernible in Fig. 8.4, where the different choice of the function h_{rs}, however, leads to changed values of $s^* = \sqrt{12/5} \approx 1.54$ and $k^* \approx 0.58\pi$ (for $\boldsymbol{k}$ directed along the x-direction).

We can summarize now the following answer to our initial question: the simple, "perpendicular" projection $\bar{\mathbf{s}}(.)$ is the optimal mapping as long as s, its maximal "error" in the vertical direction, is below a value $s^* = \sigma\sqrt{3e/2}$. In this case, apart from fluctuations, all prototypes $\boldsymbol{w}_r$ have the same value of w_{r3}. The threshold value s^* *can be interpreted as being the distance in the space* V *that corresponds to the range of the "neighborhood function"* h_{rs} *in the lattice* A. For $s > s^*$, the "perpendicular" projection can be optimized further by distortions. These distortions arise from the components w_{r3}, which now must vary with $\boldsymbol{r}$. Their variation, and, therefore, the pattern of distortions, is dominated by a wavelength of $\lambda^* = \sigma\pi\sqrt{2}$, i.e. λ^* *is also proportional to the range of the "neighborhood function"* h_{rs}.

In the previous context, V being a set of patterns, the x- and y-coordinates would correspond to two "primary" features characterized by a large variation, whereas the z-coordinate would correspond to a "secondary" feature, characterized by a smaller variation that is measured by s. Then, as long as $s < s^*$, the system converges to a topographic map of the two "primary" features only. However, when the variation of the "secondary" feature, compared to the two "primary" ones, becomes large enough, the "secondary" feature begins to be reflected in the values of the prototypes and, therefore, in the topographic map. The variation of the prototypes along the axis of the "secondary" feature is dominated by the wavelength λ^* and gives rise to an irregular pattern of "stripes" if each lattice point $\boldsymbol{r}$ is assigned a gray value that indicates the value w_{r3} of its prototype $\boldsymbol{w}_r$ along the axis of the "secondary" feature (Fig. 8.5).

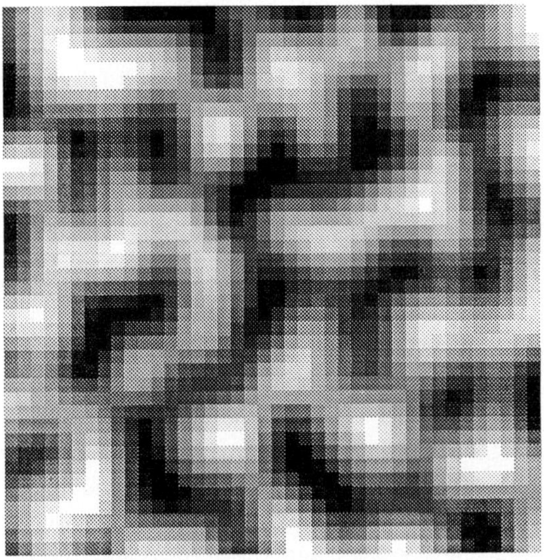

Fig. 8.5. "Striped projection". The displayed 40×40-lattice was used to obtain a "topographic map" of a 3-dimensional "feature space" given by $0 \le x, y \le 40$, $-4 \le z \le 4$ with Kohonen's algorithm ((8.4), $\sigma = 1.4$, 10^4 steps). The height (z-) dimension plays the role of the "secondary" feature, and gray values indicate its distribution over the lattice. The resulting pattern closely resembles the "ocularity stripes" found in the visual cortex. These are alternating bands of cells with stronger preference to input from either the right or the left eye (see, e.g. [8.24])

Interestingly, in the brain there are many two-dimensional arrangements of cells on which such "striped projections" seem to be realized. Prominent examples are the "ocular dominance stripes", where in addition to the "primary" two-dimensional retinal location, the additional feature "ocularity" (i.e. the degree to which a cell receives input from each eye) is mapped [8.24, 25], and the "orientation stripes", where the additional feature is line orientation [8.26, 27].

Models of the development of such "striped projections" have been previously suggested (see, e.g. [8.25, 28–31]). Here we want to emphasize that the previous analysis demonstrates that also *the particularly simple model by Kohonen can account for the phenomenon of striped projections*, a fact that has been observed already in simulations presented in [8.10] but seems to have received only little attention subsequently. This brings us to the issue of the "neural interpretation" of the model and its properties, a topic taken up in the following section.

8.4 Relation to Brain Modeling

One of the major architectural features within the brains of higher animals are *topographically organized "maps"* of various "feature spaces". They can be found in nearly all sensory and motor areas within the brain, e.g. the visual, auditive, and somatosensory fields as well as in the motor cortex, and there is both theoretical and some experimental evidence that maps of more abstract features might turn out also to play a role on higher processing levels [8.32].

In the somatosensory system the "feature space" is particularly simple. It is mapped onto a certain part of the cortex called the "somatosensory cortex". Experiments on the cortical representation of the hand surface in owl monkeys have revealed a very precise correspondence between hand locations and neurons in the cortical field [8.33]: each neuron can be excited only from receptors in some small "receptive field" in the hand surface, and the arrangement of the neurons in the cortex is a distorted, but still topographic "image" of the arrangement of their receptive fields on the skin. There is evidence that the required, very precise connectivity is not genetically prespecified but instead evolves gradually under the influence of sensory experience. Maps in different individuals show considerable variations, and they are not rigidly fixed even in adult animals. The somatotopic map can undergo adaptive changes, which have been found to be strongly driven by afferent input [8.34, 35].

The "self-organizing maps" are perhaps the simplest model that can account for the adaptive formation of such topographic representations (for other modeling approaches, see e.g. [8.30, 31, 36, 37]). In this case, the lattice A of prototypes $\boldsymbol{w}_{\boldsymbol{r}}$ corresponds to a sheet of laterally interacting adaptive neurons, one for each lattice site $\boldsymbol{r}$, that are connected to a common bundle of n input fibers from the receptors in the receptor sheet. The ith component of vector $\boldsymbol{w}_{\boldsymbol{r}}$ is interpreted as the connection strength between input fiber i and neuron $\boldsymbol{r}$.

The formation of the map is assumed to be driven by random sensory stimulation. Tactile stimuli on the receptor sheet excite clusters of receptors and thereby cause activity patterns on the input lines that are described by n-dimensional real vectors and that take the role of the input patterns $\boldsymbol{v} \in V$. The total synaptic input to each neuron $\boldsymbol{r}$ is measured by the dot product $\boldsymbol{x} \cdot \boldsymbol{w}_{\boldsymbol{r}}$. Each tactile stimulus is considered as a discrete event that leads to excitation of a localized group of neurons in the lattice A. The function $h_{\boldsymbol{rs}}$ (with $\boldsymbol{s}$ fixed and $\boldsymbol{r}$ taken as argument) is interpreted as the spatial variation of this neural excitation in A, and (8.4) can then be interpreted as a Hebbian rule together with an "activity-gated" memory loss term for the change in the synaptic strengths $\boldsymbol{w}_{\boldsymbol{r}}$ following the stimulus.

The spatial shape $h_{\boldsymbol{rs}}$ of the neural response is modeled by a Gaussian and is assumed to arise from lateral competitive interactions within the cortical sheet (for details, see e.g. [8.11]). Its center location $\boldsymbol{s}$ is assumed to coincide with the neuron receiving the largest synaptic input $\boldsymbol{w}_{\boldsymbol{r}} \cdot \boldsymbol{v}$. Strictly speaking, this is only equivalent to (8.4), where $\boldsymbol{s}$ was defined to be minimizing the Euclidean difference $||\boldsymbol{v} - \boldsymbol{w}_{\boldsymbol{s}}||$, if all vectors $\boldsymbol{w}_{\boldsymbol{r}}$ and $\boldsymbol{v}$ are assumed to be normalized. If, however, all $\boldsymbol{w}_{\boldsymbol{r}}$ are kept normalized, one can even drop the "memory loss term" $-h_{\boldsymbol{rs}(\boldsymbol{v})}\boldsymbol{v}$, as its main purpose is only to keep the vectors $\boldsymbol{w}_{\boldsymbol{r}}$ bounded. Therefore, the simulations presented below are based on the modified adaptation equation

$$w_{kli}(t+1) = (w_{kli}(t) + \varepsilon(t)h_{\boldsymbol{rs};kl}(t)v_i)/\sqrt{\Sigma_i(w_{kli}(t) + \varepsilon(t)h_{\boldsymbol{rs};kl(t)}v_i)^2}, \quad (8.17)$$

where we have also altered the notation, replacing $\boldsymbol{r}$ by (k, l) when referring to the synaptic weights of a neuron at a lattice site $\boldsymbol{r} = (k, l)$.

We still need to specify the input patterns $\boldsymbol{v}$. To this end, each input line i is taken to belong to one tactile receptor, located at a position $\boldsymbol{x}_i$ in the receptor sheet. The tactile stimuli are assumed to excite spatial clusters of receptors. As a convenient mathematical representation for a stimulus centered at $\boldsymbol{x}_{\text{stim}}$, we choose a Gaussian "excitation profile"

$$v_i = N \exp\left(-\frac{(\boldsymbol{x}_i - \boldsymbol{x}_{\text{stim}})^2}{\sigma_r^2}\right) \quad (8.18)$$

where σ_r is a measure of the "radius" of each stimulus.

With this interpretation, the algorithm, discussed in Sect. 8.2 as a means to generate a representation of a data set V that is optimal for transmission over some noisy channel, is seen as an adaptive process shaping the connectivity between two sheets of nerve cells. We can now return to the significance of the minimization of the functional $F[\boldsymbol{w}]$ in the present context.

We will assume that the primary task of a cortical map is to prepare a suitable encoding of the afferent sensory information for use in subsequent processing stages. Part of this encoded information is represented by the location of the excited neurons. Therefore, the correct transmission of this information to subsequent processing stages is equivalent to the target neurons being able to assess the origin of their afferent excitation correctly. However, such neurons typically integrate information from several different brain areas. Therefore, their "receptive fields" in these areas tend to be the larger, the higher their level in the processing hierarchy is, and their reaction to input from one source may be influenced by the current inputs from other sources. This may make it impossible to tie their response precisely to excitation in a precise location of a cortical predecessor map. However, if such neurons cannot "read" the precise source location of their input excitation, they are in a position very similar to that of a receiver at the other end of a noisy transmission channel. Therefore, *a cortical map based on minimization of the functional $F[\boldsymbol{w}]$ (8.5) might help to minimize the average transmission error between neural layers arising from fluctuations of their "functional connectivity"*.

8.5 Formation of a "Somatotopic Map"

In this section, we shall present a computer simulation for the adaptive ordering of an initially random projection between tactile receptors on the skin and neurons in a "model cortex" [8.11, 38]. The "model cortex" consists of 16 384 neurons that are arranged as a 128×128 square lattice and connected to 800 tactile receptors randomly scattered over a "receptor surface" (Fig. 8.6). The initial values w_{kli} were given independent random values chosen from the unit interval, thereby "connecting" each neuron in a random fashion to the 800 receptors.

In experiments, neurons are frequently characterized by their "receptive field properties". A "receptive field" of a neuron is the set of stimulus locations $\boldsymbol{x}_{\text{stim}}$ that lead to a noticeable excitation of the neuron. The center of the receptive field of neuron (k, l) is in the model defined as the average

$$\boldsymbol{s}_{kl} = \sum_i \boldsymbol{x}_i w_{kli} \Big/ \sum_i w_{kli} \,. \tag{8.19}$$

The mean square radius of the receptive field is a measure of the stimulus selectivity of neuron (k, l) and is defined as

$$G_{kl} = \sum_i (\boldsymbol{x}_i - \boldsymbol{s}_{kl})^2 w_{kli} \Big/ \sum_i w_{kli} \tag{8.20}$$

Figure 8.7 shows the hand-shaped receptor surface of the model used for the simulations. Each dot represents one of the 800 tactile receptors. Different regions of the hand are coded by different colors. Figure 8.8a shows the initial state of the network. Each pixel in the image corresponds to one neuron (k, l) of the neural sheet, and the pixel color simultaneously encodes two different properties of its receptive field in the hand surface: the hue indicates the location $\boldsymbol{s}_{kl}$ of the field center on the receptor surface shown in Fig. 8.7, and the saturation of the color indicates the spatial spread ($\propto G_{kl}^{1/2}$) of the receptive field. Neurons with large, diffuse receptive fields, i.e. those neurons that are connected to many, widely scattered receptors, are represented with low color saturation, while neurons with small, spatially highly specific receptive fields, i.e. connected only to receptors within a small spatial domain, are represented with bright, highly saturated colors.

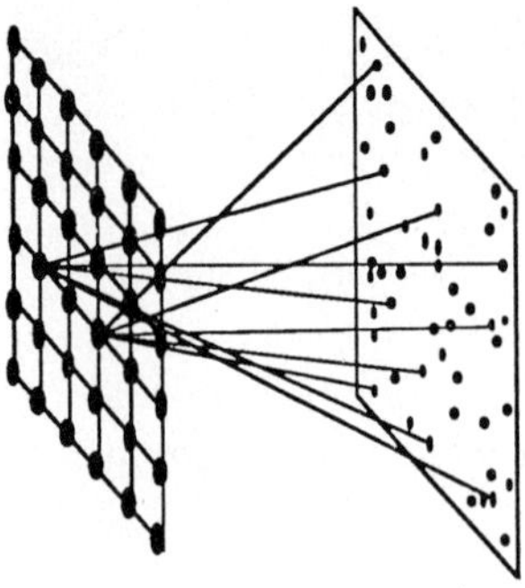

Fig. 8.6. Schematic drawing of the model system. The neuron sheet ("cortex") is represented by a square lattice of model neurons, each connected by modifiable links to all of 800 randomly located "receptors" in the receptor surface

As each neuron has its connections randomly initialized, all receptive fields, apart from statistical fluctuations, are initially very similar: they diffusely extend over the whole hand surface and, therefore, are "centered" in the middle of the receptor surface. This is reflected in Fig. 8.8a by the fairly uniform and pale color within the entire cortical sheet. Figure 8.8b shows the map after about 200 adaption steps. The brighter colors indicate the onset of a specialization of the neurons for smaller receptive fields. Finally, Fig. 8.8c shows the completed and refined map obtained after 10 000 stimuli. It shows all parts of the hand surface in their correct topographic order, and the cortical map exhibits only small fluctuations during further "stimulation". Note that each neuron is still connected to every receptor, although the connection strengths outside the neuron's receptive field are very small. The connections can, however, be "revived" if the distribution of the input pattern changes, leading to an input-driven reorganization of the cortical map [8.34].

The emergence of selectively tuned neurons with spatially restricted receptive fields in the receptor surface can also be analytically demonstrated [8.39]. For sufficiently small adaptation steps (i.e. $\varepsilon \ll 1$), one can derive the following equation for the change of the receptive-field sizes G_{kl} under one adaptation step:

$$G_{kl}(t+1) = G_{kl}(t) + \frac{\varepsilon(t) h_{rs;kl}(t)}{\sum_i w_{kli}(t)} \sum_i (\boldsymbol{x}_i - \boldsymbol{s}_{kl})^2 \left(v_i - w_{kli}(t) \frac{\sum_j v_j}{\sum_j w_{klj}(t)} \right). \tag{8.21}$$

From this relation, one can derive an equation for G_{kl} when the system has reached a stationary state,

$$G_{kl} = \frac{\int h_{rs;kl} \langle [\Gamma(\boldsymbol{x}_s) + (\boldsymbol{x}_s - \boldsymbol{s}_{kl})^2] \sum_i v_i \rangle P(\boldsymbol{x}_s) \mathrm{d}^2 \boldsymbol{x}_s}{\int h_{rs;kl} \langle \sum_i v_i \rangle P(\boldsymbol{x}_s) \mathrm{d}^2 \boldsymbol{x}_s} . \tag{8.22}$$

Here $P(.)$ denotes the probability density of the stimuli centers $\boldsymbol{x}_{\text{stim}}$ in the receptor surface, and $\langle .. \rangle$ denotes the average over all stimulus shapes (in case the stimuli are more general than the Gaussians of (8.18)). $\Gamma(.)$ represents the

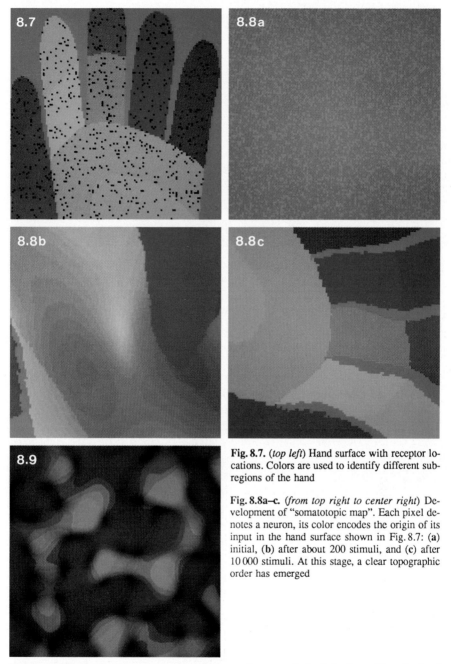

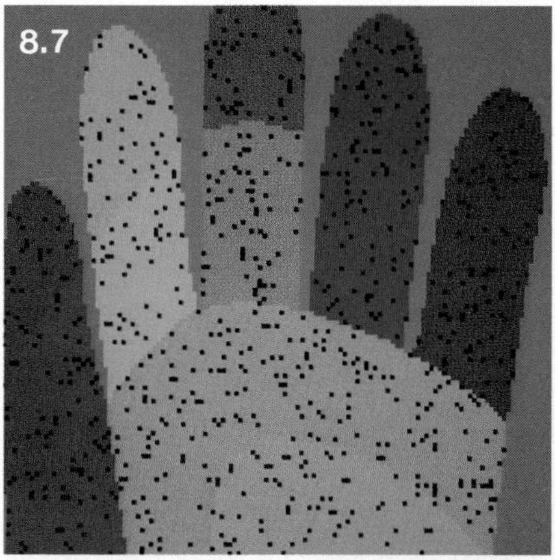

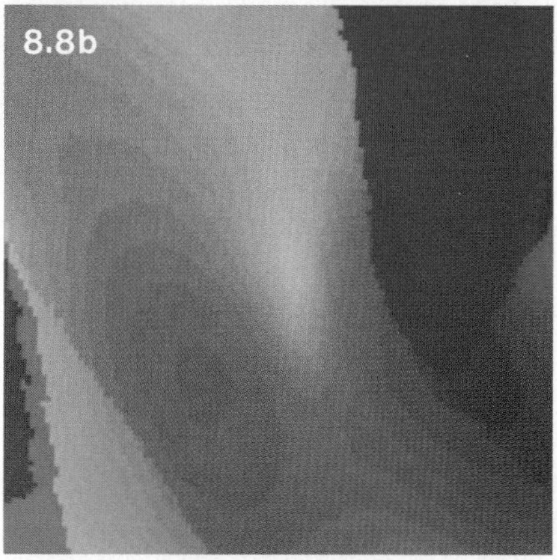

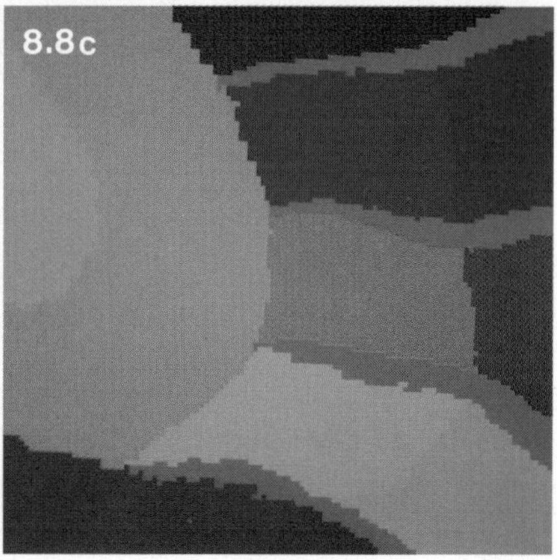

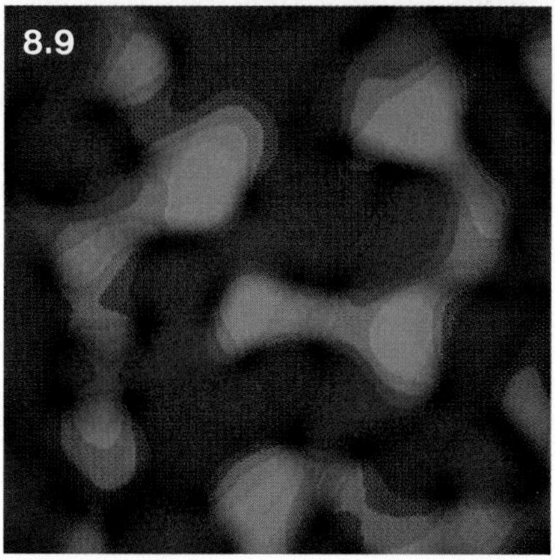

Fig. 8.7. (*top left*) Hand surface with receptor locations. Colors are used to identify different subregions of the hand

Fig. 8.8a–c. (*from top right to center right*) Development of "somatotopic map". Each pixel denotes a neuron, its color encodes the origin of its input in the hand surface shown in Fig. 8.7: (**a**) initial, (**b**) after about 200 stimuli, and (**c**) after 10 000 stimuli. At this stage, a clear topographic order has emerged

Fig. 8.9. Spatial organization of neuronal orientation preference and selectivity, formed in a simulated patch of "visual cortex" by the self-organizing process (8.17). A "rainbow" palette of colors indicates orientation preferences from 0° to 180°. Dark regions correspond to low, bright regions to high, directional selectivity

mean square radius of each stimulus, defined by

$$\Gamma(\boldsymbol{x}_s) = \frac{1}{\sum_i v_i} \sum_i (\boldsymbol{x}_i - \boldsymbol{x}_s)^2 v_i \,. \tag{8.23}$$

Equation (8.22) can be approximated well by the much simpler relation (for details see [8.40])

$$G_{kl} \approx \Gamma + M^{-1}\sigma_{\mathrm{h}}^2 \,, \tag{8.24}$$

where

$$\sigma_{\mathrm{h}}^2 = \frac{2 \sum\limits_{m,n} h_{rs;mn} \left[(r-m)^2 + (s-n)^2 \right]}{\sum\limits_{m,n} h_{rs;mn}} \tag{8.25}$$

denotes the mean square radius of the output function and M is the *local magnification factor* of the mapping from stimulus centroids $\boldsymbol{x}_{\mathrm{stim}}$ to neuron coordinates (r, s). Equation (8.24) states that the neurons develop receptive fields the area of which (proportional to G_{kl}) is the sum of two terms: the first term is essentially the area of a typical stimulus ($\propto \Gamma$) and the second term is essentially the area ($\propto \sigma^2$) of the adjustment zone in the neuron layer, but "projected back" (inverse magnification factor M^{-1}) onto the receptor sheet. Therefore, for predominantly localized tactile stimuli and narrow $h_{rs;mn}$ the neurons will develop *localized receptive fields*.

Figure 8.10 compares this theoretical result with data from a simulation (16 384 cells, 784 receptors, $h_{rs,kl} = \exp(-[r-k]^2 + [s-l]^2/\sigma_{\mathrm{h}}^2), \sigma_{\mathrm{r}} = 0.15$, 6×10^4 steps), where σ_{h} has been varied slowly between $\sigma_{\mathrm{h}} = 100$ and $\sigma_{\mathrm{h}} = 5$. The diagram shows the mean square radius of the receptive field averaged over 2 300 neurons from the center of the neural sheet plotted against the right-hand side of (8.24). The dots represent the results of the simulation and the solid line corresponds to (8.24). The agreement is very satisfactory, except for parameter values leading to large receptive fields, for which edge effects become noticeable.

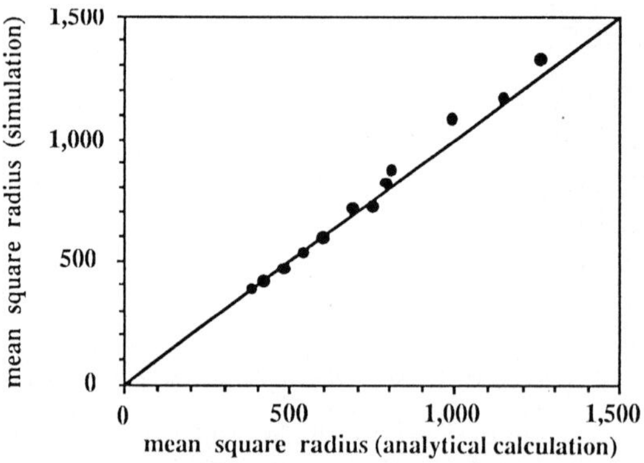

Fig. 8.10. Development of receptive field radii. The analytical result (8.24) is compared with results of a computer simulation (mean square radii are given in arbitrary, relative units)

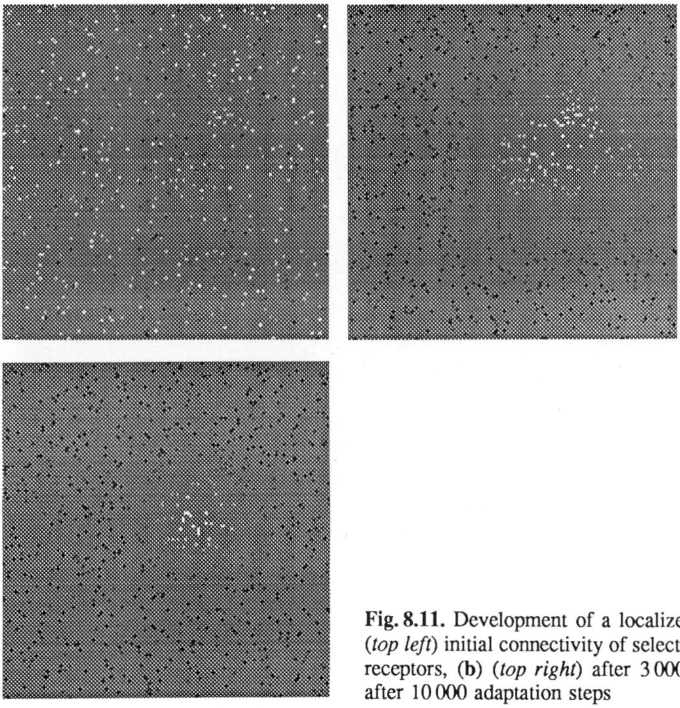

Fig. 8.11. Development of a localized receptive field: (**a**) (*top left*) initial connectivity of selected neuron with tactile receptors, (**b**) (*top right*) after 3 000 and (**c**) (*lower left*) after 10 000 adaptation steps

Figure 8.11 illustrates the state of a typical receptive field at the beginning of another simulation (Fig. 8.11a), after 3 000 iterations (Fig. 11b) and after 10 000 iterations (Fig. 8.11c) (for this run $\sigma_h = 50 \ldots 5$, $\sigma_p = 0.12$ and $t_{max} = 10^4$). The dot locations mark the positions of the tactile receptors on the receptor surface, while their brightness encodes the strength of their connection to the neuron under consideration. Initially, the field is very diffuse (Fig. 8.11a), but contracts rapidly (Fig. 8.11b), until finally it is localized well (Fig. 8.11c).

An important aspect of these simulations is the demonstration that the formation of the correct topographic map of the skin surface succeeds also in the case when the two-dimensional position information about the stimuli is provided by the very high dimensional ($d = 800$) input vectors that represent the excitation patterns of the tactile receptors. To accomplish this, the algorithm has to "detect" the relevant, two-dimensional submanifold formed by the stimuli, and has to "ignore" all remaining "orthogonal" coordinates.

The optical cortex faces a self-organization task similar to that of the somatosensory cortex. However, in contrast to the somatosensory map, the map from the eyes' retina to the optical cortex is known to represent, in addition to the "primary" features "retinal location" (pair of coordinates!), further features, such as *orientation* and *ocularity*. The spatial organization of the visual map has been studied in great detail (see e.g. [8.24, 26, 27]) and cells with similar ocularity or tuned to similar orientations were found to be arranged in *irregular "bands" or "stripes"*. Adding "ocularity" or "orientation" to the two "primary" features "retinal location" results in the need to map a three- or higher-dimensional feature space onto a two-dimensional surface [8.58, 61]. Figure 8.5 showed a self-organizing map which gave rise to a pattern of the "secondary" feature that resembles observed patterns in the visual map. A closer agreement with natural maps can be obtained if the three-dimensional input representation is replaced by high-dimensional excitation patterns on a "model retina" [8.56, 57]. Figure 8.9 shows an "orientation map" on a "model cortex" of 256×256 cells obtained in this way. "Stimuli" were of elliptic shape with randomly chosen orientations. The "stimuli" produced excitations on a "model retina" covered with 900 randomly distributed light-sensitive "receptors". Each pixel in Fig. 8.9 represents one neuron. The pixel color encodes the orientation of the stimulus to which the neuron is maximally responsive ("orientation preference"), and the brightness the degree of specificity of the neuron. Various features, such as "slabs" along which orientation selectivity changes continuously, dark "foci" of unspecific cells around which orientation changes in a clockwise or anti-clockwise fashion, and "fractures", across which orientation changes discontinuously, can be discerned and correspond well to observations from actual mapping experiments (see, e.g. [8.26, 27]).

8.6 Adaptive Orientation and Spatial Frequency Filters

In this section we consider the issue of data compression through preprocessing by local filters which select geometrically significant features from their input. A major part of the material in this section is taken from [8.40]. The input will be spatially varying patterns, for example, correlated random dot patterns or textures. The architecture of the network is very similar to that of the networks described above. It consists of an input layer with N_i neurons and an output layer with N_o neurons. Input and output units exhibit real, continuous-valued activities $\boldsymbol{i} = (i_1, .., i_{N_\mathrm{i}})$ and $\boldsymbol{o} = (o_1, .., o_{N_\mathrm{o}})$. The output layer should develop synaptic connections with the input layer to establish suitable receptive fields. The development of connections is driven by training with a set of N_π representative patterns $\{\boldsymbol{p}^\pi = (p_1^\pi, .., p_{N_\mathrm{i}}^\pi), \pi = 1, \ldots, N_\pi\}$.

The formation of the receptive fields will depend on adaptive lateral interactions between units in the output layer. These lateral interactions serve functions similar to lateral connections in Kohonen-type networks. In the latter such connections are needed to determine the output unit of maximal activity as well as to induce neighborhoodwide activities described by the functions h_{rs}, which leads to a topology-conserving neural projection between input and output layers. In the present network, lateral interactions serve the role of molding the projection such that characteristic correlations between input activities $i_1, \ldots, i_{N_\mathrm{i}}$ are detected. Such correlations are described by the the covariance matrix C of a set of characteristic patterns. This matrix has elements $C_{jk} = \langle p_j^\pi p_k^\pi \rangle$ where p_j^π denotes a sample pattern and $\langle \ldots \rangle$ denotes the average over the training set $\{\boldsymbol{p}^\pi, \pi = 1, \ldots, N_\pi\}$.

The desired filters are achieved by synaptic weights w_{jm} between input unit j and output unit m. The set of weights connecting an output unit m with all input units forms the weight vector $\boldsymbol{w}_m$, the transpose of which is the mth row of the weight matrix $\boldsymbol{W}$. Activities of the input units correspond to the presented patterns, i.e., $\boldsymbol{i} = \boldsymbol{p}^\pi$. Activities of the output units, in response to a pattern $\boldsymbol{p}^\pi$, are linear sums of the inputs weighted by the synaptic strengths, i.e., $\boldsymbol{o}^\pi = \boldsymbol{W}\,\boldsymbol{p}^\pi$. The desired filters, i.e. synaptic weights, are the eigenvectors of the covariance matrix C_{jk} defined through

$$\sum_j C_{jk} w_{km} = \lambda_m w_{jm} \ . \tag{8.26}$$

Application of such filters corresponds to the statistical technique of *principal component analysis* (see, e.g., [8.41]). The network should yield the first eigenvectors of the covariance matrix corresponding to the largest eigenvalues λ_m of C_{jk}. To render such network robust against failure of single units one may represent the mth eigenvalue by a large number of output units rather than a single unit. However, in the following we will assume for the sake of simplicity that single output units represent an eigenvector of C_{jk}.

The output units should also discern the spatial location of these characteristics in a spatially extended pattern $\boldsymbol{p}$. This latter capacity can be achieved through a translation-invariant duplication of network characteristics. We will neglect the position dependence and focus in the following only on preprocessing in a small neighborhood of input cells. We choose, therefore, two completely interconnected layers, i.e. w_{jm} initially is nonzero for all j and m.

We will describe now how the network through adjustment of its synaptic weights can adopt appropriate receptive fields when exposed to the set of training patterns. Weights between layers are adjusted upon presentation of an input pattern $\boldsymbol{p}^\pi$ according to a Hebbian rule, leading to an increase in synaptic strength if the corresponding pre- and postsynaptic potentials are of the same sign. If weight changes are small enough, the update can be performed after presentation of all patterns, i.e.

$$\Delta \boldsymbol{w}_m = \eta \langle (\boldsymbol{p}^\pi - \langle \boldsymbol{p}^\pi \rangle)(o_m^\pi - \langle o_m^\pi \rangle) \rangle \ , \tag{8.27}$$

where η is a positive parameter and where the brackets $\langle \ldots \rangle$ denote again the average over the set of patterns. The subtraction of averages in (8.27) can be interpreted as the existence of thresholds of the units. On the other hand, subtracting averages is convenient from a mathematical point of view [8.42] and allows one to assume that $\langle \boldsymbol{p}^\pi \rangle = 0$ and $\langle \boldsymbol{o}^\pi \rangle = 0$.

Let us first consider the case of a single output unit. Linsker showed that the weights of a unit that is subject to the Hebbian rule (8.27) evolve to maximize the variance of the output for a set of presented patterns [8.18]. If the weights are normalized after every update such that $\sum_i w_{i1}^2 = 1$, the Hebbian rule renders weights which characterize the direction of maximal variance of the pattern set [8.42]. Equivalently, the weight vector $\boldsymbol{w}_1$ converges to the eigenvector with the largest eigenvalue λ_1 of the covariance matrix C of the pattern set. Thus, a Hebbian learning rule for Euclidian normalized weights yields the first principal component of the input data set. The nonvanishing weights w_{i1} of the output unit define its receptive field. The output unit then acts as a feature detector which analyzes the principal feature of a presented pattern and corresponds to a so-called "matched linear filter" [8.42].

However, a single principal component usually describes only a fraction of the total information contained in a pattern. In order to transmit the complete information between the two layers, as many output cells as the rank of the covariance matrix C are required. Furthermore, in order to develop into filters of mutually orthogonal features, the output cells need to become uncorrelated. For this purpose we assume the existence of lateral, hierarchically organized connections with weights u_{lm} between output units l and m, where $l < m$. The activity of the mth output cell is then given by $o_m^\pi = \boldsymbol{w}_m \cdot \boldsymbol{p}^\pi + \sum_{l<m} u_{lm} \boldsymbol{w}_l \cdot \boldsymbol{p}^\pi$. According to (8.27), changes of synaptic connection strengths between input units and output unit m are given by

$$\Delta \boldsymbol{w}_m = \eta \left(C\boldsymbol{w}_m + \sum_{k<m} u_{km} C\boldsymbol{w}_k \right) . \tag{8.28}$$

Figure 8.12 presents the architecture of the network, in particular, the hierarchical arrangement of lateral connections. This arrangement has been chosen to guide the network to a final state in which the output units assume receptive fields corresponding to the different eigenvectors of the covariance matrix C. The cell in the output layer at the top of the hierarchy will adopt a receptive field corresponding to the eigenvector with the largest eigenvalue, the cell next in the hierarchy will represent the eigenvector with the second largest eigenvalue, and so forth.

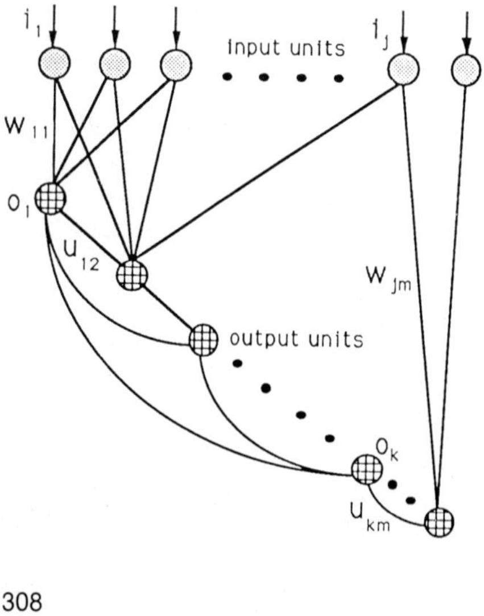

Fig. 8.12. Schematic drawing of the hierarchical network used for feature extraction

To ascertain that the cells adopt different receptive fields the lateral weights u_{lm} adapt themselves according to an *anti-Hebbian* rule: the strength of a lateral synapse is lowered if the corresponding pre- and postsynaptic activities are of the same sign. Again we assume that changes of the synaptic weights are small. The anti-Hebbian rule leads to a decrease in synaptic strength if the corresponding output units have correlated activities and is described by

$$\Delta u_{lm} = -\mu \langle \mathrm{o}_l^\pi \, \mathrm{o}_m^\pi \rangle \, . \tag{8.29}$$

Here μ is a positive learning parameter. The anti-Hebbian rule is similar to the learning rule of Kohonen's novelty filter [8.10] and to the "unlearning" rule proposed by Hopfield [8.43].

Because of the hierarchical arrangement, the cell at the top of the hierarchy will force all other output units to become uncorrelated to it; the top cell develops its receptive field in accordance with the first eigenvector and suppresses any attempt of cells lower in the hierarchy to develop the same receptive field. The second cell in the hierarchy prevents all cells below it developing a receptive field similar to its own, the latter being shaped to agree with the second eigenvector of the covariance matrix C. This chain is continued down to the cell last in the hierarchy. The selection of receptive fields in the order of decreasing eigenvalues λ_m originates from the fact that the weights w_{jm} grow fastest in the direction of the distribution of the first eigenvector, second fastest in the direction of the second eigenvector, and so on.

As a result of the proposed learning scheme, the weight vector $\boldsymbol{w}_m$ converges to the mth eigenvector of C. Convergence requires that the learning parameters η and μ governing the weights w_{jm} and u_{mn}, respectively, need to obey the inequality (we assume the ordering of eigenvalues $\lambda_1 > \lambda_2 > \ldots > \lambda_{N_\mathrm{o}}$)

$$\mu > \frac{\eta \, (\lambda_1 - \lambda_n)}{\lambda_1 \, (1 + \eta \lambda_n)} \tag{8.30}$$

for $n = 1, 2, \ldots, N_\mathrm{o}$ [8.44]. Since C is a real symmetric matrix, the weight vectors w_{jm} become orthogonal and, consequently, the output units with different receptive fields in the mature network are uncorrelated. This implies that in the mature network the lateral connections vanish after they have completed their important function of yielding orthogonal receptive fields.

Several authors have proposed inhibitory connections between output units in order to render their activities uncorrelated [8.5, 10, 46, 47]. In our scheme, lateral connections are both excitatory and inhibitory before they vanish. This results in a purely feed-forward network, which represents an important computational advantage for a parallel system. Principal component analysis has also been associated with linear feed-forward networks using optimization methods with respect to a quadratic error function, i.e., back-propagation [8.48]. The advantage of our model consists in optimal feature extraction without supervision and in the existence of biologically plausible, local adaptation rules for the weights, namely Hebbian and anti-Hebbian rules.

We will illustrate now the performance of the network for patterns of spatially varying intensity and show that the network develops feature cells with receptive fields which are similar to those of "simple cells" found in the striate cortex and which select features of different orientation and different spatial frequencies.

For this purpose, we consider a rectangular lattice of $N_{\mathrm{i}} \times N_i'$ sensory input units representing the receptive field of N_{o} output units, with $N_{\mathrm{o}} \leq N_{\mathrm{i}}\, N_i'$. We generate two-dimensional input patterns of varying intensity by first selecting random numbers s_{ij}^{π}, $\pi = 1, \ldots, N_\pi$ from the interval $[-1, +1]$. Then, in order to introduce information about the topological structure of the receptive field, the random input intensities are correlated, e.g., with their nearest neighbors in both directions. As a result, the component p_{ij}^{π} of a pattern $\boldsymbol{p}^{\pi}$ at the coordinate (i, j) of the receptive field is given by $p_{ij}^{\pi} = s_{ij}^{\pi} + s_{i-1\,j}^{\pi} + s_{i+1\,j}^{\pi} + s_{i\,j-1}^{\pi} + s_{i\,j+1}^{\pi}$. We assume vanishing boundary conditions, i.e., $s_{0j} = s_{i0} = s_{N_{\mathrm{i}}+1\,j} = s_{i\,N_{\mathrm{i}}'+1} = 0$. Note that this averaging of neighboring signals corresponds to introducing an additional layer with random activities and with fixed and restricted connections to the input layer.

Receptive fields of simple cells in cat striate cortex as recorded by Jones et al. [8.49, 50] can be described by Gabor functions which consist of an oscillatory part, namely a sinusoidal plane wave, modulated by a Gaussian, exponentially decaying part. To localize the receptive fields in our model system correspondingly, we scale the weights between layers, i.e. the weight $w(ij, m)$ between the input unit at lattice location (i, j) and the mth output unit, according to $w'(ij, m) = D(i, j) w(ij, m)$, where $D(i, j)$ is a Gaussian distribution with $D(i, j) \sim \exp[-(i - i_0)^2/\sigma_1 - (j - j_0)^2/\sigma_2]$. Here, σ_1 and σ_2 control the width of the distribution and (i_0, j_0) is the coordinate of the lattice center, i.e., $(i_0, j_0) = (N_{\mathrm{i}}/2, N_i'/2)$.

Imposing a Gaussian distribution of synaptic weights will change the eigenvalue spectrum of the covariance matrix of the input pattern. Therefore, such a network, in a strict sense, cannot develop receptive fields according to a principal component analysis. However, if the restriction to neighborhoods described by a Gaussian were not applied, the weights w_{jm} develop towards the exact eigenvectors of $\boldsymbol{C}$. The localization of w_{jm} can be exploited to prevent degeneracies between eigenvalues, which can lead to a mixing of receptive fields, resulting in asymmetrical fields. If the Gaussian distribution is chosen not to be rotationally symmetric, i.e., if $\sigma_1/\sigma_2 \neq 1$, the orientation of receptive fields is predetermined owing to imposed symmetry axes.

Figure 8.13 displays contour plots of the receptive fields of the first eight output cells after 10 000 learning cycles (from left to right and top to bottom). Solid lines correspond to positive, dashed lines to negative synaptic weights. The input lattice was a square of 20×20 units. We imposed a Gaussian distribution of synaptic weights with parameters $\sigma_1 = 12$ and $\sigma_2 = 15$. Learning parameters η and μ were equal to 0.05 and 0.1, respectively. Owing to the nonsymmetric Gaussian distribution of weights, all units have slightly elongated receptive fields. The first unit corresponds to a simple cell with all-inhibitory synaptic weights.

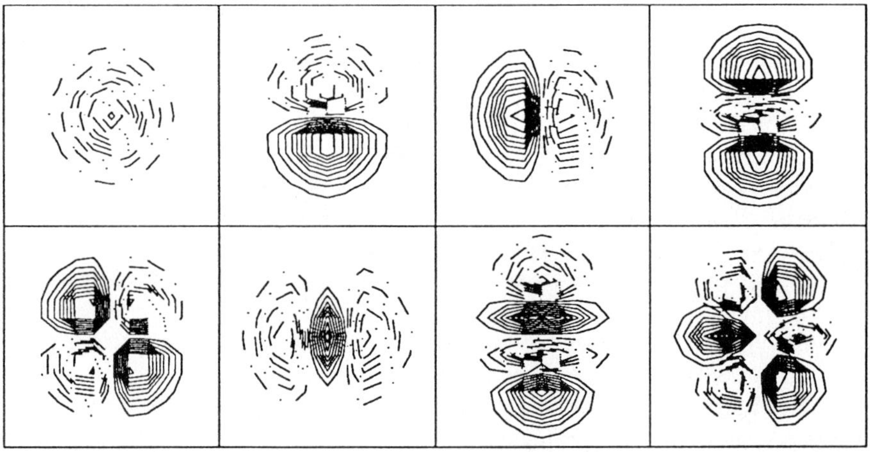

Fig. 8.13. From left to right and top to bottom: contour plots of receptive fields of output units 1–8 in the case of a square lattice of 20×20 input units. The synaptic distribution $D(i, j)$ between layers was Gaussian with $\sigma_1 = 12$ and $\sigma_2 = 15$. Solid lines indicate positive, dashed lines negative, weights. The number of learning cycles was 10 000; the learning parameters η and μ were equal to 0.05 and 0.1, respectively

The receptive fields of the second and third units display an excitatory and an inhibitory region and resemble simple cells, selective to edges of a fixed orientation. The fourth and sixth units have receptive fields with two zero crossings, corresponding to simple cells, selective to bars of a fixed orientation. In addition, the seventh unit is orientation selective, with four alternating excitatory and inhibitory regions. This unit would respond maximally to two parallel lines or bars with fixed distance and orientation.

The described units have receptive fields that resemble recorded receptive fields of simple cells in the primary visual cortex [8.16, 49, 50]. Up to now, there has not been any experimental evidence for receptive fields of the type of the fifth and eighth units, displaying four and six lobes.

8.7 Conclusion

Our previous analysis of self-organizing maps and of a hierarchical network for learning feature extraction has focused on two complementary aspects: the first aspect concerned the information-processing task carried out by each system, while the second concerned the capability of each system to account for observed phenomena of brain organization.

Regarding the first aspect, we demonstrated that both systems have in common the ability to *compress data*. This ability is realized in different ways for the two systems: self-organizing maps achieve data compression by a nonlinear mapping of their input patterns onto a lower-dimensional manifold. The points of this manifold can be considered "code labels" requiring less storage space than, and allowing an approximate reconstruction of, the original data. The self-organizing maps lead to an encoding that is a compromise between minimization of the reconstruction error for the original data and preservation of their similarity relationships under the encoding transformation.

The hierarchical network for feature extraction achieves data compression by performing a principal component analysis of its input data. The available cells organize their connectivity such that they automatically extract the principal components with the largest eigenvalues of the signal correlation matrix. Their output values represent the amplitudes of these principal components and, therefore, provide a lower-dimensional signal from which the original signal can be reconstructed with minimal expected square error.

With regard to the second aspect, we showed that self-organizing maps can explain several properties of the organization of cortical areas, such as the ubiquitous "striped projections" and the hierarchical feature maps found in the visual cortex, as a consequence of a *single* principle. This principle is related to the minimization of the functional $F[w]$ and interpretable as a process of adaptive synaptic modification. The hierarchical feature-extraction network complements this ability by explaining the formation of the small-scale structure observed in the various receptive fields encountered in cells of the visual cortex. The receptive-field properties of these cells resemble Gabor filters, and very similar receptive fields are developed by the cells of the artificial network.

A better understanding of the operation of the brain involves the investigation of several levels of organization. The research presented in this contribution was meant to be a small step towards this goal.

Acknowledgement. We would like to thank R. Kufrin and G. Quinn for their help and support in all technical matters concerning the use of the Connection Machine system. The authors are grateful to the Boehringer-Ingelheim Fonds for providing a fellowship to K. Obermayer. This research has been supported by the University of Illinois at Urbana-Champaign. Computer time on the Connection Machine CM-2 has been made available by the National Center for Supercomputer Applications.

References

8.1 D.J. Amit: *Modeling Brain Function* (Cambridge University Press 1989)
8.2 D.J. Amit, H. Gutfreund, H. Sompolinsky: Storing infinite number of patterns in a spin-glass model of neural networks, Phys. Rev. Lett. **55**, 1530–1533 (1985)
8.3 D.J. Amit, H. Gutfreund, H. Sompolinsky: Information storage in neural networks with low level of activity, Phys. Rev. A **35**, 2293–2303 (1987)
8.4 S. Grossberg: Adaptive pattern classification and universal recoding: I. Parallel development and coding of neural feature detectors, Biol. Cybern. **23**, 121–134 (1976)
8.5 D.E. Rumelhart, D. Zipser: Feature discovery by competitive learning, Cognitive Science **9**, 75–112 (1985)
8.6 J. Buhmann, R. Divko, K. Schulten: Associative memory with high information content, Phys. Rev. A **39**, 2689–2692 (1989)
8.7 E. Gardner, B. Derrida: Optimal storage properties of neural network models, J. Phys. A **21**, 271–284 (1988)
8.8 T. Kohonen: Self-organized formation of topologically correct feature maps, Biol. Cybern. **43**, 59–69 (1982)
8.9 T. Kohonen: Analysis of a simple self-organizing process, Biol. Cybern. **44**, 135–140 (1982)
8.10 T. Kohonen: *Self-Organization and Associative Memory*, Springer Series in Information Sciences 8 (Springer, Berlin, Heidelberg 1984)
8.11 K. Obermayer, H. Ritter, K. Schulten: Large-scale simulation of self-organizing neural network on parallel computers: Application to biological modelling, Parallel Computing **14**, 381–404 (1990)
8.12 H. Ritter, T. Martinetz, K. Schulten: Topology conserving maps for learning visuomotor-coordination, Neural Networks **2**, 159–168 (1989)
8.13 T. Martinetz, H. Ritter, K. Schulten: Three-dimensional neural net for learning visuomotor-coordination of a robot arm, IEEE Transactions on Neural Networks **1**, 131–136 (1990)
8.14 H. Ritter, T. Martinetz, K. Schulten: *Neuronale Netze – Eine Einführung in die Neuroinformatik selbstorganisierender Netzwerke* (Addison-Wesley, Bonn 1990) (in German); *Neural Computation and Self-organizing Maps: An Introduction* (Addison-Wesley, New York 1992)
8.15 D.H. Hubel, T.N. Wiesel: Receptive fields of single neurones in cat's striate cortex, J. Physiol. **148**, 574–591 (1959)
8.16 J.P. Jones, L.A. Palmer: An evaluation of the two-dimensional Gabor filter model of simple receptive fields in cat striate cortex, J. Neurophysiol. **58**, 1133–1258 (1987)
8.17 W.M. Cowan: The development of the brain, Sci. Am. **241**, 107–117 (1979)
8.18 R. Linsker: Self-organization in a perceptual network, IEEE Computer **21**, 105–117 (1988)
8.19 Y. Linde, A. Buzo, R.M. Gray: An algorithm for vector quantizer design, IEEE Trans. Comm. **28**, 84–95 (1980)
8.20 J. Makhoul, S. Roucos, H. Gish: Vector quantization in speech coding, Proc. IEEE (1985), 73-1551-1558
8.21 H. Ritter, K. Schulten: Kohonen's self-organizing maps: Exploring their computational capabilities, *Proc. IEEE ICNN 88*, San Diego (IEEE Computer Society press 1988), Vol. I, pp. 109–116
8.22 S.P. Luttrell: "Self-organisation: A derivation from first principles of a class of learning algorithms", in: *Proc. IJCNN 89*, Washington DC (IEEE Computer Society Press, 1989), Vol. II pp. 495–498
8.23 H. Ritter, K. Schulten: Convergence properties of Kohonen's topology conserving maps: Fluctuations, stability and dimension selection, Biol. Cybern. **60**, 59–71 (1989)
8.24 S. LeVay, T.N. Wiesel, D.H. Hubel: The development of ocular dominance columns in normal and visually deprived monkeys, J. Comp. Neurol. **191**, 1–51 (1974)
8.25 K.D. Miller, J.B. Keller, M.P. Stryker: Ocular dominance column development: analysis and simulation, Science **245**, 605–615 (1989)
8.26 G.G. Blasdel, G. Salama: Voltage-sensitive dyes reveal a modular organization in monkey striate cortex, Nature **321**, 579–585 (1986)
8.27 D.H. Hubel, T.N. Wiesel: Sequence regularity and geometry of orientation columns in the monkey striate cortex, J. Comp. Neurol. **158**, 267–294 (1974)
8.28 C. von der Malsburg: Development of ocularity domains and growth behavior of axon terminals, Biol. Cybern. **32**, 49–62 (1979)

8.29 C. von der Malsburg, J. Cowan: Outline of a theory for the ontogenesis of iso-orientation domains in visual cortex, Biol. Cybern. **45**, 49–56 (1982)
8.30 A. Takeuchi, S. Amari: Formation of topographic maps and columnar microstructures, Biol. Cybern. **35**, 63–72 (1979)
8.31 D.J. Willshaw, C. von der Malsburg: How patterned neural connections can be set up by self-organization, Proc. R. Soc. London B **194**, 431–445 (1976)
8.32 H. Ritter, T. Kohonen: Self-organizing semantic maps, Biol. Cybern. **61**, 241–254 (1989)
8.33 J.H. Kaas, R.J. Nelson, M. Sur, C.S. Lin, M.M. Merzenich: Multiple representations of the body within the primary somatosensory cortex of primates, Science **204**, 521–523 (1979)
8.34 J.H. Kaas, M.M. Merzenich, H.P. Killackey: The reorganization of somatosensory cortex following peripheral nerve damage in adult and developing mammals, Annual Rev. Neurosci. **6**, 325–256 (1983)
8.35 M.M. Merzenich et al.: Somatosensory cortical map changes following digit amputation in adult monkeys, J. Comp. Neurol. **224**, 591 (1984)
8.36 C. von der Malsburg, D.J. Willshaw: How to label nerve cells so that they can interconnect in an ordered fashion, Proc. Natl. Acad. Sci. USA **74**, 5176–5178 (1977)
8.37 J.C. Pearson, L.H. Finkel, G.M. Edelman: Plasticity in the organization of adult cerebral maps: A computer simulation based on neuronal group selection, J. Neurosci. **12**, 4209–4223 (1987)
8.38 K. Obermayer, H. Ritter, K. Schulten: Large-scale simulation of a self-organizing neural network: Formation of a somatotopic Map, *Parallel Processing in Neural Systems and Computers*, edited by R. Eckmiller, G. Hartmann, and G. Hauske (North-Holland, Amsterdam 1990) 71–74
8.39 K. Obermayer, H. Ritter, K. Schulten: A neural network model for the formation of topographic maps in the CNS: Development of receptive fields, IJCNN-90, Conf. Proceedings II, 423–429, San Diego (1990)
8.40 J. Rubner, K. Schulten: A self-organizing network for complete feature extraction, Biol. Cybern. **62**, 193–199 (1990)
8.41 D.N. Lawlwy, A.E. Maxwell: *Factor Analysis as a Statistical Method* (Butterworths, London 1963)
8.42 E. Oja: A simplified neuron model as a principal component analyzer, J. Math. Biology **15**, 267–272 (1982)
8.43 J.J. Hopfield, D.I. Feinstein, R.G. Palmer: "Unlearning" has a stabilizing effect in collective memories, Nature **304**, 158–159 (1983)
8.44 J. Rubner, P. Tavan: A self-organizing network for principal component analysis, Europhys. Lett. **10**, 693–698 (1989)
8.45 H. Ritter: Asymptotic level density for a class of vector quantization processes, Internal Report A9, Helsinki Univ. of Technology (1989)
8.46 C. von der Malsburg: Self-organization of orientation sensitive cells in the striate cortex, Kybernetik **14**, 85–100 (1973)
8.47 A.L. Yuille, D.M. Kammen, D.S. Cohen: Quadrature and the development of orientation selective cortical cells by Hebb rules, Biol. Cybern. **61**, 183–194 (1989)
8.48 P. Baldi, K. Hornik: Neural networks and principal component analysis: leraning from examples without local minima, Neural Networks **2**, 53–58 (1989)
8.49 J.P. Jones, L.A. Palmer: The two-dimensional spatial structure of simple receptive fields in cat striate cortex, J. Neurophysiol. **58**, 1187–1211 (1987)
8.50 J.P. Jones, A. Stepnoski, L.A. Palmer: The two-dimensional spectral structure of simple receptive fields in cat striate cortex, J. Neurophysiol. **58**, 1112–1232 (1987)
8.51 M. Cottrell, J.C. Fort: A stochastic model of retinotopy: A self-organizing process, Biol. Cybern. **53**, 405–411 (1986)
8.52 N.G. van Kampen: *Stochastic Processes in Physics and Chemistry* (North-Holland, Amsterdam 1981)
8.53 E.I. Knudsen, S. du Lac, S.D. Esterly: Computational maps in the brain, Ann. Rev. Neurosci. **10**, 41–65 (1987)
8.54 T. Kohonen: *Clustering, Taxonomy and Topological Maps of Patterns*, Proc 6th Int. Conf. on Pattern Recognition, Munich pp. 114–128 (1982)
8.55 H. Ritter, K. Schulten: On the stationary state of Kohonen's self-organizing sensory mapping, Biol. Cybern. **54**, 99–106 (1986)

8.56 K. Obermayer, H. Ritter, K. Schulten: A Principle for the Formation of the Spatial Structure of Cortical Feature Maps, Proc. Natl. Acad. Sci. USA **87**, 8345–8349 (1990)
8.57 K. Obermayer, H. Ritter, K. Schulten: A Model for the Development of the Spatial Structure of Retinotopic Maps and Orientation Columns, IEICE Trans. Fund. Electr. Comm. Comp. Sci., in press (1992)
8.58 K. Obermayer, G.G. Blasdel, K. Schulten: A Statistical Mechanical Analysis of Self-Organization and Pattern Formation during the Development of Visual Maps, Phys. Rev. A, in press (1992)
8.59 E. Erwin, K. Obermayer, K. Schulten: Self-Organizing Maps: Ordering, Convergence Properties and Energy Functions, Biol. Cybern., in press (1992)
8.60 E. Erwin, K. Obermayer, K. Schulten: Self-Organizing Maps: Stationary States, Metastability and Convergence Rate, Biol. Cybern., in press (1992)
8.61 K. Obermayer, K. Schulten, G.G. Blasdel: A Comparison of a Neural Network Model for the Formation of Brain Maps with Experimental Data, in: Advances in Neural Information Processing Systems 4, Eds. D.S. Touretzky, R. Lippman, Morgan Kaufmann Publishers, in press (1992)

9. Layered Neural Networks

Eytan Domany and Ronny Meir

With 8 Figures

Synopsis. Some of the recent work done on layered feed-forward networks is reviewed. First we describe exact solutions for the dynamics of such networks, which are expected to respond to an input by going through a sequence of preassigned states on the various layers. The family of networks considered has a variety of interlayer couplings: linear and nonlinear Hebbian, Hebbian with Gaussian synaptic noise and with various kinds of dilution, and the pseudoinverse (projector) matrix of couplings. In all cases our solutions take the form of layer-to-layer recursions for the mean overlap with a (random) key pattern and for the width of the embedding field distribution. Dynamics is governed by the fixed points of these recursions. For all cases nontrivial domains of attraction of the memory states are found. Next we review studies of unsupervised learning in such networks and the emergence of orientation-selective cells. Finally the main ideas of three supervised learning procedures, recently introduced for layered networks, are outlined. All three procedures are based on a search in the space of internal representations; one is designed for learning in networks with fixed architecture and has no associated convergence theorem, whereas the other two are guaranteed to converge but may require expansion of the network by an uncontrolled number of hidden units.

9.1 Introduction

We review here work done, mainly in the physics community, on layered feed-forward neural networks [9.1].

First we describe results obtained for the dynamics of multilayered networks that store random patterns on all layers [9.2–6]. These networks can be viewed as an associative memory that produces a "wave" of activity, passing from one layer to the next, in response to inputs presented to the first layer. As will be explained, these networks bridge a gap between conventional physicists' models and those studied by the "connectionist" school of computer scientists. Solutions of their dynamics are readily obtainable, and provide the first example of analytically obtained nontrivial basins of attraction for a model neural network.

Next, studies of unsupervised learning in such networks are briefly reviewed. In particular, we outline the work of Linsker [9.7–12] on the development of orientation-selective cells. The central idea and observation of this body of work is that when the input layer is submitted to random noise, and the interlayer connections are allowed to develop in response to the activity thus generated, nontrivial "ordered" structures emerge, which can perhaps be related to prenatal development of orientation-selective cells. A related unsupervised learning procedure [9.13] produces correlated responses to presentation of uncorrelated random inputs. The appearance of these correlations is abrupt, describable as a discontinuous transition when the number of presented patterns increases above a critical value.

Finally we present the general ideas and methods used in three recently introduced *supervised* learning algorithms [9.14–16]. The common feature of these algorithms is that the internal representations play a central role in the learning process. The main difference is that one algorithm [9.14] tries to solve the learning problem for a fixed architecture, and is not guaranteed to converge, whereas the other two [9.15, 16] always find a solution, but at the price of increasing the network size during learning in an uncontrolled manner, which closely resembles similar algorithms developed earlier [9.17].

9.2 Dynamics of Feed-Forward Networks

9.2.1 General Overview

The models discussed in this section lie in a "no man's land" between what we can loosely term physicists' and nonphysicists' models.

Contributions of the theoretical-physics community to the field of neural networks concentrated primarily [9.1] on the Hopfield model [9.18, 19]. and extensions thereof [9.1, 20, 21]. The typical physicist's network is characterized by three general features [9.22]. *First,* it is *homogenous* and uniform. By this we mean that all elements of the network are functionally similar. A *second* characteristic of these networks is the kind of task they attempt to perform; in most cases the networks studied by physicists deal with *random patterns*. That is, the network is to be designed in such a way that some preset, randomly selected states play a special role in its dynamics. Physicists are interested primarily in typical or average properties of an ensemble of such networks; these properties are usually calculated in the thermodynamic limit, i.e. when the number of basic units $N \to \infty$. The *third* attribute of physicists' networks is that in general a "cognitive event", such as recall of a memory, is associated with a *stable state* of the network's dynamics. For some applications, one is interested in embedding *stable cycles* in the network [9.23] In general, physicists' networks are *recurrent*, even when feedback does not play a central role in the dynamics.

Physicists' networks usually model an associative memory. Such a memory arrives at the correct stable state in response to presentation of an erroneous input (initial state). The memory is distributed and therefore robust; considerable "damage" to the network will leave its retrieval characteristics nearly unaffected.

Questions asked about these networks concern their storage capacity [9.24, 25], the size and shape of the domains of attraction of the stable states [9.26–30], and the time it takes for the network to relax to such a state [9.30, 31].

Another issue addressed more recently by physicists is that of learning [9.32–37]. Here the aim is to generate couplings and thresholds that ensure that the desired states are indeed stable attractors of the dynamics, with sizeable domains of attraction. The dynamic process associated with learning has also been investigated [9.38–40].

On the other hand, neural networks were studied by nonphysicists (in particular computer scientists) for many years [9.41–43]. The networks considered by this community differ, by and large, from physicists' networks in all the above-mentioned respects. These networks are usually not functionally uniform: for example, some units receive input only from the external world and not from other units of the network. Whereas these elements constitute the *input*, a different set of units passes information out of the network and serves to generate its *output*. There may also be processing units that have no direct contact with the external world: all their inputs come from, and all their outputs go to other units of the network itself.

More importantly, most computer scientists are interested in *finite* networks, which perform *specific*, well-defined tasks, such as determining whether a given geometrical figure is singly or multiply connected [9.14]. The answer of the network to a posed problem is read off its output units at some preset time. Hence stable attractors do not necessarily play a special role in the dynamics of such networks. In many of the most popular and widely studied examples the network architecture is purely feed-forward [9.43]. The task of these networks is usually to implement a desired mapping of input onto output space. The standard learning procedure used by the connectionist community is back propagation [9.45–47].

The family of networks studied by us belongs to the nonphysicists' class with respect to some of the attributes listed above. It has an input layer and can be viewed as producing an output on some remote layer; hence it is *inhomogeneous*. Being feed-forward, it does not have stable attractors of its dynamics. Instead, a wave of activity that passes the layers sequentially may be interpreted [9.48] as a "cognitive event". On the other hand, ours is a physicist's network in that *random patterns* are assigned to each layer (including the internal or "hidden" ones) and average properties are calculated in the thermodynamic limit.

The main analytic result obtained for our model addresses the following question: given an initial state, which has sizeable overlap with one key pattern on the input layer, what is the overlap with the corresponding stored pattern on layer l? If this overlap approaches a value close to 1, the memory has recalled the stored pattern successfully. Successful recall occurs if the initial state is not too far from the stored pattern, i.e. is in its *domain of attraction*. This problem is first solved for the case of Hebbian interlayer couplings. The solution is based on the observation [9.6] that the fields generated by cells of layer l on the units of layer $l+1$ are Gaussian-distributed independent random variables. The solution takes the form of layer-to-layer recursions of the mean overlap and of the width of the field distribution. The solution is extended to a number of more general cases; that of static synaptic noise [9.4, 20] added to the Hebbian couplings, the problem of diluted bonds [9.20], and the model with nonlinear synapses (i.e. the bonds are nonlinear functions of the Hebbian couplings [9.49]). Finally we present a solution of a layered network in which the interlayer couplings were obtained by the pseudoinverse method [9.26, 42, 50]. All the above-mentioned solutions appear in the form of layer-to-layer recursions, from which *domains of attraction* are calculated.

9.2.2 The Model: Its Definition and Solution by Gaussian Transforms

The network is composed of binary-valued units (cells, spins) arranged in layers:

$$S_i^l = \pm 1; ,$$

where $l = 0, 1, 2, \ldots L$ is a layer index, and each layer contains N cells. The state of unit S_i^{l+1} is determined by the state of the units on the previous layer l, according to the conditional probability

$$P(S_i^{l+1} | S_1^l, S_2^l, \ldots S_N^l) = \mathrm{e}^{\beta S_i^{l+1} h_i^{l+1}} / 2 \cosh(\beta h_i^{l+1}) , \tag{9.1}$$

$$h_i^{l+1} = \sum_j J_{ij}^l S_j^l .$$

Here J_{ij}^l is the strength of the connection from cell j on layer l to cell i on layer $l + 1$ (see Fig. 9.1). The quantity h_i^{l+1} is the field produced by the entire layer l on site i of the next layer. The parameter $\beta = 1/T$ controls stochasticity; the $T \to 0$ limit reduces to the deterministic form

$$S_i^{l+1} = \mathrm{sgn}(h_i^{l+1}) .$$

Dynamics of such a network can be defined as follows. Initially the first (input) layer is set in some fixed state externally. In response to that, all units of the

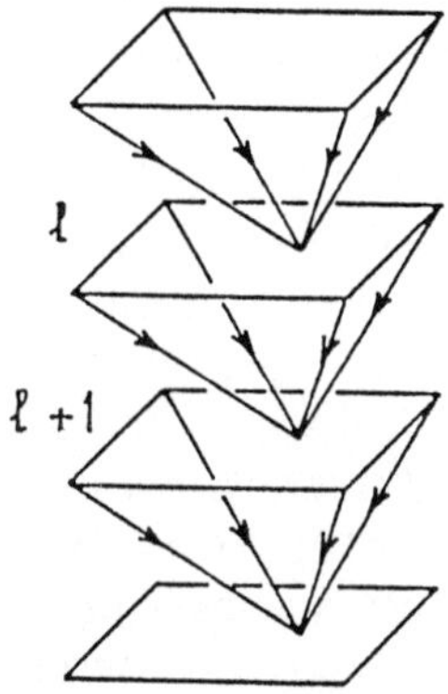

Fig. 9.1. Layered feed forward network. The state of cell i in layer $l+1$ is determined by the states of all cells of layer l. Input is presented to the first layer and output read from the last

second layer are set synchronously at the next time step, according to the rule (9.1), the next layer follows at the next time step, and so on. Thus the response of the network to an initial state is an "avalanche" of coherent activity that produces the appearance of a sequence of states, on layer l at *time* l.

For an alternative interpretation of this dynamics, imagine a single-layer recurrent network with couplings K_{ij}, in which units update their states synchronously (such as is the case of the Little model [9.51]). If we set in the layered network all $J^l_{ij} = K_{ij}$, i.e. independent of the layer index, the resulting layered network dynamics will be precisely identical to the dynamics of the recurrent network, with the layer index of the former playing the role of time for the latter. *Hence every recurrent network is equivalent to a properly defined layered feed-forward network* [9.45]. Letting the bonds depend on and vary with the layer index is, therefore, completely equivalent to allowing the couplings of a recurrent network to vary with time. Therefore one can interpret solutions of these layered networks as an "annealed approximation" [9.52] to the dynamics of networks with feedback.

Returning to our feed-forward network, we have to specify the bonds J^l_{ij}. These are chosen so that the network performs a desired task. A reasonable task for a layered network is to require that, in response to a particular input, a preset sequence of states develops on subsequent layers. We consider the problem of embedding in the network different random sequences of patterns, associating pattern $\xi^l_{i\mu}$ with layer l; the pattern index μ takes one of $\mu = 1, 2, ...\alpha N$ possible values. These states, $\xi^l_{i\mu}$, are the key patterns of the network. The interlayer couplings can be chosen according to any one of a variety of standard learning procedures. The simplest choice is that of outer product, or *Hebbian*, couplings [9.18, 51],

$$J^l_{ij} = \frac{1}{N} \sum_{\nu=1}^{\alpha N} \xi^{l+1}_{i\nu} \xi^l_{i\nu} \,. \tag{9.2}$$

Other choices are also discussed below.

By "solving the model" we mean that for a given initial state, i.e. the state of the first layer, we can predict the state on layer l that results from the network's dynamics. Of course we predict the state in the sense of *averages*, over the thermal noise associated with the dynamics, as well as over the choice of key patterns. An initial state is characterized by its overlap M^0_μ with the key patterns on the *first* ($l = 0$) layer; here we present results for initial states that have finite overlap with *only one* key pattern, i.e.

$$M^0_\mu = \frac{1}{N} \sum_i S^0_i \xi^0_{i\mu} = \begin{cases} O(1) & \text{for } \mu = 1 \\ O(1/\sqrt{N}) & \text{for } \mu > 1 \end{cases} \,. \tag{9.3}$$

With this initial condition we let the network develop in time according to the stochastic dynamic rule (9.1). Denote by M^l_μ the overlap of any particular state $\{S^l_i\}$, obtained on layer l, in the course of this dynamic process, for some particular choice of the key patterns $\{\xi^l_{i\mu}\}$. Our aim is to calculate the *average overlap* for $l > 0$:

$$m_\mu^l = \overline{\langle M_\mu^l \rangle} = \frac{1}{N} \sum_i \overline{\langle S_i^l \xi_{i\mu}^l \rangle} = \overline{\langle S_i^l \xi_{i\mu}^l \rangle} . \tag{9.4}$$

In this expression the brackets $\langle \cdot \rangle$ denote the thermal average over the stochastic dynamic process; the bar denotes the average over the key pattern assignments. With the initial conditions specified above, we expect $M_\mu^l = O(1/\sqrt{N})$ for all $\mu > 1$, whereas M_1^l may be of order unity. A network that corrects errors of the input is expected to start with a low but finite initial overlap with one key pattern and yield increasingly larger overlaps on subsequent layers. In order to compress notation we suppress the layer index and prime all variables associated with layer $l+1$ (i.e. we use S_i' and $\xi_{i\mu}'$). Unprimed variables refer to layer l. Another simplifying notation is the following: all brackets and bars will refer to averages taken over *primed* variables. The fact that unprimed variables have also to be averaged over is implicitly assumed everywhere; as we will demonstrate, all unprimed averages are taken care of by the law of large numbers, and only primed variables have to be averaged explicitly. Consider therefore

$$m_1' = \overline{\langle \xi_{i1}' S_i' \rangle} . \tag{9.5}$$

First we perform the thermal average over the dynamics of the last (primed) layer. From (9.1) we immediately get

$$m_1' = \overline{\tanh(\beta \xi_{i1}' h_i')} = \overline{\tanh(\beta H_{i1}')} , \tag{9.6}$$

where H_{i1}' is the "embedding field" associated with pattern $\nu = 1$, given by

$$H_{i1}' = \frac{1}{N} \sum_j \xi_{j1} S_j + \frac{1}{N} \sum_{\nu > 1} \xi_{i1}' \xi_{i\nu}' \sum_j \xi_{j\nu} S_j = M_1 + \sum_{\nu > 1} \xi_{i1}' \xi_{i\nu}' M_\nu . \tag{9.7}$$

With respect to averaging over the patterns, we explicitly perform averages over ξ', keeping in mind that thermal and configurational averages are to be taken (if needed; see below) for previous layers as well.

The essence of the method used throughout this paper is the treatment of the embedding field H_{i1}' as the sum of a "signal" m_1 and a Gaussian "noise" x. The signal arises from the key pattern $\nu = 1$, with which the initial state has non negligible overlap. The noise is generated by the other patterns, contained in J_{ij}, as well as by thermal fluctuations. One has to show that, as the thermal and configurational averages implied in (9.6) are taken, the embedding field $H_{i1}' = m_1 + x$ is Gaussian distributed. If so, averages such as (9.6) can be performed by integrating over x, with the correct Gaussian weight.

The stochastic variable M_1 in (9.7), which gives rise to the "signal", is the average of N stochastic variables $\xi_{j1} S_j$. Therefore using the law of large numbers we have

$$M_1 = m_1 + O(1/\sqrt{N}) . \tag{9.8}$$

Since we assume that $M_1 = \mathrm{O}(1)$, deviations of M_1 from m_1 can be neglected. Hence with respect to the first term in (9.7), all thermal and configurational averaging has been taken into account. Turning now to the second term,

$$x = \sum_{\nu>1} \xi'_{i1}\xi'_{i\nu}M_\nu \, , \tag{9.9}$$

note first that even though we also have $M_\nu = m_\nu + \mathrm{O}(1/\sqrt{N})$, we cannot replace M_ν by its average value and neglect its fluctuations, since $m_\nu = 0$ for $\nu > 1$. Keeping in mind, however, the fact that for $\nu > 1$ all $M_\nu = \mathrm{O}(1/\sqrt{N})$, we note that the Lindeberg condition [9.53] is satisfied for x, and therefore we can use the central-limit theorem, according to which the stochastic variable x is Gaussian distributed, with mean $\bar{x} = 0$ and variance given by

$$\Delta^2 = \sum_{\nu,\mu>1} \overline{\xi'_{i\nu}\xi'_{i\mu}}M_\nu M_\mu = \sum_{\nu>1} M_\nu^2 = \alpha N\overline{\langle M_\nu^2\rangle} + \mathrm{O}\left(\frac{1}{\sqrt{N}}\right) , \tag{9.10}$$

where we have used $\overline{\xi'_{i\nu}\xi'_{i\mu}} = \delta_{\nu\mu}$ and self-averaging of the M_ν^2. In (9.10) $\overline{\langle M_\nu^2\rangle}$ denotes the average over *unprimed* variables. Since averaging over the ξ' is equivalent to averaging over the Gaussian distributed variable x, we can express (9.6) in the form

$$m'_1 = \int \mathrm{d}x \tanh[(\beta(m_1 + x)] \frac{\mathrm{e}^{-(x/\Delta)^2/2}}{\sqrt{2\pi\Delta^2}} \, . \tag{9.11}$$

This *exact* relation constitutes a recursion that determines the average overlap on layer l+1 in terms of the average overlap m_1 and the width Δ^2, both characteristic of layer l. To complete the solution, a recursion for the width is also needed, in order to express

$$(\Delta')^2 = \sum_{\mu>1}^{\alpha N} \overline{\langle (M'_\mu)^2\rangle} \tag{9.12}$$

in terms of m_1 and Δ^2. We must evaluate, for $\mu > 1$,

$$\overline{\langle (M'_\mu)^2\rangle} = \frac{1}{N} + \frac{1}{N^2}\sum_{i\neq j} \overline{\tanh(\beta H'_{i\mu})\tanh(\beta H'_{j\mu})} \, , \tag{9.13}$$

$$H'_{i\mu} = \xi'_{i\mu}\xi'_{i1}m_1 + M_\mu + \sum_{\nu\neq 1,\mu} \xi'_{i\mu}\xi'_{i\nu}M_\nu \, . \tag{9.14}$$

Again we have replaced M_1 by m_1, neglecting fluctuations, but have kept M_ν for $\nu > 1$. Averages over ξ' are done, as before, by noting that (for $i \neq j$) the variables x_i and x_j, defined as

$$x_i = \sum_{\nu\neq 1,\mu} \xi'_{i\mu}\xi'_{i\nu}M_\nu \, , \tag{9.15}$$

are independent [9.54] and Gaussian distributed, with mean $\bar{x}_i = 0$ and width Δ.

If we write $\xi'_{i\mu}\xi'_{i1} = \eta, \quad \xi'_{j\mu}\xi'_{j1} = \eta'$, (9.13) takes the form

$$\overline{\langle (M'_\mu)^2 \rangle} = \frac{1}{N} + \frac{1}{4} \sum_{\eta,\eta'=\pm 1} \int \frac{\mathrm{d}x\,\mathrm{d}y}{2\pi\Delta^2} \mathrm{e}^{-(x^2+y^2)/2\Delta^2}$$
$$\times \tanh \beta(\eta m_1 + M_\mu + x) \tanh \beta(\eta' m_1 + M_\mu + y)\,. \tag{9.16}$$

Noting that $M_\mu = \mathrm{O}(1/\sqrt{N})$ we expand the integrand and find to leading order in M_μ that

$$\overline{\langle (M'_\mu)^2 \rangle} = 1/N + (1-q)^2 \beta^2 M_\mu^2\,, \tag{9.17}$$

where

$$q = \overline{\langle S_i \rangle^2} = \int \frac{\mathrm{d}x}{\sqrt{2\pi\Delta^2}} \mathrm{e}^{-(x/\Delta)^2/2} \tanh^2[\beta(m_1 + x)]\,. \tag{9.18}$$

Finally, substituting (9.17) in (9.12) we find that the new width is given by

$$(\Delta')^2 = \alpha + (1-q)^2 \beta^2 \Delta^2\,. \tag{9.19}$$

At this point it is important to note that the difference between the layered network treated here and the diluted model of Derrida et al. [9.21], on the one hand, and the Hopfield model [9.18], on the other, lies entirely in the expression (9.19) for the width Δ. For the dilute model the embedding fields H_i and H_j of any pattern μ on two distinct sites $i \neq j$ are completely uncorrelated, whereas for the layered network their correlation is $M_\mu^2 = \mathrm{O}(1/N)$. This correlation gives rise to the second term in (9.19), which leads to nontrivial domains of attraction for the layered network, as opposed to the extremely diluted model.

In summary, the solution of a layered network with random key patterns on each layer and Hebbian interlayer couplings is given by recursions of the form

$$m^{l+1} = \frac{1}{\sqrt{2\pi}} \int \mathrm{d}y\; \mathrm{e}^{-y^2/2} \tanh[\beta(m^l + \Delta^l y)] \tag{9.20a}$$
$$(\Delta^{l+1})^2 = \alpha + \beta^2 (1-q^l)^2 (\Delta^l)^2 \tag{9.20b}$$

with

$$q^l = \int \frac{\mathrm{d}x}{\sqrt{2\pi}} \mathrm{e}^{-y^2/2} \tanh^2[\beta(m^l + y\Delta^l)]\,.$$

In the deterministic limit, $\beta \to \infty$, these recursions become

$$m^{l+1} = \mathrm{erf}(m^l/\sqrt{2}\Delta^l)\,, \quad (\Delta^{l+1})^2 = \alpha + \frac{2}{\pi} \mathrm{e}^{-(m^l/\Delta^l)^2}\,. \tag{9.21}$$

In order to find the overlap on layer $l+1$ we have to iterate these recursions. The initial state determines m_1^{o}, the overlap on the first layer, and $(\Delta^{\mathrm{o}})^2 = \alpha$. These recursions allow us to predict the overlap on any layer l; in particular, whether for large l the overlap is close to its maximal value of 1 (interpreted as recall of the appropriate memory), or close to 0 (i.e. no recall). To answer

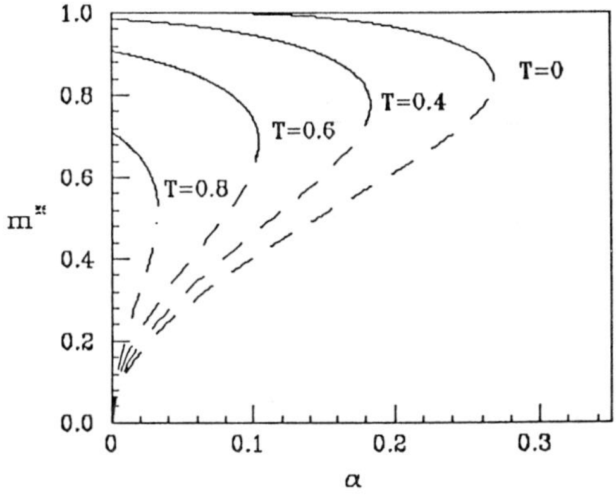

Fig. 9.2. Fixed points of (9.20), as a function of the storage capacity α for different temperatures T. The upper (*solid*) branch of each curve, as well as the $m^* = 0$ line, are stable fixed points. The lower (*dashed*) branches are unstable. For $T < 1.0$ the two branches merge at $\alpha_c(T)$

this question we look for the fixed points $m^*(\alpha, T)$ of the recursions, plotted in Fig. 9.2. First note that $m^* = 0$ is always a fixed point that corresponds to no recall. Only this fixed point exists for $T > 1$. At temperatures $T < 1.0$ we find two more fixed points. These form, as a function of α, a stable upper branch (solid lines in Fig. 9.2) and an unstable lower branch, which merge at $\alpha_c(T)$.

Understanding the fixed point structure immediately leads to the resulting "phase diagram". At temperatures $T > 1$ every initial state flows to $m^* = 0$. That is, too much noise destroys the memory completely. On the other hand, for $T < 1$ recall *is* possible, provided α is not too large. If $\alpha > \alpha_c(T)$, all initial states flow to the $m^* = 0$ fixed point. On the other hand, for $\alpha < \alpha_c(T)$, the overlap develops in a manner that depends on its initial value. If the initial overlap is large enough,

$$m^{\mathrm{o}} > m_{\mathrm{c}}^{\mathrm{o}}(\alpha, T) ,$$

we obtain $m^l \to m^* \simeq 1$ for large l, whereas $m_1^l \to 0$ for $m_1^{\mathrm{o}} < m_{\mathrm{c}}^{\mathrm{o}}(\alpha, T)$. A convenient way to summarize these statements is in terms of R, the radius of the domain of attraction of the memory state, defined as

$$R = 1 - m_{\mathrm{c}}^{\mathrm{o}}(\alpha, T) . \tag{9.22}$$

R measures how close the initial configuration must be to one of the key patterns in order to guarantee convergence to its "images" on subsequent layers. Plots of R v. α, for different values of T, are shown in Fig. 9.3.

For completeness we present in Fig. 9.4 the layer-to-layer development of the overlap with a key pattern, as obtained by iterating (9.21), with $\alpha = 0.20$ and different initial overlaps. For initial overlaps $m^{\mathrm{o}} = 0.8, 1.0 > m_{\mathrm{c}}^{\mathrm{o}}(\alpha = 0.20)$ the recursions take the system to $m^* \simeq 1$, whereas for $m^{\mathrm{o}} = 0.20$ the system flows to $m^* = 0$. The circles are obtained from simulations; deviations from the analytic solution are due to finite-size effects ($N = 200$) and to statistical fluctuations.

Various other aspects of the model described and solved above, such as the response of the network to initial states with nonvanishing overlap and with more

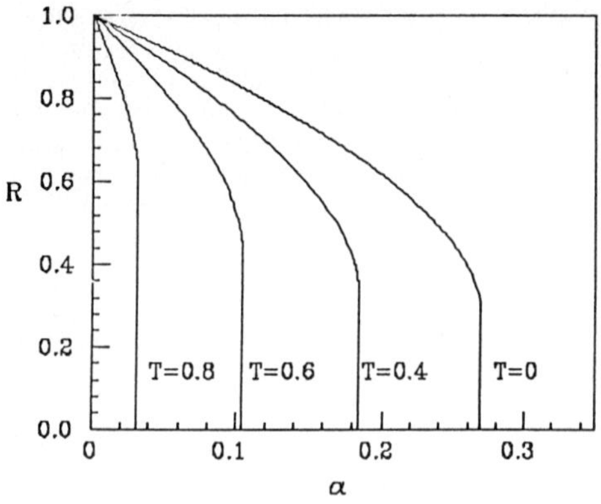

Fig. 9.3. The size R of the basin of attraction of the memory phase v. α, for various values of the temperature T, defined in (9.22)

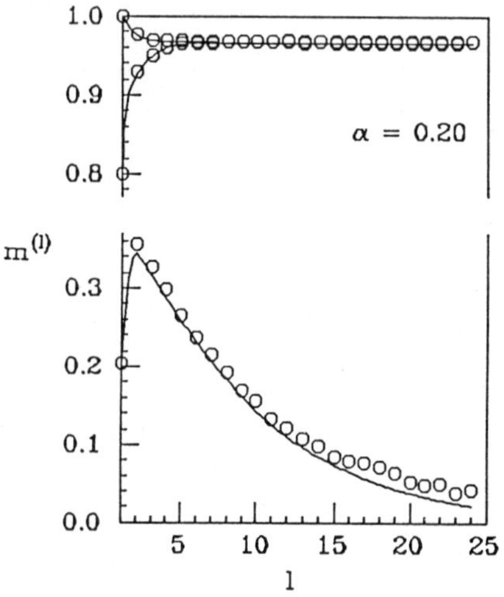

Fig. 9.4. The overlap m^l v. the layer index (or time) l, obtained by iterating (9.21), with α = 0.20, and different initial overlaps m°. For m° = 0.8, 1.0 $>$ $m_c^{\circ}(\alpha = 0.20)$ the recursions take the system to $m^* \simeq 1$, whereas for m° = 0.20 the system flows to m^* = 0. The circles are obtained from simulations; deviations from the analytic solution are due to finite-size effects (N = 200) and to statistical fluctuations

than one pattern, and details of the role played by finite size effects can be found in [9.3].

We turn now to describe briefly another interesting question, which, again, can be answered analytically for our layered network. Consider two different initial states on the first layer of the network that have nonvanishing overlap and let them evolve according to the same dynamical rule. A quantity of interest is Q^l, the overlap between the states obtained on subsequent layers. It is of interest to find out how the manner in which Q^l develops depends on the phase (recall or no recall) in which the dynamics takes place, and on the initial overlap. Similar questions have been addressed in the context of diluted networks by Derrida et al. [9.21] and by Derrida [9.55].

We consider only the deterministic T = 0 case. Clearly if the two configurations are initially identical, they will remain identical on all layers. Without going into the details of the calculation (see Derrida and Meir 1988), which are based on the Gaussian method described above, we summarize our results, at zero temperature, as follows. The recursion relations (9.20) are supplemented by one for the overlap Q^l. For $\alpha > \alpha_c$ the only stable fixed point is at $Q = 0$; another fixed point, at $Q = 1$, is unstable. This means that in the phase of no recall any two distinct states, no matter how close initially, will diverge away from each other. For $\alpha < \alpha_c$, however, one finds an additional stable fixed point, at $0 < Q^* < 1$, which corresponds to the two configurations having a fixed (independent of l) nonzero limiting overlap.

An interesting question concerns the evolution of two initially close configurations in the memory phase, i.e. $\alpha < \alpha_c$. We know that each of the initial configurations evolves to a state that has a large overlap with the same key pattern. Moreover, we just found that the two states have nonvanishing limiting overlap Q^*. Nevertheless, even if the initial configurations start off arbitrarily close to each other (i.e., $Q^0 \simeq 1$), they always begin to diverge away from one another, even in the good memory phase, as $Q^l \longrightarrow Q^* < 1$ for large l. This initial divergence of arbitrarily close initial configurations is reminiscent of the behavior of chaotic dynamical systems, and has been observed also in the diluted Hopfield model [9.21] and in diluted spin glasses [9.55].

9.2.3 Generalization for Other Couplings

We now derive layer-to-layer recursions for the overlap, for networks whose interlayer couplings differ from the simple Hebbian, or *outer product*, case with full connectivity that was treated above. We use the same initial conditions on the first layer as before; i.e., significant overlap with key pattern 1, whereas for all $\nu > 1$ we have $M_\nu = O(1/\sqrt{N})$.

Static Synaptic Noise. We consider the same problem as in Sect. 9.2.2, but with a noise term added [9.49] to the Hebbian couplings,

$$J_{ij}^l = \frac{1}{N} \sum_{\nu}^{\alpha N} \xi_{i\nu}^{l+1} \xi_{j\nu}^l + \frac{1}{\sqrt{N}} z_{ij}^l \, . \tag{9.23}$$

Here z_{ij}^l are independent, Gaussian-distributed random variables, with mean and variance given by

$$[z_{ij}^l] = 0 \, , \quad [z_{ij}^l z_{km}^{l'}] = \alpha \Delta_0^2 \delta_{ik} \delta_{jm} \delta_{ll'} \, , \tag{9.24}$$

where $[\cdot]$ denotes the average over the synaptic noise. Such noise represents a degrading of the "correct" couplings; the performance of the network in its presence provides a measure of its robustness. Following the notation of Sect. 9.2.2, we have

$$\begin{aligned} m_1' &= [\overline{\langle \xi_{i1}' S_i' \rangle}] = [\overline{\tanh(\beta H_{i1}')}] \\ H_{i1}' &= M_1 + \frac{1}{\sqrt{N}} \sum_j z_{ij} S_j \xi_{i1}' + \sum_{\nu > 1} \xi_{i1}' \xi_{i\nu}' M_\nu = m_1 + y + x \, . \end{aligned} \tag{9.25}$$

As before, H'_{i1} has a "signal" term, m_1, but now there are two distinct Gaussian noise terms, with vanishing means $\overline{x} = [\overline{y}] = 0$, and variances $\overline{x^2} = \Delta^2 = \sum_{\nu>1} M_\nu^2$ and $[\overline{y^2}] = \alpha\Delta_0^2$. Hence the total "noise" in H'_{i1} is Gaussian distributed, with mean zero and variance

$$\delta^2 = \Delta^2 + \alpha\Delta_0^2 , \tag{9.26}$$

and we get

$$m_1' = \int \frac{du}{\sqrt{2\pi}} e^{-u^2/2} \tanh \beta(m_1 + \delta u) , \tag{9.27a}$$

complemented by the recursion for the width, obtained using the same steps and arguments that led to (9.19):

$$\begin{aligned} (\Delta')^2 &= \alpha + (1-q)^2\beta^2\Delta^2 , \\ q &= \int \frac{du}{\sqrt{2\pi}} e^{-u^2/2} \tanh^2 \beta(m + u\delta) . \end{aligned} \tag{9.27b}$$

The deterministic ($\beta \to \infty$) limit of these recursions [9.4], is given by

$$\begin{aligned} m_1^{l+1} &= \operatorname{erf}(m^l/\sqrt{2}\delta^l) , \quad (\delta^l)^2 = (\Delta^l)^2 + \alpha\Delta_0^2 , \\ (\Delta^{l+1})^2 &= \alpha + \frac{2}{\pi}\left(\frac{\Delta^l}{\delta^l}\right)^2 e^{-(m^l/\delta^l)^2} . \end{aligned} \tag{9.28}$$

By iterating (9.28) for different Δ_0 one observes [9.6] that $\alpha_c(\Delta_0)$, the storage capacity, decreases with increasing static synaptic noise. An additional degrading effect of the noise is the decrease of m^*, the limiting overlap with the recalled pattern, with increasing Δ_0. Here we present only the effect of the static noise on the basin size; Fig. 9.5 shows R v. Δ_0 for various values of α. As can be expected the basin size decreases with increasing α and static noise Δ_0.

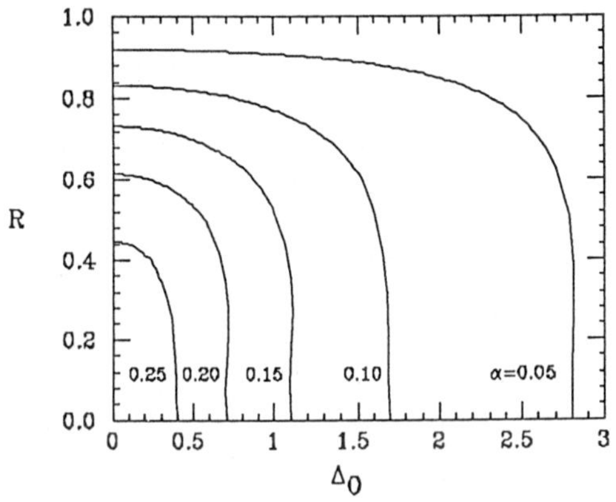

Fig. 9.5. Size of basin of attraction R v. static noise width Δ_0 for various values of the storage capacity α

Dilution. A different potentially degrading action on the network is one of removing some of the bonds; e.g. setting the associated couplings to zero. This goes under the name of *dilution* [9.20]. Consider the layered network with randomly diluted Hebbian couplings

$$J_{ij}^l = c_{ij}^l \frac{1}{cN} \sum_\nu \xi_{i\nu}^{l+1} \xi_{j\nu}^l , \tag{9.29}$$

where c_{ij}^l are independent random variables, chosen from the distribution

$$P(c_{ij}) = c\delta(c_{ij}, 1) + (1-c)\delta(c_{ij}, 0) . \tag{9.30}$$

The concentration of nonzero bonds is c; the bond from i to j can be removed independently of the removal of the bond from j to i. Using again the notation of Sect. 9.2.2, and denoting by $[\cdot]$ averages over the variables c_{ij}^l, we want to calculate $m_1' = [\overline{\tanh(\beta H_{i1}')}]$, with

$$H_{i1}' = \tilde{M}_1 + \sum_{\nu>1} \xi_{i1}' \xi_{i\nu}' \tilde{M}_\nu , \quad \tilde{M}_\mu = \frac{1}{cN} \sum_{j=1}^N c_{ij} \xi_{j\mu} S_j . \tag{9.31}$$

Note that $\tilde{M}_\mu$, as defined here, should carry also a site index i, since it depends on c_{ij}. Fluctuations of $\tilde{M}_\mu$ are due to three causes: thermal, configurational (ξ) and dilution effects. Since S_j does not depend on the c_{ij} (which determine the effect of S_j on S_i') we have $[c_{ij}\xi_{j\mu}S_j] = [c_{ij}]\xi_{j\mu}S_j$ and find that the embedding field H_{i1} again contains a "signal", m_1, and a noise term $x = \sum_{\nu>1} \xi_{i1}' \xi_{i\nu}' \tilde{M}_\nu$ with mean $[\overline{x}] = 0$ and variance δ^2,

$$\delta^2 = \left[\overline{x^2}\right] = \left[\sum_{\nu>1} \tilde{M}_\nu^2\right] = \alpha \frac{1-c}{c} + \sum_{\nu>1} M_\nu^2 = \alpha \Delta_0^2 + \Delta^2 . \tag{9.32}$$

This variance has exactly the form (9.26), obtained for the case of static synaptic noise. The average overlap m_1' and the width $(\Delta')^2$ are also given by the same expression (9.27) as in the case of static noise. Therefore the problem of dilution maps exactly onto that of static synaptic noise, with the effective variance of the static noise given by $\Delta_0^2 = (1-c)/c$. Diluting one half of the bonds at random, i.e. $c = 1/2$, corresponds to $\Delta_0 = 1$; the corresponding critical capacity (at $T = 0$) is $\alpha_c \simeq 0.16$ (see [9.6]). The effect of $c = 1/2$ dilution on the size of the basin of attraction is shown in Fig. 9.6.

Nonlinear Synapses. We turn now to the case of couplings of the form [9.20, 49]

$$J_{ij}^l = \frac{\sqrt{\alpha N}}{N} F\left(\frac{1}{\sqrt{\alpha N}} \sum_{\nu=1}^{\alpha N} \xi_{i\nu}^{l+1} \xi_{j\nu}^l \right) , \tag{9.33}$$

where $F(x)$ is a (generally nonlinear) function of x. This general class of models includes some interesting cases, such as clipped synapses (i.e. $J_{ij} = \mathrm{sgn}\left[\sum \xi_{i\nu}^{l+1} \xi_{j\nu}^l\right]$), and selective dilution (see below). Derivation of the recursion

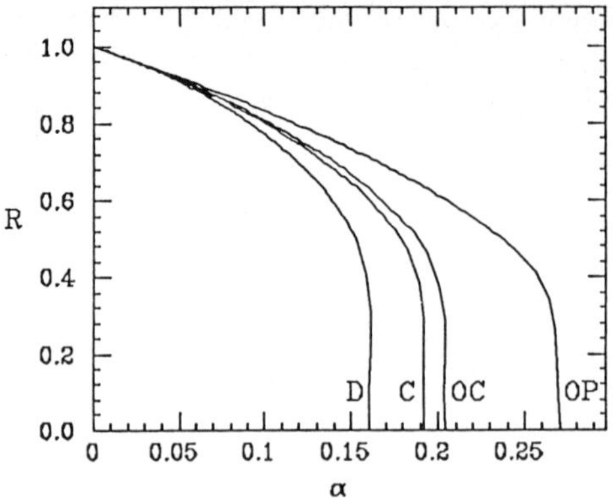

Fig. 9.6. Size of basin of attraction R v. storage capacity α, corresponding to models discussed: fully connected Hebbian, or outer product (OP); randomly diluted outer product with $c = 0.5$ bond concentration (D); clipped (C); and optimally clipped (OC)

for this case is along the same lines as that of the previous sections and of the similar problem for fully connected Hopfield networks [9.20, 49]. The restrictions on F are also similar to those given in that case [9.20]. As before, we wish to calculate the thermal and configurational average of the overlap $m_1' = \overline{\langle \xi_{i1}' S_i \rangle} = \overline{\tanh[\beta H_{i1}']}$. The embedding field is now given by

$$H_{i1}' = \xi_{i1}' \frac{\sqrt{\alpha N}}{N} \sum_j F\left(\frac{1}{\sqrt{\alpha N}} \sum_\nu \xi_{i\nu}' \xi_{j\nu}\right) S_j \,, \tag{9.34}$$

To perform the average over ξ', we treat H_{i1}' as a Gaussian-distributed random variable. Again we obtain similar recursions as for the case of static synaptic noise (9.27) with an effective inverse temperature $\overline{\beta} = \beta\overline{F'}$ and the static width parameter given by $\Delta_0^2 = \overline{F^2}/(\overline{F'})^2 - 1$. Here the bar over F and its derivative represents the average of these functions over their (Gaussian-distributed) argument, $x = 1/\sqrt{\alpha N}\Sigma\xi_{i\nu}'\xi_{j\nu}$. We give here results for two choices of F: clipped (C), and optimally clipped (OC) couplings. The corresponding $F(x)$ take the forms $F^{\mathrm{C}}(x) = \mathrm{sgn}(x)$ and $F^{\mathrm{OC}}(x) = \mathrm{sgn}(x)$, if $|x| > x_0$ and $F^{\mathrm{OC}} = 0$ otherwise, with x_0 chosen so that the resulting Δ_0 is minimal. The basin size R for each of these cases is plotted v. α in Fig. 9.6. As can be expected, the best performance (largest basin) is achieved for the original outer-product couplings.

Pseudoinverse. The network with Hebbain couplings (9.2) had the disadvantage that for $\alpha > 0$ the internal fields $h_{i\nu}^{l+1} = \sum_j J_{ij}^l \xi_{j\nu}^l$ of the patterns $\xi_{i\nu}^l$ had a broad distribution. Hence the exact patterns are not stable fixed points of the dynamics (9.1), owing to the negative Gaussian tail of the distribution of $\xi_{i\nu}^{l+1} h_{i\nu}^{l+1}$. But there is a matrix J_{ij}^l that gives a sharp distribution

$$\xi_{i\nu}^{l+1} h_{i\nu}^{l+1} = 1 \tag{9.35}$$

for all ν, l, and i. This holds not only for random patterns but for any set $\{\xi_{i\nu}^l\}$ of linearly independent patterns too. Since there are at most N such patterns, the network has a maximal capacity of $\alpha_c = 1$. This matrix can be calculated from the pseudoinverse [9.26, 42] of (9.35); for the layered structure one obtains

$$J_{ij}^{l} = \frac{1}{N} \sum_{\mu\nu} \xi_{i\nu}^{l+1} \left[C_l^{-1}\right]_{\nu,\mu} \xi_{j\mu}^{l} \tag{9.36}$$

with the correlation matrix

$$[C_l]_{\nu,\mu} = \frac{1}{N} \sum_{i} \xi_{i\nu}^{l} \xi_{i\mu}^{l} \,. \tag{9.37}$$

J_{ij}^{l} has two properties that are important for an associative memory. (i) It is a projector onto the linear space spanned by the p patterns $\xi_{j\nu}^{l}$, i.e. the orthogonal space is projected out. (ii) Of all matrices for which (9.35) holds, J_{ij}^{l} has the minimal norm $\sum_j (J_{ij}^{l})^2$; hence one expects a maximal basin of attraction [9.57]. Historically, the completely connected feedback network was studied first [9.26]. A feed-forward network with one layer of couplings (9.36) has recently been considered [9.50], and the extremely diluted anisotropic network [9.21] with these couplings was solved exactly [9.28, 58].

Here we want to solve the layered network with the pseudoinverse couplings (9.36). For simplicity we consider an initial state S_i^0 of layer $l = 0$ that has a nonvanishing overlap m_1^0 with pattern ξ_{i1}^{0}, only, i.e.

$$\langle S_i^0 \rangle = \xi_{i1}^{0} m_1^0 \,. \tag{9.38}$$

Here the brackets denote the average over initial states. The internal fields

$$h_i^{l+1} = \sum_{\nu} \xi_{i\nu}^{l+1} \sum_{\mu} [C_l^{-1}]_{\nu\mu} m_{\mu}^{l} \tag{9.39}$$

are Gaussian distributed, since $\xi_{i\nu}^{l+1}$ is uncorrelated to C_l and m_{μ}^{l}. A fairly straightforward analysis yields the following recursion for the overlap:

$$m^{l+1} = \int \frac{\mathrm{d}z}{\sqrt{2\pi}} \mathrm{e}^{-z^2/2} \tanh\left[\beta(m^l + \Delta^l z)\right] \,. \tag{9.40}$$

At $T = 0$ this reduces to

$$m^{l+1} = \operatorname{erf}\left[\frac{m^l}{\sqrt{2}\Delta^l}\right] \,. \tag{9.41}$$

These recursions are supplemented by that of the width;

$$(\Delta^l)^2 = \frac{\alpha}{1-\alpha}\left(1 - (m^l)^2\right) \,. \tag{9.42}$$

The width of the field distribution is zero at $m^0 = 1$, and hence at $T = 0$ $m^l = 1$ is a fixed point as it should be. For $m^0 \neq 1$ the width diverges for $\alpha_c = 1$.

For the corresponding extremely asymmetrically diluted network one obtains [9.28] the same equations as in the layered model (9.40–42). We present in

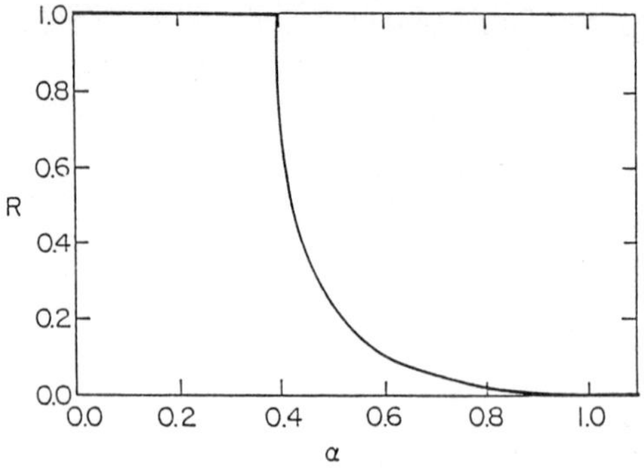

Fig. 9.7. Size of basin of attraction R v. storage capacity α for pseudoinverse interlayer couplings

Fig. 9.7 R v. α for the case $T = 0$ and pseudoinverse couplings between the layers. This situation is intermediate between layered Hebbian and strongly diluted [9.21] networks. For $\alpha < \alpha_u \simeq 0.39$ all initial states with nonvanishing overlap flow to the correct pattern (as is the case for the diluted model), whereas for $\alpha_u < \alpha < 1$ the memory state has a nontrivial basin of attraction (as in the layered network with Hebbian couplings).

9.3 Unsupervised Learning in Layered Networks

In this section we review studies of unsupervised learning. We first explain what this means. In a supervised learning procedure the network is presented with examples; presentation of an input is accompanied by presentation of the correct corresponding output. The input is used to generate the response of the system, using the current weights. A learning algorithm provides a recipe to modify the couplings in a manner that leads towards realization of the desired input–output association. The outcome of the learning step depends on the actual result of the dynamic process just described, *as well as on the correct output.* In an unsupervised learning procedure, on the other hand, *no correct output is provided to the network.* Presentation of an input is accompanied, as before, by a dynamic process that utilizes the current weights; changes of these weights, however, depend only on the actual state that the system arrived at, and *not on any externally provided correct output.*

A very interesting development, which offers a simple mechanism for explaining the formation of feature-analyzing cells and their columnar organization in a layered self-adaptive network, was presented by R. Linsker in a series of papers [9.7–12]. Work by neurobiologists over the past twenty years has demonstrated unequivocally that layers of orientation-selective cells exist in the visual cortex of many mammals, such as cats and monkeys (for a recent review see [9.59]). Moreover, these cells are organized into columns of cells of the same or similar orientation. A surprising fact is that, in some cases, they are present at birth.

Linsker's work will be discussed in two phases, corresponding to the development of his own ideas. In the first phase, Linsker studies a simple layered network that develops according to an unsupervised Hebbian learning rule to yield the biologically plausible results discussed above. The second phase then goes on to define an information-theoretic principle, which attempts to "explain" why it is advantageous to the system to develop in such a way.

9.3.1 Hebbian Learning and the Development of Orientation-Sensitive Cells and Columns

The network studied by Linsker [9.7–9] consists of several layers $(A, B, \ldots)$ of cells. Each cell in layer M $(M = B, C, \ldots)$ receives inputs from some neighborhood of cells in the previous layer L. The positions (in layer L) of the inputs to each M-cell are chosen randomly according to a probability distribution whose magnitude decreases monotonically with distance from the point overlying the cell's location. Linsker uses a Gaussian distribution in his simulations. The response of an M-cell to a stimulus coming from the L-layer is linear, i.e

$$M_n = a_1 + \sum_i c_{ni} L_i \ , \tag{9.43}$$

where a_1 is a constant and c_{ni} is the connection strength. The connections themselves change according to the following Hebbian rule:

$$\Delta c_{ni} = a_2 M_n L_i + a_3 L_i + a_4 M_n + a_5 \ , \tag{9.44}$$

where the as are constants $(a_2 > 0)$.

The network responds to presentation of random input patterns to the first layer. With the assumption that the connections change slowly with each presentation, Linsker casts the above equation in a differential form, by averaging it over many presentations. Thus he obtains the equation

$$\dot{c}_{ni} = \sum_j Q^L_{i,j} c_{nj} + \left[k_1 + (k_2/N) \sum_j c_{nj} \right] \ , \tag{9.45}$$

where $k_{1,2}$ are functions of the parameters $a_1 - a_5$ and $Q^L_{i,j}$ is proportional to the two-point autocorrelation function of layer-L activity at cells i and j. The M-layer correlation function can then be calculated [9.8] from Q^L and c_{ni} by the formula

$$Q^M_{n,m} \propto \sum_i \sum_j Q^L_{i,j} c_{ni} c_{mj} \ . \tag{9.46}$$

Using the above two equations one can now investigate the development of the layered system. Starting with white noise on the first layer A, one calculates $Q^A_{i,j}$ (which is just a delta function for the case where the noise is chosen independently on each site). Using (9.45) with random initial conditions one proceeds to calculate the asymptotic value of the connections between layer A and B. This process can be proved to converge under certain conditions [9.7]. Having determined the A-to-B connections, one calculates the activities of the B cells (which are no longer uncorrelated), and so forth for all layers.

Several points should be added to the above brief account. (1) There are several adjustable parameters. (2) The synaptic positions of each cell are chosen randomly and independently. (3) A hard constraint is imposed on the weights so that they do not increase without limit under (9.45). With a suitable choice of parameters, Linsker demonstrates that the possible mature morphologies of cells in each layer are of a limited variety. Next he demonstrates that the connections between layer A and B develop in such a way as to reach their maximum (or minimum) allowed values. Choosing the parameters such that the all-excitatory regime is obtained, one can easily see that Q^B_{nm} is of a Gaussian nature. Proceeding to evaluate the connection between layers B and C, one finds several possible regimes [9.7, 11]. Without going into the details, it is possible to show that there is a choice of parameters such that cells in layer C are of a center-surround nature (i.e circularly symmetric cells with excitatory core and inhibitory surround, or vice versa). An ON-center cell, for example, responds maximally to an input that has a centered spot of light against a dark background. It should be noted that such cells are observed in mammalian retinae.

Having chosen the parameters such that the cells in layer C are of an ON-center nature, one finds that the correlation function Q^C_{ij} is of a Mexican-hat shape. Cells developing according to (9.45), with a Mexican-hat type of Q-function, can be divided into two broad classes. (1) Centrally symmetric cell types, and (2) cells whose input field has a preferred orientation. Which of these options is chosen depends on the depth of the minimum of the Mexican hat. Linsker has argued, using "energy" considerations [9.8], that if the minimum is sufficiently deep, the cells should develop an oriented receptive field, thus breaking the symmetry inherent in the problem. Thus the scenario for the formation of orientation-selective cells is as follows: cells in succesive layers have Mexican-hat-type correlation functions with increasingly deep minima. At a certain layer (Linskers' layer F) the depth is sufficient to produce orientation sensitivity in the next layer. This behavior is analagous to breaking the symmetry in a ϕ^4 field theory [9.60]. It should be noted that the preferred orientations of the cells in layer G are independent of each other.

Up to this stage no lateral connections (within the layers) were allowed. Adding excitatory connections between layer-G cells, formed prior to the feed-forward connections, one finds that the cells in layer G organize into regions of similar orientation preference. These regions are remarkably similar to the regions found experimentally in macaque monkeys. A detailed discussion of the possible morphologies is given in [9.9, 11]. Having produced complex of results from a very simple learning rule and architecture, Linsker [9.10] proceeds to ask what the principle behind this self-organization of the network is. For example, what is the information-theoretic meaning of the cells developing as they do?

9.3.2 Information-Theoretic Principles Guiding the Development of the Perceptual System

To understand the behavior of the development better, note that (9.45) can be written in the form (dropping the index n for the moment, i.e considering a single cell)

$$\dot{c}_i \propto -(\partial E/\partial c_i)\,, \quad E = E_Q + E_k \tag{9.47}$$

with

$$E_Q = (-1/2)\sum_{ij} Q_{ij}c_ic_j\,, \quad E_k = -k_1 - (k_2/2N)\left(\sum_j c_j\right)^2 .$$

Thus the connections develop in such a way as to minimize the cost function E. Alternatively one may say that the learning rule (9.45) minimizes E_Q, subject to the constraint E_k = const, which implies that the total connection length $\sum_j c_j$ should be fixed. It is easy to see that

$$E_Q = -(1/2)\langle(M - \bar{M})^2\rangle \tag{9.48}$$

so that the Hebbian rule maximizes the output variance, given the constaint mentioned. Linsker then notes that had the constraint been of the form $\Sigma_j c_j^2$ = const, the development would give rise to connections that maximize the output variance, subject to the constraint of having a fixed (Euclidean) length. This is a special case of a well-known statistical method, called principal component analysis (PCA), which aims at identifying the salient features of a high-dimensional set of data points [9.61]. This particular form of Hebbian rule had been studied previously [9.62]. The generalization to the case of more than one M-cell is also discussed in [9.11].

Linsker further notes that under certain conditions (but not always) the PCA rule discussed above leads to the principle of optimal inference. By optimal inference we mean the following. We are given a particular value of the output M and some values for the connections c_i. We are then asked to estimate the input activities, and minimize the mean square error between the estimated and actual values. It turns out [9.11] that in certain cases PCA and minimal square error lead to the same set of c values.

Motivated by this observation, Linsker proceeds to postulate a principle for the development of a perceptual system, subject to noise and other constraints (such as geometrical ones for example). Before describing the principle, we define a quantity R, the information rate, which is a well-known quantity in information theory (see for example [9.63]). This quantity, also called the mutual information, essentially measures how much information one random variable (in our case the output M) conveys about another random variable (the input L). The larger R the more information the outputs convey about the input. The expression for R is

$$R = \langle \ln[P(L|M)/P(L)] \rangle\,, \tag{9.49}$$

where $P(L)$ is the input probability distribution, and $P(M|L)$ is the probability distribution of the output, given the input. The average is taken with respect to the full probability distribution $P(M, L)$. Having introduced the mutual information function R, Linsker then goes on to postulate it as an organizing principle in the development of a perceptual system. Details about the consequences of this rule can be found in [9.11].

Before concluding our review of his work, we mention that Linsker has demonstrated recently [9.12] that, using the principle of maximal information preservation described above, one obtains the result that, under certain conditions, center-surround cells are preferably formed. This result is very interesting since it may shed some light on questions such as the reason for the appearance of such cells in the visual systems of many species. It would be interesting to see if such a principle can explain other types of cell response observed in mammals. An open question at this stage is whether one can devise biologically plausible local learning rules (such as the Hebb rule), which can be demonstrated to maximize the information flow.

9.3.3 Iterated Hebbian Learning in a Layered Network

Another unsupervised learning scheme, iterated learning (IL), was introduced independently and has been studied recently [9.13]. This scheme shares some of the features used by Linsker but differs in some respects.

IL was applied to a layered network with binary-valued elements (versus continuous-valued, as used by Linsker). Also, full interlayer connectivity (versus local) was assumed. On the other hand, here learning was also unsupervised and Hebbian; in both schemes one waits for the couplings incident on layer l to mature before modification of the couplings between l and $l+1$ starts.

Conforming with Sect. 9.2, let us denote the key patterns on layer l by $\xi_{i\nu}$ and those of layer $l+1$ by $\xi'_{i\nu}$. The Linsker-type Hebbian learning has the following form in this notation:

$$J_{ij}(t+\delta t) = J_{ij}(t) + k\sum_{\nu} S'_{i\nu}(t)\xi_{j\nu} \; , \tag{9.50}$$

where

$$S'_{i\nu}(t) = \mathrm{sgn}\left[\sum_{j} J_{ij}(t)\xi_{j\nu}\right] \tag{9.51}$$

is the state of unit i of layer $l+1$, obtained, with couplings $J_{ij}(t)$, in response to the appearance of pattern ν on layer l. It is assumed implicitly that between t and $t+\delta t$ all stored patterns, $\nu = 1, 2, \ldots, \alpha N$, are presented.

As opposed to (9.50), the IL procedure uses

$$J_{ij}(t+\delta t) = k\sum_{\nu} S'_{i\nu}(t)\xi_{j\nu} \; . \tag{9.52}$$

That is, the old value of $J_{ij}(t)$ is discarded after every presentation cycle.

The procedure is repeated until convergence, i.e.

$$S'_{i\nu}(t+\delta t) = S'_{i\nu}(t) \,, \tag{9.53}$$

for all i and ν: at that point layer $l+1$ has "matured", one sets $\xi'_{i\nu} = S'_{i\nu}$, and the process is started for the next layer.

Iterated learning treats the *internal representations*, rather than the couplings, as the basic object to be modified in the course of the learning process. In effect one searches for such second-layer key patterns, $\xi_{i\nu}$, as are reproduced *precisely* with the Hebbian couplings

$$J_{ij} = \sum_{\nu} \xi'_{i\nu} \xi_{j\nu} \,. \tag{9.54}$$

An interesting aspect of IL is a *phase transition associated with the learning process*. When $\alpha > \alpha_c \simeq 0.18$, a "correlation catastrophy" occurs: all (initially random) key patterns collapse onto a *single state*. That is, beyond a critical capacity, no matter in which key pattern the first layer is initialized, the network flows to one common high-layer output state. The nature and causes of this collapse are not completely understood; for details see [9.13].

9.4 Supervised Learning in Layered Networks

We turn now to describe briefly several learning algorithms introduced recently by physicists. We choose to concentrate on learning rules designed for feed-forward networks with binary-valued (McCulloch–Pitts) threshold elements [9.14–16] (see also [9.64]). This excludes the much more widely used back-propagation algorithm [9.45–47]. Before turning to the first algorithm described here [9.14], which learns by *choosing internal representations* (CHIR), we state the problem of learning in concrete terms. Consider the particularly simple network architecture presented in Fig. 9.8. This is a feed-forward network with N inputs, a single output, and H hidden units. There are 2^N possible input patterns $(S_1^{\text{in}}, S_2^{\text{in}}, \ldots, S_N^{\text{in}})$. The states of hidden layer units S_i^{h} and the output unit, S^{out} are determined by the input according to the rules

$$\begin{aligned} S_i^{\text{h}} &= \operatorname{sgn}\left(\sum_{j=1}^{N} W_{ij} S_j^{\text{in}} + \theta_i\right) \,, \\ S^{\text{out}} &= \operatorname{sgn}\left(\sum_{i=1}^{H} w_i S_i^{\text{h}} + \theta\right) \,. \end{aligned} \tag{9.55}$$

Here W_{ij} are the weights assigned to the connections from input unit j to hidden unit i and w_i the weights assigned to the connections from the latter to the output unit. The parameters θ_i (θ) are thresholds associated with the hidden-

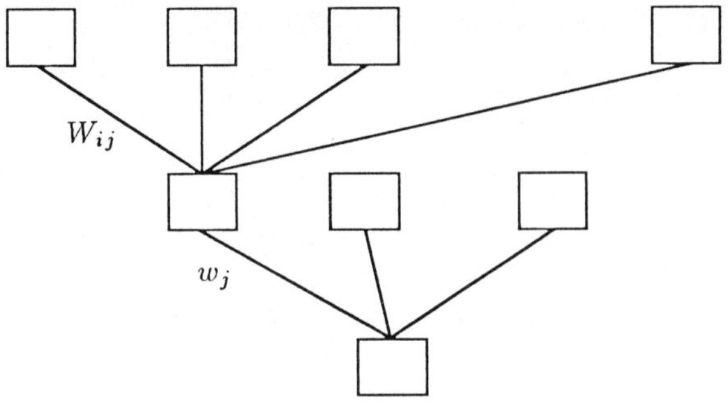

Fig. 9.8. Network with N input units, one hidden layer with H units, and a single output

layer (output) cells. They may be regarded as weights that connect the units to a constant input ($S_0 = 1$); from here on, reference to the "weights" implies thresholds as well.

A typical task of such a network is *classification*: divide the input space into two categories A and B; whenever a pattern that belongs to A is presented as input, the network is required to respond by setting $S^{\text{out}} = +1$, whereas an input from category B gives rise to $S^{\text{out}} = -1$. There are 2^{2^N} possible distinct input–output mappings (predicates). Obviously an arbitrary set of weights and thresholds, when used in the dynamic rule (9.55), will not produce the mapping required for our A versus B classification task. Supervised learning takes place in the course of a training session; inputs are presented sequentially to the network, and the outputs generated in response to them are compared to the (known) correct classification. Weights are modified according to some rule, in a manner that reduces the discrepancy between the desired and actual outputs. *The basic problem of learning is to find an algorithm, i.e a synaptic modification rule, that produces a set of connections and thresholds that enables the network to perform a preassigned task.*

In some instances hidden units are not necessary; the desired input–output association can be realized by directly connecting the input (source) units to the output (target). In these instances a proper set of weights can be found in a simple, elegant, and local fashion by the *perceptron learning rule* (PLR), which we now briefly explain.

Consider $j = 1, \ldots, N$ source units that are directly connected to a single target unit i. When the source units are set in any one of $\mu = 1, \ldots, M$ patterns, i.e. $S_j = \xi_j^\mu$, we require that the target unit (determined using (1)) takes preassigned values ξ_i^μ. A set of weights, W_{ij}^*, θ_i^*, for which this input-output relationship is satisfied constitutes a solution of the problem. Starting from any arbitrary initial guess for the weights, an input ν is presented, resulting in the output taking some value S_i^ν. Now modify every weight according to the rule

$$\begin{aligned} W_{ij} &\to W_{ij} + \Delta W_{ij} \,, \\ \Delta W_{ij} &= \eta(1 - S_i^\nu \xi_i^\nu)\xi_i^\nu \xi_j^\nu \,, \end{aligned} \tag{9.56}$$

where $\eta > 0$ is the step-size parameter. The threshold θ gets modified by the same rule, with $\xi_j^\nu = 1$. Now another input pattern is presented, and so on, until all inputs draw the correct output. Note that the PLR modifies weights only when presentation of input ν produces an erroneous output. When that happens, each weight is changed, in a Hebbian fashion, towards values that correct the error.

The most impressive aspect of this learning rule is the existence of an associated *convergence theorem*, which states that if a solution to the problem exists, the PLR will find a solution in a finite number of steps [9.44, 65]. However, single-layer perceptrons can solve only a very limited class of problems. The reason for this is that of the 2^{2^N} possible partitions of input space only a small subset (less than $2^{N^2}/N!$) is linearly separable [9.66]. One of the most widely known examples of a problem that cannot be solved by a single-layer perceptron is *parity* [9.44, 67]: assign output +1 to inputs that have an odd number of +1 bits, and -1 otherwise. Parity (and any other) classification task can be solved once a single hidden layer is inserted beween input and output. This, however, makes learning difficult; in particular, the PLR cannot be implemented. The reason for this is that in order to determine the corrective increment of a weight W_{ij}, one has to know the correct state of both presynaptic and postsynaptic cells, i and j. For multilayer perceptrons only the states of the input "cells" and output are known; no information regarding the correct state of the hidden layer is a priori available. Therefore, when in the course of the training session the network errs, it is not clear which connection is to blame for the error and what corrective action is to be taken.

The CHIR algorithm circumvents this "credit assignment" problem by guessing *internal representations*. Whereas other learning rules view the *weights* as the central independent variables of the learning process, CHIR views the internal representations (the states taken by the hidden layers of the network in response to presentation of inputs from the training set) as the fundamental entities to be determined. This is a conceptually plausible assumption; in the course of learning, a biological or artificial system should form various maps and representations of the external world. Once such representations are chosen, simple and local Hebbian learning rules, such as the PLR, can be trivially implemented, and the problem of learning becomes one of *choosing proper internal representations*. Failure of the PLR to converge to a solution is used as an indication that the current guess of internal representations needs to be modified. To demonstrate the idea, imagine that we are not supplied with the weights that solve the problem; however, the correct internal representations are revealed. That is, we are given a *table* with M rows, one for each input. Every row has H bits $\xi_i^{\mathrm{h},\mu}$, for $i = 1, \ldots, H$, specifying the state of the hidden layer obtained in response to input pattern μ. One can now view each hidden-layer cell i as the target cell of the PLR, with the N inputs viewed as source. Given sufficient time, the PLR will converge to a set of weights $W_{i,j}$, connecting input unit j to hidden unit i, so that indeed the input–output association that appears in column i of our table will be realized. In a similar fashion, the PLR will yield a set of weights

w_i, in a learning process that uses the hidden layer as source and the output unit as target. Thus, in order to solve the problem of learning, all one needs is a search procedure in the space of possible internal representations for a table that can be used to generate a solution. This sounds, a priori, a rather hopeless undertaking; if the training set contains all possible inputs (i.e. $M = 2^N$), there are 2^{H2^N} possible distinct internal representations (i.e. tables). The CHIR algorithm actually searches a much smaller space of tables; the reason for this, details of the algorithm, and tests of its performance on a number of classification tasks are presented in [9.14]. The method has been extended to multiple outputs and more than one hidden layer [9.68], to networks with binary weight [9.69], and to schemes that minimize cost functions [9.70]. Most recently, a version of CHIR that *learns without storing the table of internal representations* has been introduced [9.71].

In a recent paper Rujan and Marchand [9.15] also used the basic idea of viewing the internal representations as the object of central importance. Their approach is basically geometrical. Obviously all possible inputs constitute the 2^N corners of the unit hypercube embedded in N dimensions. Each corner can be colored black or white, depending on its desired classification. Hidden units are needed when the black and white colored corners are not linearly separable. Rujan and Marchand use a "greedy algorithm" to scan a particular subset of hyperplanes that chop off corners of the same color from the hypercube, until the remaining (convex) body contains only corners of the opposite color. Each hidden unit corresponds to one of these hyperplanes. The value the hidden unit takes in response to an input depends on which side of the hyperplane is the corner that corresponds to the input presented. Every input maps onto one of $H+1$ corners of a hypercube in the H-dimensional space of internal representations; only similarly colored corners of input space map onto the same hidden-space corner. Rujan and Marchand show that the resulting set of colored corners in this hidden space is linearly separable, and hence the weights that produce the correct output can be found.

There are $N3^{N-1}$ hyperplanes in the subset scanned in the course of the learning procedure. For every hyperplane scanned one has to check to which side all (remaining) input corners belong. Furthermore, whenever a plane is accepted, all previously accepted planes must be examined for intersections; only nonintersecting ones are kept. The process is convergent by definition, but the architecture varies in the course of learning, hidden units are being added and removed, and there is no guaranteed bound on the number of hidden units one will end up with. At worst, if the number of hidden units reaches the number of inputs, one obtains the well-known "grandmother cell" solution of the learning problem: every hidden unit responds positively to one and only one input. In the examples reported, however, this does not seem to happen; for the problem of parity the algorithm generated the optimal architecture of N hidden units.

The idea of first generating internal representations and only then learning the weights by the PLR was adopted by Mezard and Nadal [9.16], in the context of a "tiling algorithm". This algorithm also incorporates the Rujan–Marchand approach in that it allows a varying number of hidden units within each layer, as well as a variable number of layers [9.17].

In each layer there is a "master unit" that will eventually develop into the output unit of the networks. When a new layer is added, the initial weights incident on the new master unit are chosen so that it has a smaller number of errors than the previous one. If the subsequently used PLR generates more errors, the networks reverts back to the initial weights. This ensures that eventually all errors get eliminated by a network whose number of layers is less than the initial number of errors (which, however, may be exponential in the number of input units).

This growth in "depth" is preceded by lateral growth. Units are added to a layer until the internal representations it provides are "faithful", meaning that no two patterns of distinct desired output are mapped onto the same internal representation.

The algorithm always converges to a solution, but again there is no guarantee against the lateral growth getting out of control, in a way that generates hidden units in a number comparable to the number of inputs.

9.5 Summary and Discussion

We have reviewed some of the work done by physicists on layered feed-forward neural networks. First we presented, in a fairly detailed fashion, a basic, simple model which is designed to take the network through a sequence of states, with each layer generating the state of the next one. When the embedded states are random, and the interlayer couplings are obtained from them in a Hebbian manner, the problem is exactly soluble. That is, given an initial state, development of the states on all subsequent layers can be calculated. This model is the layered version of the Hopfield model, and in terms of simplicity and solvability it is indeed on a par with the latter. As can be done for the Hopfield model, here also the solution can be extended to a variety of generalized Hebbian (outer product) connections. These generalizations include adding static noise, diluting the bonds in various ways, and introducing couplings whose strength is a nonlinear function of the outer product. The case of pseudoinverse couplings between neighboring layers can also be solved; these couplings take the network with no error through a sequence of random patterns assigned to the sequence of layers.

This family of neural networks is unique in that its dynamics is exactly soluble and is controlled by attractors with nontrivial domains of attraction. The only other model with exactly soluble dynamics is the extremely diluted asymmetric model of Derrida et al. We found that many of our results exhibit qualitative agreement with those obtained for the dilute model and also with the fully connected Hopfield model. For all three models one finds that the stationary states (and their layered analogs) satisfy similar equations for the overlap with the recalled key pattern (9.20a). This equation is easily interpreted as the Gaussian average of the embedding field. The only difference between the three models is in the expressions for the width of this distribution. One should keep in mind that the embedding field is Gaussian distributed for the layered and dilute networks, whereas in the fully connected case this result is incorrect, and is based on approximations (such as using a replica-symmetric solution).

Next we reviewed some of the work done on learning in layered networks. First we reviewed unsupervised learning procedures, introduced by Linsker, that lead to the emergence of orientation-selective cells on deep layers. Attempts to frame the development of such units in an information-theoretic context were also briefly reviewed.

Finally some newly introduced supervised learning algorithms were mentioned. These algorithms address the issue of learning in networks composed of binary-valued linear threshold elements. They treat the internal representations as the fundamental entity with which the learning process has to deal. In our opinion the main thrust of future work will be concerned with the development of efficient learning rules that yield good solutions with high probability in reasonable learning times, for networks of fixed, or at least well-bound size.

Many of the results of our own research reviewed here were obtained in a most fruitful collaboration with W. Kinzel, whom we thank. We also thank T. Grossman for useful advice. This research was partially supported by a grant from Minerva, and by the US–Israel Binational Science Foundation (BSF).

References

9.1 For a recent review of physicists' contributions see W. Kinzel, Physica Scripta **T25**, 144 (1989)
9.2 E. Domany, R. Meir and W. Kinzel: Europhys. Lett. **2**, 175 (1986)
9.3 R. Meir and E. Domany: Phys. Rev. Lett. **59**, 359 (1987); Europhys. Lett. **4**, 645 (1988); Phys. Rev. A **37**, 608 (1988)
9.4 R. Meir: J. Phys. (Paris) **49**, 201 (1988)
9.5 B. Derrida and R. Meir: Phys. Rev. A **38**, 3116 (1988)
9.6 E. Domany, W. Kinzel and R. Meir: J. Phys. A **22**, 2081 (1989)
9.7 R. Linsker: Proc. Nat. Acad. Sci. USA **83**, 7508–7512 (1986)
9.8 R. Linsker: Proc. Nat. Acad. Sci. USA **83**, 8390–8394 (1986)
9.9 R. Linsker: Proc. Nat. Acad. Sci. USA **83**, 8779–8783 (1986)
9.10 R. Linsker, in: *Computer Simulation in Brain Science*, edited by R. Cottrill (Cambridge University Press, Cambridge 1988)
9.11 R. Linsker: IEEE Computer (March 1988) 105–117
9.12 R. Linsker: to be published in the proceedings of the 1988 Denver Conference on Neural Information Processing Systems (Morgan Kauffman)
9.13 R. Meir and E. Domany: Phys. Rev. A **37**, 2660 (1988)
9.14 T. Grossman, R. Meir, and E. Domany: Complex Systems, **2**, 555 (1988)
9.15 P. Rujan and M. Marchand: Complex Systems **3**, 229 (1989)
9.16 M. Mézard and J.P. Nadal: J. Phys. A **22**, 2191 (1989)
9.17 See for example J.E. Hopcroft and R.L. Mattson, Synthesis of Minimal Threshold Logic Networks, IEEE Trans. Electronic Computers, **EC-14**, 552 (1965)
9.18 J.J. Hopfield: Proc. Natl. Acad. USA **79**, 2554 (1982)
9.19 D.J. Amit, H. Gutfreund, and H. Sompolinsky: Ann. Phys. **173**, 30 (1987)

9.20 H. Sompolinsky, in: *Heidelberg Colloquium on Glassy Dynamics* edited by J.L. van Hemmen and I. Morgenstern, Lecture Notes in Physics Vol. **275** (Springer, Berlin, Heidelberg 1987)
9.21 B. Derrida, E. Gardner, and A. Zippelius: Europhys. Lett. **4**, 167 (1987)
9.22 See for example E. Domany, J. Stat. Phys. **51**, 743 (1988)
9.23 See H. Ritter, K. Obermayer, K. Schulten, and J. Rubner, this volume, Chap. 8
9.24 E. Gardner: J. Phys. A **21**, 257 (1988); E. Gardner and B. Derrida, J. Phys. A **21**, 271 (1988)
9.25 W. Krauth and M. Opper: J. Phys. A **22**, L519 (1989)
9.26 I. Kanter and H. Sompolinsky: Phys. Rev. A **35**, 380 (1987)
9.27 A.C.C. Coolen, J.J. Denier van der Gon, and Th.W. Ruijgrok: Proc. nEuro88; A.C.C. Coolen, H.J.J. Jonker, and Th.W. Ruijgrok, Utrecht preprint (1989)
9.28 M. Opper, J. Kleinz, H. Köhler, and W. Kinzel: J. Phys. A **22**, L407 (1989)
9.29 T.B. Kepler and L.F. Abbott: J. Phys. (Paris) **49**, 1657 (1988)
9.30 H. Horner, D. Bormann, M. Frick, H. Kinzelbach, and A. Schmidt: Z. Phys. B **76**, 381 (1989)
9.31 I. Kanter: Phys. Rev. A **40**, 2611 (1989)
9.32 S. Diedrich and M. Opper: Phys. Rev. Lett. **58**, 949 (1987); E. Gardner, N. Stroud, and D.J. Wallace: J. Phys. A **22**, 2019 (1989)
9.33 W. Krauth and M. Mézard: J. Phys. A **21**, L745 (1987)
9.34 P. Peretto: Neural Networks **1**, 309–322 (1988)
9.35 J.F. Fontanari and R. Meir: Caltech preprints (1989)
9.36 L.F. Abbott and T.B. Kepler: J. Phys. A **22**, L711 (1989)
9.37 F.J. Pineda: Phys. Rev. Lett. **59**, 2229 (1987)
9.38 M. Opper: Europhys. Lett. **8**, 389 (1989)
9.39 J.A. Hertz, G.I. Thorbergson, and A. Krogh: Physica Scripta **T25**, 149 (1989)
9.40 W. Kinzel and M. Opper: this volume, Chap. 4
9.41 For reviews see: J.D. Cowan and D.H. Sharp, Quarterly Reviews of Biophysics, **21** 365 (1988); R.P. Lippmann, IEEE ASSP Magazine, **4**, 4 (1987)
9.42 T. Kohonen: *Self Organization and Associative Memory* (Springer, Berlin, Heidelberg 1984)
9.43 D.E. Rumelhart and J.L. McClelland: *Parallel Distributed Processing: Explorations in the Microstructure of Cognition*, 2 vols. (MIT Press, Cambridge, Mass. 1986)
9.44 M. Minsky and S. Papert: *Perceptrons*, expanded edition (MIT Press, Cambridge, Mass. 1988)
9.45 D.E. Rumelhart, G.E. Hinton, and R.J. Williams: in *Parallel Distributed Processing: Explorations in the Microstructure of Cognition* edited by D.E. Rumelhart and J.L. McClelland, (MIT Press, Cambridge, Mass. 1986) Vol. 1, p. 318
9.46 Y. Le Cun: Proc. Cognitiva, **85**, 593 (1985); P.J. Werbos, Ph.D. thesis, Harvard University (1974)
9.47 D.B. Parker, MIT Technical Report TR-47 (1985)
9.48 M. Abeles: *Local Cortical Circuits* (Springer, Berlin, Heidelberg 1982)
9.49 H. Sompolinsky: Phys. Rev. A **34**, 2571 (1986); J.L. van Hemmen and R. Kühn, Phys. Rev. Lett. **57**, 913 (1986); J.L. van Hemmen, Phys. Rev. A **36**, 1959 (1987)
9.50 W. Krauth, M. Mézard, and J.P. Nadal: Complex Systems, **2**, 387 (1988)
9.51 W.A. Little: Math. Biosci. **19**, 101 (1975)
9.52 S. Amari and K. Maginu: Neural Networks, **1**, 63 (1988)
9.53 See for example W. Feller, *An Introduction to Probability Theory and its Applications* (Wiley, New York 1966) Vol. II p. 256
9.54 It is very important to realize that the *embedding fields* H_i and H_j are *not* independent. Their correlation is $M_\mu^2 \sim 1/N$. This correlation gives rise to the layer-to-layer recursive variation of the width parameter Δ^l, which in turn, causes the appearance of non trivial domains of attraction
9.55 B.Derrida: J. Phys. A **20**, L721 (1987)
9.56 W. Kinzel: Z. Physik B **60**, 205 (1985)
9.57 W. Krauth, J.P. Nadal, and M. Mézard: J. Phys. A: Math. Gen. **21**, 2995 (1988)
9.58 J. Kleinz: diploma thesis, Justus-Liebig University Giessen (1988)
9.59 For a recent review see D. H. Hubel, Los Alamos Science **16**, 14 (1988)
9.60 D. Kammen and A. Yuille: Biol. Cybern. **59**, 23 (1988)
9.61 P. Huber: Ann. Statistics **13**, 435 (1985)
9.62 E. Oja: J. Math. Biol. **15**, 267 (1982)
9.63 R.E. Blahut: *Principles and Applications of Information Theory*, (Addison-Wesley, 1987)
9.64 B. Widrow and R. Winter: Computer **21**, 25 (1988)

9.65 F. Rosenblatt: Psych. Rev. **62**, 386 (1958); *Principles of Neurodynamics* (Spartan, New York 1962)
9.66 P.M. Lewis and C.L. Coates: *Threshold Logic* (Wiley, New York 1967)
9.67 J. Denker, D. Schwartz, B. Wittner, S. Solla, J.J. Hopfield, R. Howard, and L. Jackel: Complex Systems **1**, 877–922 (1987)
9.68 T. Grossman, Complex Systems **3**, 407 (1989)
9.69 T. Grossman, in *Advances in Neural Information Processing Systems 2*, edited by D. Touretzky (Morgan Kaufman, San Mateo 1990) p. 516
9.70 R. Rohwer, in *Advances in Neural Information Processing Systems 2*, edited by D. Touretzky (Morgan Kaufman, San Mateo 1990) p. 538); A. Krogh, G.I. Thorbergsson, and J.A. Hertz, *ibid*, p. 773; D. Saad and E. Marom, Complex Systems **4**, 107 (1990)
9.71 D. Nabutovsky, T. Grossman, and E. Domany: Complex Systems **4**, 519 (1990)

Elizabeth Gardner – An Appreciation

Elizabeth Gardner was born on 25 August 1957 in Cheshire, where her father Douglas worked as a chemical engineer with ICI. Her mother Janette was also a scientist, having gained a Ph.D. in physical chemistry at the University of Edinburgh. The family moved to Scotland the following year, on the expansion of BP Chemicals at Grangemouth.

Her interest in science extended beyond her school activities at St. Hillary's in Edinburgh, and from a remarkably early age: when she was two years old she was overheard pointing out to granny "that's the Plough, and that's Orion ...". The fascination with astronomy, which has drawn so many towards science, continued through her school years and found encouragement at home, with many a cold evening spent at the 4-inch reflector telescope her parents gave her at the age of 12. Around the same time, it also became clear that experimental work was probably not going to be her strongest bent: when she was 14, she and her brother Richard surveyed Blackford Hill using surrounding landmarks and came to the conclusion that its summit was 560 feet below sea level!

Elizabeth applied to Cambridge and Edinburgh for undergraduate entry. Her application to Cambridge was unsuccessful and her interview there was followed by the suggestion: "Don't you think you'd be better at a Scottish university?" – remarkably sound and disinterested advice. She came to Edinburgh University in 1975 with the intention of studying for the astrophysics degree, but, on learning that the mathematical physics courses were strongly recommended for that degree, she registered in her first year for the mathematical physics degree and continued in it for her four years as an undergraduate. She had an outstanding record, gaining more than 90% in every degree examination in every year. In her first year she was in the top three in all of her courses: mathematics, applied mathematics, and physics. In her second year she was awarded the Class Medal as the top student in all three subjects. Again in third year she was awarded the Class Medal, and graduated with first class honours, sharing the Tait Medal and Robert Schlapp Prize.

Elizabeth's quiet and reserved manner was well known and never more noticeable than during formal interviews, but in scientific discussions she showed a vividly different side. This emerged early on during her D.Phil. studies, about half way through her first year at Oxford when she met with her supervisor Dr I.J.R. Aitchison to talk about possible research projects. Till then her time had, as is usual, been mostly taken up with the graduate course work, in which she had naturally performed very well, though without making any special impact. But when, that day, the discussion with her turned to some problems in gauge theory, it became apparent that she was positively bursting with excitement: she was intensely eager to begin research work, for which indeed she proved to have a burning, though rarely visible, passion.

By the time she was in her third year as a graduate student she had acquired an unusual authority, and her opinion was much sought by those who knew her and recognized her quality. When a difficult point came in coffee-time discussions and people were stuck, they would turn to her. Until asked a specific question, she would probably not have joined in the discussion, but, when asked, she would often turn out to have thought about the problem already and to have worked out the answer – which she whould explain precisely and economically. Otherwise, not having much taste for small talk or thinking out loud, she would simply say that she hadn't thought about it and didn't know.

Elizabeth had extremely high intellectual standards, both as regards the sort of problem she felt she wanted to work on and as regards the style of the work itself. This was reflected in her personal dissatisfaction with the main work in her thesis, which concerned an attempt to learn about non-perturbative aspects of non-Abelian gauge theories. Naturally approximations had to be made – and they violated gauge invariance. Within the context of the approach adopted, her work was original and inventive and marked by an easy-seeming technical competence. But she was not happy with an approximation which, she said, she could neither justify, nor control, nor improve upon systematically. In expressing these serious reservations about her own work, she showed how central to her inner self was the pursuit of the highest possible standards of scientific endeavour.

Although she fulfilled all the requirements, Elizabeth in fact never formally graduated from Oxford; she always said she could leave that until she was 90.

After her D.Phil. at Oxford, Elizabeth spent 2 years as a postdoc in the theoretical group at Saclay with the support of a Royal Society Fellowship. It was here that she started to work on problems of field theory defined on random lattices. This work was motivated by the idea that field theories on random lattices have the advantage of being defined on lattices and yet of keeping the continuous translational invariance.

It was this which provided for her the bridge between field theory and the theory of disordered systems, on which she was going to devote all her future scientific work. Following this change of direction, she contributed to several topics in the theory of disordered systems, in the localization problem for which a perturbative method was developed for the calculation of Lyapunov exponents, in the effect of disorder in the critical behaviour of weakly disordered systems, and in stereological properties of random materials.

During this period Elizabeth also started to work on spin glasses using both renormalization techniques and replica methods. Her first works on spin glasses were calculations on hierarchical lattices, on extensions of the Sherrington–Kirkpatrick model to p-spin interactions, and on random energy models. Her deep understanding of the modern ideas developed in the theory of spin glasses allowed Elizabeth to become very quickly one of the best world experts on the theory of neural networks.

She returned to the Physics Department at the University of Edinburgh in October 1984. This was her base for the remainder of her brief research career, although she continued a close collaboration with colleagues at Saclay and latterly became overwhelmed with invitations elsewhere, to conferences and workshops, and for scientific visits. She started with a two-year postdoctoral position supported by the Science and Engineering Research Council (SERC) for work in her D.Phil. area of theoretical particle physics. At the end of this period she applied unsuccessfully for an SERC Advanced Fellowship to study disordered systems and neural-network models; one may speculate that again her modesty and diffidence at formal interviews was a crucial factor. It was clear to the Department however that she was a scientist of outstanding potential, and a package of support was quickly put together, involving a short period under a contract for neural-network studies from the Royal Signals and Radar Establishment at Malvern and a Dewar Fellowship from the University of Edinburgh. This enabled her to continue at Edinburgh, pending what was a successful second application for an Advanced Fellowship, for five years from October 1987.

Soon after her return to Edinburgh in 1984 it became clear that her scientific curiosity remained driven by the study of disordered systems. Her entry into the study of neural-network models was remarkably incisive. At that time numerical simulations were being performed at Edinburgh to try to understand the storage capacity in the Hopfield net and the fraction of states perfectly stored in the model. That this quantity depends upon the ratio of the number of patterns stored to the number of nodes in the net can be readily understood on the basis of a simple "signal and interference" argument, but the functional form obtained from numerical simulations can not. Within three or four days of being first exposed to the problem Elizabeth came up with the exact result which explained the data perfectly. Shortly thereafter she became interested in the training algorithms which were being rediscovered and developed and made the subject her own. There then followed a flood of papers of the highest scientific quality, elucidating the dynamics of network models, the effect of multi-spin interactions, the role of asymmetry and dilution, and above all the maximum storage capacity. For this last work she developed a beautiful technique exploring the space of the interactions themselves (the connection strengths in the network model). In her approach, in contrast to the usual statistical mechanics in which the spin variables are replicated to deal with the average over the quenched disorder, the statistical properties in the space of interactions are explored by replicating the interactions to deal with the average over the possible spin states.

Elizabeth bore her illness privately, with dignity and with remarkable strength of character. The first diagnosis of cancer was made in 1986; her colleagues at Edinburgh were aware only of the physical changes resulting from the treatment, the flowering of her scientific talent, and her selfless support for her graduate students. The disease recurred in the subsequent two years and she succumbed to it on 18 June 1988, only 9 months into her 5-year Advanced Fellowship. She had been interviewed for a lectureship in the Department of Physics at Edinburgh less than three weeks before. The seriousness of her illness was even then not fully known to her colleagues, and Elizabeth was pleased to be reassured that the University expected to be able to appoint her to a permanent position at the end of her Advanced Fellowship.

Gerhard Toulouse has offered what is surely a fitting epitaph: in retrosepct, the brilliant creativity and the admirable dignity of her last years are the best apology for pure research and the ethics of knowledge.

This appreciation was prepared by D.J. Wallace (Edinburgh), in consultation with B. Derrida (Saclay) and I.J.R. Aitchison (Oxford); it originally appeared in a memorial issue of Journal of Physics A **22**, 1955–2273 (1989).

Subject Index

Printing: COLOR-DRUCK DORFI GmbH, Berlin
Binding: Buchbinderei Lüderitz & Bauer, Berlin

Physics of Neural Networks

Models of Neural Networks I 2nd Edition
E. Domany, J.L. van Hemmen, K. Schulten (Eds.)

Models of Neural Networks II: Temporal Aspects of Coding and Information Processing in Biological Systems
E. Domany, J.L. van Hemmen, K. Schulten (Eds.)

Neural Networks: An Introduction 2nd Edition
B. Müller, J. Reinhardt and M. T. Strickland